Philosophical Concepts in Physics discusses advances in science against the historical and philosophical backgrounds in which they occurred. Readers are given an opportunity to reflect upon the nature of a scientific enterprise that they may previously have come to know only from the perspective of traditional accounts of science or retrospectively from course work in the sciences. A major goal is to impress upon the reader the essential and ineliminable role that philosophical considerations have played in the actual practice of science and in the construction of scientific theories.

Scientific knowledge, because of its putative certainty and objective method of discovery, is often seen as being essentially different from other types of knowledge. As popularly understood, physics and philosophy might seem about as far removed from each other as two intellectual disciplines could be. However, this book illustrates the formative mutual influences that physics and philosophy have had, and continue to have, on each other. The discussion of central philosophical issues is anchored in the specific historical context and in the actual content of relevant scientific activity.

Some necessary introduction to the history of ancient and early modern science is presented first, but major emphasis is given to the watersheds of twentieth century physics: namely relativity and, especially, quantum mechanics. The reader is assumed to have some background in elementary classical physics, but no knowledge of relativity or quantum mechanics is assumed.

The book will be of particular interest to science, engineering, philosophy and general humanities students who have had an introductory course in physics, to scientists with an interest in the relation of philosophy to physics and to philosophers of science. It could also be used as a text in a junior/senior level undergraduate course in the philosophy of science.

Philosophical Concepts in Physics

Philosophical Concepts in Physics

THE HISTORICAL RELATION BETWEEN

PHILOSOPHY AND SCIENTIFIC THEORIES

JAMES T. CUSHING
DEPARTMENT OF PHYSICS, UNIVERSITY OF NOTRE DAME

PUBLISHED BY THE PRESS SYNDICATE OF THE UNIVERSITY OF CAMBRIDGE
The Pitt Building, Trumpington Street, Cambridge, United Kingdom

CAMBRIDGE UNIVERSITY PRESS
The Edinburgh Building, Cambridge CB2 2RU, UK   http://www.cup.cam.ac.uk
40 West 20th Street, New York, NY 10011-4211, USA   http://www.cup.org
10 Stamford Road, Oakleigh, Melbourne 3166, Australia

© James T. Cushing 1998

This book is in copyright. Subject to statutory exception
and to the provisions of relevant collective licensing agreements,
no reproduction of any part may take place without
the written permission of Cambridge University Press.

First published 1998
Reprinted 1998

Printed in Great Britain at the University Press, Cambridge

Typeset in Times 10/13 pt [VN]

A catalogue record for this book is available from the British Library

Library of Congress Cataloguing in Publication data

Cushing, James T., 1937–
    Philosophical concepts in physics: the historical relation
between philosophy and scientific theories / James T. Cushing.
        p.   cm.
    Includes bibliographical references and index.
    ISBN 0 521 57071 9 (hardcover). – ISBN 0 521 57823 X (pbk.)
    1. Physics – Philosophy.   2. Physics – History.   3. Science –
Philosophy.   4. Science – History.   I. Title.
QC6.C85 1997
530′.01–dc21   97-3059 CIP

ISBN 0 521 57071 9 hardback
ISBN 0 521 57823 X paperback

For Nimbilasha, Christine and Patricia

Contents

Preface xv
Copyright acknowledgments xviii

PART I **The scientific enterprise** 1

1. Ways of knowing 3
   1.1 Philosophy
   1.2 Logical deduction
   1.3 Self-evident first principles
   1.4 Rationalists versus empiricists
   1.5 The status of scientific knowledge
      1.A Descartes' *Regulae*
      Further reading

2. Aristotle and Francis Bacon 15
   2.1 Aristotle
   2.2 Observation versus experimentation
   2.3 The universe as organism
   2.4 Aristotle on motion
   2.5 Francis Bacon
   2.6 *The New Organon*
   2.7 Bacon and Aristotle compared
      2.A Aristotle's *On the Heavens* and *Physics* on motion
      Further reading

3. Science and metaphysics 29
   3.1 Origins of scientific method
   3.2 A popular view of science
   3.3 Hume and Mill on induction
   3.4 Popper on observation and hypothesis
   3.5 Justification for hypotheses
   3.6 Scientific knowledge and truth
      3.A Hume, Mill and Popper on scientific knowledge
      Further reading

PART II **Ancient and modern models of the universe** 41

4. Observational astronomy and the Ptolemaic model 43
   4.1 Elementary observations
   4.2 The celestial sphere

Contents

    4.3 Eratosthenes' determination of the earth's size
    4.4 Aristarchus' heliocentric model of the universe
    4.5 The planets
    4.6 Ptolemy's geocentric model of the universe
    4.7 To save the phenomena
        4.A Determination of the absolute sizes of planetary orbits
        Further reading

5 The Copernican model and Kepler's laws   59
    5.1 Copernicus and his heliocentric model
    5.2 Advantages of the Copernican theory
    5.3 Shortcomings of the Copernican theory
    5.4 Kepler's laws
        5.A Conic sections
        Further reading

6 Galileo on motion   74
    6.1 The impetus theory
    6.2 Galileo's naturally accelerated motion
    6.3 Projectile motion
    6.4 Inertia
    6.5 Galileo on Aristotle
        6.A Galileo's *Dialogues Concerning Two New Sciences*
        Further reading

PART III   The Newtonian universe   87

7 Newton's *Principia*   89
    7.1 Isaac Newton
    7.2 Newton's philosophy of science
    7.3 Outline of Newton's argument in the *Principia*
    7.4 Newton's three laws of motion
    7.5 The logical structure of classical mechanics
        Further reading

8 Newton's law of universal gravitation   103
    8.1 Newton's astronomical data and deductions
    8.2 An inverse-square law
    8.3 The moon's centripetal acceleration
    8.4 The law of gravitation for point masses
    8.5 Gravitation for extended bodies
    8.6 Inertial and gravitational masses
        Further reading

9 Some old questions revisited   114
    9.1 An illustration of Newton's geometrical proofs
    9.2 Kepler's first and third laws
    9.3 Perturbations
    9.4 The ocean tides prior to Newton
    9.5 The earth–moon system and tidal bulges

9.A Newton and Young on wave interference
Further reading

PART IV A perspective 133

10 Galileo's *Letter to the Grand Duchess* 135
- 10.1 The background
- 10.2 A basic issue
- 10.3 The *Letter to the Grand Duchess*
- 10.4 Galileo and Urban VIII
- 10.5 Religion *vis-à-vis* natural philosophy
  - 10.A Galileo's *Letter to the Grand Duchess*
  - Further reading

11 An overarching Newtonian framework 148
- 11.1 A revolution
- 11.2 A broad coherence
- 11.3 Views on space prior to Newton
- 11.4 Newton's absolute space
- 11.5 Physical versus mathematical spaces
- Further reading

12 A view of the world based on science: determinism 164
- 12.1 The belief in simple laws
- 12.2 The meaning of determinism
- 12.3 Why the clockwork universe?
- 12.4 An unwarranted optimism
- 12.5 Two maps as examples
- Further reading

PART V Mechanical versus electrodynamical world views 181

13 Models of the aether 183
- 13.1 Emergence of the optical aether
- 13.2 The elastic solid aether
- 13.3 The electromagnetic aether
- 13.4 Thomson's and Maxwell's models
- 13.5 Maxwell's arguments for the aether
  - 13.A Maxwell on the aether versus action at a distance
  - Further reading

14 Maxwell's theory 195
- 14.1 Maxwell's equations
- 14.2 The displacement current
- 14.3 The final classical theory
- 14.4 The Michelson–Morley experiment
- 14.5 Precursors to relativity
  - 14.A Maxwell's equations in mathematical form
  - Further reading

Contents

15 The Kaufmann experiments 208
- 15.1 Rival theories of electromagnetic mass
- 15.2 Kaufmann's experiments
- 15.3 Planck's analysis of Kaufmann's work
- 15.4 Subsequent determinations of e/m_0
- 15.5 Conclusions
- 15.A Some technical details
- Further reading

PART VI The theory of relativity 223

16 The background to and essentials of special relativity 225
- 16.1 Albert Einstein
- 16.2 Einstein's skepticism about classical physics
- 16.3 The postulates
- 16.4 Time dilation and length contraction
- 16.5 The Lorentz transformations
- 16.A A technical detail on Einstein's *Gedankenexperiment*
- Further reading

17 Further logical consequences of Einstein's postulates 241
- 17.1 Relativistic Doppler effect
- 17.2 Mass–energy equivalence
- 17.3 The twin paradox
- 17.4 Simultaneity and coexistence
- 17.A Some calculational details
- Further reading

18 General relativity and the expanding universe 252
- 18.1 The basic principles
- 18.2 Experimental tests
- 18.3 The stability of the classical universe
- 18.4 The Einstein and Friedmann universes
- 18.5 Hubble's law
- 18.6 A modern model of our universe
- 18.A A derivation of Hubble's law
- Further reading

PART VII The quantum world and the completeness of quantum mechanics 271

19 The road to quantum mechanics 273
- 19.1 Some historical background
- 19.2 Planck's hypothesis
- 19.3 Bohr's semiclassical model
- 19.4 Actual discoveries versus rational reconstructions
- 19.5 Two routes to quantum mechanics
- 19.6 Forging the Copenhagen interpretation
- Further reading

20   Copenhagen quantum mechanics   290
   20.1   Some simple quantum-mechanical systems
   20.2   Interpretations of the wave function
   20.3   A fundamental distinction between large and small
   20.4   The uncertainty relation
   20.5   Photon interference – the double slit
   Further reading

21   Is quantum mechanics complete?   305
   21.1   The completeness of quantum mechanics
   21.2   The Bohr–Einstein confrontations
   21.3   The measurement problem
   21.4   Schrödinger's cat paradox
   21.5   Dirac on the effect of measurement
   Further reading

PART VIII Some philosophical lessons from quantum mechanics   317

22   The EPR paper and Bell's theorem   319
   22.1   The EPR paradox
   22.2   An analysis of the EPR paper
   22.3   Bell's theorem
   22.4   A derivation of Bell's theorem
      22.A   A calculation of the EPRB correlations
   Further reading

23   An alternative version of quantum mechanics   331
   23.1   An overview
   23.2   The Copenhagen interpretation
   23.3   A logically possible, empirically viable alternative
   23.4   The value of an alternative interpretation
   23.5   Explanation versus understanding
   23.6   Attempts at understanding quantum mechanics
      23.A   Some mathematical details of Bohm's theory
   Further reading

24   An essential role for historical contingency?   345
   24.1   Underdetermination
   24.2   A dilemma for the realist
   24.3   An alternative historical scenario?
   24.4   Internal versus external explanations
   Further reading

PART IX   A retrospective   357

25   The goals of science and the status of its knowledge   359
   25.1   Einstein on science and its goals
   25.2   A reductionist program
   25.3   Styles of scientific inference
   25.4   A paradox of confirmation

Contents

    25.5 The paradigm model of science
    25.6 An eclectic description of science
    25.7 One modern world view based on science
         Further reading

Notes 378
General references 398
Bibliography 400
Author index 412
Subject index 415

Preface

This book has grown out of an elective, one-semester junior/senior level interdisciplinary course I have taught for several years to students in arts and letters, science, and engineering at the University of Notre Dame. It allows one to examine a selection of philosophical issues in the context of specific episodes of the development of physical laws and theories. Many students with science and engineering backgrounds find this exercise informative – for some unsettling, but still rewarding. Although a major goal of the exposition is to impress upon the reader the essential and ineliminable role that philosophical considerations have played in the actual practice of science, more space is devoted to the history and content of science than to philosophy *per se*. The reason for this is that I believe that meaningful and useful philosophy of science can only be done within the context of the often tortuous historical route to new insights. Another way to put this is that it takes a lot of history of science to anchor even a little philosophy of science.

Some necessary background from the history of ancient and early modern science is presented first, but major emphasis is given to the immediate precursors to and the content of the watersheds of twentieth-century physics: relativity and, especially, quantum mechanics. This is not a systematic exposition of either the history or the philosophy of science, but an individualistic, perhaps to some even an idiosyncratic, selection of topics and episodes from the history and philosophy of physics. Developments in science are presented against the historical and philosophical backgrounds in which they occurred. At times the term 'construction' may seem more appropriate than 'discovery' for the way theories have been developed and, especially in the later chapters, the question of the influence of historical, philosophical and even social factors on the very form and content of scientific theories is discussed. Quantum mechanics proves to be a particularly rich source of material on this topic.

The reader is assumed to have some knowledge of elementary classical physics at the level of an introductory one-year course in that subject. Since relativity and quantum physics will still be new to such an audience, enough background is provided on those subjects to make them accessible to the readers for whom this book is intended. The quotations at the beginning of each of the nine major divisions (or Parts) of this book, and the supplemental material included at the ends of the chapters, form integral parts of the presentation and are meant to be read. The footnotes (that are gathered together in the *Notes* section after the last chapter of this book) contain literature citations and sometimes expository comments on the text material, including in some cases mathematical details. For these reasons, it

Preface

is important to take cognizance of the information provided in the footnotes. In these notes, GB stands for *Great Books of the Western World* (Robert M. Hutchins, ed.). A few chapters have an appendix in which mathematical details have been isolated in order to make the main body of the text suitable for a wider audience. (The basic ground rule is that advanced mathematics does not appear in the text proper.) Additional references are given at the end of each chapter to facilitate further study on points that may be of particular interest to the reader. Moreover, in the *General References* section near the end of this book, several works are suggested as sources for overall background information. All references cited in this book are listed together in the *Bibliography* section before the *Index*.

Although this book emphasizes an interdisciplinary approach to physics, history and philosophy, and uses some primary source material in the text, a disclaimer (a type of 'truth in advertising') is in order here. I am by no means an expert in the history and philosophy of science and have relied heavily on secondary sources. As someone trained as a scientist, I am interested in broadening the view of science students on the connections between philosophy and fundamental science, and in doing this within the context of physics as they recognize it. My hope is that the reader will become sufficiently taken with that question to pursue it further in more formal historical and philosophical treatises. In a classroom situation, the material in the text serves as a basis for discussion and for additional reading on the part of the students. There are surely places where I have been unable to escape from the thrall of the physicist's folklore 'knowledge' and skewed view of history and of philosophy. The style is often informal and it would make the text unreadable to enter all of the caveats that might be necessary. Proper historians and philosophers of science will no doubt be unhappy about some of what I say. I can only apologize in advance for the infelicities that will doubtless stand out to them in these pages and hope that the general reader will, nevertheless, find something of interest and value. I must thank my friends and colleagues here at Notre Dame and elsewhere who have, over many years, helped to reduce somewhat my own ignorance in these subjects. Professor Ernan McMullin first sparked my interest in the history and philosophy of science some twenty or so years ago and for that I thank him – although he is not to be held in any way responsible for the shortcomings of his 'student'. He, Professors Samir K. Bose, Gerald L. Jones, William D. McGlinn, Stephen M. Fallon and Robert E. Kennedy have been good enough to read through some sections of my manuscript and, thus, save me from certain embarrassments. The remaining flaws are mine alone.

For assistance with the redaction of the manuscript, I am especially indebted to Ms Alisa N. Ellingson and to Dr Yuri V. Balashov, both for the exceptional care they exercised in reading through preliminary drafts of the manuscript and for extremely helpful suggestions that substantially improved the text and references. Their work was supported by the Department of Philosophy and by the Program in the History and Philosophy of Science here at Notre Dame. Mr Neal Nash is responsible for all of the line drawings in the text. I also thank the Department of Physics and the College of Science at the University of Notre Dame – more particularly, the Chairman Gerald L. Jones and the Dean Francis J. Castellino – for the freedom over the years to be able to teach interdisciplinary courses in physics and philosophy and to do research in an area that overlaps often disparate academic turfs. Finally, my wife Nimbilasha has for many years shown the patience, support and encouragement without which it would not have been possible to complete a project whose beginnings go back nearly two decades.

James T. Cushing

Copyright acknowledgments

A book of this nature, with much quoted material from primary sources, would not be possible without the permission granted by many publishers and individuals to use material on which they currently hold, or previously held, the copyright. Even though the fair-use provision of the copyright laws covers many of my citations and some others are now in the public domain, I nevertheless contacted all of the publishers whose material I cite and I wish to list them below in acknowledgment. Every quotation in this text is cited by publisher in an accompanying footnote and any figure based on copyrighted material is so indicated (as 'adapted from') in a footnote keyed to the first reference that appears to that figure. Although I am grateful to all of these publishers and individuals, I must recognize the extraordinary generosity of The American Physical Society, Encyclopaedia Britannica, University of California Press, Cambridge University Press, The University of Chicago Press, Dover Publications, Mrs Melitta Mew, Northwestern University Press, Open Court Publishing Co. and Oxford University Press for releases that went far beyond anything supported merely by the concept of fair use of copyrighted material.

Here, then, are the names of all of those publishers and individuals whom I acknowledge for use of copyrighted material. All references are to the Bibliography at the end of this book.

The American Physical Society: Bloch (1976); Bohm (1952); Cushing (1981, 1982); Einstein et al. (1935); Frank (1949); Goldberg et al. (1967).
American Scientist: Jensen (1987).
Archive for History of Exact Sciences: Klein (1962).
Bantam Doubleday Dell Publishing Group: Burtt (1927); Drake (1957); *The Jerusalem Bible* (1966).
C. H. Beck'sche Verlagsbuchhandlung: Caspar (1937).
The Benjamin/Cummings Publishing Co.: Watson (1970).
The British Journal for the History of Science: Westfall (1962).
University of California Press: Galilei (1967); Newton (1934).
Cambridge University Press: Archimedes (1897); Bell (1987); Bohr (1934); Descartes (1977a,

1977b); Eddington (1926); Kepler (1992); Maxwell (1890); Moore (1989); Schrödinger (1944); Ziman (1978).
Carol Publishing Group: Einstein (1934).
Leopold Cerf: Descartes (1905).
The University of Chicago Press: Blackwell (1977); Cushing (1994); Duhem (1969); Heilbron (1985); Kuhn (1970); de Santillana (1955); Westfall (1980a).
C. J. Clay & Sons: Kelvin (1904).
Cornell University Press: Cooper (1935).
Crown Publishers: Einstein (1954a).
Daedalus: Holton (1968).
Dover Publications: Born (1951); Copernicus (1959); Heath (1981b); Jevons (1958); Lorentz (1952); Lorentz et al. (n.d.); Maxwell (1954); More (1934); Poincaré (1952).
Encyclopaedia Britannica: Bacon (1952); Copernicus (1952); Faraday (1952); Harvey

Copyright acknowledgments

(1952); Jefferson (1952); Kepler (1952a, 1952b); Lucretius (1952); Newton (1952); Ptolemy (1952).

Harper–Collins Publishers: Boorse and Motz (1966); Heisenberg (1958, 1971); Mill (1855).

Harvard University Press: Cohen (1971); Cohen and Drabkin (1948); Kuhn (1957); Manuel (1968); Quine (1990).

Historical Studies in the Physical Sciences: Hanle (1979).

The Johns Hopkins University Press: Koyré (1957).

Humanities Press International: Whittaker (1973).

Professor Max Jammer: Jammer (1989).

The Journal of Philosophy: Fine (1982b).

Kluwer Academic Publishers: Heilbron (1988); Huygens (1934).

Alfred A. Knopf: Frank (1947); Monod (1971).

Mrs Robert B. Leighton: Leighton (1959).

Literistic Ltd.: Clark (1972).

Longman, Rees, Orme, Brown & Green: Herschel (1830).

Macmillan Publishing Co.: Hertz (1900); Plato (1892); Thayer (1953).

The McGraw-Hill Companies: White (1934).

Mrs Melitta Mew: Popper (1965, 1968).

The University of Michigan Press: Donne (1959).

The MIT Press: Feynman (1965).

John Murray Publishers: Fahie (1903); Peacock (1855).

The New York Review of Books: vann Woodward (1986).

North-Holland Publishing Co.: van der Waerden (1967).

Northwestern University Press: Galilei (1946).

W. W. Norton & Co.: Butterfield (1965); Freud (1965).

University of Notre Dame Press: Cushing and McMullin (1989).

Open Court Publishing Co.: Mach (1960); Schilpp (1949).

Oxford University Press: Aristotle (1942a, 1942b, 1942c); Dirac (1958); Gell-Mann (1981); Hume (1902).

Penguin Books UK: Smith (1974).

Pergamon Press: Landau and Lifshitz (1977).

Philosophical Library: Planck (1949); Przibram (1967).

The Philosophical Magazine: Bohr (1913).

The Philosophical Review: Quine (1951).

Princeton University Press: Duhem (1974); von Neumann (1955).

Reports on Progress in Physics: Clauser and Shimony (1978).

Routledge, Chapman & Hall: Russell (1917).

Science History Publications: Cardwell (1972).

Serif Publishing Co.: Klein (1964).

Simon & Schuster: Luria (1973); Whitehead (1967).

Sterling Lord Literistic: Koestler (1959).

Walker Publishing Co.: Born (1971).

Professor Richard S. Westfall: Westfall (1971).

John Wiley & Sons: Eisberg (1961); Laplace (1902).

PART I

The scientific enterprise

I hold that in all cases of inductive inference we must invent hypotheses, until we fall upon some hypothesis which yields deductive results in accordance with experience. Such accordance renders the chosen hypothesis more or less probable, and we may then deduce, with some degree of likelihood, the nature of our future experience, on the assumption that no arbitrary change takes place in the conditions of nature.
William Jevons, *The Principles of Science: A Treatise on Logic and Scientific Method*

Pure logical thinking cannot yield us any knowledge of the empirical world; all knowledge of reality starts from experience and ends in it. Propositions arrived at by purely logical means are completely empty as regards reality. Because Galileo saw this, and particularly because he drummed it into the scientific world, he is the father of modern physics – indeed, of modern science altogether.

. . .

A complete system of theoretical physics is made up of concepts, fundamental laws which are supposed to be valid for those concepts and conclusions to be reached by logical deduction. It is these conclusions which must correspond with our separate experiences.

. . .

The structure of the system is the work of reason; the empirical contents and their mutual relations must find their representation in the conclusions of the theory. In the possibility of such a representation lie the sole value and justification of the whole system, and especially of the concepts and fundamental principles which underlie it ... [T]hese latter are free inventions of the human intellect, which cannot be justified ... *a priori*.
Albert Einstein, *On The Method of Theoretical Physics*

Science is not a system of certain, or well-established, statements; nor is it a system which steadily advances towards a state of finality. Our science is not knowledge (*epistēmē*): it can never claim to have attained truth, or even a substitute for it, such as probability.

. . .

We do not know: we can only guess.

. . .

The advance of science is not due to the fact that more and more perceptual experiences accumulate in the course of time. Nor is it due to the fact that we are making ever better use of our senses. Out of uninterpreted sense-experiences science cannot be distilled, no matter how industriously we gather and sort them. Bold ideas, unjustified anticipations, and speculative thought, are our only means for interpreting nature: our only organon, our only instrument, for grasping her. And we must hazard them to win our prize. Those among us who are unwilling to expose their ideas to the hazard of refutation do not take part in the scientific game.
Sir Karl Popper, *The Logic of Scientific Discovery*

PART I **The scientific enterprise**

From impacts on our sensory surfaces, we in our collective and cumulative creativity down the generations have projected our systematic theory of the external world. Our system is proving successful in predicting subsequent sensory input. How have we done it?

. . .

Observation sentences are the link between language, scientific or not, and the real world that language is all about.

. . .

Traditional epistemology sought ground in sensory experience capable of implying our theories about the world, or at least of endowing those theories with some increment of probability. Sir Karl Popper has long stressed, to the contrary, that observation serves only to refute theory and not to support it.

. . .

Pure observation lends only negative evidence, by refuting an observation categorical that a proposed theory implies.

Willard Quine, *Pursuit of Truth*

1

Ways of knowing

Since we hope to learn something about science and its operation and since science concerns itself with a certain type of knowledge and its attainment, let us begin with a brief consideration of how we arrive at knowledge. A common type of knowledge is that based on opinion or on the acceptance of another's authority. Most of our everyday knowledge used for dealing with the practicalities and necessities of life is gained in either of these fashions. Much of what one learns in the course of reading a book is taken as true simply because it has been presented on the printed page, although hopefully you will be more critical than that. Thus, we can merely have an opinion about a proposition and then decide to accept it as true or we can appeal to the authority of another, as 'Einstein says' or 'Aristotle says,' or to the authority of a text, as 'The Bible says,' or we can assert a fact to be 'obvious'. Consider the following example:

> We hold these truths to be self-evident, that all men are created equal; that they are endowed by their Creator with certain unalienable rights; that among these are life, liberty, and the pursuit of happiness. That, to secure these rights, governments are instituted among men, deriving their just powers from the consent of the governed; that, whenever any form of government becomes destructive of these ends, it is the right of the people to alter or to abolish it, and to institute a new government, laying its foundation on such principles, and organizing its powers in such form, as to them shall seem most likely to effect their safety and happiness.[1†]

In writing the Declaration of Independence, Thomas Jefferson (1743–1826) did not adduce evidence for the correctness of these givens, but circumvented all argument on that point by stating them to be self-evident. While such may be a neat rhetorical ploy, it may or may not, depending upon your predilections, seem to offer a solid basis for the truth of the statement.

1.1 PHILOSOPHY

As with so much in our Western tradition of thought, we find some of the earliest systematic treatments of the nature of human knowledge among the discourses of Socrates (c. 470–399 B.C.), Plato (428/427–348/347 B.C.) and Aristotle (384–322 B.C.), those ancient Greeks who contributed so greatly to the philosophical foundations of Western culture. *Philosophy* in its etymological roots comes from two Greek words meaning friendly or loving (φιλος, pronounced philos) and knowledge

† *Notes* begin on p. 378 and sources may be located via the *General References* (pp. 398–9) and the *Bibliography* (pp. 400 ff.)

(σοφια, sophia). Originally, it connoted love of wisdom, a desire for fresh experience, or the pursuit of intellectual culture. It later came to denote the disciplined study of reality and of human nature. Socrates lived in Athens during the period of the Peloponnesian War between that city and Sparta. After the defeat of Athens by Sparta, the Athenian court of 500 citizens found Socrates guilty of impiety and of corrupting the minds of the young with his philosophical questioning that led them to deny the existence of the gods. For this he was condemned to die by drinking hemlock. Although Socrates wrote nothing himself, we know of his dialogues through the teaching of his most famous pupil, Plato. Born of a distinguished Athenian family, Plato founded the Academy in Athens in 387 B.C. This was in some ways the counterpart of a modern university. Plato's most outstanding student by far was Aristotle, a prodigious writer and thinker whom we shall meet again several times in these beginning chapters.

Plato distinguished between mere *opinion* and *science*. For him science was the ideal of human knowledge and was seen as necessarily true and unchanging. We might take, as a model of such a system of knowledge, axiomatic geometry, as later found in Euclid's *Elements* (about 300 B.C.). In his *Republic*, Plato discussed the ideal state and the philosopher kings fit to rule it. He tells us that these philosophers are to desire a knowledge of the truth and of reality, not just of superficialities. According to him, true knowledge must be real, stable and unchanging. It must be of unique, immutable objects (his *Forms*), whereas belief has to do with appearances; knowledge is infallible, while belief may be true or false. Plato distrusted the world of sense experience because of its constant change and posited the existence of another world of changeless forms accessible only to the intellect. In his conception these forms were the true reality and existed independently of the sensible world. Experience and sense data give us only an approximate picture of these forms. As an example, he would suppose that the concepts of mathematics (more particularly of plane geometry), such as a circle, line and triangle, exist only in the abstract and that knowledge of them is not gained through sense experience, since a true circle, line or triangle does not exist in the sense world. We cannot draw a mathematical point or line because these will have extension when produced by a pen or pencil. This was Plato's answer to the quest for the immutable amidst the change of the sensible world.

Although both Plato and Aristotle were realists in that they believed in the existence of an external world independent of the mind of the observer, we may briefly characterize their differences by saying that Plato posited a reality far removed from immediate sense experience (realism of ideas or of forms), while Aristotle assigned primary reality to the objects of sense experience (realism of natures or of substances). Aristotle believed that form and matter were intellectually distinguishable, but not in fact separable in the real world of experience. He taught that all genuine knowledge is gained through logical demonstration proceeding from true and necessary first principles that are abstracted from sense experience or observation. His treatise on natural (or material) bodies was titled *Physica* (*Physics*) since the Greek word φυσικα (physika) was originally derived from the adjective for 'natural' (things) and, as used by Aristotle, had the meaning of natural science. It was in this sense that scientists were once referred to as natural philosophers. In his *Physics* he tells us that we must start from the things that are more knowable and obvious to us and proceed toward those that are clearer and more knowable by their nature.

> When the objects of an inquiry, in any department, have principles, conditions, or elements, it is through acquaintance with these that knowledge, that is to say scientific knowledge, is attained. For we do not think that we know a thing until we are acquainted with its primary conditions or first principles, and have carried our analysis as far as its simplest elements. Plainly therefore in the science of Nature, as in other branches of study, our first task will be to try to determine what relates to its principles.
>
> The natural way of doing this is to start from the things which are more knowable and obvious to us and proceed towards those which are clearer and more knowable by nature; for the same things are not 'knowable relatively to us' and 'knowable' without qualification. So in the present inquiry we must follow this method and advance from what is more obscure by nature, but clearer to us, towards what is more clear and more knowable by nature.
>
> Now what is to us plain and obvious at first is rather confused masses, the elements and principles of which become known to us later by analysis.[2]

Aristotle, unlike Plato (with his innate forms), holds that there is no such thing as innate (or inborn) knowledge. For Aristotle all knowledge must begin with input from the external world from which we may eventually generalize, abstract or induce the overarching schemes and unifying principles that please the mind and that produce in us a sense of finally having understood an area of knowledge or a collection of individual sense experiences.

1.2 LOGICAL DEDUCTION

Once we have arrived at these general principles (and we later discuss at length how this is done), we can then deduce further results from these by logical reasoning (that is, by valid inference). By *deduction* we mean the process of reasoning from the general to the particular, from a given premise or set of premises to necessary conclusions. A standard example of a deductive argument is that, given all men to be mortal and Socrates to be a man, it necessarily follows that Socrates is mortal. The syllogism, in its various forms, is one of the most basic and elementary tools used in logic. Aristotle sometimes uses the term *syllogism* loosely to mean a form of argument or discussion. He gave the definition: 'A syllogism is a discourse in which, certain things being stated, something other than what is stated follows of necessity from their being so.'[3] More precisely, 'syllogism' has a technical meaning as a structured form of arguing from two premises to a conclusion.

Although there are several forms of syllogisms and of deductive reasoning in general, we consider here only what are referred to as *hypothetical propositions* that will prove useful later. These statements are of the form 'If p, then q,' where p and q stand for two propositions. As a fanciful example consider the statement, 'If you are good, then I will give you a lollipop.' (Or, an instructor might say to a student, 'If you do well in my course, then I will give you an A for a final grade.') Once we accept this statement as binding, it necessarily follows that, provided you are good, I must give you the promised reward.[4] Let us now look at some possible outcomes of this reward–punishment scheme and see what we can logically conclude. If you are good, then, of course as already stated, you do get your lollipop. Suppose you do in fact get a lollipop. Does it necessarily follow that you were good? Of course not, since I could have given you a lollipop even if you were not good. The statement was not 'If and only if you are good, then I will give you a lollipop', but merely 'If you are good.' Finally, what follows if in fact you do not get a lollipop? You must not have been good since, if you had been good, you would certainly have received a lollipop.

In these hypothetical propositions, we refer to the *if* clause as the *antecedent* and the *then* clause as the *consequent*. Two important points are these: once the antecedent condition is fulfilled, the consequent always obtains; and, when the consequent does not occur, then the antecedent condition cannot have been fulfilled. That is, if we negate the consequent, then we must negate the antecedent. We introduce a shorthand or symbolic notation that can simplify the analysis of the structure of an argument. As before, let p and q denote propositions and $\sim p$ (non p) and $\sim q$ (non q) represent their negations. Finally, let the symbol $\Rightarrow$ stand for 'implies'. Hence we can state 'If p, then q' equally as well as $p \Rightarrow q$. The general logical rule that negating the consequent negates the antecedent becomes simply $\sim q \Rightarrow \sim p$.

Let us apply these results to the type of argument we meet in our study of physical theories. Suppose we take as given the statement (that we refine in Chapter 8) 'If Newton's law of gravitation governs a planet orbiting the sun, then the orbit of the planet is an ellipse.' We can make the following identification: p – 'Newton's law of gravitation governs a planet orbiting the sun' and q – 'The orbit of the planet is an ellipse.' The proposition itself is $p \Rightarrow q$. What if careful observation shows that the planetary orbit is not exactly an ellipse? In other words, we have established $\sim q$. Having negated the consequent, we must then negate the antecedent. That is, $\sim p$ follows so that Newton's law of gravitation does not govern the planet. As we shall see much later, one of the reasons that Newton's law of gravitation was replaced by Einstein's general theory of relativity is that a planetary orbit in fact departed slightly from elliptical shape.

Suppose, however, that (as seemed to be the case for a long time) the orbits of the planets were ellipses. Would this necessarily establish the validity of Newton's law of gravitation? Certainly not, since another agent could be responsible for keeping the planets in their orbits – say, angels beating with their wings against one side of the planets, as might be facetiously alleged some people believed at one time. Finally, if we know that Newton's law of gravitation is not correct ($\sim p$), does it follow that the planets do not go around the sun in elliptical orbits ($\sim q$)? No, it does not, since some other theory that is correct might yield the same result regarding elliptical orbits.

Let us summarize these important results as follows. Given the conditional statement (or hypothetical proposition) $p \Rightarrow q$, it necessarily follows that $\sim q \Rightarrow \sim p$, but it does not necessarily follow that $q \Rightarrow p$ or that $\sim p \Rightarrow \sim q$. Notice that there is a profound asymmetry between being able to prove an hypothesis and being able to disprove one. As a simple application of these general logical rules, criticize for yourself the following argument as to its logical validity. 'Because science, based on the belief in simple, objective laws of nature, has been so prodigiously successful for the past few centuries, we can necessarily conclude that these assumed laws of nature are correct and do exist as part of an external, objective reality.' We leave it to the reader both to identify properly propositions p and q, so that the basic argument can be put into the form $p \Rightarrow q$, and then to decide the validity of drawing the stated conclusion.

1.3 SELF-EVIDENT FIRST PRINCIPLES

So far we have given a few useful rules about logically valid relations between a given statement – what will sometimes be termed first principles – and consequences deducible from it. The critical question now is how we arrive at the truth of these givens (the antecedents). On this point let us examine the writing of René Descartes (1596–1650), the French mathematician and philosopher who is often considered the father of modern philosophy. Descartes was born in a rural area of France to a father who was a lawyer and judge in the province of Brittany. His early education (1604–1614) was by Jesuits at the Royal College in La Flèche. Descartes at an early age directed his studies and investigations to establishing an absolutely certain foundation for human knowledge. He spent most of his working years in Holland. In 1629, Descartes wrote his *Regulae* (*Rules for the Direction of the Mind*), although this was not published until after his death. He proposed that the method of mathematics could be generalized to science so that absolutely certain knowledge could be attained.

His plan was to start with self-evident propositions as his first principles. These first principles with which science was to begin were to be obtained through *intuition*, by which Descartes meant the direct apprehension of those truths that a clear and disciplined mind saw as certain. To these absolutely self-evident first principles one was then to apply logical deduction in order to obtain valid conclusions. Descartes' quest for certain knowledge was similar to Aristotle's in that both would begin with evident truths and then proceed from there by deduction. For Descartes the knowledge gained by intuition was even more secure than that obtained by deduction since intuition is immediate and simpler. A few quotations from the *Regulae* concerning these issues are given in Section 1.A at the end of this chapter.

In such a program Descartes saw an infallible means of adding to human knowledge:

> [I]f a man observe [these simple rules] accurately, he shall never assume what is false as true, and will never spend his mental efforts to no purpose, but will always gradually increase his knowledge and so arrive at a true understanding of all that does not surpass his powers.[5]

PART I   The scientific enterprise

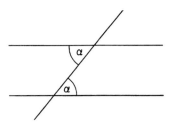

FIGURE 1.1   Parallel lines cut by a third straight line

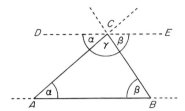

FIGURE 1.2   Interior angles theorem for a triangle

As a simple example of the type of argument Descartes had in mind, suppose we take as essentially a matter of definition the fact that a straight line subtends 180° and as an axiom (or postulate) that, when two parallel straight lines are cut by a third straight line, then the opposite interior angles (α of Figure 1.1) are equal. (This property can also be obtained from Euclid's famous parallelism axiom that through any point not on a straight line one and only one parallel to the original line can be drawn.) From this definition and this axiom let us now obtain the well-known result that the sum of the interior angles of a plane triangle is 180°. In Figure 1.2 we have drawn the line DE parallel to the base AB of triangle ABC. Using the property illustrated in Figure 1.1, we see that angle DCA equals α, while angle ECB equals β. Since $\alpha + \beta + \gamma = 180°$, we obtain the stated result.

It was this type of argument (or proof) from 'obvious' first principles that Descartes took as his model. Euclidean geometry with its self-evident axioms and scheme of logical deductions should lead to absolutely certain results. However, as we shall see in later chapters, it was found in the nineteenth century that non-Euclidean geometries could be constructed in which the parallel-line postulate is false. The problem with Descartes' scheme is that certain principles that appear to be self-evident to us may in fact be incorrect in the real world. That is, if the parallel-line axiom is accepted, then the stated result about the sum of the interior angles of a triangle does follow. However, mathematicians have discovered logical alternatives to Euclidean geometry. There are spaces in which a three-sided figure bounded by straight lines will have its interior angles sum to more or less than 180°. An example is a triangle drawn on the surface of a sphere. Without observation, we cannot be certain which of these geometries is that of the space we actually live in. This is the sense in which Albert Einstein (1879–1955) meant that logical thinking alone cannot give us knowledge of the real world. (See the quotations for Part I (*The scientific enterprise*).) We must very often appeal to sense experience to decide which general and logically possible principles are in fact respected by nature. All human knowledge is fallible. In his later work, such as *Discourse on Method* (1637), Descartes did retreat somewhat from this purely intuitive, deductive approach to physical laws. He continued to believe that the general principles of natural science could be demonstrated in this fashion, but admitted that these had to be combined with data from sense experience in order to derive the more specific laws of science. He also saw the need to do experiments to decide among these alternative laws. This was the germ of the scientific method as it has since come to be practiced.

In the mid-nineteenth century, the astronomer Sir John Herschel (1792–1871) believed in the existence of a straightforward and reliable procedure of beginning from unproblematic observational data and using deduction to gain certain knowledge:

> [O]ne preliminary step to make... is the absolute dismissal and clearing the mind of all prejudice ... and the determination to stand and fall by the result of a direct appeal to facts in the first instance, and of strict logical deduction from them afterwards.[6]

While Descartes began with principles self-evident to the mind, whereas Herschel started with empirical evidence, each believed he had an unassailable foundation for knowledge.

1.4   RATIONALISTS VERSUS EMPIRICISTS

This basic difference in emphasis on reason or on experience recurs throughout the ages among philosophers. *Rationalists*, such as Descartes, rely mainly upon reason as the source of knowledge, while *empiricists*, such as John Herschel (cited above) or David Hume (1711–1776) whom we discuss in Chapter 3, rely more upon experience, through observation and experiment, as the basis of knowledge. This latter approach brings us to the problem of induction. By *induction* we mean a process of reasoning from particulars to the general, from the individual to the universal. For example, from the fact that all men of past ages who have ever existed have died, we might induce the general proposition that all men, including those still alive and even those yet to be born, are mortal. The question of how one goes from specific instances or observations to those general principles we need to begin a deductive argument is the subject of the next chapter.

We end this discussion by considering a quotation from the writings of the theoretical physicist Max Planck (1858–1947), one of the pioneers of modern physics and a founder of the quantum theory of light, for which he received the Nobel Prize in physics in 1918. Planck, like Aristotle, lays emphasis on the primacy of sense knowledge that has been essential for the eventual development of natural science. This is reflected in the following excerpt from *The Meaning and Limits of Exact Science*.

> If we seek a foundation for the edifice of exact science which is capable of withstanding every criticism, we must first of all tone down our demands considerably. We must not expect to succeed at a stroke, by one single lucky idea, in hitting on an axiom of universal validity, to permit us to develop, with exact methods, a complete scientific structure. We must be satisfied initially to discover some form of truth which no skepticism can attack. In other words, we must set our sights not on what we would like to know, but first on what we do know with certainty.
>
> Now then, among all the facts that we do know and can report to each other, which is the one that is absolutely the most certain, the one that is not open even to the most minute doubt? This question admits of but one answer: 'That which we experience with our own body.' And since exact science deals with the exploration of the outside world, we may immediately go on to say: 'They are the impressions we receive in life from the outside world directly through our sense organs, the eyes, ears, etc.' If we see, hear or touch something, it is clearly a given fact which no skeptic can endanger.[7]

Even here, though, Planck attaches too much certitude to the reliability of what we observe with our senses. Not only can we misinterpret the stimuli that our senses send to the brain, as in many common varieties of optical illusions for instance, but there remains the more serious question of interpreting the data we gather from observation and experiment. There have been cases in the history of science in which skilled scientists of the hightest repute have 'seen' or 'verified', through observation and experiment, the prediction of some hypothesis, even though this prediction subsequently turned out not to correspond to reality and could not be reproduced by other observers. For example, Sir William Herschel (1738–1822), discoverer of the planet Uranus, the father of John Herschel and the most famous astronomer of the eighteenth century, was able with the powerful telescopes he manufactured to resolve into individual stars several nebulae that had previously appeared to be milky luminous patches in the sky. In the mid 1780s, he conjectured that all nebulae were composed of individual stars so that none were made of a luminous fluid. In 1790 he did observe a nebula that he was forced to interpret as a central star surrounded by a cloud of luminous fluid. In the interim period, however, Herschel claimed to resolve into individual stars both the Orion and Andromeda nebulae. In fact, though, Orion is a gaseous cloud containing a continuous distribution of matter, not just individual stars, while Andromeda is a galaxy of stars.

In more recent times, there was a dispute of several years' duration between the American Robert Millikan (1868–1953), who eventually received the 1923 Nobel Prize in physics for his observations of the electron as the fundamental and indivisible unit of negative electric charge, and Felix Ehrenhaft (1879–1952) of Vienna, who, along with his colleagues, obtained data that seemed to show the existence of electric charges much smaller than that of Millikan's electron.[8] Subsequent attempts by others to reproduce Ehrenhaft's results were unsuccessful, so that Millikan's view prevailed. It seems as though Ehrenhaft misinterpreted his data, perhaps largely because he believed that electrical charge should be continuously divisible. There have also been suggestions that Ehrenhaft was not always as careful as he ought to have been in his experiments. The point is not that either Herschel or Ehrenhaft was a dishonest man. Furthermore, especially in Herschel's case, one unquestionably had a highly skilled observer and careful worker. The difficulty arose in the interpretation of data and with the influence that one's expectations had on this interpretation.

Nevertheless, having entered these cautions against Planck's claim of the importance and reliability of sense data, we must still appreciate that such data are what remain more or less intact as theories come and go. This attitude is nicely exemplified in the introduction to an advanced text on quantum physics written in the second quarter of the twentieth century. The book itself abounds with pictures of the physical data that the then-current theory attempted to explain. As a justification for the large amount of laboratory data presented, the author tells us:

> That photographs are an extremely important feature of any book on atomic spectra may be emphasized by pointing out that, of all the theories and knowledge concerning atoms, the spectrum lines will remain the same for all time.[9]

1.5 THE STATUS OF SCIENTIFIC KNOWLEDGE

The basic problem we framed in this chapter is that of providing a sound foundation for scientific knowledge – really, for the hypotheses or assumptions that form the core of our fundamental theories of the physical universe. We indicated that throughout history there has been a recurrent tension between intuition and observation as the basis for such a foundation. As the introductory quotations for Part I (*The scientific enterprise*) illustrate, an essential and ineliminable role is played by human creativity in fashioning a scientific theory. More than logic and empirical evidence alone is required in order to arrive at our successful, overarching theories that account for the phenomena we observe. In the following chapters we shall see how we finally arrived at an understanding of the corrigibility of scientific knowledge. To appreciate our position today, we must know where we came from intellectually and how we attained our present vantage.

In the process of doing this, we have to consider whether or not our present knowledge depends sensitively upon the historical path by which we reached it – that is, what role historical contingency plays in the evolution of ideas (perhaps much as it does in the evolution of living organisms).[10] For example, the 'path dependence' of a dominant technology is a well-documented and familiar fact. As illustrations, consider the case of video cassette recorders in which the technologically superior Betamax format has been nearly universally supplanted by the VHS one, essentially because a powerful group of Japanese manufacturers of video components decided to push the marketing of that format. Similarly, Apple MAC computers are in many ways innovatively superior to and more user friendly than the much more dominant IBM PCs that gained the upper hand through better marketing strategies. A more mundane example is provided by the standard arrangement of characters on typewriter and computer keyboards. This layout was originally employed in order to slow down the typist (to prevent mechanical levers from jamming in early model typewriters), yet remains even though more efficient arrangements of the characters are known. Once a dominant technology becomes ensconced (even if largely on the basis of 'chance' factors), it is too difficult and costly to retool and replace it. The 'best' competitor does not always win here. Similarly, English has become the virtually universal language for international scientific and business conferences, not because of any intrinsic superiority of that language over some other (such as French, German or Japanese), but simply because of the historical contingency of the political and military history of great nations. Could this be so in the realm of scientific ideas as well? Are the laws and theories[11] of our science nothing more than one contingent way to describe the world, or do they yield truth? For now, this is posed as an open question, to which we return often in this book.

I.A DESCARTES' *REGULAE*

In his *Regulae* Descartes laid out the rules that were to lead man to certain knowledge. Here are a few excerpts from that work.

Rule II

Only those objects should engage our attention, to the sure and indubitable knowledge of which our mental powers seem to be adequate.

Science in its entirety is true and evident cognition.... Thus in accordance with the above maxim we reject all such merely probable knowledge and make it a rule to trust only what is completely known and incapable of being doubted.

...

Consequently if we reckon correctly, of the sciences already discovered, Arithmetic and Geometry alone are left, to which the observance of this rule reduces us.

...

But one conclusion now emerges out of these considerations, viz. not, indeed, that Arithmetic and Geometry are the sole sciences to be studied, but only that in our search for the direct road towards truth we should busy ourselves with no object about which we cannot attain a certitude equal to that of the demonstrations of Arithmetic and Geometry.

...

Rule III

In the subjects we propose to investigate, our inquiries should be directed, not to what others have thought, nor to what we ourselves conjecture, but to what we can clearly and perspicuously behold and with certainty deduce; for knowledge is not won in any other way.

...

Now I admit only two mental operations, viz., intuition and deduction.

By *intuition* I understand, not the fluctuating testimony of the senses, nor the misleading judgement that proceeds from the blundering constructions of imagination, but the conception which an unclouded and attentive mind gives us so readily and distinctly that we are wholly freed from doubt about that which we understand. Or, what comes to the same thing, *intuition* is the undoubting conception of an unclouded and attentive mind, and springs from the light of reason alone; it is more certain than deduction itself, in that it is simpler, though deduction, as we have noted above, cannot by us be erroneously conducted. Thus each individual can mentally have intuition of the fact that he exists, and that he thinks; that the triangle is bounded by three lines only, the sphere by a single superficies, and so on. Facts of such a kind are far more numerous than many people think, disdaining as they do to direct their attention upon such simple matters.

. . .

This evidence and certitude, however, which belongs to intuition, is required not only in the enunciation of propositions, but also in discursive reasoning of whatever sort. For example, consider this consequence: 2 and 2 amount to the same as 3 and 1. Now we need to see intuitively not only that 2 and 2 make 4, and likewise 3 and 1 make 4, but further that the third of the above statements is a necessary conclusion.

Hence now we are in a position to raise the question as to why we have, besides intuition, given this supplementary method of knowing, viz., knowing by *deduction*, by which we understand all necessary inference from other facts that are known with certainty. This, however, we could not avoid, because many things are known with certainty, though not by themselves evident, but only deduced from true and known principles by the continuous and uninterrupted action of a mind that has a clear vision of each step in the process. It is in a similar way that we know that the last link in a long chain is connected with the first, even though we do not take in by means of one and the same act of vision all the intermediate links on which that connection depends, but only remember that we have taken them successively under review and that each single one is united to its neighbour, from the first even to the last. Hence we distinguish this mental intuition from deduction by the fact that into the conception of the latter there enters a certain movement or succession, into that of the former there does not. Further deduction does not require an immediately presented evidence such as intuition possesses; its certitude is rather conferred upon it in some way by memory. The upshot of the matter is that it is possible to say that those propositions indeed which are immediately deduced from first principles are known now by intuition, now by deduction, i.e. in a way that differs according to our point of view. But the first principles themselves are given by intuition alone, while, on the contrary, the remote conclusions are furnished only by deduction.

These two methods are the most certain routes to knowledge, and the mind should admit no others. All the rest should be rejected as suspect of error and dangerous.[12]

FURTHER READING

At the end of each chapter you will find a few references suggested for more discussion of certain topics raised in the text. The reader should appreciate, though, that chapter footnotes (that are collected together in the *Notes* section at the end of the book) also contain many relevant references that are not repeated in these *Further Reading* sections. Moreover, the *General References* section at the end of the book contains brief descriptions of several reference works that will frequently be useful. Again, these will not usually be repeated under *Further Readings*. It is assumed that the interested reader will be aware of these background references and consult them occasionally, as the need arises.

Philipp Frank's *Philosophy of Science* provides a useful view, from a physicist–philosopher's perspective, of the relation between science and philosophy. Ernan McMullin's essay 'Cosmic Order in Plato and Aristotle' gives a helpful guide to what proved to be an influential view of the universe, one that forms the foundation of much of what we discuss in the following chapters. Finally, Gerald Holton's *The Scientific Imagination* contains a valuable set of case studies in the history of science (including the Millikan–Ehrenhaft encounter mentioned in this chapter).

2

Aristotle and Francis Bacon

In order to contrast by means of specific examples two different ways of reaching general conclusions about the phenomena of reality, we compare the work of Aristotle with that of Sir Francis Bacon (1561–1626). As we shall see, the difference is one of emphasis and execution rather than of principle. Aristotle favors the discussion of general principles that can be arrived at on the basis of an intuitive grasp of natures, while Bacon stresses the importance of slow and careful induction from specific cases.

2.1 ARISTOTLE

Aristotle was born in 384 B.C. at Stagira, a Greek colonial town on the peninsula of Chalcidice in Macedonia on the northern shores of the Aegean. His father Nicomachus (*d. c.* 374 B.C.) was a physician to Amyntas II of Macedon (*d.* 370/369 B.C.), grandfather of Alexander the Great (356–323 B.C.). In 367 B.C. Aristotle, at age seventeen, came to Athens and began his studies as a pupil in Plato's Academy. Plato was at that time a man of sixty. Aristotle's life falls rather naturally into three major periods. He spent twenty years at the Academy until Plato's death (348/347 B.C.). The anti-Macedonian mood prevalent in Athens may have been partly responsible for his leaving that city and traveling for the next twelve years. In this second period he went to Macedon in 342 B.C. at the behest of Philip II (382–336 B.C.) to tutor that king's son, Alexander. During his thirteenth through sixteenth years, Alexander was taught by Aristotle. Thereafter, Alexander began to share more of his father's responsibilities and had less time and need for Aristotle's instruction. Upon Philip's assassination in 336 B.C., Alexander at age twenty ascended the throne and no longer required a mentor. There was then little to hold Aristotle in Macedon.

The third period of Aristotle's life began in 335 B.C., when he returned to Athens once that city had come under the Macedonian rule of Alexander. During the next twelve years, Aristotle headed the Peripatetic School in the Lyceum outside Athens. This institution was somewhat like a modern college and most of the finished works of Aristotle as we have them today were set down by him (or by his students and colleagues) as lectures given in the Lyceum. As a thinker and teacher, Aristotle showed himself to be a philosopher, psychologist, logician, moralist, political thinker, biologist and the founder of literary criticism. When Alexander the Great died in 323 B.C. at age 33, resentment against the Macedonian rule and Aristotle's

Macedonian background made life in Athens uncomfortable for the philosopher. At this time, a charge of impiety was brought against him for an elegy he had delivered twenty years earlier on behalf of one of Alexander's allies. Aristotle recalled the fate of Socrates before him in that same city and left saying that he would not let the Athenians offend twice against philosophy.[1] He retired in voluntary exile to the Ionian city of Chalcis in Euboea, the native district of his mother. He died there a few months later in 322 B.C.

2.2 OBSERVATION VERSUS EXPERIMENTATION

We saw in Chapter 1 that in his *Physics* Aristotle emphasized the need to begin with sense experience, or data from the real world, and from this to proceed to general laws or rules. For him there must always be an experiential part to the examination of any general question about the sensible world. We shall see a concrete example of Aristotle's method in the next section when we study his views on motion. A later age would criticize Aristotle for his characteristic of passing from a preliminary examination of specific cases to sweeping general conclusions. For the present, we distinguish the fine powers of *observation*, that some of the early Greeks had, from *experimentation*, that belongs by and large to a later age. In observation, one carefully notices and takes account of the phenomena nature happens to present to the senses. (For example, looking at people in everyday circumstances and seeing how they function, what motivates them, and what makes them happy are observations.) In experimentation, one creates new or controlled situations in nature and exacts a response from nature. (For example, isolating people or animals from their normal environments to study their responses to specific and carefully controlled positive or negative stimuli is experimentation.)

When we say that some of the early Greeks were careful observers, we do not mean to imply that their conclusions were uniformly reliable in all fields. Aristotle's works on ethics and even in the science of biology stood up to subsequent scrutiny and parts of them remain valid even in modern times, whereas others are rather poor, such as those on falling bodies and projectile motion (as we shall see). Aristotle, like a number of other ancient Greek thinkers, had fine powers of observation coupled with a penchant for cosmological speculation. He often considered large questions and proposed solutions to them, as opposed to studying finite, tractable problems. (There is some indication, though, that later in his career he gradually shifted from an essentially speculative approach to knowledge to one with a considerable inductive component as well.) It seems a fact, unfortunate or otherwise, that the most evident, pressing and meaningful questions for us are horribly complex and 'ultimate' ones. The Greeks, and many thinkers in our Western tradition since that time, have participated in what is often referred to as the great dialogue (the writings contained in the so-called *Great Books*). Much of this might be characterized as a structure of grand schemes that are restatements of and unsuccessful attempts at answers to these difficult questions. We say that many of these attempts have been unsuccessful in the sense that they have not provided the definitive and final answers sought, although they have given rise to continuing discussion and, hopefully, to progress.

2.3 THE UNIVERSE AS ORGANISM

As an example of a development of thought that culminated in the teachings of one great intellect, let us consider the classical Greek view of the physical universe as most fully formulated and articulated by Aristotle. The early Greek world as represented in the epics of the *Iliad* and the *Odyssey* of Homer (about 850 B.C.) abounds with gods, both major and minor. These are directly responsible for the functioning of the physical universe (for example, Aeolus, the wind god; Poseidon, the sea god; and Demeter, goddess of the fruitful soil and of agriculture). They also decided the fortunes of men (for example, Ares (Mars in Roman mythology), the god of war; the three goddesses known as the Fates: Clotho, who spun the thread of life, Lachesis, who determined its length, and Atropos, who cut it off). Since the Greek gods were fickle and often cruel, nature appeared to the early Greeks as capricious and beyond their control or comprehension. A transition began with the Ionian philosophers represented by Thales (*c.* 624–*c.* 546 B.C.), Anaximander (610–546/545 B.C.) and Anaximenes (*fl. c.* 545 B.C.) for whom there was just one permanent element (or perhaps only a few of them) in the universe from which all was built. They presented a more or less depersonalized interaction of things in the physical universe. But even Democritus (*c.* 460–*c.* 370 B.C.), with his rigidly mechanistic interpretation of nature, still used analogies from the organic world. Although we return to these philosophers in Chapter 12, our purpose here is to indicate that there was a school of Greek thought that was developing toward an impersonal or mechanistic view of the physical universe.

However, Socrates spoke out very strongly against such mechanistic philosophies. His teachings come to us through the writings of his pupil Plato. In Plato's *Phaedo*, a group of Socrates' disciples visits the old philosopher in prison after he has been condemned to die. The dialogue that follows is Socrates' last before he drinks the hemlock in the presence of his companions. In it Socrates discusses man, the soul and the question of immortality. He searches for a science satisfying the needs and aspirations of man. For him, as for his disciple Plato, the mechanistic philosophies proved to be inadequate for this purpose and had to be rejected since they separated man from nature and nature from the realm of the good and the beautiful. In his *Timaeus* Plato presents what we may refer to as his scientific view of the operation of the universe and of the creatures in it and states that 'the world became a living creature truly endowed with soul and intelligence by the providence of God.'[2] Plato found repulsive the explanation that the mechanistic philosophies gave of the living world as a product of pure chance through the mechanisms of inanimate beings. (We return to this question, from a modern perspective, in Chapters 12 and 25.)

In his treatises *Physics* and *De Caelo* (*On the Heavens*), Aristotle presented his *organismic* view of the universe, a view that dominated much of Western thought until the sixteenth century. By 'organismic' we mean here only that Aristotle viewed and attempted to explain the behavior of inanimate bodies by analogy with living organisms. In using this term, we are not implying that Aristotle saw inanimate objects as living, but rather that the former, like the latter, appear to be directed toward a goal. Each object had its own nature, tendency or goal. Thus, it was the nature of a light element, such as fire, to go up and of a heavy element, such as earth, to go down. These were their natural motions. Since it was the nature of earth to seek the center of the universe, it followed that the earth itself must have its center at the center of the universe and be spherical in shape. As all the parts of the earth attempt to fall toward the center of the universe from various directions, they must finally impinge upon one another and form a spherical body. Similarly, since a circle was considered an unchanging (and, hence, potentially eternal) figure and since the heavens and their element, aether, were also seen as unchanging, the natural motion of heavenly bodies (for example, the stars) was taken to be circular motion. Because change and corruption were evident phenomena in the world of the earth, the earth was assigned a nature quite different from that of the heavens. Natural motion for the terrestrial elements (earth, air, fire, water) was straight-line motion (for instance, straight up for fire and straight down for earth). This fundamental distinction between the laws that governed the heavens and those that held on the earth remained until the time of Galileo Galilei (1564–1642). (See the quotations for Part III (*The Newtonian universe*).) Any motion that was not natural was violent or forced motion and required the constant action of an external agent to effect it.

In Aristotle's world system, the universe was spherical with the earth at the center and with the planets occupying successive layers. These outer spherical shells (those above the orb of the moon) consisted of a fifth element – the aether. His universe was finite and there was nothing beyond the farthest edge of the outermost sphere. The sublunar region, that included the earth, was composed of four elements: fire, air, water and earth. Fire and earth were the 'extremes' (highest and lowest), while air and water were the 'intermediates' (in the region between the extremes). Each element had its own natural place and, once there, it was quiescent. Of central importance were the notions of change and becoming: the realization of potentiality into actuality, and this extended to Aristotle's understanding of the motion of physical objects. An element was fully actual only in its natural place and each thing tended toward the good for its nature. In this scheme there was no self-sustaining, uncaused motion, since an agent was required for all motion. The motive efficacy of the heaviness of a body composed of earth increased as the body came closer to its natural place. (Lightness was the corresponding causal principle for fire.) As we shall see in more detail below, a true vacuum was impossible because there could be no natural motion there. A medium served both as a cause of motion and as providing resistance to it.

Final causes played an essential role in Aristotle's explanation of the physical universe. His was a unified, overarching world view of a whole. Since living beings appear to act, or to be directed, toward an end or goal, Aristotle sought to explain the behavior of inanimate objects in terms of final causes (or in terms of that for which or toward which a thing acts in virtue of its nature).

> The necessity that each of the simple bodies should have a natural movement may be shown as follows. They manifestly move, and if they have no proper movement they must move by constraint; and the constrained is the same as the unnatural. Now an unnatural movement presupposes a natural movement which it contravenes, and which, however many the unnatural movements, is always one. For naturally a thing moves in one way, while its unnatural movements are manifold. The same may be shown from the act of rest. Rest, also, must either be constrained or natural, constrained in a place to which movement was constrained, natural in a place movement to which was natural. Now manifestly there is a body which is at rest at the centre. If then this rest is natural to it, clearly motion to this place is natural to it.
>
> . . .
>
> But since 'nature' means a source of movement within the thing itself, while a force is a source of movement in something other than it or in itself *qua* other, and since movement is always due to either nature or to constraint, movement which is natural, as downward movement is to a stone, will be merely accelerated by an external force, while an unnatural movement will be due to the force alone.[3]

Some of Aristotle's commentators characterized this by explaining that a heavy object in seeking to be nearer the center of the earth (and of the universe) naturally fell down and its tendency or urge (and therefore its speed) increased as it got closer to its goal. This apparent purposefulness of action by which living organisms act for an end or move toward a goal is termed *teleonomy* (from the Greek words *telos* for end or goal and *nomos* for law) or a teleonomic principle. In using this analogy to illustrate the concept of teleonomy, we do not mean that only living things can be goal directed. Such an organismic or teleonomic outlook did not, in fact, finally prove useful in advancing our understanding of the physical universe.

Aristotle's views on the subject of physics are often represented as essentially freezing the thinking of those who followed him until the time of the Renaissance. We point out in Chapter 6 that this was not universally true, since there were major thinkers, both in antiquity and in the Middle Ages, who differed with Aristotle on the subject of motion. Nevertheless, many took his work as dogma. After him, questions, even of fact, were often resolved by appeal to his principles rather than by observation. Such a priori[4] reasoning does not allow one to question premises and hence modify them readily. As we shall see, the rise of science as we know it today was marked by a change from organism to a mechanistic view of nature.

2.4 ARISTOTLE ON MOTION

We now summarize Aristotle's teaching on the motion of bodies and, as usual, relegate most of the direct quotations to the end of this chapter. In his *On the Heavens* Aristotle states that bodies have their natural motions produced by weight (or by lightness) and that the distance a given body will cover in a fixed time increases with its weight. On this point we find in his *Physics*:

> We see that bodies which have a greater impulse either of weight or of lightness, if they are alike in other respects, move faster over an equal space, and in the ratio which their magnitudes bear to each other.[5]

This is often summarized by saying that Aristotle held that heavier bodies fall more rapidly than less heavy ones, and this in direct proportion to their weight. But, it is important to realize that some of the terms Aristotle employed may not have had the same technical meaning for him as they do for us today. To understand what an author from another age meant, we should attempt to put ourselves into his mind set. We think of the weight of a body as its 'heaviness' (that can be measured, for example, by an ordinary scale) and that, near the surface of the earth, is essentially a constant (proportional to the intrinsic mass of the body). It appears that for Aristotle the speed with which a body fell was a measure of its weight. From this point of view, it is conceivable that the weight of a body might vary as the body moved. It is also possible that Aristotle had in mind the average velocity (over the distance of motion), as opposed to our modern notion of instantaneous velocity.[6] His theory or description of motion was largely qualitative, or at best semiquantitative, in nature and simply cannot be pressed too far quantitatively, as we shall see. None of this is said to deny that there were serious inadequacies in Aristotle's account of motion, but only to indicate that it may not have been manifestly inconsistent and wholly without merit.

The natural motion proper to a body was determined by the proportion of the four basic elements in it. (Thus, a body composed largely of the element earth would have a greater natural downward speed than one that had a sizable proportion of the element fire in it.) A body consisting of earth moved more quickly the closer it was to its natural place (the center of the earth and of the universe).[7] Hence, its weight (in Aristotle's sense) might have increased as it moved (and thus its speed too). In the original Greek text of the passage quoted above, Aristotle nowhere uses the verb for fall, so that he need not be talking there about the vertical motion of free fall.[8] In any case, by the Middle Ages, and into the Renaissance, people generally believed that his doctrine stated that the rate of fall of a body was in direct proportion to its weight (or heaviness in our sense). One reason for this may be that many of the ideas of classical Greek antiquity came to the people of Europe prior to the Renaissance through the works of Roman authors written in Latin. In his *De Rerum Natura* (*On the Nature of the Universe*), the Roman poet and philosopher Lucretius (*c*. 96–*c*. 55 B.C.) tells us that:

> For whenever bodies fall through water and thin air, they quicken their descents in proportion to their weights, because the body of water and subtle nature of air cannot retard everything in equal degree, but more readily give way, overpowered by the heavier: on the other hand empty void cannot offer resistance to anything in any direction at any time, but must, as its nature craves, continually give way; and for this reason all things must be moved and borne along with equal velocity though of unequal weights through the unresisting void.[9]

Lucretius does use the verb 'fall' and that may account in part for the Renaissance belief that Aristotle had done the same.

In discussing natural motion through a medium (say, air or water), Aristotle assumes that the speed with which a body moves is inversely proportional to the density of (or to the resistance offered by) the medium. We can express his belief that the natural speed v of a body is directly proportional to its weight W and inversely proportional to the resistance R of the medium as

$$v = \frac{W}{R} \tag{2.1}$$

Three comments are in order about Eq. (2.1) The first is that Aristotle's laws of motion are not expressed in equations in his writings. That is a modern convenience we have employed in an attempt to make the central concepts of his ideas on motion intelligible to a modern reader. The second is that one ought not be concerned with the units in which Eq. (2.1) is written since, again, we do not mean that equation to be used for quantitative calculations. Finally, it is unclear whether, to best represent Aristotle's ideas with Eq. (2.1), one should take v to be the average or the instantaneous speed.

In his *Physics* he argues that a body would move with infinite speed through a void that offers no resistance to motion ($R = 0$ in Eq. (2.1)). Since he considers this impossible, Aristotle concludes that there can be no void. The existence of a void also runs counter to his concept of natural motion as directed toward a place, because Aristotle feels that one cannot define place in a (universal) void. In considering hypothetical motion in a void, he states:

> [N]o one could say why a thing once set in motion should stop anywhere; for why should it stop *here* rather than *here*? So a thing will either be at rest or must be moved *ad infinitum*, unless something more powerful get in its way.[10]

Modern physical theory would conclude from this that a body once set into motion in a void would continue forever in its natural state of motion (that would be a straight line). Instead, Aristotle rejects the possibility of a void (since infinite rectilinear motion appears untenable in a finite universe). This impossibility of a void is central to Aristotelian physics and to the Aristotelian world view in general. In the previous section we saw that Plato rejected – as did Aristotle later – the mechanistic and materialistic philosophies associated with atomism in which atoms moved about chaotically in a void. By denying even the logical possibility of a void, Aristotle foreclosed any such materialistic philosophy. Furthermore, the Aristotelian universe was finite in extent, giving it an absolute center in relation to which the place and motion of any body could be uniquely defined.[11] Of course, such arguments do not produce necessary truths, but are based, ultimately, on an opinion about what the world must (surely) be like.

As we stated earlier, and as Eq. (2.1) makes clear, (sublunary) motion, even when constant, requires a cause (weight or force). Since every motion that was not natural required an external motive agent and hence an explanation, projectile motion posed a serious problem. Once an arrow has left the bowstring, it does not fall straight down (as it should according to its natural tendency to seek the center of the earth), but instead follows a curved trajectory in its return to the ground. Aristotle met this challenge by assuming that the surrounding air closes in behind the arrow (or behind any projectile) and urges it along its curved path. This explanation is sufficiently implausible or counterintuitive that his position on projectile motion was one of the weak points of Aristotelian physics and was often attacked.

In modern philosophy of science the term *ad hoc* (Latin: literally, 'for this') is often used pejoratively for explanations or devices (like Aristotle's in the preceding paragraph) that are introduced into a theory or model simply to produce agreement with some troublesome observations or data. Their use can signal that a theory is inconsistent or in serious trouble. Several times in the coming chapters we refer to ad hoc moves in just this spirit. Of course, one should appreciate that it is often not a simple matter to decide whether or not a modification made in response to a difficulty encountered by a theory is truly ad hoc or, instead, a creative and progressive development that will lead to fruitful new research. Frequently an adjustment can be seen as convincingly and pejoratively ad hoc only retrospectively with the aid of hindsight, once a complete and successful theory is already in place.

2.5 FRANCIS BACON

In order to present in outline an approach to knowledge with an emphasis very different from that of Aristotle's, we discuss briefly the *Novum Organum* (*The New Organon*) of Sir Francis Bacon. Since the collection of Aristotle's treatises on logic as an instrument (*organon*) of science and scientific reasoning had become known as *The Organon* and since Bacon's work was to be very critical of Aristotle and his school, it is significant that Bacon should title his own treatise *The New Organon*. As with any summary, the following is a poor substitute for actually reading Bacon. Our main interest in this will be as a vehicle for introducing some valuable ideas rather than to set this work up as a guide to scientific reasoning.

A few comments about Bacon and his times may help furnish some perspective. A man of letters and a prolific essayist, he lived in England during the reigns of Elizabeth I (1533–1603) and of James I (1566–1625) in the age of William Shakespeare (1564–1616). His scientific contemporaries in England were William Gilbert (1540–1603), who was physician to Elizabeth I and whose *De Magnete* (*On the Magnet*) was the beginning of the modern history and theory of both magnetism and electricity, and William Harvey (1578–1657), whose *De Motu Cordis* (*On the Motion of the Heart*) was a classic of inductive science and laid the foundations for our understanding of the heart and of the circulatory system in humans and animals. During the same period Galileo in Italy was studying the motion of falling bodies and paving the way for *mechanics* (the study of motion in general) as we know it today and for much of modern scientific method. In Germany Johannes Kepler (1571–1630) was struggling with the problem of the nature of the orbits of the planets around the sun and would eventually announce his three laws of planetary motion, without which Isaac Newton (1642–1727), two generations later, might never have been able to formulate his universal law of gravitation.

Francis Bacon was born January 22, 1561, the son of Sir Nicholas Bacon (1509–1579), Lord Keeper of the Great Seal under Elizabeth I. His mother was a zealous Puritan. Early in his education he became disenchanted with Aristotle and syllogistic learning as practiced since ancient times. This had a great impact on his later thinking. In 1579 he turned to law to get ahead quickly after his father's death left Francis with neither wealth nor income. At age twenty-three, and for the next thirty years, he held a seat in the House of Commons. For nearly twenty-five years he lived in the shadow of the great legal scholar Sir Edward Coke (1552–1634) (his bitter rival) and of the politically influential Robert Devereux (1566/67–1601), the second Earl of Essex (his friend and patron). Elizabeth appointed Bacon to prepare a case against Essex for the way in which Essex had settled a rebellion in Ireland. Bacon sent Essex to the scaffold and thereby gained power through the Queen's favor. In 1603 James I became king and Bacon's fortunes improved considerably, although he was never able to obtain James' support for his grand project (*The Great Instauration* that included *The New Organon*) of restructuring natural philosophy. Bacon defended James in Parliament on the divine right of kings issue and in 1613 Coke, with whom Bacon had repeated conflicts over the question of this royal prerogative, was dismissed from all offices. Then, in succession, Bacon became Attorney General (replacing Coke), Lord Keeper of the Great Seal in 1617, prosecuted Sir Walter Raleigh (1554–1618) in 1618 on a previous treason charge, for which Raleigh was executed and Bacon was rewarded with the Chancellorship, prosecuted Thomas Howard (1561–1626), the first Earl of Suffolk, in 1619 as an embezzler of government funds, for which Bacon became Baron Verulam, published *The New Organon* in 1620 (at age sixty), and was made Viscount St. Albans in 1621. But in 1621 Coke indicted Bacon on twenty-eight charges of accepting bribes and this led to Bacon's own downfall. He spent the remaining five years of his life working out schemes for the advancement of science. While attempting an experiment to preserve a hen by freezing it in the snow, he contracted pneumonia and died.

Bacon was an enormously gifted, complex person with many interests and with corresponding demands upon his talents. Not surprisingly, some of his major undertakings failed to achieve the goals he had hoped they might. His *The New Organon* (that was never completed) did not actually influence the great contemporary scientific minds of his time mentioned above (or even those immediately succeeding him). Partisans of Bacon have at times claimed too much for *The New Organon*. He was isolated from and quite critical of the developments of his scientific contemporaries. Bacon rejected the Copernican 'hypothesis' and treated with contempt the researches of Galileo and of Gilbert. He underrated the importance of mathematics and of exact measurement. Nevertheless, some of his observations and comments in *The New Organon* are telling.

2.6 *THE NEW ORGANON*

Now what of Bacon's work itself? He urged that men resist the temptation to go immediately from a few specifics to the broadest generalities, but rather to proceed with caution. He noted that philosophical systems often flourish most in the hands of the first author and afterwards degenerate. Men should attempt to examine and to interpret the facts of nature rather than to anticipate the results. He argued that philosophers err by studying the broadest generalities, thus very likely starting at the wrong end of the problem. Although we may find this obvious today, it was not evident to most people in his time. While Bacon, by his own admission, proposed no universal or complete theory of knowledge, he stressed the central role of induction and of verification by controlled observations and by experiments.

Bacon not only believed it was a moral imperative that science be used for the betterment of mankind, but also that the fruits, or practical results, that it produced were an important indication of the correctness of its starting premises or general principles. Once again, as in Descartes' *Discourse on Method*, we see the germ of the idea, so important for the progress of modern science, that correct predictions serve as a warrant for the soundness of an hypothesis. In a famous metaphor Bacon urges that science must steer a mean course between the extremes of the empirical (empiricists, mere indiscriminate fact gatherers – ants) and the dogmatical (rationalists, pure abstract thinkers – spiders) and blend both modes (hypothesis and observation to produce fruits – bees).

> Those who have treated of the sciences have been either empirics or dogmatical. The former like ants only heap up and use their store, the latter like spiders spin out their own webs. The bee, a mean between both, extracts matter from the flowers of the garden and the field, but works and fashions it by its own efforts. The true labor of philosophy resembles hers, for it neither relies entirely nor principally on the powers of the mind, nor yet lays up in the memory the matter afforded by the experiments of natural history and mechanics in its raw state, but changes and works it in the understanding. We have good reason, therefore, to derive hope from a closer and purer alliance of these faculties (the experimental and rational) than has yet been attempted.[12]

Above all, Bacon advocated a slow and careful advance from particulars to generalities. He urged use of a combination of induction and deduction in arriving at knowledge. In Bacon's ladder of axiom, one is to make modest generalizations based on specific observations and data, check these modest theories by comparing their predictions with facts once again, then combine these generalizations into more general ones, check their predictions against observations, and in this way carefully proceed to the most general axioms, theories or laws. This ascent to axioms–descent to works scheme is depicted diagrammatically in Figure 2.1.

2.7 BACON AND ARISTOTLE COMPARED

Aristotle and Bacon both believed that one must begin with experience as input in order to arrive finally, through reason, at general laws and principles. A major difference between them, however, lies in the emphasis each placed on the need for and extent of the observational phase of the program. In certain areas, Aristotle had a penchant for going quickly from a few observations to the most universal hypotheses that were thereafter taken as absolute and immutable truths. At the

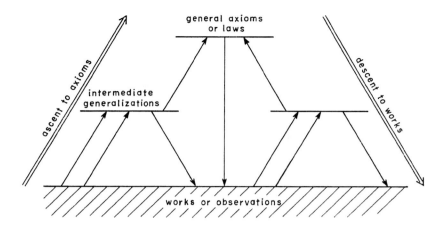

FIGURE 2.1 **Bacon's ladder of axiom**

other extreme, Bacon's philosophy is sometimes criticized because he placed too much emphasis on the almost blind gathering of all types of uncorrelated facts. This is actually somewhat unjustified as the aphorism quoted above indicates since Bacon castigated the empirics (who indiscriminately gather facts) as well as the a priori philosophers. One cannot simply take a mass of factual data and use logic alone to pry general laws from it. As we discuss in the next chapter, judgment is necessary both in deciding which facts to consider and in abstracting a scientific theory from these. The successors of Bacon are not the scientists of the seventeenth and later centuries, but the inductive logicians like David Hume and John Mill (1806–1873), men more concerned with the philosophical problems associated with induction than with constructing specific scientific theories. It may be reasonable to see Descartes, rather than Bacon, as a true precursor of modern philosophy and of scientific reasoning, since his theory of knowledge based on a rigid deductive system seems closer to modern scientific method than Bacon's, that at times appeared to be preoccupied with induction.

2.A ARISTOTLE'S *ON THE HEAVENS* AND *PHYSICS* ON MOTION

In his *On the Heavens* Aristotle argues that bodies have their natural motions produced by weight.

> Of necessity, we assert, ... a moved thing which has no natural impetus cannot move either towards or away from the center. Suppose a body *A* without weight, and a body *B* endowed with weight. Suppose the weightless body to move the distance *CD*, while *B* in the same time moves the distance *CE*, which will be greater since the heavy thing must move further. Let the heavy body then be divided in the proportion *CE* : *CD* (for there is no reason why a part of *B* should not stand in this relation to the whole). Now if the whole moves the whole distance *CE*, the part must in the same time move the distance *CD*. A weightless body, therefore, and one which has weight will move the same distance, which is impossible.[13]

PART I   The scientific enterprise

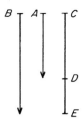

FIGURE 2.2   **Natural motions caused by weight**

Figure 2.2 may help you in following this passage. Body *A* that is weightless is assumed to move in natural motion a distance *CD* in the same time that body *B* that has weight moves the greater distance *CE*. Aristotle divides the body *B* into two parts, one of whose weights compared to the original body *B* is in the ratio *CD* : *CE*. He then states that this less weighty body will move a distance *CD* in the same time that *B* moves *CE*. His conclusion is that a weightless body (one having no natural impetus) cannot move in natural motion since, if it did, it would move just as did one with natural impetus (or weight). This he considers impossible since there is a cause for the latter motion, but not for the former.

Later in the same treatise, Aristotle tells us:

> By absolutely light, then, we mean that which moves upward or to the extremity, and by absolutely heavy that which moves downward or to the centre. By lighter or relatively light we mean that one, of two bodies endowed with weight and equal in bulk, which is exceeded by the other in the speed of its natural downward movement.
>
> . . .
>
> For any two portions of fire . . . the upward movement of the greater is quicker than that of the less, just as the downward movement of a mass of gold or lead, or of any other body endowed with weight, is quicker in proportion to its size.[14]
>
> . . .
>
> A given weight moves a given distance in a given time; a weight which is as great and more moves the same distance in a less time, the times being in inverse proportion to the weights. For instance, if one weight is twice another, it will take half as long over a given movement.[15]

These passages could be taken as giving a measure of weight in terms of the speed (or rate of fall) of a body (in natural motion).

Aristotle allows the speed of a moving object to increase as it gets closer to its natural place:

> [E]arth moves more quickly the nearer it is to the center, and fire the nearer it is to the upper place.[16]

Does this increased urge result in an increased weight as a cause of motion?

In his *Physics* Aristotle attempts to establish the impossibility of the existence of a vacuum.

> Now the medium causes a difference because it impedes the moving thing, most of all if it is moving in the opposite direction . . .

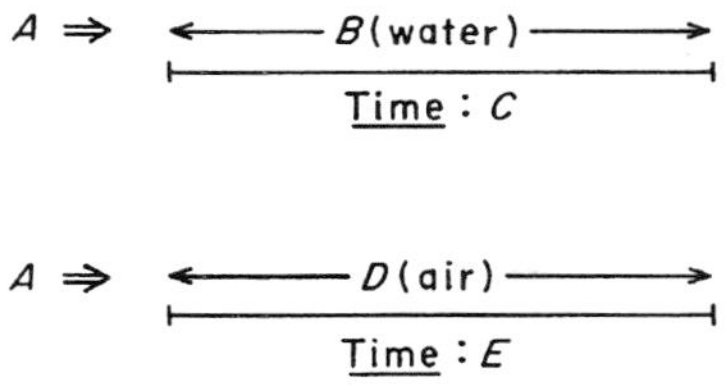

```
A ⇒   ←———— B(water) ————→
      |————————————————————|
              Time : C


A ⇒   ←———— D(air) ————→
      |————————————————|
              Time : E
```

FIGURE 2.3 **Motion through a medium**

> *A*, then, will move through *B* in time *C*, and through *D*, which is thinner, in time *E* (if the length of *B* is equal to *D*), in proportion to the density of the hindering body. For let *B* be water and *D* air; then by so much as air is thinner and more incorporeal than water, *A* will move through *D* faster than through *B*. Let the speed have the same ratio to the speed then, that air has to water. Then if air is twice as thin, the body will traverse *B* in twice the time that it does *D*, and the time *C* will be twice the time *E*. And always, by so much as the medium is more incorporeal and less resistant and more easily divided, the faster will be the movement.
>
> . . .
>
> These are the consequences that result from a difference in the media; the following depend upon an excess of one moving body over another. We see that bodies which have a greater impulse either of weight or of lightness, if they are alike in other respects, move faster over an equal space, and in the ratio which their magnitudes bear to each other. Therefore they will also move through the void with this ratio of speed. But that is impossible; for why should one move faster? (In moving through *plena* [space filled with matter, as opposed to a void] it must be so; for the greater divides them faster by its force. For a moving thing cleaves the medium either by its shape, or by the impulse which the body that is carried along or is projected possesses.) Therefore all will possess equal velocity. But this is impossible.[17]

Notice that in Aristotelian physics, the medium plays a dual role: (1) to provide a cause for motion (toward a proper or natural place) and (2) to resist the motion (lest it be infinite). That is, in terms of our representation via Eq. (2.1), both factors on the right (*W* and *R*) are causal.

In Figure 2.3 we have a graphic representation of Aristotle's argument. Body *A* moves (in natural motion) equal distances, once through water (*B*) in a longer time *C* and once through air (*D*) in a shorter time *E*. In the second sentence of the second paragraph above Aristotle assumes that the speeds are in the (inverse) ratio of the densities of the two media. He goes on (in a passage not reproduced here) to conclude that a body would move infinitely rapidly through a void since it would encounter no resistance. Because he considers this impossible, he deduces that a void or vacuum cannot exist. In the last paragraph just quoted he attempts to strengthen his conclusion that there can be no void by arguing that, since a body moves more rapidly in direct proportion to its weight, bodies would move at different rates through a void without any apparent cause for this difference in speed.[18] Rejecting this alternative, he states that all bodies would move at equal rates in a void, an equally unacceptable result for him. Hence, Aristotle rejects the possibility of the existence of a void. He holds that the medium is necessary for the motion of the body.

PART I   **The scientific enterprise**

FURTHER READING

Books I and II of Aristotle's *Nicomachean Ethics* give the reader some picture of the interrelatedness of various components of his unified teleonomic world view: although this part of the *Ethics* is concerned with human affairs, it is pervaded by the overarching teleonomic aspect of Aristotle's thought and makes for an interesting comparison with parts of his *Physics* and *On the Heavens* that we have considered in this chapter. Chapter 1 of Stanley Jaki's *The Relevance of Physics* discusses the Aristotelian view of the physical world as organism. Adwin Green's *Sir Francis Bacon* is a readable and informative biography of this complex person.

3

Science and metaphysics

In this chapter we discuss a few of the classic views of science that were popular from the time of the Renaissance until the early part of the twentieth century and then indicate some later changes in the conception of science. We return to a more complete overview of the current status of the philosophy of science as a retrospective in Chapter 25.

3.1 ORIGINS OF SCIENTIFIC METHOD

In the previous chapter we used Bacon as an example of a proponent of what developed into one important aspect of the modern scientific method. We also referred to Descartes as the father of modern philosophy and scientific reasoning. Galileo is often credited as being the first working scientist to apply modern scientific method in his investigations. (Chapter 6 will discuss the scientific writings and research of Galileo.) Although it is simplest for purposes of exposition to focus on the works of specific individuals such as Bacon, Descartes or Galileo to illustrate the rise of modern scientific thought and practice, these seventeenth-century thinkers were not the first to break with Aristotelian tradition. They did have predecessors. For instance, in the thirteenth century an experimental dimension for science was already advocated by the English Franciscan friar Roger Bacon (c. 1220–1292). And, as we shall see in more detail in Chapter 6, some of the fourteenth-century Ockhamists in Paris applied mathematical methods to the problem of motion and obtained results that contributed to the foundations of modern mechanics and of calculus. (This group was named after William of Ockham (c. 1285–1347/1349), famous for his often-quoted razor (or criterion for assessing an explanation), sometimes paraphrased as 'That is best that is simplest and works.') John Buridan (1300–1358) formulated a concept of impetus and treated the effect of gravity on a falling body in terms of what we would today call uniformly accelerated motion. Nicholas Oresme (1325–1382) anticipated some of the elements of analytic geometry by employing coordinates to represent arbitrary functions and reasoned to a distance–time formula for uniformly accelerated motion. Leonardo da Vinci (1452–1519) knew of the work of this Parisian School and his own investigations in mechanics were influenced by it. He stressed the importance of using mathematics in science, the need for observation and experiment, and the reciprocal roles played by (in modern terminology) both induction and deduction in discovering the laws of nature. The point is that, even though we often present new ideas as represented by one person, it is necessary to realize that the roots of revolutionary insights usually go far into the past and have involved histories. Here we can give only a brief outline of their origins.

Underlying the work of Bacon, of Galileo and of other proponents of the 'scientific method' is a belief in the objective validity of science. This means that nature really exists outside of and independently of us, that there are laws of nature that function without regard for our expectations, and, moreover, that we are capable of discovering those laws. This is an act of faith on the part of the scientist. Science itself cannot prove the correctness or truth of this basic assumption of an external world governed by knowable laws. Perhaps this nondemonstrable element of belief at the base of the scientific enterprise strikes you as alien to some preconceived notions you may have about the nature of science. If that is the case, then the present chapter could be a bit disquieting.

So, before turning in succeeding chapters to our study of concrete physical laws, we make a few remarks on the nature of the scientific enterprise.

3.2 A POPULAR VIEW OF SCIENCE

A current popular conception of science, that bears some similarity to the type of program proposed by Francis Bacon, can be set down in outline form as:

a observation
b hypothesis
c prediction
d confirmation

This simple model of science and how it operates depicts the process as a purely objective one in which careful observations are made, or controlled experiments performed, after which rules or laws are extracted from these empirical data and general hypotheses are formed. Next, these general hypotheses or conjectured theories are employed to predict new observational consequences. These predictions are used to confront reality and, if the predictions are verified, then the theory has been confirmed. By this process science amasses more and more knowledge about the physical world and brings us closer to the truth.

Let us now examine this model a bit more critically. Certainly nearly all physical scientists have agreed that one must begin with the data of sense experience. Not only men like Bacon, Galileo and Newton were of this persuasion. Harvey, in his revolutionary and pioneering treatise on the circulation of the blood in the human body, urged, as Aristotle had, that we proceed from things more known to those less known. This approach is elegantly and concisely summarized in his statement 'that the facts cognizable by the senses wait upon no opinions, and that the works of nature bow to no antiquity...'.[1] But what are we to do with these facts and data?

3.3 HUME AND MILL ON INDUCTION

Previously we termed induction the process of abstracting general laws or principles from specific observations. Can one ever know with certainty that a law gained from induction is correct? David Hume, the eighteenth-century British philosopher, historian, economist and essayist, held that philosophy is basically an inductive, experimental science of human nature. In his *An Enquiry Concerning Human Understanding* (1758), he discusses the notion of causality, the validity of induction and the necessity of sense data preceding ideas. Here we simply summarize the quotation from Hume's work given in Section 3.A. In analyzing the concepts of cause and effect, Hume points out that what we actually observe is only one event following another, rather than a necessary connection between them. There is no direct observation of a cause (or connection) for the supposed effect. Since what we see are events conjoined (or associated) but never connected, the terms *cause* and *effect* are words without operational meaning. If we know just one instance of two events being conjoined (one following another), then we do not state a general law governing their connection in terms of cause and effect. However, if we repeatedly see the same pair of events conjoined, as a body initially at rest being struck by another moving object so that after the collision the first body is also in motion, we do refer to cause and effect. (In this example, the collision of the moving object with the body at rest would be termed the cause for the effect of that body being set into motion.) That is, from the occurrence of constant conjunction, we infer a necessary connection. However, the difference between one such pair of events and many is that the mind is carried by habit to feel the existence of a necessary connection (in our thought).

In essence, Hume asked what reason we have to suppose that future observations would resemble past ones. He challenged placing unlimited trust in universal, predictive laws induced from observed facts. As a simple example, from the fact that the sun has risen every day in the past since the beginning of recorded history, we might induce the 'law' that the sun necessarily rises every day without exception. Can we, however, be absolutely certain that the sun will rise tomorrow? No. There is no way around the provisional character of physical laws gained by induction. Hume saw the need for induction, but felt its validity to be unprovable.

The later British philosopher, economist and ethical theorist John Mill also wrote extensively on the question of induction. Mill was a child prodigy educated by his father so that he read Greek by age eight, when he started Latin, studied Euclid and then Aristotle at age twelve. He was not only a philosopher, but a political activist as well. In his *System of Logic* (1843), Mill discusses induction and the assumed uniformity of the laws of nature. He argues that the main problem for the science of logic is a justification of the process of induction that is the operation of discovering and proving general propositions. Mill sees the question to be one of discovering the conditions under which an induction is logically legitimate. Involved in all inductions is the assumption that the course of nature is uniform or that the universe is governed by general laws. However, this is itself an example of induction since we assume it to be true because it has always turned out to be so in individual instances. This uniformity of the course of nature is the premise (or first principle) that can be used to justify induction. Unfortunately, neither Mill nor anyone else has proven the correctness of this premise. It remains an assumption, albeit one that few would dispute and without which we cannot do science.

3.4 POPPER ON OBSERVATION AND HYPOTHESIS

You may feel that a concern for the nature of the cause–effect relationship, the validity of induction and the status of physical laws is just philosophical nit-picking that is best left for some idle Saturday afternoon and that we can get on with the serious business of science without being concerned about such questions. But even if we gloss over these issues, consider the relation between observation and hypothesis. How independent are the operations of gathering data and of discovering theories? The fact-gathering necessary for science is rarely random. As Charles Darwin (1809–1882), whose *On the Origin of Species* (1859) laid the foundation of the theory of evolution, pointed out, no one can be a good observer unless he is an active theorizer. That is, the discovery of new facts and the evolution of a theory are reciprocal processes. The experiments a scientist undertakes do not yield immediate and obvious sense data but rather those collected with considerable difficulty. She must have a theory to guide her in this search, else she would not know which of the myriad phenomena in nature to examine. As we shall see in our study of physics, those phenomena that one age sees as crucial in light of a then-current theory often appear less important at a later time when a different theory has become more popular.

Perhaps you object that these are still mere details of the relationship between observation and hypothesis. One way or another, scientists do arrive at candidates for physical laws and then from these deduce new consequences that can be looked for. What do we conclude when a theory predicts a result that is found to correspond with our observations? Has the theory been verified? No, we simply know that it has not yet been refuted. Sometimes we put a more positive gloss on this by speaking of corroboration or of confirmation. That is, if a theory makes a prediction that is definitely contradicted by sufficiently careful observation or experimentation, then the theory is necessarily logically incorrect. It has been refuted. A given theory (p) implies or makes a specific prediction (q). If the prediction turns out to be false ($\sim q$) as indicated by further analysis or by experimental results, then it necessarily follows that the theory is false ($\sim p$). (Unfortunately, it is not always a simple matter to know when the data are sufficiently good to refute an hypothesis, as we shall see in Chapters 4 and 5 when we discuss stellar parallax.) However, if the prediction is correct (q), we cannot be certain that the theory (p) is correct.

One recent British philosopher of science, Sir Karl Popper (1902–1994), stressed that the hallmark of a scientific theory is not that it can be verified but rather that it is capable of being refuted or falsified. According to this criterion, a theory can qualify as scientific only if it makes specific predictions that can be subjected to comparison with the real world to find out whether or not these predictions are indeed true in physical reality. In his *Conjectures and Refutations*, an excerpt of which is given at the end of this chapter, Popper asks what the criterion is by which one can distinguish a scientific theory from a nonscientific one. As a young man (around 1919), he contrasted Karl Marx's (1818–1883) theory of history, Sigmund Freud's (1856–1939) psychoanalysis and Alfred Adler's (1870–1937) individual psychology with Einstein's general theory of relativity. Since the former three all had great explanatory power in that whatever happened could always be interpreted after the fact as confirming those theories, Popper concluded that verifications alone are not of central importance since in these cases they are so easily come by. In the case of general relativity, or of any truly scientific theory, definite predictions are made that are subsequently tested by experiment. There is no guarantee ahead of time that the observational facts will not disagree with these predictions. Such theories are refutable. Only predictions that have this element of risk (in that they might be refuted) should count as meaningful supportive evidence for scientific theories. A good scientific theory prohibits certain outcomes from occurring in nature and a stringent test of such a theory is an attempt to falsify or refute it by actually observing those prohibited results. Hence, for Popper, the hallmark of scientific theories is that they are (in principle) refutable or falsifiable. (This is not the same as saying that they are in fact constantly refuted. A successful scientific theory survives many serious attempts to refute it.) This is similar to the position of the American philosopher Willard Quine (1908–). (Recall the quotations for Part I (*The scientific enterprise*).)

3.5 JUSTIFICATION FOR HYPOTHESES

Therefore, as long as a scientific theory holds up under comparison with observations, it may be correct. We can never be certain that a theory is correct, only that refuted theories are incorrect. Still, scientists do use theories to make predictions and, when these predictions accord with nature, then scientists continue to work with such theories. This model of postulating a theory and then making specific predictions with it is termed the *hypothetico-deductive* method. Although that method does not prove a theory as necessarily correct, it does give us a *warrant* for accepting the theory (always provisionally of course). *Retroduction*, as we define it here, signifies arguing for the plausibility of a theory on the basis of its successful specific predictions. From our modern perspective, we can recognize that this is the method that Newton advocates in his *Rules of Reasoning in Philosophy*.

> In experimental philosophy we are to look upon propositions inferred by general induction from phenomena as accurately or very nearly true, notwithstanding any contrary hypotheses that may be imagined, till such time as other phenomena occur, by which they may either be made more accurate, or liable to exceptions.[2]

So far we have seen three types of warrants for the axioms or postulates from which we then deduce logically certain implications or predictions:

i *axiomatic* – Here the axiom is claimed to be self-evident, obvious or immediate. As we saw in Chapter 2, Aristotle often arrived at his general cosmological principles in this fashion after a rather cursory examination of the data of experience. Descartes attempted to base the science of mechanics on such first principles seen to be true in their own right, once properly understood.

ii *inductive* – Here generalizations are made from similarities perceived in a large group of particular events or observations. This is the method proposed by Bacon and later espoused by Hume and by Mill.

iii *retroductive* – Here we go from consequences back to hypotheses as in the hypothetico-deductive method above.

Let us return now to the falsification process that Popper feels is essential for the operation of science. Is this process quite so simple and certain? No, since observational data alone are never sufficient to force scientists to abandon a model or theory that has previously been successful. Such theories are often modified or reinterpreted. As a case in point, Planck was involved in a long and vitriolic controversy with several distinguished colleagues over the interpretation and relation of some of the laws of heat and of mechanics. Even though he had sound arguments on his side, it was many years before his point of view was generally accepted. In his *Scientific Autobiography* he states:

> This experience gave me also an opportunity to learn a fact – a remarkable one in my opinion: a new scientific truth does not triumph by convincing its opponents and making them see the light, but rather because its opponents eventually die, and a new generation grows up that is familiar with it.[3]

For example, we shall see that observation alone did not decide the issue between Claudius Ptolemy's (*fl.* 127–145 A.D.) model of the universe that placed the earth at its center and Nicholas Copernicus' (1473–1543) model that placed the sun at its center. If not observation alone, then what are the other criteria by which scientists decide among rival theories? A certain economy of explanation, a symmetry of form and a beauty of formulation have always been important factors. These are aesthetic matters that cannot be quantified and are not agreed upon universally. For some, discovering beauty and a sense of symmetry is the end of science and the reason for doing it. We return to these questions several times throughout this book.

3.6 SCIENTIFIC KNOWLEDGE AND TRUTH

Even though the absolute objectivity and certitude of science have now been opened to some doubt, you may still hold that nevertheless science both seeks and finds truth. Is this obvious, though? Since we can never be certain that a given model or theory of the physical world is correct, can we say that such theories are true? As we emphasized previously, there is an essential asymmetry between the confirmation and the refutation of a theory or of a proposition. Refutation is a logically certain and valid inference in that, once the prediction (or consequent) has been denied, then the theory (or antecedent) must also be denied. However, there is no logically valid means of establishing the truth of a theory from the agreement of its predictions with observed physical phenomena. (An exception to this would occur if one were able to show that there existed only a finite number of logically possible explanations and all but one were refuted, say by observation. However, this is rare in complex, realistic situations.) Even if a theory were true (in the sense that it did actually embody and uniquely account for an absolutely and universally valid law of nature), we would have no way to know that it was true. Are not such theories rather constructs of the human mind by which we represent nature? One can take scientific theories to be just a means of simplifying and codifying the observed facts about nature, and not an explanation of them.

In his 1941 lecture *The Meaning and Limits of Exact Science*, Planck discussed the task of science as introducing order and regularity into the multitude of diverse sense experiences. He observed that, even though the individual data of sense experience have great solidity, they are of limited significance. We feel compelled to postulate the existence of a real world that underlies and unifies these diverse surface phenomena. This belief constitutes an irreducible, nondemonstrable element of science, but one we feel unable to dispense with. According to Planck, science studies the phenomenological world in order to construct a scientific world picture of our experiences and from this to gain an understanding of what he calls the real world of objects, the ultimate, metaphysical reality that can never be completely known. We approach the metaphysically real world through improvement of our phenomenological world picture. Planck's real metaphysical world of objects is similar to Plato's world of forms or essences that we mentioned in Chapter 1.

If we take a brief retrospective glance at the evolution of the concept and status of scientific knowledge as we have presented it in these first three chapters, we can associate the major developments with the following sequence of philosophers.

Plato, Aristotle → Bacon, Descartes → Galileo, Newton → Hume, Mill → Popper, Quine

What began as a quest for certain knowledge and understanding of the phenomena of nature, based either on deduction from self-evident first principles (Aristotle, Descartes) or on careful induction to lead to unassailable general laws (Bacon), has today been abandoned as unattainable. One modern philosopher has referred to this goal of attainable certain knowledge as the Bacon–Descartes ideal.[4] The basic problem with that program is that there is no logically valid way to arrive at first principles (or general laws), either on the basis of intuition, of induction or of a combination of the two. A content-increasing logic (or a logic of induction) does not exist. Deductive logic merely allows us to search out those statements already contained (implicitly) in the premises. The conclusion contains no more facts or information than the premises (even though it may be more useful or easily recognizable there than in the premises). The trend of these philosophical developments has been a decrease in the certainty we are able to claim for scientific knowledge. We can only hope to refute candidates for scientific theories, not to prove them as true. From this perspective our successful, accepted scientific theories are consistent, but not necessarily 'true', stories about the way the world might 'really' be. In the next chapter we begin to examine how the history of actual scientific practice leads us to such an evaluation of the status of scientific theories.

3.A HUME, MILL AND POPPER ON SCIENTIFIC KNOWLEDGE

In his *An Enquiry Concerning Human Understanding* Hume discussed the concepts of cause and effect.

> We have sought in vain for an idea of power or necessary connexion in all the sources from which we could suppose it to be derived. It appears that, in single instances of the operation of bodies, we never can, by our utmost scrutiny, discover anything but one event following another, without being able to comprehend any force or power by which the cause operates, or any connexion between it and its supposed effect.... All events seem entirely loose and separate. One event follows another; but we never can observe any tie between them. They seem *conjoined*, but never *connected*. And as we can have no idea of any thing which never appeared to our outward sense or inward sentiment, the necessary conclusion *seems* to be that we have no idea of connexion or power at all, and that these words are absolutely without any meaning, when employed either in philosophical reasonings or common life.
>
> . . .
>
> Even after one instance or experiment where we have observed a particular event to follow upon another, we are not entitled to form a general rule, or foretell what will happen in like cases; it being justly esteemed an unpardonable temerity to judge of the whole course of nature from one single experiment, however accurate or certain. But when one particular species of event has always, in all instances, been conjoined with another, we make no longer any scruple of foretelling one upon the appearance of the other, and of employing that reasoning, which can alone assure us of any matter of fact or existence. We then call the one object, *Cause*; the other, *Effect*. We suppose that there is some connection between them; some power in the one, by which it infallibly produces the other, and operates with the greatest certainty and strongest necessity.
>
> It appears, then, that this idea of a necessary connexion among events arises from a number of similar instances which occur of the constant conjunction of these events; nor can that idea ever be suggested by any one of these instances, surveyed in all possible lights and positions. But there is nothing in a number of instances, different from every single instance, which is supposed to be exactly similar; except only, that after a repetition of similar instances, the mind is carried by habit, upon the appearance of one event, to expect its usual attendant, and to believe that it will exist. This connexion, therefore, which we *feel* in the mind, this customary transition of the imagination from one object to its usual attendant, is the sentiment or impression from which we form the idea of power or necessary connexion. Nothing farther is in the case. Contemplate the subject on all sides; you will never find any other origin of that idea. This is the sole difference between one instance, from which we can never receive the idea of connexion, and a number of similar instances, by which it is suggested.
>
> . . .
>
> [T]herefore, we may define a cause to be *an object, followed by another, and where all the objects similar to the first are followed by objects similar to the second*. Or in other words *where, if the first object had not been, the second never had existed*. The appearance of a cause always conveys the

mind, by a customary transition, to the idea of the effect. Of this also we have experience. We may, therefore, suitably to this experience, form another definition of cause, and call it, *an object followed by another, and whose appearance always conveys the thought to that other.*[5]

In *A System of Logic* Mill turned to the problem of induction.

What Induction is, therefore, and what conditions render it legitimate, cannot but be deemed the main question of the science of logic – the question which includes all others.

. . .

For the purposes of the present inquiry, Induction may be defined: the operation of discovering and proving general propositions.

. . .

Induction properly so called, as distinguished from those mental operations, sometimes, though improperly, designated by the name, which I have attempted in the preceding chapter to characterize, may, then, be summarily defined as generalization from experience. It consists in inferring from some individual instances in which a phenomenon is observed to occur, that it occurs in all instances of a certain class; namely, in all which *resemble* the former, in what are regarded as the material circumstances.

In what way the material circumstances are to be distinguished from those which are immaterial, or why some of the circumstances are material and others not so, we are not yet ready to point out. We must first observe that there is a principle implied in the very statement of what Induction is; an assumption with regard to the course of nature and the order of the universe: namely, that there are such things in nature as parallel cases; that what happens once, will, under a sufficient degree of similarity of circumstances, happen again, and not only again, but always. This, I say, is an assumption involved in every case of induction. And, if we consult the actual course of nature, we find that the assumption is warranted; the fact is so. The universe, we find, is so constituted, that whatever is true in one case, is true in all cases of a certain description; the only difficulty is, to find *what* description.

This universal fact, which is our warrant for all inferences from experience, has been described by different philosophers in different forms of language: that the course of nature is uniform; that the universe is governed by general laws; and the like.

. . .

[T]he principle which we are now considering, that of the uniformity of the course of nature, will appear as the ultimate major premiss of all inductions; and will, therefore, stand to all inductions in the relation in which, as has been shown at so much length, the major proposition of a syllogism always stands to the conclusion; not contributing at all to prove it, but being a necessary condition of its being proved; since no conclusion is proved for which there cannot be found a true major premiss.[6]

In his *Conjectures and Refutations* Popper addressed the concept of falsification.

I therefore decided to do what I have never done before: to give you a report on my own work in the philosophy of science, since the autumn of 1919 when I first began to grapple with the problem, '*When should a theory be ranked as scientific?*' or '*Is there a criterion for the scientific character or status of a theory?*'

. . .

I knew, of course, the most widely accepted answer to my problem: that science is distinguished from pseudo-science – or from 'metaphysics' – by its *empirical method*, which is essentially *inductive*, proceeding from observation or experiment. But this did not satisfy me. On the contrary, I often formulated my problem as one of distinguishing between a genuinely empirical method and a non-empirical or even a pseudo-empirical method – that is to say, a method which, although it appeals to observation and experiment, nevertheless does not come up to scientific standards. The latter method may be exemplified by astrology, with its stupendous mass of empirical evidence based on observation – on horoscopes and on biographies.

But as it was not the example of astrology which led me to my problem I should perhaps briefly describe the atmosphere in which my problem arose and the examples by which it was stimulated. After the collapse of the Austrian Empire there had been a revolution in Austria: the air was full of revolutionary slogans and ideas, and new and often wild theories. Among the theories which interested me Einstein's theory of relativity was no doubt by far the most important. Three others were Marx's theory of history, Freud's psycho-analysis, and Alfred Adler's so-called 'individual psychology'.

. . .

I found that those of my friends who were admirers of Marx, Freud, and Adler were impressed by a number of points common to these theories, and especially by their apparent *explanatory power*. These theories appeared to be able to explain practically everything that happened within the fields to which they referred. The study of any of them seemed to have the effect of an intellectual conversion or revelation, opening your eyes to a new truth hidden from those not yet initiated. Once your eyes were thus opened you saw confirming instances everywhere: the world was full of *verifications* of the theory. Whatever happened always confirmed it. Thus its truth appeared manifest; and the unbelievers were clearly people who did not want to see the manifest truth; who refused to see it, either because it was against their class interest, or because of their repressions which were still 'unanalyzed' and crying aloud for treatment.

. . .

With Einstein's theory the situation was strikingly different. Take one typical instance – Einstein's prediction, just then confirmed by the findings of Eddington's expedition. Einstein's gravitational theory had led to the result that light must be attracted by heavy bodies (such as the sun), precisely as material bodies were attracted. As a consequence it could be calculated that light from a distant fixed star whose apparent position was close to the sun would reach the earth from such a direction that the star would seem to be slightly shifted away from the sun; or, in other words, that the stars close to the sun would look as if they had moved a little away from the sun, and from one another.

. . .

Now the impressive thing about this case is the *risk* involved in a prediction of this kind. If observation shows that the predicted effect is definitely absent, then the theory is simply refuted. The theory is *incompatible with certain possible results of observation* – in fact with results which everybody before Einstein would have expected. This is quite different from the situation I have previously described, when it turned out that the theories in question were compatible with the most divergent human behaviour, so that it was practically impossible to describe any human behaviour that might not be claimed to be a verification of these theories.

. . .

One can sum up all this by saying that *the criterion of the scientific status of a theory is its falsifiability, or refutability, or testability.*[7]

FURTHER READING

Ralph Blake *et al.*'s *Theories of Scientific Method* gives a nice overview of this subject from the Renaissance through the nineteenth century. Karl Popper's *Conjectures and Refutations* contains essays and lectures on the theme 'that we can learn from our mistakes' (via falsifiability) and sets this thesis against a broad historical background of philosophical positions on scientific knowledge. Baruch Brody's *Readings in the Philosophy of Science* is a collection of papers, by prominent modern philosophers of science, on scientific explanation and prediction, on the structure and function of scientific theories and on the confirmation of scientific hypotheses.

PART II

Ancient and modern models of the universe

[I]n the investigation of naturally accelerated motion we were led, by hand as it were, in following the habit and custom of nature herself, in all her various other processes, to employ only those means which are most common, simple and easy.

...

[W]e shall not be far wrong if we put the increment of speed as proportional to the increment of time; hence the definition of motion which we are about to discuss may be stated as follows: A motion is said to be uniformly accelerated when starting from rest, it acquires, during equal time-intervals, equal increments of speed.

...

The present does not seem to be the proper time to investigate the cause of the acceleration of natural motion concerning which various opinions have been expressed by various philosophers, some explaining it by attraction to the centre, others to repulsion between the very small parts of the body, while still others attribute it to a certain stress in the surrounding medium which closes in behind the falling body and drives it from one of its positions to another. Now, all these fantasies, and others too, ought to be examined; but it is not really worthwhile. At present it is the purpose of our Author merely to investigate and to demonstrate some of the properties of accelerated motion (whatever the cause of this acceleration may be).

Galileo Galilei, *Dialogues Concerning Two New Sciences*

Now just as all the parts of the earth mutually cooperate to form its whole, from which it follows that they have equal tendencies to come together in order to unite in the best possible way and adapt themselves by taking a spherical shape, why may we not believe that the sun, moon, and other world bodies are also round in shape merely by a concordant instinct and natural tendency of all their component parts? If at any time one of these parts were forcibly separated from the whole, is it not reasonable to believe that it would return spontaneously and by natural tendency? And in this manner we should conclude that straight motion is equally suitable to all world bodies.

...

[T]he circular motion of the projector [on the earth] impresses an impetus upon the projectile to move, when they separate, along the straight line tangent to the circle of motion at the point of separation.... [T]he projectile would continue to move along that line if it were not inclined downward by its own weight, from which fact the line of motion derives its curvature.

Galileo Galilei, *Dialogue Concerning the Two Chief World Systems*

4

Observational astronomy and the Ptolemaic model

A central assumption of this book is that general philosophical discussions about science should be based upon specifics from the actual history of science. Therefore, we begin by considering the model of the universe accepted by the ancients and then we analyze some of the philosophical implications of this episode. In this chapter we present a summary of the naked-eye observations available to the ancients and outline the model of the universe that they built using this data.[1] The next chapter then gives a description of the present-day theory that overthrew it. In Chapter 12 we discuss in more depth the profound shift in philosophical perspective necessitated by this change, as well as the contributions of several of the major precursors and shapers of that revolution.

4.1 ELEMENTARY OBSERVATIONS

In order to get into the proper frame of mind to appreciate a discussion of an ancient model of the universe in which the earth was pictured as being at rest, the modern reader who 'knows' that the earth moves around the sun and also spins on its axis might ask the following question. How would you demonstrate that the earth makes an orbit about the sun and that it turns on its own axis? Just use naked-eye observations – no photographs from satellites allowed. You will find that this is not such a simple exercise. Perhaps then the view the ancients had of their universe will seem less bizarre to you.

The most obvious fact evident to even an unsophisticated observer – modern or ancient – is the daily rising and setting of the sun and the recurrent change of the seasons. If one drives a stake or a long pole into the ground along a vertical or plumb line, one can observe several important features of the shadow cast by the sun. As the day progresses, both the direction and the length of the shadow vary, the length being longest early in the morning and late in the evening and shortest at midday. (The specific shape of the curve traced out by the tip of this shadow need not concern us here.) However, the direction of this shortest shadow is the same each day. That is, if a marker is placed on the ground at the tip of the shortest shadow one day and then the same is done on succeeding days, all of these markers lie on one straight line. This allows the observer to define two reference points: *north* can be taken as the invariant direction in which this shortest shadow points (with east, south and west as the conventional compass directions relative to north) and *noon* can be taken as the instant at which this shortest shadow occurs. (These and several other results we present are what an observer in the northern hemisphere (but not too near the equator) would see. But, most of the ancient peoples whose records we have did live in such regions of the northern hemisphere.) The elapsed time between successive noons is a convenient means of defining the solar day that we take as the *mean solar day*. A little more observation shows that the shadow of the vertical pole progresses in a general direction from west toward east during the day. Yet, on just two days during the year does the tip of this shadow move along a perfectly straight line from west to east. These days are the *vernal equinox* (around March 21 on our modern calendars) and the *autumnal equinox* (around September 23). These days mark, respectively, the beginning of spring and of fall. On these two dates the lengths of the days and nights are always equal. Furthermore, there is a day on which the noon shadow is longer than on any other day. This is the *winter solstice* (around December 22). It marks the beginning of winter and on this date the night is longest and the daylight hours fewest. The *summer solstice* (around June 22) is the start of summer and on that date the noon shadow is shorter than at any other time, the night being the shortest and the daylight hours most.

The year can then be taken as the period of time between one vernal equinox (spring) and the next. Unfortunately, the integral number of days in such a year is not exactly the same from year to year since the vernal equinox does not always occur on the same date. This problem of the solar calendar plagued astronomers from the beginning of recorded history up through the sixteenth century. Various ancient societies coped with this problem in different ways. A natural starting point for some was a year of 360 days (following, perhaps, from the roughly thirty-day period of the moon and the choice of twelve months in a year). However, with such an inaccurate calendar, the seasons drifted noticeably from year to year. To help correct that, the Egyptians added a five-day holiday season to make a 365-day year. This calendar eventually became part of the heritage of the Roman Empire. In 46 B.C. Julius Caesar (100–44 B.C.) further reformed the calendar to $365\frac{1}{4}$ days by having three years of 365 days followed by a leap year of 366 days. Since the solar year is only about 11 minutes shorter than $365\frac{1}{4}$ days, the slippage of the seasons was slight with the Julian calendar. Still, by the sixteenth century the vernal equinox had moved backwards from March 21 to March 11. This was a matter of concern to Pope Gregory XIII (1502–1585) because Easter is defined as the first Sunday after first full moon after the vernal equinox and the dates of all other movable Church feasts depend on that of Easter. In a society with civil and ecclesiastical calendars and functions so closely intertwined, it would not do to have Easter slip into summer and Christmas into spring or summer. In 1582 the Gregorian calendar was introduced into Christian Europe when Gregory the Great decreed that the day after October 4, 1582, be October 15, 1582. In this calendar, leap year is suppressed in every centesimal year (ending in 00) whose number cannot be divided by 400. Thus it was dropped in 1700, 1800, and 1900, but will not be dropped in 2000. This pressing need for calendar reform played a role in precipitating the Copernican revolution in the medieval conception of the universe.

4.2 THE CELESTIAL SPHERE

In addition to the diurnal, or daily, motion of the sun through the sky, there is the diurnal motion of the stars in the sky and the motion of the sun relative to the background provided by these stars. We begin with the stars and consider what one sees while watching the stars on a clear night from a vantage point in the middle-northern latitudes. There is one star, Polaris (the Pole Star or North Star), whose position in the heavens remains fixed during any one night and from night to night. Latitude is defined as the elevation of the Pole Star above the horizon at the place of observation on the earth. Let us take as a representative elevation in northern latitudes about 45°, since, for example, Paris is 49°, London 51°, New York City 41°, Chicago 42° and Los Angeles 34°. All of the stars in the heavens move with a westward rotation on arcs of circles all centered about Polaris and at the rate of 15° per hour. Those stars within a circle centered at Polaris and tangent to the observer's horizon are termed circumpolar stars and are always present in the sky (although visibility during daylight hours is made difficult due to the brightness of the sun). The stars beyond this circle rise and set each evening. The size of the circumpolar circle increases as one moves north. The most important single fact about the stars is that their positions relative to the North Star, or relative to each other, remain fixed. For example, the Big Dipper rotates or turns during the night, but its shape remains the same. The pattern of stars is like a photograph that turns at a uniform rate (15° per hour) about Polaris. It is for this reason that we refer to the *fixed stars*. Since the pattern of the stars and their journeys across the sky repeat themselves without variation night after night, it is evident why the ancients naturally assumed that the stars were fixed in a pattern on an enormous sphere (the *celestial sphere*) that rotated at a uniform rate about the earth as center (see Figure 4.1). We say that the rotation of the celestial sphere points from N to S since a right-threaded screw turned in the direction of rotation of the celestial sphere would advance from N to S along the axis. The celestial sphere makes one revolution per day (actually, per 23 hours and 56 minutes, as we show shortly).

The apparent motion of the sun through the heavens is not quite as simple as that of the stars. From day to day the position of the sun changes relative to the fixed stars. A great circle is a circle drawn on a sphere and centered about the center of the sphere. The great circle that the projection of the sun onto the celestial sphere follows on its eastward journey through the various constellations of stars is known as the *ecliptic*. This is the apparent annual path of the sun through the background of the fixed stars in the heavens. It takes the sun just one year to travel around this ecliptic fixed on the celestial sphere. The celestial sphere (and the sun) was pictured as rotating westward at the rate of one complete revolution per day, while the sun also had a much slower eastward motion along the ecliptic. Since noon (and hence the day, hour and minute) is defined by the position of the sun in the sky and since the sun falls behind the stars on the celestial sphere each day by

$$\frac{24 \text{ hours}}{365 \text{ days}} \times \frac{60 \text{ minutes}}{1 \text{ hour}} \cong 4 \text{ minutes/day}$$

PART II Ancient and modern models of the universe

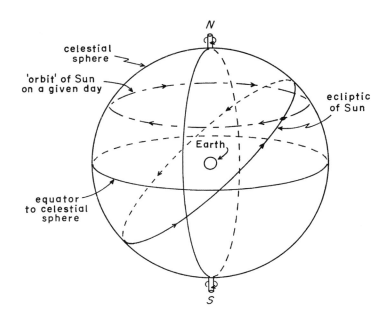

FIGURE 4.1 The two-sphere model: a spherical earth at the center of the celestial sphere

the celestial sphere appears to execute one complete revolution about the earth in 4 minutes less than one day. In fact, a *sidereal day* is defined as the interval between two successive transits of the first point of Aries over the upper meridian (the highest point on the star's transit) of any place. To sufficient accuracy for our purposes, we have one sidereal day equals 23 hours 56 minutes of mean solar time. On any given day the sun pursues a circular path through the heavens, but the size of the circle varies from day to day as the sun changes its position on the celestial sphere along the ecliptic (see the dashed circular orbit for the sun in Figure 4.1). This variation in elevation of the sun in the sky accounts for the variation in the length of a noon shadow, corresponding to the change in the seasons.

Since the earliest history, constellations of stars have been identified. *Constellation* is the term used for a group of stars to which the name of an object or creature has been given. In ancient times the twelve constellations of the zodiac were used to divide the ecliptic into twelve equal segments. The numbers and names of these are given in Table 4.1, along with the dates that the sun enters these constellations in its eastward journey along the ecliptic. The correlation between the position of the sun along its ecliptic (that is, its being in a particular constellation) and the seasons of the year was interpreted as a control of the seasons by the various constellations and was one of the origins of astrology. By the second century A.D., a catalogue of 48 constellations, comprising some 1,022 stars, had been established. The list included the twelve constellations of the zodiac. This catalogue remained authoritative until the time of the Renaissance. In 1930 the International Astronomical Union defined 88 constellations that cover the entire sky.

Table 4.1. *The constellations of the zodiac*

No.	Name	Date	No.	Name	Date
1	Aries (Ram)	March 21	7	Libra (Balance)	September 23
2	Taurus (Bull)	April 20	8	Scorpio (Scorpion)	October 24
3	Gemini (Twins)	May 21	9	Sagittarius (Archer)	November 22
4	Cancer (Crab)	June 22	10	Capricorn (Goat)	December 22
5	Leo (Lion)	July 23	11	Aquarius (Water Bearer)	January 20
6	Virgo (Virgin)	August 23	12	Pisces (Fishes)	February 19

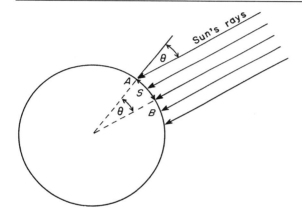

FIGURE 4.2 **Eratosthenes' argument for the size of the earth**

4.3 ERATOSTHENES' DETERMINATION OF THE EARTH'S SIZE

Before we confront the problem of the planets, let us consider the question of the shape of the earth itself. We saw in Chapter 2 that Aristotle believed that the earth was spherical. However, in the third century B.C. the Greek astronomer and geographer Eratosthenes (*c.* 276–*c.* 194 B.C.) actually determined the circumference of the earth as follows. On a day when the noon sun was directly overhead at the Egyptian city of Syene (*B*) (today known as Aswan), he measured the angle between the vertical and the sun's rays at Alexandria (*A*), a city directly to the north (see Figure 4.2). Simple geometry tells us that the arc length s from A to B is given in terms of the circumference C of the earth by the proportion $\theta/360° = s/C = s/2\pi r_e$, where r_e is the radius of the earth itself. Since Eratosthenes knew the distance from Syene to Alexandria, his direct measurement of θ as about 1/50 of a complete circle (or 7.2°) allowed him to determine C within about 5 percent of its presently known value of 25,000 miles [4.03×10^4 km].[2]

PART II Ancient and modern models of the universe

4.4 ARISTARCHUS' HELIOCENTRIC MODEL OF THE UNIVERSE

The ancient model of the universe shown in Figure 4.1 is often referred to as the *two-sphere universe*. It is in many ways very successful, since it collates much data and presents a coherent picture of the earth, the sun and the stars. It is still used for purposes of celestial navigation. Although this two-sphere model was the most widely accepted until quite modern times, there were alternatives proposed even by some of the ancients. Aristarchus of Samos (c. 310–230 B.C.) postulated that the sun was at the center of the universe and that the earth moved around the sun. We learn of this theory of Aristarchus' in the writing of Archimedes (c. 290/280–212/211 B.C.) of Syracusae (in Sicily), the greatest mathematician, physicist and inventor of antiquity. In his work *The Sand-Reckoner*, Archimedes offers a proof that, no matter how large the number of particles of matter in the universe, there exists a number that exceeds it. In passing, the introduction states:

> Now you are aware that the 'universe' is the name given by most astronomers to the sphere whose centre is the centre of the earth and whose radius is equal to the straight line between the centre of the sun and the centre of the earth. This is the common account, as you have heard from astronomers. But Aristarchus of Samos brought out a book consisting of some hypotheses, in which the premisses lead to the result that the universe is many times greater than that now so called. His hypotheses are that the earth revolves about the sun in the circumference of a circle, the sun lying in the middle of the orbit, and that the sphere of the fixed stars, situated about the same centre as the sun, is [very] great....[3]

However, there was one observational fact that mitigated strongly against this sun-centered theory: namely, *stellar parallax*. A simple example of the phonomenon of parallax is the following. Hold two objects (say, your two index fingers) vertically, one somewhat closer to you than the other and both roughly an arm's length in front of you. Without moving these objects, move your head from side to side. The two objects appear to move relative to each other. Similarly, if, from the earth, we view two fixed stars (shown as A and B in Figure 4.3), once from position C and then from the position D on the earth's orbit about the sun, the angle ACB is smaller than angle ADB so that the separation of the stars A and B will appear to vary. Since no such effect of stellar parallax can be observed with the naked eye, the ancients had little reason to accept Aristarchus' theory. His model offered no gain in simplicity and seemed to be refuted by the facts. Of course, there is one obvious way to dispense with the parallax objection. If the fixed stars (or the surface of the celestial sphere) are sufficiently far away from the earth, then the angles ACB and ADB will be equal for all practical purposes and no effect of stellar parallax will be detectable. That was the route that Copernicus finally took by demanding that the distance from the earth to the celestial sphere be at least 1,500,000 earth radii. This was immensely larger than the 20,000 or so earth radii that, as we see below, the ancients were prepared to accept.

Stellar parallax measurements are extremely difficult to make. Only in modern times have such observations been successful. In terms of Figure 4.3, the parallax angle for a given star (A) is defined as α when the two positions of the earth (C and D)

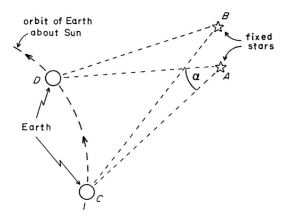

FIGURE 4.3 **Stellar parallax**

are on opposite sides of the diameter of the earth's orbit around the sun and the line drawn from the star to the sun is perpendicular to this diameter. Even for the star closest to the earth, Alpha Centauri, a triple star in the constellation Centaurus (not one of the zodiacal constellations), this parallax angle is very small, being on the order of one second of arc (1″). Among early unsuccessful attempts at such measurements were those of the Englishmen James Bradley (1693–1762) in 1725 and William Herschel at the end of the same century. Although they were unable to detect the effect sought, Bradley was led to the discovery of stellar aberration (see Section 13.1) announced in 1728 and Herschel to the discovery of binary stars in the early 1800s. The first direct observations of stellar parallax were made in 1837–1839 by the German astronomer Friedrich Struve (1793–1864) on Vega (also known as Alpha Lyrae), the brightest star in the northern constellation Lyra (lyre); by the German astronomer and mathematician Friedrich Bessel (1784–1846) on 61 Cygni in the constellation Cygnus (swan); and by the Scottish astronomer Thomas Henderson (1798–1844) on Alpha Centauri in the constellation Centaurus (centaur).

4.5 THE PLANETS

So far we have presented the two-sphere model of the universe without the planets, except for the sun. To the ancients the known planets were the sun, the moon, Mercury, Venus, Mars, Jupiter and Saturn. The planets presented a genuine problem to any theory of the universe, as their root from the Greek word 'wanderer' implies. They all have a westward diurnal motion with the stars and gradual eastward motion through the background of the stars, as does the sun. During their journey through the heavens, the planets stay near the ecliptic, moving sometimes north and sometimes south of it, but nearly always staying within the zodiac, a strip of the heavens 16° wide and centered on the ecliptic. In the two-sphere model of the universe, there was really no way to order the distances of the various planets from the earth. The rule of thumb used was to assume that the longer a planet's period (the time for one complete revolution around the earth), the greater was its distance from the earth.[4] By the fourth century B.C. or so, the planets were ordered outward from the earth as: moon, Mercury, Venus, sun, Mars, Jupiter and Saturn. However, Venus and Mercury presented a special problem since their angular separation in the heavens is never very great, a fact for which the Ptolemaic model provided no natural explanation. (We return to this difficulty in Section 5.2.)

The outstanding problem of ancient cosmology was to explain the irregularities of the motions of the planets. Plato is said to have asked what are the uniform and ordered movements that can account for the apparent movements of the planets. His colleague Eudoxus ($c.400$–$c.350$ B.C.) proposed the first in a long series of models that compounded various circular motions. In that *homocentric-sphere model*, each of the seven heavenly bodies was assigned a shell of some finite thickness. (This alone would lead to eight spherical shells, counting the celestial sphere for the stars. In fact, the model had many more shells to account for the complicated motions of the planets.) The ancients appealed to a belief in the impossibility of a void to assign a size to their universe (that, in modern parlance, would be only the solar system).[5] Assuming the ancients' simple proportionality between the period and the radius of an orbit for a planet, we can estimate the size of their universe. Since the period of Saturn (the outermost planet then known) is about twenty-nine years, one would conclude that its orbit is twenty-nine times as large as that of the sun about the earth. Given the earth–sun distance (see Section 4.A), it follows that the distance from the earth to the celestial sphere (that was to be just at the outermost edge of the region occupied by Saturn) was estimated to be about $20,000 r_e$.

Let us summarize what we have so far presented in this chapter and see how successively more complex models of the universe came to be proposed. Plato, Aristotle and most other ancients felt that circular motion was eternal and therefore the only type suitable for the heavenly bodies. A certain egocentrism and anthropocentrism demanded the earth be at the center of the universe and at absolute rest. (After all, we experience no sense of motion on the earth when we are at rest with respect to it. Think about yourself looking at these facts.) The stars appear fixed with respect to one another, although they all seem to revolve about the earth. The planets, on the other hand, do not maintain fixed positions relative to each other and appear to pursue erratic paths across the sky. This led the ancients to place the stars on a large, common celestial sphere that revolved about the earth as center. It was assumed that all heavenly bodies moved in circular orbits with uniform speed, reflecting a type of perfection that befit their eternal and unvarying nature. Although Aristarchus attempted to place the sun at the center of the universe and have the earth revolve about it, his inability to measure stellar parallax was considered evidence against his theory. Hence, the Greeks chose to have the earth stand still. A criterion of simplicity of explanation is evident in all of this.

There were, however, even from ancient times, two pieces of data that were taken as opposing the geocentric theory of circular planetary orbits. First, the planets

Observational astronomy and the Ptolemaic model

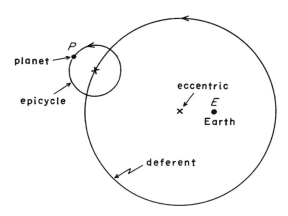

FIGURE 4.4 Eccentrics, deferents and epicycles

varied in brightness throughout the year and, second, there was the *retrograde motion* of the planets. This latter effect refers to the fact that a planet, such as Mars, in its eastward journey through the stars, seems at times to slow down, stop, reverse its direction for a short while, and then proceed eastward again.

Eccentric circles were postulated to explain the variations in brightness, the basic assumption being that the closer a planet is the brighter it is. The center of the planet's orbit is displaced from the earth to a point called the eccentric, as shown in Figure 4.4. (The earth remains at the center of the celestial sphere.) Retrograde motion is accounted for by the *epicycle* that is a smaller circle on which the planet moves uniformly while the center of this epicycle moves at a uniform rate around the larger eccentric circle, known as the *deferent*. In this theory, the planet moves on the epicycle in the same direction that the epicycle moves on the deferent. Therefore, the motion of the planet (P) relative to the earth (E) is a composition of these uniform circular motions. The exact shape of the planet's orbit depends upon the relative radii of the epicycle and of the deferent and upon the relative speeds of motion around these circles. A possible path is shown in Figure 4.5. At points 1 and 3 the planet will progress eastward through the background of fixed stars, whereas at point 2 its apparent motion will reverse itself or retrogress. As with the motion of the sun along its ecliptic, this eastward motion of the planets through the stars is in addition to (or is superimposed upon) the overall westward diurnal rotation of the celestial sphere that 'carries' the planets with it. In this model the rate of westward diurnal rotation of the celestial sphere is greater than that of the deferents of the planets so that the positions of the planets generally 'drift' eastward relative to the background of the celestial sphere.

4.6 PTOLEMY'S GEOCENTRIC MODEL OF THE UNIVERSE

Around 150 A.D. Claudius Ptolemy, an Alexandrian astronomer, mathematician and geographer, refined (or further complexified, depending upon your point of

PART II **Ancient and modern models of the universe**

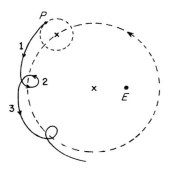

FIGURE 4.5 *Epicyclic motion according to Ptolemy*

view) this model of the universe to fit the data then available. In our previous discussion, we saw that the two-sphere universe, based on uniform circular motion and combined with eccentrics, deferents and epicycles for each planet, could give a qualitative fit to the known facts of observational astronomy. These basic ideas were perfected by Ptolemy's predecessor Hipparchus (*fl.* 146–127 B.C.). Ptolemy presented the details of this model in his *Almagest* (literally, the greatest (work)), as the Arabs later referred to it. He summarized this as follows.

> And so, in general, we have to state that the heavens are spherical and move spherically; that the earth, in figure, is sensibly spherical also when taken as a whole; in position, lies right in the middle of the heavens, like a geometrical centre; in magnitude and distance, has the ratio of a point with respect to the sphere of the fixed stars, having itself no local motion at all.[6]

He remained an Aristotelian in his acceptance of uniform circular motion for the planets:

> [I]t is first necessary to assume in general that the motions of the planets in the direction contrary to the movement of the heavens are all regular and circular by nature, like the movement of the universe in the other direction. That is, the straight lines, conceived as revolving the stars or their circles, cut off in equal times on absolutely all circumferences equal angles at the centres of each; and their apparent irregularities result from the positions and arrangements of the circles on their spheres through which they produce these movements, but no departure from their unchangeableness has really occurred in their nature in regard to the supposed disorder of their appearances.
>
> But the cause of this irregualr appearance can be accounted for by as many as two primary simple hypotheses. For if their movement is considered with respect to a circle in the plane of the ecliptic concentric with the cosmos so that our eye is the centre, then it is necessary to suppose that they make their regular movements either along circles not concentric with the cosmos, or along concentric circles; not with these simply, but with other circles borne upon them called epicycles. For according to either hypothesis it will appear possible for the planets seemingly to pass, in equal periods of time, through unequal arcs of the ecliptic circle which is concentric with the cosmos.[7]

However, in order to obtain quantitative agreement between this model and the data (here, the exact locations of the planets at any past or future date), the deferent, epicycle and eccentric were not sufficient. Ptolemy introduced a device known as the *equant*. Figure 4.6 shows the earth located at *F* (still the center of the celestial

Observational astronomy and the Ptolemaic model

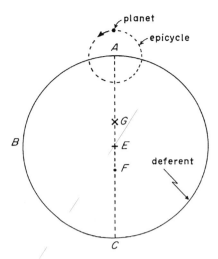

FIGURE 4.6 **The equant (G) in Ptolemy's model**

sphere) and E as the center of the deferent's eccentric circle. Ptolemy slips the equant (G) into his system as follows.

> The epicycles do not have their centres borne on the eccentric circles whose centres are those with respect to which the epicycles' centres revolve in a regular eastward motion and cut off equal angles in equal times.... [T]he epicycles' centres are borne on circles equal to the eccentrics effecting the anomaly, but described about other centres.
>
> ...
>
> For about centre E and diameter AEC, let there be the eccentric circle ABC on which the epicycle's center is borne. On this diameter let F be taken as the ecliptic's centre, and G ... as the centre around which we say the epicycle's mean passage is regularly effected.[8]

That is, in terms of Figure 4.6, the center of the epicycle (A) no longer moves with a constant speed (or sweeps out equal angles in equal time) with respect to the center (E) of the deferent but now with respect to the equant point (G). It is only with respect to this off-center point that the rate of motion of a point on the deferent would appear uniform. With the expediency of the equant, Ptolemy departed from the Aristotelian ideal he had subscribed to and almost all pretense of uniform circular motion was lost.

Figure 4.7 represents the essentials of Ptolemy's universe as presented in his *Almagest*. (Only one set of circles for a single planet is shown but, of course, there is one set for each of the planets, as well as for the sun and moon.) As the data became more accurate, epicycles on epicycles were required to obtain the necessary numerical precision. In fact the Ptolemaic theory never quite worked, in that more and more adjustments were required as the observational data improved in accuracy. The theory became essentially a matter of simply fitting the data rather than of providing a fundamentally correct representation of the physical universe.

The equant provides a nice example of an ad hoc device whose only function in a theory or model is to produce agreement with some troublesome data. (Recall the

PART II Ancient and modern models of the universe

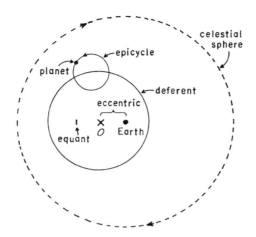

FIGURE 4.7 **The Ptolemaic universe**

discussion at the end of Section 2.4.) In the present case, the concept of the equant even violated the spirit of the underlying Aristotelian doctrine of uniform circular motion. With all of its geometrical complexity, the Ptolemaic model of the universe was still unable to fit the data and remain faithful to its conceptual foundation. It was not internally consistent.[9] The entire system became so cumbersome that, when Alfonso X of Castile (1221–1284), the king of Spain from 1252–1284, was introduced to the Ptolemaic model, he is alleged to have said that, if the Lord Almighty had consulted him before embarking upon the Creation, then he, Alfonso, would have recommended something better.

4.7 TO SAVE THE PHENOMENA

As we mentioned in passing in Section 2.3, Plato gave his description of the universe in his *Timaeus*.[10] In his opinion the earth was created as a perfect sphere and set into natural uniform circular motion. This Platonic doctrine was passed on to later ages through a series of commentaries on the works of Plato and Aristotle. In the *Commentary on Aristotle's De Caelo* by the Greek philosopher Simplicius (*fl. c.* 530 A.D.), this tradition is formulated as:

> Plato lays down the principle that the heavenly bodies' motion is circular, uniform, and constantly regular. Thereupon he sets the mathematicians the following problem: What circular motions, uniform and perfectly regular, are to be admitted as hypotheses so that it might be possible to save the appearances presented by the planets?[11]

Ptolemy remained faithful to this Platonic goal when he stated in his *Almagest* that:

> Now, since our problem is to demonstrate, in the case of the five planets as in the case of the sun and moon, all their apparent irregularites as produced by means of regular and circular motions (for these are proper to the nature of divine things which are strangers to disparities and disorders) the successful accomplishment of the aim as truly belonging to mathematical theory in

philosophy is to be considered a great thing, very difficult and as yet unattained in a reasonable way by anyone.[12]

In this Platonic tradition astronomy was not meant to attempt an explanation of physical reality, but merely to give a mathematical representation of it ('to save the phenomena' in Plato's phrase). Since the heavenly bodies are of a divine nature, they obey laws different from those found on the earth. There is no connection between the two and that makes it impossible for us to know anything about the physics of the heavens.

While this attitude toward physical models may strike the modern reader as strange, it does bring up nicely the issue of *realism* versus *instrumentalism*. The basic question there is how seriously ('realistically') a theory is to be taken.[13] For us this usually depends upon the evidence for a theory. At first a theory (or explanatory 'story') is taken tentatively and then more seriously as it continues to be successful in more cases and we become acclimated to it. In a sense, there is a psychological component in this, perhaps similar to that in Hume's explanation of how we form a habit of the mind to 'see' causal connections after we have observed a sufficiently large number of constant conjunctions of one type of event followed by another of a given kind. So, we naturally ask when it is reasonable to accept a theory as literally true. With respect to early astronomical models, we can observe the following.

The ancient Babylonians discovered empirical rules for calculating the positions and recurrences of the planets, but constructed no geometrical picture or causal explanation to account for these regularities. Their scheme was a purely instrumental one. Similarly, Plato and Eudoxus took astronomical constructions instrumentally. Thus, in the homocentric-sphere model, the cause driving the thick spherical shells within which the planets wandered was not sought. Astronomy was seen as a discipline between mathematics and physics and this created some tension or uncertainty about its character. It was not necessarily to be taken realistically. Aristotle had quite a different attitude and for him the outermost (or celestial) sphere became the genuine cause of driving this set of nested spheres. In turn each of these driven spherical shells drove the adjacent shell interior to it. Ptolemy's position appears to have been ambivalent on this realism issue. Actually, the later Ptolemaic model did not lend itself as readily to a causal explanation as had the much earlier homocentric-sphere one of Eudoxus and Aristotle. The causal aspect of the later models was not much in evidence.

The writings of Aristotle and of Ptolemy dominated cosmological and astronomical thinking for many centuries. In Chapter 6 we shall look at some of the criticisms of and challenges to this ancient corpus of knowledge prior to the Renaissance. For now, though, we simply acknowledge that by 1500 A.D. a new conception of the universe was required. This would be the era of Copernicus and of Kepler in which a key transition was made. Copernicus held his picture, or model, to be literally true (that is, he did not deny its reality) and, as we shall see in the next chapter, he explicitly stated his belief in the actual motion of the earth. Initially one might not be inclined to think that Copernicus took his model at face value, since the preface to his *De Revolutionibus Orbium Caelestium* (*On The Revolutions of the*

PART II Ancient and modern models of the universe

Heavenly Spheres) contains a disclaimer about the literal truth of his own representation:

> Since the newness of the hypotheses of this work – which sets the earth in motion and puts an immovable sun at the centre of the universe – has already received a great deal of publicity, I have no doubt that certain of the savants have taken grave offense and think it wrong to raise any disturbance among liberal disciplines which have had the right set-up for a long time now. If, however, they are willing to weigh the matter scrupulously, they will find that the author of this work had done nothing which merits blame. For it is the job of the astronomer to use painstaking and skilled observation in gathering together the history of the celestial movements, and then – since he cannot by any line of reasoning reach the true causes of these movements – to think up or construct whatever causes or hypotheses he pleases such that, by the assumption of these causes, those same movements can be calculated from the principles of geometry for the past and for the future too. This artist is markedly outstanding in both of these respects: for it is not necessary that these hypotheses should be true, or even probable; but it is enough if they provide a calculus which fits the observations.... For it is sufficiently clear that this art is absolutely and profoundly ignorant of the causes of the apparent irregular movements. And if it constructs and thinks up causes – and it has certainly thought up a good many – nevertheless it does not think them up in order to persuade anyone of their truth but only in order that they may provide a correct basis for calculation. But since for one and the same movement varying hypotheses are proposed from time to time, as eccentricity or epicycle for the movement of the sun, the astronomer much prefers to take the one which is easiest to grasp. Maybe the philosopher demands probability instead; but neither of them will grasp anything certain or hand it on, unless it has been divinely revealed to him.[14]

However, that preface was not written by Copernicus, but by the Lutheran theologian Andreas Osiander (1498–1552) who was in charge of the printing of the book due to Copernicus' failing health. Even more than simply taking this new model of the universe seriously, Kepler actually sought causes for the motions of the planets about the sun. We turn to these new developments in Chapter 5.

4.A DETERMINATION OF THE ABSOLUTE SIZES OF PLANETARY ORBITS

We saw how Eratosthenes determined the size of the earth's radius r_e (see Figure 4.2 and the accompanying discussion). In this note we outline how Aristarchus was able to express the radius of the earth's orbit R_e about the sun in terms of the already known value of r_e. This was done in his treatise *On The Sizes and Distances of the Sun and Moon*.[15] Incidentally, the argument given below remains equally valid in either the earth-centered or sun-centered model of the universe.

The distance R_e from the earth to the sun can be related to the distance R_m from the earth to the moon by observing the moon when it is half full, as indicated in Figure 4.8. The orbits of the earth and of the moon are taken to be circular. When the moon is exactly half full, the upper angle in the triangle of Figure 4.8 must be a right angle. At that time one measures the angle α between R_m and R_e as seen from the earth. Simple trigonometry relates R_m and R_e as $\cos\alpha = R_m/R_e$. Aristarchus took α to be about $87°$ so that he found $R_e = 19 R_m$.[16] This is a very difficult measurement to make because it is not easy to determine when the moon is

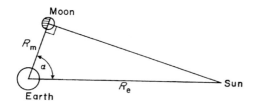

FIGURE 4.8 Earth–sun–moon relation at half-moon

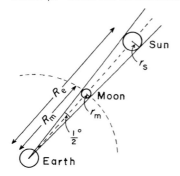

FIGURE 4.9 The angle subtended by the sun or moon at the earth

precisely half full and then to measure the angle α very accurately. Recall that Aristarchus had neither a telescope nor sophisticated instruments to measure angles. Today we know that α is $89°\,51'$ so that this relation becomes $R_e = 382 R_m$. Aristarchus' underestimation of the size of the earth's orbit eventually led to an underestimation of the size of the solar system.

Next, one can measure the angle subtended at the eye of an observer on the earth by the moon or by the sun, as shown in Figure 4.9. In both cases this angle is nearly $\frac{1}{2}°$ of arc. From this we can learn two facts. First, we can set up the ratio $(2r_m)/(2\pi R_m) = (\frac{1}{2})/(360°)$, where r_m is the radius of the moon itself and $2\pi R_m$ is the circumference of the orbit of the moon about the earth. From this we obtain $R_m = 229 r_m$. Second, from the similar triangles of Figure 4.9, we can write $r_m/r_s = R_m/R_e$. If we use Aristarchus' value for the ratio on the right, we obtain $r_s = 19 r_m$; if we use the modern value, we have $r_s = 382 r_m$.

Finally, Aristarchus was able to relate these quantities to the radius of the earth r_e by observing the moon when it was eclipsed by the earth, as illustrated in Figure 4.10. Here the moon passes through the shadow cast by the earth. By measuring the time it took the moon to enter this shadow completely (that is, to travel a distance $2r_m$) and then the time between the entrance of the front edge of the moon into this shadow and the emergence of the front edge of the moon from it, Aristarchus could set up a simple proportion to determine the distance Y of the moon's path through the shadow. He found the second time to be approximately twice the first time so that he took Y to be $Y = (2)2r_m = 4r_m$. Again, since he did not have an accurate means to measure time intervals, he was in error in this determination. A more accurate measurement yields $Y = (1.3)4r_m = 5.2 r_m$.

PART II   Ancient and modern models of the universe

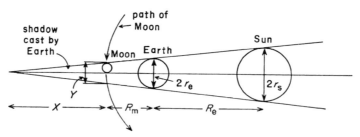

FIGURE 4.10   Aristarchus' argument for the earth–moon distance

From the similar triangles of Figure 4.10 we can set up the (approximate) ratios[17]

$$\frac{X}{Y} = \frac{X + R_m}{2r_e} = \frac{X + R_m + R_e}{2r_s} \tag{4.1}$$

Since we do not need the value of X (the distance that the earth's shadow extends beyond the moon, a distance we cannot observe without traveling into space beyond the moon), we can solve for X from the first and third terms of Eq. (4.1) as

$$X = \frac{Y(R_m + R_e)}{2r_s - Y} \tag{4.2}$$

If we now substitute this into the equation consisting of the second and third terms of Eq. (4.1), we find

$$r_s \left[\frac{Y(R_m + R_e)}{2r_s - Y} + R_m \right] = r_e \left[\frac{Y(R_m + R_e)}{2r_s - Y} + R_m + R_e \right] \tag{4.3}$$

If we insert Aristarchus' values for Y, R_e and r_s into Eq. (4.3), we obtain $r_m = 0.35 r_e$, $R_m = 80 r_e$ and $R_e = 19 R_m = 1520 r_e$. If we repeat the same calculations using the more modern values (mainly the better value for the angle α of Figure 4.8), we find $R_e = 24{,}400 r_e$.

FURTHER READING

The classic locus for most of the material in this chapter is Thomas Kuhn's *The Copernican Revolution*. Otto Neugebauer's *The Exact Sciences in Antiquity* and Michael Crowe's *Theories of the World from Antiquity to the Copernican Revolution* are also useful references. Arthur Koestler's *The Sleepwalkers* provides fascinating insights into Copernicus and Kepler. Pierre Duhem's *To Save the Phenomena* follows the development of physical theory from Plato to Galileo. Charles Gillispie's *The Edge of Objectivity* exposes the reader to a grand sweep of the history of scientific ideas from the time of Galileo to the twentieth century.

5

The Copernican model and Kepler's laws

The form of the Ptolemaic system, as depicted in Figure 4.7, was in accord with the general thinking of the day, fit past observational data and predicted fairly well the future positions of the then-known planets. In his *De Revolutionibus*, published at his death in 1543, and in his earlier *Commentariolus* (*Sketch of the Hypotheses for the Heavenly Motions*), about 1514, Nicolaus Copernicus seriously attacked the Ptolemaic model. He did this largely because he felt that some of the devices (in particular, the equant) used to compound circular motions in Ptolemy's system produced motions that were not uniform enough.

> [I]n setting up the solar and lunar movements and those of the other five wandering stars, [mathematicians] do not [all] employ the same principles, assumptions, or demonstrations for the revolutions and apparent movements. For some make use of homocentric circles only, others of eccentric circles and epicycles, by means of which however they do not fully attain what they seek.... But even if those who have thought up eccentric circles seem to have been able for the most part to compute the apparent movements numerically by those means, they have in the meanwhile admitted a great deal which seems to contradict the first principles of regularity of movement.[1]
>
> ...
>
> Yet the planetary theories of Ptolemy and most other astronomers, although consistent with the numerical data, seemed likewise to present no small difficulty. For these theories were not adequate unless certain equants were also conceived; it then appeared that a planet moved with uniform velocity neither on its deferent nor about the center of its epicycle. Hence a system of this sort seemed neither sufficiently absolute nor sufficiently pleasing to the mind.[2]

Copernicus suggested putting the earth, rather than all the heavens, in motion.

> Why therefore should we hesitate any longer to grant to it [the earth] the movement which accords naturally with its form, rather than put the whole world in a commotion – the world whose limits we do not and cannot know? And why not admit that the appearance of daily revolution belongs to the heavens but the reality belongs to the Earth?[3]

We mentioned in Chapter 4 that he overcame the stellar parallax objection by increasing the size of the universe tremendously so that the effect became too small to observe with the naked eye. Still, he was basically an Aristotelian in accepting the necessity of uniform circular motion.

> After this we will recall that the movement of the celestial bodies is circular. For the motion of a sphere is to turn in a circle; by this very act expressing its form, in the most simple body, where beginning and end cannot be discovered or distinguished from one another, while it moves through the same parts in itself.

PART II **Ancient and modern models of the universe**

. . .

> We must however confess that these movements are circular or are composed of many circular movements, in that they maintain these irregularities in accord with a constant law and with fixed periodic returns: and that could not take place, if they were not circular.[4]

5.1 COPERNICUS AND HIS HELIOCENTRIC MODEL

Copernicus was in no sense a revolutionary by nature, but a rather timid person. He was born at Torun, Poland, in 1473. At age eleven, Nicolaus lost his father and was henceforth raised by an uncle who was a scholarly priest and later a bishop. This rather autocratic uncle decided that the young man should be trained for the Church and obtained an appointment for Nicolaus as Canon of Frauenburg in 1497. This sinecure provided him with an income, but had very few duties attached to it. He remained on leave from his post at Frauenburg until 1512 while he completed his education and traveled. He was a lawyer and physician, trained also in Greek, mathematics and astronomy. He became aware of Aristarchus' sun-centered theory of the universe and maintained an active interest in astronomy for the rest of his life. At Frauenburg he established an observatory. By 1514 his reputation as an astronomer was such that he was invited to the Lateran Council to discuss calendar reform. In the same year he presented a short outline of his heliocentric theory in the *Commentariolus*. This was not published, but was circulated as a manuscript. In 1539 Georg Rheticus (1514–1576), a young professor of mathematics at Wittenberg, visited Copernicus and remained for two years. Copernicus completed his major work, *De Revolutionibus*, about 1530 but would not publish it. In 1540 Rheticus was allowed to publish the *Narratio Prima* (*The First Account of the Revolutionary Book by Nicolaus Copernicus*) that was a general account of the Copernican system. Finally, Rheticus and other friends prevailed upon him to publish the *De Revolutionibus*. Tradition has it that, on the last day of his life, May 24, 1543, as he lay dying of a brain hemorrhage, Copernicus was given a copy of his only published work.

It may be helpful to situate Copernicus' work in the context of the intellectual currents of his time. The Renaissance, that lasted in Europe from the fourteenth to the seventeenth century or so, was an age of rediscovery of much of the ancient Greek and Roman traditions in learning, as well as of new discovery (for example, the great voyages, including those of Christopher Columbus (1451–1506)) and of calendar reform. (Recall that the Gregorian calendar was instituted in 1582.) There was a loosening of the hold of Ptolemaic astronomy. It was an age of change and the Roman Church took no stand (for or against) on the Copernican system at first. The Reformation of the sixteenth century, led by Martin Luther (1483–1546) and John Calvin (1509–1564), had as a central tenet of Protestantism that a pure and direct interpretation of the Bible was to be the fundamental source of Christian knowledge. Therefore, it is not surprising that Protestant Churches were against the Copernican model as opposing Holy Scripture. (Remember Osiander's preface to

The Copernican model and Kepler's laws

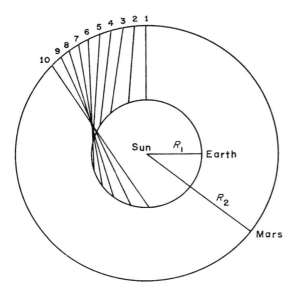

FIGURE 5.1 **Mars–earth configuration in the Copernican model**

De Revoltionibus disclaiming the literal truth of Copernicus' system.) For about sixty years after its publication, the Roman Church kept 'hands off', as it were, Copernicus' great book. It was with the Counter-Reformation, a reaction against the Protestant charges of laxness in religious and political matters in the late sixteenth and early seventeenth centuries, that the Roman Church took a hard line against Copernicanism and in 1616 *De Revolutionibus* was placed on the Index of books forbidden to be read by Catholics. We return to these issues in Chapter 10 when we discuss the trial of Galileo.

We now consider the Copernican model in some detail by examining how the phenomenon of apparent retrograde motion is accounted for in this heliocentric, or sun-centered, universe. In this theory the earth and all of the planets revolve around the sun in the same direction. This direction is also that in which the earth spins on its own axis. Figures 5.1 and 5.2 demonstrate qualitatively the heliocentric explanation of the retrograde motion of Mars as seen from the earth. We take a specific case $R_2 = 2.5 R_1$ and $\omega_1 = \omega_2/4$. This qualitatively (but not quantitatively) resembles the earth–Mars configuration in that the size of Mars' orbit is larger than the earth's and Mars also has the longer period of revolution. Here we use the Greek symbol ω (omega) to denote the angular velocity (usually measured in degrees per second or radians per second). The relation among the linear speed v, the radius of the circle R, the period τ (the time taken for one complete revolution around the sun) and the angular velocity ω (measured in radians per second) is, for uniform circular motion, $v = (2\pi R)/\tau = R\omega$. Figure 5.1 shows the positions of the earth and of Mars at ten equally spaced time intervals. This is what an observer at rest with respect to the sun would see. In Figure 5.2 we have translated these data to a reference point fixed on the earth. The retrograde motion is evident. We 'see' only the projection of the

61

PART II Ancient and modern models of the universe

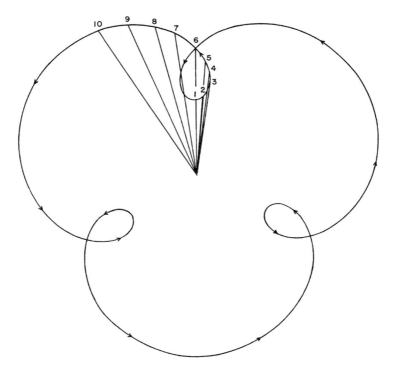

FIGURE 5.2 **Epicyclic motion according to Copernicus**

(tangential) velocity perpendicular to the line of sight. The retrograde motion is an (apparent) effect due to the relative motion of the earth and planet. Copernicus appreciated this aspect of his system.

> For if the annual revolution were changed from being solar to being terrestrial, and immobility were granted to the sun, the risings and settings of the signs and of the fixed stars – whereby they become morning or evening stars – will appear in the same way; and it will be seen that the stoppings, retrogressions, and progressions of the wandering stars are not their own, but are a movement of the Earth and that they borrow the appearances of this movement.[5]

5.2 ADVANTAGES OF THE COPERNICAN THEORY

At the qualitative, or semiquantitative, level, Copernicus' model has several aesthetic and practical advantages over Ptolemy's. In the heliocentric model only two parameters (or fudge factors), namely the radius of the orbit and the speed of the planet, are required for each planet. Then, both the variation in brightness and the retrograde motion follow 'naturally'. For the geocentric model at least five parameters are required for each planet (the radius of its orbit (or deferent) and its speed, the radius of the epicycle and its speed and the offset of the eccentric) for even a qualitative fit.

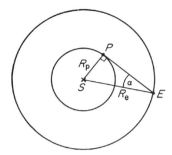

FIGURE 5.3 Determination of the size of the orbit of an inner planet

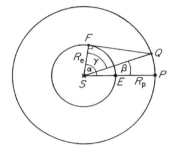

FIGURE 5.4 Determination of the size of the orbit of an outer planet

The Copernican system also resolves the problem of the order of the sun, Mercury and Venus. Since the sun is placed at the center of the universe, only the order of Mercury and Venus need be settled. For either of these inferior or inner planets (those whose orbits are smaller than that of the earth), the size of the orbit relative to that of the earth's can be determined as indicated in Figure 5.3. When the angle α between the earth–sun line (ES) and the earth–planet line (EP) reaches its maximum value, then the angle SPE is a right angle (since PE is tangent to the circle of radius R_p) and simple trigonometry yields $\sin\alpha = R_p/R_e$. The radius of the inner planet's orbit R_p is thus given in terms of that of the earth's orbit R_e, once this maximum angle α has been measured by observation.[6]

A similar, but somewhat more complicated, argument can be employed to determine orbital radii of the superior or outer planets as indicated in Figure 5.4.[7] We begin with the sun (S), the earth (E) and the outer planet (P) all in the collineation SEP. At some later time, when the earth has advanced to point F and the planet to Q and when angle SFQ is a right angle (that will always occur at some instant since the earth moves more rapidly in its orbit than does an outer planet), we can argue as follows. Knowing the time t_{EF} it takes the earth to travel from E to F and assuming uniform circular motion, we can determine the angle α (angle PSF) as $\alpha/360° = t_{EF}/365$. Similarly, the angle β (that is, angle QSP) is simply $\beta/360° = t_{EF}/$(period of outer planet). Since $\gamma = \alpha - \beta$, we have for the right triangle SFQ that $R_e = R_p \cos\gamma$. In the geocentric model there was no such method for uniquely ordering the planets. We can also appreciate now that these triangulation techniques in the Copernican model, plus Aristrachus' method outlined at the end of

PART II Ancient and modern models of the universe

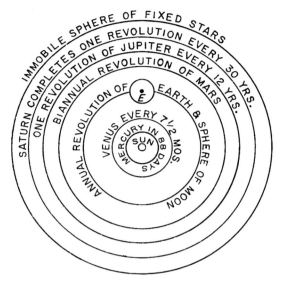

FIGURE 5.5   The Copernican model of the universe

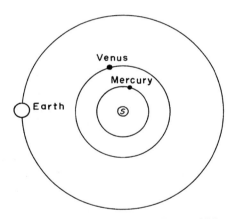

FIGURE 5.6   The 'morning stars' Venus and Mercury

Chapter 4, allow us to determine the absolute magnitudes of the sizes of the orbits of the planets. Figure 5.5 shows the heliocentric model of the universe as Copernicus conceived of it.[8] Notice that the moon has been placed in its own circular orbit about the earth.

Another impressive qualitative feature of the Copernican model is that it quite naturally explains why Venus and Mercury must appear fairly close together, often as morning and evening stars. Since the radii of the orbits of Mercury and of Venus are each much smaller than that of the earth, we see from Figure 5.6 that they must always have a small angular separation as seen from earth. On the other hand, in the older Ptolemaic model, the sun, Mercury and Venus each had its own orbit about the earth and there would not seem to be any a priori reason why they should not at times appear very far apart in the heavens. The geocentric model had to adjust their orbits in a completely ad hoc fashion.

5.3 SHORTCOMINGS OF THE COPERNICAN THEORY

However, the precise quantitative success of Copernicus' original model was quite another matter. Ptolemy required a system of over thirty circles as deferents and epicycles, as well as eccentrics and equants, to obtain an acceptable quantitative fit to astronomical data. Copernicus, too, was forced to use minor epicycles and some eccentrics to attain quantitative success. For a degree of accuracy equal to that of Ptolemy's model, Copernicus also required over thirty circles. In fact, his *De Revolutionibus* even employs the equivalent of equants in treating the moon and Mercury.[9] From the point of view of practical economy and simplicity, there was little to choose between the two schemes. By this exacting criterion, Copernicus also failed to solve the problem of the planets. His long delay in publishing *De Revolutionibus* may have been due in part to the fact that his early expectations of the *Commentariolus* (for a simple explanation of planetary motion) were not fulfilled upon detailed examination of the problem.

Observations made by the ancient Chaldeans and Greeks, along with some more recent ones by the Arabs, formed most of the data that Copernicus used in fitting his model of the universe. He made very few observations himself and these were not particularly accurate. Although Copernicus was not aware of it at the time, some of these reported observations were incorrect. Not only was his model of the universe not quite right, but he was attempting to fit it to data some of which were in error. Extensive, accurate and reliable observations were required before significant progress could be made.

It was the Dane Tycho Brahe (1546–1601) who provided these. Born near the Elsinore of Shakespeare's *Hamlet*, he was raised by a cantankerous vice-admiral uncle and himself developed into an eccentric. Tycho's appearance was striking since he had a silver and gold nose that replaced the real one lost in a duel as a young man. While still a student in 1560 at the University of Copenhagen, he saw a partial eclipse of the sun that had been forecast. What impressed him most was the fact that such an event could be predicted by astronomy. Thus began a lifelong passion for exact observation. Against his uncle's wishes, he continued his study of the stars by constructing huge, precise instruments with which to make the best naked-eye observations then possible. Not only did Tycho gather data more accurate than any before his time, but he also accumulated a vast compendium of continuous positions of the planets and reliable locations of nearly 1000 stars.

In 1572 he became famous by discovering a new star where there had not been one previously. It remained visible for over a year before it faded, but its position in the heavens did not vary so that it was definitely a star. This fact contradicted Platonic and Aristotelian doctrine that the heavens above the sphere of the moon were unchangeable. Subsequently, in 1577 he demonstrated that the great comet seen moving across the heavens that year was also beyond the lunar sphere. The Platonic/Aristotelian universe, to which Ptolemy and Copernicus had subscribed, was now shattered. In 1576 King Frederick II of Denmark (1534–1588) gave Brahe an island between Copenhagen and Elsinore where Tycho had a great observatory constructed. For the next twenty years he remained there and gathered his treasure of data. Here at Uraniburg on this island of Hveen, Brahe observed the heavens, frequently gave clamorous banquets and treated the tenants there miserably. In 1597 he was forced to leave Hveen and finally two years later came to Prague as imperial mathematician to Emperor Rudolph II (1552–1612) and settled at nearby Benatek Castle.

5.4 KEPLER'S LAWS

It was to Benatek Castle that Johannes Kepler came in 1600. Kepler was born into an impoverished noble family at Weil in southwest Germany. Throughout his life he existed on the brink of financial ruin and was harried by family problems. As a brilliant student he won entrance to the University of Tübingen in 1587, where the astronomer Michael Maestlin (1550–1631) introduced him to the work of Copernicus. Throughout his life Kepler was a mystic. He originally intended to enter the ministry, but in 1594 accepted a position as mathematician and astronomer at Graz in Austria. One of his responsibilities there was to prepare an annual calendar of astrological forecasts. These proved to be quite accurate.

In 1595 Kepler became convinced that the structure of the universe was intimately connected with the five regular solids that can be constructed with identical faces made from equilateral figures. These are the tetrahedron (pyramid), the cube, the octahedron (eight equilateral triangles), the dodecahedron (twelve pentagons) and the icosahedron (twenty equilateral triangles). Kepler's idea was to inscribe, successively, each of these regular solids within spheres, one inside the other. This would allow six concentric spheres with the sun at the center. Each sphere contained one planetary orbit corresponding to the six planets then known (Mercury, Venus, earth, Mars, Jupiter and Saturn). In fact, this purported explanation fails (since it does not give correctly the ratios of the planetary orbits and since there are more than six planets). Nevertheless, this was an idea to which Kepler returned many times. In 1596 he published his first major work, *Mysterium Cosmographicum* (*The Cosmographic Mystery*), in which he expounded this theory. The book attracted much attention and both Brahe and Galileo corresponded with him about it. Many years later, in 1619, Kepler returned to a similar theme in his *Harmonices Mundi* (*Harmonies of the World*) and discovered his third law of planetary motion that we discuss later. Here, as so often elsewhere, Kepler had mystical reasons for his convictions about the nature of the universe. One factor in his early acceptance of the Copernican theory was his belief that the sun should be at the center of the universe by virtue of its dignity and power, being a place where God would reside as prime mover.

The Copernican model and Kepler's laws

When Tycho Brahe died in 1601, Kepler was appointed his successor at Benatek Castle. From 1600 to 1606 Kepler worked on the orbit of Mars, first attempting to fit it exactly with combinations of circular motions. He obtained agreement with Tycho's data to within eight minutes of arc, but realized that, since the data were more accurate than this, his theory must be incorrect. In his *Astronomia Nova* (*New Astronomy*) of 1609 Kepler states:

> [F]or us who, by divine kindness were given an accurate observer such as Tycho Brahe, for us it is fitting that we should acknowledge this divine gift and put it to use.... Henceforth I shall lead the way toward that goal according to my own ideas. For, if I had believed that we could ignore these eight minutes, I would have patched up my hypothesis accordingly. But since it was not permissible to ignore them, those eight minutes point the road to a complete reformation of astronomy: they have become the building material for a large part of this work....[10]

Here we see one of the hallmarks of modern science – the need for accurate quantitative agreement between theory and experiment. Prior to Brahe, an astronomical model was taken as adequate if it agreed with observations to within about ten minutes of arc. His observations were at least twice as accurate.

Kepler temporarily abandoned this problem of Mars' orbit. He took as an (incorrect) working assumption that the speed of a planet varied inversely as the distance from the sun (because he believed that rays emanating from the sun – his *anima motrix* – pushed the planets around in their orbits). By a long and tortuous argument he derived his second law of equal areas in equal times. He finally returned to finding the orbit of Mars. In all it took him six years of unbelievable labor to discover that it is an ellipse. He gave all the details of these false starts in his *New Astronomy*. In 1618–1621 he published the *Epitome Astronomiae Copernicus* (*Epitome of Copernican Astronomy*) that summarized for the general public his work, including the three laws of planetary motion that now bear his name. His last major work, the *Rudolphine Tables*, appeared in 1627 and was based largely on the observations made by Brahe. These astronomical tables that Kepler constructed allowed one to calculate the positions of the planets for any date in the past or future.

We now state the three planetary laws that took Kepler nearly twenty years to formulate. These laws were discovered largely empirically and by trial and error, with no coherent theory behind them. They were simply concise mathematical summaries of regularities that Kepler found by studying an incredible amount of data over a period of many years.

I   *Kepler's first law*: The planets move on ellipses about the sun at one focus.

In Section 5.A we develop some of the mathematical properties of the ellipse. Technically, the classical definition of an ellipse is the locus of all points (P) such that the sum of the distances (d_1 and d_2) from two fixed points (the foci, F and F')

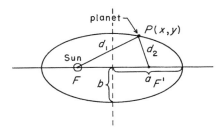

FIGURE 5.7   The geometry of an ellipse

remains constant (see Figure 5.7). All of the orbits of the planets are nearly in the plane of the earth's orbit about the sun (also known as the *ecliptic*). Once elliptical orbits are used in the Copernican model, that model does become much superior in simplicity and accuracy to that of the Ptolemaic system. An ellipse is characterized by two parameters that we take to be the semimajor axis a and the semiminor axis b, as shown in Figure 5.7. The standard modern form for the equation of an ellipse is

$$\frac{x^2}{a^2} + \frac{y^2}{b^2} = 1 \tag{5.1}$$

Kepler's first statement of this law is in the *New Astronomy*:

> Therefore, the orbit of a planet is an ellipse.[11]

In his *Epitome of Copernican Astronomy*, he demonstrates this in a section headed:

> It is left ... to prove that an elliptic orbit is constituted, concerning which you have said the observations bear witness.[12]

2   *Kepler's second law*: A radius vector drawn from the sun to the planet sweeps out equal areas in equal times.

We illustrate this law in Figure 5.8. Suppose that during a two-month period a planet moves from point 1 to point 2 along its orbit so that its radius vector drawn from the sun sweeps out the area A_{12}. Then, according to Kepler's second law, during any other two-month period (say in moving from 3 to 4) the radius will sweep out an equal area (A_{34}). Since the distance of the planet from the sun varies as the planet moves around its orbit, we see that the planet must move more rapidly when it is closer to the sun than when it is farther away if it is to sweep out equal areas in equal times. This statement that the rate at which area is swept out is a constant for a given planet gives a precise relation between the instantaneous speed of a planet and its distance from the sun. It can be written as a mathematical equation, but we do not need that at present.

Kepler first stated his second law in Section 40 of his *New Astronomy*, but the discussion is very involved and the result is obtained incorrectly assuming eccentric circular motion for the planet. It is stated concisely for an elliptical orbit in his

The Copernican model and Kepler's laws

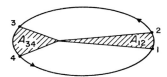

FIGURE 5.8 **Kepler's second law of planetary motion**

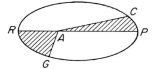

FIGURE 5.9 **The 'delay' of a planet**

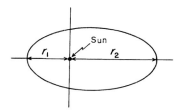

FIGURE 5.10 **The perihelion (r_1) and aphelion (r_2) distances**

Epitome of Copernican Astronomy:

> Wherefore the delay of the planet in arc PC is to the delay in the equal arc RG as the area of triangle PCA is to the area of triangle RGA.[13]

By 'delay' here Kepler means the transit time along the arc in question. Figure 5.9 labels the quantities referred to in this quotation. The sun is located at A. In this statement of his law, Kepler took the arc lengths PC and RG to be equal, so that neither the areas PCA and RGA nor the transmit times (delays) are equal. From the proportionality stated, we see that if the areas are taken to be equal, then so are the transit times (although, of course, the arc lengths would no longer be equal). Appreciate that Kepler's second law states that the rate at which area is swept out is constant for a given planet, not that it is the same for all planets. For example, the area swept out by the earth in any one day is the same as the area swept out by the earth in any other day, but different from the area swept out by Mars on any given day. That is, the specific numerical value of this rate of sweeping out area varies from planet to planet in our solar system.

3. *Kepler's third law*: The ratio of the cube of the mean radius R of a planet's orbit to the square of its period τ is a fixed constant for all planets in the solar system.

$$\frac{R^3}{\tau^2} = const \qquad (5.2)$$

PART II Ancient and modern models of the universe

Table 5.1. *Astronomical data for the planets*

Planet	Mean radius of orbit (km)	Mean period (days)	Eccentricity $\sqrt{1-(b/a)^2}$	$(R^3/\tau^2)/10^{19}$
Mercury	5.79×10^7	88	0.206	2.51
Venus	1.08×10^8	225	0.007	2.49
Earth	1.50×10^8	365	0.017	2.53
Mars	2.27×10^8	687	0.093	2.48
Jupiter	7.78×10^8	4,333	0.048	2.51
Saturn	1.43×10^9	10,759	0.056	2.53
Uranus	2.87×10^9	30,685	0.047	2.51
Neptune	4.49×10^9	60,190	0.009	2.50
Pluto	5.91×10^9	90,737	0.249	2.51

Table 5.2. *Data on the earth, moon and sun*

	Earth	Moon	Sun
Mass (kg)	5.98×10^{24}	7.34×10^{22}	1.99×10^{30}
Mean radius (km)	6.37×10^3	1.72×10^3	6.97×10^5
Mean distance from earth (km)		3.80×10^5	1.49×10^8

The mean radius of an elliptical orbit is defined as one-half the sum of the distance of closest approach of the planet to the sun r_1 (the *perihelion*), and the greatest distance between the planet and the sun r_2 (the *aphelion*) (see Figure 5.10) as

$$R = \frac{1}{2}(r_1 + r_2) \qquad (5.3)$$

(Notice that this mean radius R is just the semimajor axis of the ellipse in Eq. (5.1) and in Figure 5.7.) The period τ is the time it takes the planet to make one complete journey on its orbit around the sun. Specifically, if R_1 and τ_1 are the mean radius and period of one planet (say, the earth) and R_2 and τ_2 those of another (say, Mars), then Kepler's third law states

$$\frac{R_1^3}{\tau_1^2} = \frac{R_2^3}{\tau_2^2} \qquad (5.4)$$

It is essential to realize that Kepler's third law holds only among orbits about a common central body (for example, the planets around the sun or a set of satellites around the earth). However, it would give no relation, for instance, between the radius and period of the earth about the sun and those of the moon about the earth.

This law is found in his *Harmonies of the World*, in which he returned to the theme of his earlier *Cosmographic Mystery* and attempted to obtain the laws of the universe from a combination of geometry, astronomy, music and astrology.

> But it is absolutely certain and exact that the ratio which exists between the periodic times of any two planets is precisely the ratio of the 3/2th power of the mean distances....[14]

In Table 5.1 we list some astronomical data for the planets. The last column in that table shows the validity of Kepler's third law, Eq. (5.2). Table 5.2 contains some useful data on the earth, moon and sun.

5.A CONIC SECTIONS

Let us begin with the special case of an ellipse, since it figures in Kepler's first law. Reference to Figure 5.7 of the text will aid us in our discussion. The *eccentricity e* is defined as $e = \sqrt{a^2 - b^2}/a$ and is a measure of the departure of the ellipse from a circle. When $e = 0$, then $a = b$ and we are back to a circle. The orbits of earth and of Mars are nearly circular since $e_{\text{earth}} = 0.0167$ and $e_{\text{Mars}} = 0.0934$, whereas the orbit of Mercury is quite elongated with $e_{\text{Mercury}} = 0.2056$. The verbal definition of the ellipse that accompanied Figure 5.7 in the text can be stated mathematically in terms of that figure as $PF + PF' = const$ for all points P on the ellipse. If we let O denote the center of the ellipse and take $x = a$, $y = 0$, then we have $(a + OF + (a - OF')) = const = 2a$, since $OF = OF'$. Therefore, it follows that $PF + PF' = 2a$. From Pythagoras' theorem we can write

$$(PF)^2 = y^2 + (OF + x)^2, \quad (PF')^2 = y^2 + (OF - x)^2$$

If we let $x = 0$, $y = b$, then we have $PF = PF' = a$ so that $a^2 = ((OF)^2 + b^2)$ and $OF = OF' = \sqrt{a^2 - b^2}$. Straightforward algebra then leads to the standard equation of an ellipse given in Eq. (5.1).

Around 350 B.C. Menaechmus, a pupil of Eudoxus', discovered conic sections as plane sections of right circular cones. Apollonius of Perga (*c.* 262–*c.* 190 B.C.) also defined conic sections in terms of cutting a right circular cone. (That is, a conic section is simply the curve generated by the intersection of a plane and a right circular cone.) However, Pappus (*fl. c.* 320 A.D.) cited Euclid as having known a directrix property of conic sections and stated the following theorem that established an equivalent definition for conic sections.

> [I]f the distance of a point from a fixed point [the focus] is in a given ratio [the eccentricity] to its distance from a fixed line [the directrix], the locus of the point is a conic section which is an ellipse, a parabola, or a hyperbola according as the given ratio is less than, equal to, or greater than unity.[15]

The meaning of this definition is illustrated in Figure 5.11 (where the vertical y-axis is parallel to the directrix and F is the fixed point (here, the origin) mentioned in the theorem) and the equation corresponding to this is

$$\frac{r}{d + x} = e \tag{5.5}$$

or, equivalently

$$x^2 + y^2 = e^2(x + d)^2 \tag{5.6}$$

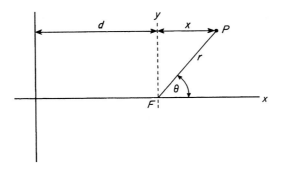

FIGURE 5.11 The focus–directrix definition of a conic section

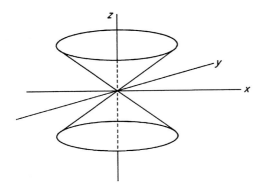

FIGURE 5.12 A right circular cone

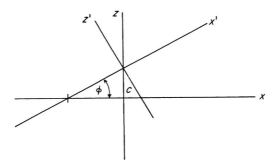

FIGURE 5.13 A plane in three dimensions

Now, let us return to Menaechmus' definition of a conic section. The equation of a right (circular) cone in three dimensions is (see Figure 5.12)

$$x^2 + y^2 = z^2 \tag{5.7}$$

and that of a plane parallel to the y-axis is (see Figure 5.13)

$$z = x \tan\phi + c \tag{5.8}$$

The (projection onto the x–y plane of the) intersection of this plane with the cone (a *conic* section) is simply obtained by eliminating z between Eqs. (5.7) and (5.8). This produces a result of the form of Eq. (5.6). We can also obtain the equation of the intersection of the plane and the cone in the intersecting plane itself ($z' = 0$ in Figure 5.13). The result turns out to have the same form as Eq. (5.6) so that it, too, is a conic section. This means that the intersection of a plane with a cone produces a conic section in the plane of this intersection (Menaechmus' definition). We have, therefore, established the equivalence of these two classic definitions of conic sections (in terms of slicing a cone with a plane and in terms of a directrix and a focus). One can also make explicit the reduction of the general expression (Eq. (5.6)) for a conic section to the three well-known special cases of the ellipse, parabola and hyperbola by rewriting (with a bit of care) Eq. (5.6) for the three cases $0 \leq e < 1$, $e = 1$ and $e > 1$.

Finally, we can easily obtain the polar form of this conic section by recalling Eq. (5.5) above and noting, from the geometry of Figure 5.11, that $x = r\cos\theta$ so that

$$\frac{r}{d + r\cos\theta} = e$$

or

$$r = \frac{de}{1 - e\cos\theta} \tag{5.9}$$

Equation (5.9) will prove useful when we discuss Newton's work on planetary orbits in Chapter 9.

FURTHER READING

Thomas Kuhn's *The Copernican Revolution* and Arthur Koestler's *The Sleepwalkers* remain essential reading for the story of this chapter, while Judith Field's *Kepler's Geometrical Cosmology* discusses Kepler's cosmological theories, and their origins, in detail.

6

Galileo on motion

In this chapter we present a prelude to the foundations of modern *mechanics* that is basically the study of the descriptions and of the causes of the motion of bodies. Aristotle and Galileo are the historical characters on whose work we focus. As we pointed out in Chapter 3, it is important to appreciate that all developments in science take place against the historical, philosophical and social backgrounds in which the scientists find themselves. All scientists, no matter how singularly gifted, build upon the work of their predecessors, even when they overturn old beliefs and theories. Unlike the Athena of Greek mythology who emerged full grown from the head of Zeus, new theories in science do not materialize in final form from one mind working in isolation, but are developed as part of a larger movement.[1] That theme, to which we return often in this book, is well illustrated in the following.

6.1 THE IMPETUS THEORY

Over the centuries there were numerous transcriptions and translations of Aristotle's works, as well as endless commentaries upon them. Not all those who studied the Aristotelian tradition were uncritical of it. This was especially true for Aristotle's position that 'unnatural' motion required the action of an external agent. Hipparchus, perhaps the greatest astronomical observer of antiquity, expressed somewhat vaguely the concept of an impressed force that was transmitted to a moving body. This impressed force was gradually dissipated by the surrounding medium so that the body eventually came to rest.

In 533 A.D. the sixth-century Greek Christian philosopher, John Philoponus, whose writings contributed to both Arabic culture and medieval Western thought, commented on Aristotle's *Physics*:

> Here is something absolutely false, and something we can better test by observed fact than by any demonstration through logic. If you take two masses greatly differing in weight, and release them from the same elevation, you will see that the ratio of times in their movements does not follow the ratio of weights, but the difference in time is extremely small; so that if the weights do not greatly differ, but one, say, is double the other, the difference in times will be either none at all or imperceptible.[2]

Here Philoponus takes exception to the received Aristotelian dogma that a heavier body will fall more rapidly than a lighter one. He also denied that the medium through which an object moved was a causal factor in the way that Aristotle held it was. (Recall Eq. (2.1) and the discussion in Section 2.A on the dual causal role of the medium for Aristotle.) Philoponus saw no difficulty with motion through a void. In modern notation we might attempt to render his idea on the influence of the medium more as $v = (W - R)$, rather than the inverse proportion suggested by Aristotle (as represented in Eq. (2.1)). (Again, we emphasize that this equation is not to be taken as any expression of a quantitative law.) Notice that motion in a void ($R = 0$) would no longer require an infinite speed as it had for Aristotle. Furthermore, Philoponus argued for the existence of an impressed force, similar to that of Hipparchus'. In making a case against Aristotle's explanation of projectile motion, he asks rhetorically and then responds:

> When one projects a stone by force, is it by pushing the air behind the stone that one compels the latter to move in a direction contrary to its natural direction? Or does the thrower impart a motive force to the stone, too?
>
> ...
>
> From these considerations and from many others we may see how impossible it is for forced motion to be caused in the way indicated. *Rather is it necessary to assume that some incorporeal motive force is imparted by the projector to the projectile*, and that the air set in motion contributes either nothing at all or else very little to this motion of the projectile. If, then, forced motion is produced as I have suggested, it is quite evident that if one imparts motion 'contrary to nature' or forced motion to an arrow or a stone the same degree of motion will be produced much more readily in a void than in a plenum. And there will be no need of any agency external to the projector .... [3]

Avicenna (980–1037), an important Muslim philosopher–scientist and the leading Muslim Aristotelian, had a doctrine of impressed force similar to that of Philoponus. Avempace (c 1095–1138/1139), who integrated Islamic and Greek philosophy into his own system, also supported views consistent with Philoponus' impressed force. These ideas became known in the West during the Middle Ages through Latin translations made of the commentaries of the Spanish Arab Averroes (1126–1198) on the works of Aristotle. A significant aspect of these theories of impressed force is that they all saw this impetus as the cause, rather than as an effect, of motion.

In the fourteenth century an important conceptual change took place about this impressed force. The English Franciscan William of Ockham, perhaps the most influential philosopher of his time, held that motion, once it existed, did not require a continuing cause to maintain it. This disagreed with all previous conjectures that required a force for sustained motion. John Buridan, a French Aristotelian philosopher who studied under Ockham at the University of Paris, formulated an *impetus theory*. In it the mover of an object transmits to the object a power proportional to the product of the amount of matter (or mass) in the object times the speed of the object. In modern terminology, this would mean that impetus is identified with mv. This impetus was a permanent impressed force that remained indefinitely in the body, unless diminished by another external agent. It did allow the possibility that a moving body left to itself would continue in motion. We must not read too much into this from our modern vantage point. It is unclear whether the subsequent motion would be rectilinear, circular or some other type. This impetus theory was pursued by one of Buridan's students, Nicholas Oresme, a French Roman Catholic bishop and Aristotelian scholar who studied uniformly accelerated motion. He proved the Merton theorem ($x = \frac{1}{2}v_f t$) relating the distance x traveled in a time t by a body starting from rest and attaining a final velocity v_f. This theorem was first discovered at Merton College, Oxford, in the 1330s. This is essentially the same law that Galileo would establish 300 years later.

We have outlined this development of views of motion contrary to Aristotle's teachings on the subject in order to indicate that there did exist a long and gradual change of concepts on motion during the Middle Ages. Furthermore, the typical representation of science in these 'Dark Ages' as devoid of original thinking and as completely dominated by Aristotle is overly simplistic. This medieval tradition in science was only rediscovered around the turn of the twentieth century by the pioneering work of Pierre Duhem (1861–1916), a French physicist and philosopher. In fact, the impetus theory was known to the next major figure to whom we turn.

6.2 GALILEO'S NATURALLY ACCELERATED MOTION

On February 15, 1564, Galileo Galilei was born in Pisa. His father was a musician. Galileo received his early education at the ancient monastery of Vallombrosa near Florence. In 1581 he entered the University of Pisa to study medicine, but withdrew in 1585 for lack of funds and never took a degree. His scientific reputation was such, however, that in 1589 he was appointed a lecturer in mathematics at the University of Pisa. It was during this period that Galileo is supposed to have conducted his public demonstration at the Leaning Tower to disprove Aristotle's statement about the rate at which bodies fall. Although there seems to be no solid evidence that Galileo actually dropped two bodies from the Leaning Tower of Pisa, he did claim to perform experiments on the motion of falling bodies by using inclined planes, as we discuss later. After resigning from Pisa over conflicts with the Aristotelians, he moved to the University of Padua in 1592 where he remained for eighteen years. There he set up a workshop to make scientific and mathematical instruments to supplement his income that was inadequate for his family needs. Although Galileo never married, he did have a mistress by whom he had two daughters and a son. In 1609 Galileo made marked improvements in the crude telescopes that had recently been invented and he began his astronomical observations that did much to undermine the Ptolemaic, earth-centered model of the universe in favor of the Copernican, sun-centered one. We discussed both of these in Chapters 4 and 5. In 1610 he returned to Florence, having both an appointment to the University of Pisa and a position as Philosopher and Mathematician to the Grand Duke Cosimo II de' Medici (1590–1621). Galileo had his first encounter with the Inquisition in 1616. He wrote his *Dialogue Concerning the Two Chief World Systems* during the years 1625–1630. This defense of the Copernican system against the Ptolemaic one was published in 1632. Later that year, sales and publication were halted by order of the Holy Office and Galileo was summoned to Rome by the Inquisition. He was convicted in 1633 and made to abjure his heretical teachings concerning the Copernican system. From 1633 until his death in 1642, Galileo was confined to his Villa at Arcetri, near Rome. It was during this period that he wrote his great *Dialogues Concerning Two New Sciences*, that was published in 1638 in Leiden. Although we discuss some of Galileo's writings on scientific method in Chapter 12, we confine ourselves now to the *Dialogues*.

The first new science treated in the *Dialogues* is what we, today, would refer to as the strength of materials. Even though we are not concerned with this subject here, it is interesting to note that his discussion of the height to which a column of water rises in a suction pump provided the basis for the development of the barometer by his pupil Evangelista Torricelli (1608–1647). With the proper understanding of the functioning of this instrument came an end to the old Aristotelian belief in *horror vacui* (the claim that nature abhorred a vacuum and would not tolerate one).

On the Third Day of the *Dialogues* Galileo introduces his treatment of the second new science, that of motion.

> My purpose is to set forth a very new science dealing with a very ancient subject. There is, in nature, perhaps nothing older than motion, concerning which the books written by philosophers are neither few nor small; nevertheless I have discovered by experiment some properties of it which are worth knowing and which have not hitherto been either observed or demonstrated. Some superficial observations have been made, as, for instance, that the free motion of a heavy falling body is continuously accelerated; but to just what extent this acceleration occurs has not yet been announced; for so far as I know, no one has yet pointed out that the distance traversed, during equal intervals of time, by a body falling from rest, stand to one another in the same ratio as the odd numbers beginning with unity.
>
> It has been observed that missiles and projectiles describe a curved path of some sort; however, no one has pointed out the fact that this path is a parabola. But this and other facts, not few in number or less worth knowing, I have succeeded in proving; and what I consider more important, there have been opened up to this vast and most excellent science, of which my work is merely the beginning, ways and means by which other minds more acute than mine will explore its remote corners.
>
> This discussion is divided into three parts; the first part deals with motion which is steady or uniform; the second treats of motion as we find it accelerated in nature; the third deals with the so-called violent motions and with projectiles.[4]

Here Galileo is concerned with *kinematics*, that is the quantitative description of the motion of bodies. He steadfastly prescinded from inquiring into the cause of motion (*dynamics*), since he felt that the time was not yet ripe for such an undertaking. (See the third quotation for Part II (*Ancient and modern models of the universe*).) One of his great discoveries was that in natural free-fall a body is accelerated toward the earth at a constant rate. That is, when we release an object near the surface of the earth, it falls with a constant acceleration. Galileo referred to such motion with constant acceleration as naturally accelerated motion. In his *Dialogues* Galileo states that naturally accelerated motion should be defined in a way that not only agrees with the facts but is also such that the velocity increases in as simple a manner as possible. Since he had previously defined uniform motion as that in which equal distances are covered in equal increments of time, he proposed that for the free fall of an object the acceleration is such that equal increments of velocity are gained in equal intervals of time.

> A motion is said to be equally or uniformly accelerated when, starting from rest, its [speed] receives equal increments in equal times.[5]

The change in velocity Δv is proportional to the increment of time elapsed Δt so that $\Delta v \propto \Delta t$, or $\Delta v = $ const Δt. His key insight was to conceive of time t as, what we today would term, the independent variable so that the position, $x = x(t)$, and the velocity, $v = v(t)$, became functions of this variable that simply flows on at its own (uniform) rate. This would also be important for Newton as the 'fluent' variable when he created his 'fluxions' (known to us as the differential calculus).

Galileo used the criterion of simplicity in seeking a correct description of naturally accelerated motion.[6] This and agreement with experiment provided warrant for the truth of his description.[7] For him naturally (or uniformly) accelerated motion included not only the vertical free-fall of a body, but the motion of an object down a smooth incline as well. Galileo argued that, from his definition of uniformly accelerated motion, it followed that an object starting from rest would cover a distance proportional to the square of the elapsed time. As confirmation of this prediction, he cited the results of many experiments he performed with inclined planes. Because Galileo could not measure very small time intervals with appreciable accuracy, he could not study directly the vertical free-fall of a body, but had to 'dilute' the effect of gravity by using an inclined plane of gentle slope to decrease the rate of descent of the ball. He cut a smooth groove in a long board and lined it with parchment.[8] A perfectly round, polished brass ball (starting from rest) was allowed to roll down the incline (as indicated in Figure 6.1). In about a hundred trials for various inclinations and lengths of incline, Galileo found that the distance of descent (along the incline face) always varied closely as the square of the time, although the final velocity of the ball depended only upon the height of the incline, not upon the incline angle. Since accurate clocks did not exist then, time intervals were measured by weighing the amount of water that flowed at a steady rate out of a large vessel with a small hole at its base. Today these classic experiments can be repeated easily using a common stop watch. Much of Galileo's greatness in conceiving and executing these experiments lay in his being able to extract a great deal of useful data by means of such crude instruments. Often, especially in the early history of science, clever people made important discoveries using simple apparatus.

We can summarize the results of such experiments by stating that the distance x traveled is proportional to the square of the time t (that is, $x \propto t^2$). This, of course, assumes that the object starts from rest at the top of the incline. Since algebra had yet to be adapted to the application of these types of problems, Galileo had only geometry and ordinary language as tools in his proofs. We examine one of these in Section 6.A.

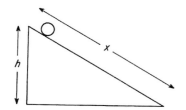

FIGURE 6.1 Galileo's inclined-plane experiment

6.3 PROJECTILE MOTION

Having discussed uniformly accelerated motion, we now consider *projectile motion* in which a body undergoes a constant downward acceleration and maintains a constant horizontal velocity. Galileo discovered that the trajectory or path of the projectile could be found by treating each of these motions (the vertical and the horizontal) independently and then combining them. In other words, he learned that this assumption of independent horizontal and vertical motions led to results that agreed with observation. We denote the horizontal coordinate by x and the vertical one by y. For the present, we discuss only a particular example in which the projectile is launched in a horizontal direction with a velocity v_0. Since there is no acceleration in the horizontal direction, we have for this uniform motion (taking advantage of modern notation).

$$x = v_0 t \tag{6.1}$$

Let us choose as our origin ($x = 0$, $y = 0$) the starting point of the projectile and agree to measure y positive downward. Then we have

$$y = \frac{1}{2}gt^2 \tag{6.2}$$

We use the symbol g to denote the acceleration due to gravity at the surface of the earth. Figure 6.2 represents the trajectory of the projectile. We obtain an explicit expression for the curve $y = f(x)$ by eliminating the time t between Eqs. (6.1) and (6.2) as

$$y = \left(\frac{g}{2v_0^2}\right)x^2 \tag{6.3}$$

This is the equation of a parabola and that confirms Galileo's claim made in an excerpt given in the previous section. Galileo's actual geometrical argument for this result is reproduced in Section 6.A.

6.4 INERTIA

As we stated previously, at no time did Galileo seek the cause of the natural downward acceleration that all bodies experience near the surface of the earth. In

PART II Ancient and modern models of the universe

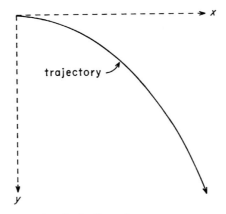

FIGURE 6.2 **Projectile motion**

Chapter 7 we shall see that it was Isaac Newton who stated precisely the quantitative laws of motion, of mechanics and of gravitation. However, Galileo did employ a quantity termed momentum that he defined as the product of the weight of a body times its velocity, and he also formulated a concept of the *inertia* of a body, a property by which a body tended to persist in its state of motion. Galileo's concept of inertia was not quite correct and our modern idea of inertia is due instead to Descartes, Christiaan Huygens (1629–1695) and Newton.

In an exchange between Simplicio and Salviati, two characters in the *Dialogues*, Galileo argues that since a smooth ball projected down an incline would accelerate and continually increase its speed and since one projected up an incline would be decelerated and continually slow down until it stopped, then a ball projected along a horizontal surface would continue to move along this surface with undiminished speed. He used as confirmation of this the fact that a heavy ball dropped from the mast of a moving ship lands on the deck at the base of the mast (not behind the mast), as indicated in Figure 6.3. So, a body left to itself will persist in its state of motion. However, it is not clear whether this will be straight-line motion or motion in a circle parallel to the surface of the earth. In some passages it seems as though for him natural, or inertial, motion was one that neither rose nor fell but that always remained equidistant from the center of the earth.[9]

Descartes did properly state the principle of inertia: that every body tends to continue its motion in a straight line (or to remain at rest), unless it is under some constraint. In his *Principles of Philosophy* (1644) Descartes states this as follows.

> XXXVII. *The first law of nature: that each thing as far as in it lies, continues always in the same state; and that which is once moved always continues to move.*
>
> [E]very thing, in so far as it is simple and undivided, remains always in the same state as far as in it lies and never changes unless caused by an external agent.
>
> . . .
>
> XXXIX. *The second law of nature: that all motion is of itself in a straight line; and thus things which move in a circle always tend to recede from the centre of the circle they describe.*

80

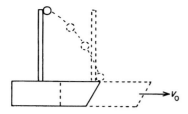

FIGURE 6.3   **An example of inertial motion**

> [E]very part of a body, left to itself, continues to move, never in a curved line, but only along a straight line... and all curvilinear motion is always constrained.[10]

In the process of so defining inertia, Descartes introduced the notion of an infinite universe to which we return in Chapter 11.

In his *The Motion of Colliding Bodies*, Huygens took the following as his first hypothesis.

> Any body already in motion will continue to move perpetually with the same speed and in a straight line unless it is impeded.[11]

Although this work was not published until 1703, eight years after Huygens' death, it was written in the mid 1650s. This concept of inertial motion must have been very much 'in the air' at that time since this axiom is simply stated without any argument or justification on its behalf. This statement of inertia is quite similar to the one subsequently given by Newton as Law I of his *Philosophiae Naturalis Principia Mathematica* (*Mathematical Principles of Natural Philosophy*) – the famous *Principia*. (See Chapter 7.)

This line of thought from Galileo through Newton produced a new definition of natural motion. No longer did each body have its own state of natural motion (such as straight-line down for earth, straight-line up for fire, uniform circular for heavenly bodies), but the universal motion for all free bodies was uniform straight-line motion (or rest). That is the law of inertia. This modern concept of inertia did not begin only with Galileo and occur as a sharp break with the Aristotelian concept of different natural motions for different types of bodies. As we saw in Section 6.1, there was a gradual evolution from Aristotle's belief that a uniform velocity required the continuous action of a uniform force to a concept of an impetus (having some affinity to modern inertia).[12] This does not detract from Galileo's greatness, but simply places him in some historical perspective.

6.5   GALILEO ON ARISTOTLE

We now discuss an attempt by Galileo to refute the Aristotelian dogma that the rate of fall of a body is proportional to its weight. (Notice that in the following Galileo uses weight in the sense of heaviness, much as we do today. See Section 2.4 on this.) Although Aristotle is often criticized for his a priori form of reasoning, Galileo himself offered the following argument as a proof that Aristotle's doctrine on falling

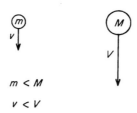

FIGURE 6.4 Galileo's representation of Aristotle on free fall

$v < v' < V$

FIGURE 6.5 Two connected weights

bodies is self-contradictory. If m and M represent, respectively, light and heavy stones, then, according to Aristotle, M should fall more rapidly than m ($v < V$, as indicated in Figure 6.4). Now tie M and m together to form one body (as shown in Figure 6.5) that falls at the rate v'. Then m should retard M while M should make m fall more rapidly ($v < v' < V$). However, since ($m + M$) is heavier than M, it should fall more rapidly than M alone ($v' > V$). Galileo claims that this results in an inconsistency ($v' < V$ and $v' > V$ simultaneously) in Aristotle's hypothesis. This can be avoided if $v = v' = V$: the statement that all bodies fall at the same rate. If Galileo is correct and we can conclude that m and M must fall at the same rate, why need we do any experiments to observe that this is the case? How could it happen that a penny and a feather dropped in air do not fall at the same rate?

We must analyze some of Galileo's unstated, but implicit, assumptions here, such as that the rate of fall of a body is independent of its shape (not strictly true) and of its speed and depends only upon its weight. Even if we were to accept Galileo's argument as correct, all we could conclude is that the assumption that different bodies fall at different (constant) speeds is incorrect. This is not the same as saying that all bodies fall at the same (constant) speed. They do not, and Galileo knew it. This argument was largely a polemical device to undercut the Aristotelian position and, by inference, to support Galileo's. Furthermore, an Aristotelian might have countered Galileo by contending that the nature of body ($m + M$) is different from those of m and M combined or that the weight (as a measure of speed in Aristotle's sense) of ($M + m$) was the same as that of M or of m alone so that all would fall with a common speed. (In fact, Galileo goes on in this argument to make a distinction of just that sort for his own purposes.)

However, Galileo does not eschew appeal to experiment but later in the same passage has his spokesman Salviati relate that:

> Aristotle says that 'an iron ball of one hundred pounds falling from a height of one hundred cubits reaches the ground before a one pound ball has fallen a single cubit.' I say that they arrive at the

> same time. You find, on making the experiment, that the larger outstrips the smaller by two finger-breadths, that is, when the larger has reached the ground, the other is short of it by two finger-breadths;....[13]

Incidentally, Galileo was really being unjust to Aristotle since the first sentence of this quotation, that was itself claimed to be taken directly from Aristotle, does not seem to appear in the extant writings of Aristotle.[14]

The famous Leaning Tower of Pisa 'experiment' is often cited as proof that all bodies fall at the same rate. Let us attempt to judge whether or not Galileo could possibly have gotten the results commonly attributed to him for the legendary leaning tower demonstration. Of course, we are here using the 'hindsight' of modern theoretical mechanics.[15] When proper account is taken of air resistance, one finds the following for the case of two spheres made of the same material (say steel) and released simultaneously. The larger one will have fallen a greater distance than (and hence be ahead of) the smaller one at any time t (assuming, of course, that both begin at rest at the same elevation). For balls of roughly 100 lb [equivalent to a mass of 45.4 kg] and 1 lb [0.454 kg], that are released 200 ft [61.0 m] above the ground, we can calculate directly that the larger ball would hit the ground 3.55 s after they had been released simultaneously and that the smaller ball would then be about three feet [0.91 m] above the ground. So, it appears that Galileo could not possibly have obtained the results he claimed: namely, that the two balls were separated vertically by no more than 'two finger-breadths' upon impact. One can also show by direct calculation that the time for the smaller ball to fall 200 ft [61.0 m] is 3.58 s. Hence, an error in the release time of only 0.03 s could mask this effect of a three-foot [0.91 m] separation upon impact. This means that great care would have to be taken to insure that both balls are released simultaneously. (Realize that a typical human reaction time is about 0.10 s.)

So far, we have considered the case of two objects of the same material (or density), but of different radii. Suppose, however, we have two bodies of the same size, as there is some indication Galileo would have done.[16] In particular, let us take a 16-lb [7.3 kg] steel shot and a softball (that is about the same size as the shot). Each turns out to have a radius of around 6.0×10^{-2} m, but their respective densities are in the ratio of about twenty. Now, when these two bodies of greatly different masses and approximately the same size are dropped simultaneously from atop a high tower (again, 200 ft [61.0 m]), they do not in fact hit the ground nearly together, but are separated vertically by over twenty feet [6.1 m]. Not only is this supported by calculations based on modern mechanics for the fall of a body through a medium (such as air) that offers resistance to motion through it,[17] but this has been found in direct observation.[18] Also, the time it would take the softball to fall the 200 ft [61.0 m] would be 3.81 s (whereas the steel ball takes only 3.55 s). In this case, then, an error in release time of 0.26 s would be required to mask the effect. This is far less likely than in the former case of two objects of the same material but different sizes.

It is these facts, among others, that indicate Galileo could never actually have performed his famous Leaning Tower of Pisa experiment and gotten the alleged result that a heavy body and a light one had the same time of fall from the top of the tower.[19]

6.A GALILEO'S *DIALOGUES CONCERNING TWO NEW SCIENCES*

We begin with the passage in which Galileo attempted to refute Aristotle's views on motion. Here, as elsewhere in the *Dialogues*, Galileo used the dialogue format. The three characters are Salviati, Galileo's spokesman modeled after his friend Filippo Salviati (1582–1614); Sagredo, the epitome of the leisured Italian intellectual, a practical and broad-minded man, willing to judge a case on its merits, and modeled after Galileo's deceased friend, the Venetian nobleman, Giovanfrancesco Sagredo (1571–1620); and Simplicio, the caricature of a pedantic Aristotelian, a slave to dogma who can no longer think for himself.

> *Salviati.* But, even without further experiment, it is possible to prove clearly, by means of a short and conclusive argument, that a heavier body does not move more rapidly than a lighter one provided both bodies are of the same material and in short such as those mentioned by Aristotle. But tell me, Simplicio, whether you admit that each falling body acquires a definite speed fixed by nature, a velocity which cannot be increased or diminished except by the use of force or resistance.
>
> *Simplicio.* There can be no doubt but that one and the same body moving in a single medium has a fixed velocity which is determined by nature and which cannot be increased except by addition of momentum or diminished except by some resistance which retards it.
>
> *Salviati.* If then we take two bodies whose natural speeds are different, it is clear that on uniting the two, the more rapid one will be partly retarded by the slower, and the slower will be somewhat hastened by the swifter. Do you not agree with me in this opinion?
>
> *Simplicio.* You are unquestionably right.
>
> *Salviati.* But if this is true, and if a large stone moves with a speed of, say, eight while a smaller moves with speed of four, then when they are united, the system will move with a speed less than eight; but the two stones when tied together make a stone larger than that which before moved with a speed of eight. Hence the heavier body moves with less speed than the lighter; an effect which is contrary to your supposition. Thus you see how, from your assumption that the heavier body moves more rapidly than the lighter one, I infer that the heavier body moves more slowly.
>
> *Simplicio.* I am all at sea because it appears to me that the smaller stone when added to the larger increases its weight and by adding weight I do not see how it can fail to increase its speed or, at least, not to diminish it.
>
> *Salviati.* Here again you are in error, Simplicio, because it is not true that the smaller stone adds weight to the larger.
>
> *Simplicio.* This is, indeed, quite beyond my comprehension.[20]

On the Fourth Day of the *Dialogues Concerning Two New Sciences*, Galileo shows that projectile motion is parabolic motion.

> I now propose to set forth those properties which belong to a body whose motion is compounded of two other motions, namely, one uniform and one naturally accelerated;....
>
> . . .

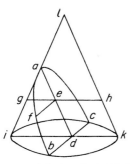

FIGURE 6.6 Appolonius' definition of a parabola

A projectile which is carried by a uniform horizontal motion compounded with a naturally accelerated vertical motion describes a path which is a semi-parabola.[21]

Galileo must first recall a basic property of a parabola that he takes as characterized by Appolonius' definition in terms of a plane cutting a cone (see Section 5.A).

> [I]magine a right cone, erected upon the circular base *ibkc* [Figure 6.6] with apex at *l*. The section of this cone made by a plane drawn parallel to the side *lk* is the curve which is called a *parabola*. The base of this parabola *bc* cuts at right angles the diameter *ik* of the circle *ibkc*, and the axis *ad* is parallel to the side *lk*; now having taken any point *f* in the curve *bfa* draw the straight line *fe* parallel to *bd*; then, I say, the square of *bd* is to the square of *fe* in the same ratio as the axis *ad* is to the portion *ae*. Through the point *e* pass a plane parallel to the circle *ibkc*, producing in the cone a circular section whose diameter is the line *geh*. Since *bd* is at right angles to *ik* in the circle *ibk*, the square of *bd* is equal to the rectangle formed by *id* and *dk*; so also in the upper circle which passes through the points *gfh* the square of *fe* is equal to the rectangle formed by *ge* and *eh*; hence the square of *bd* is to the square of *fe* as the rectangle *id·dk* is to the rectangle *ge·eh*. And since the line *ed* is parallel to *hk*, the line *eh*, being parallel to *dk*, is equal to it; therefore the rectangle *id·dk* is to the rectangle *ge·eh* as *id* is to *ge*, that is as *da* is to *ae*; whence also the rectangle *id·dk* is to the rectangle *ge·eh*, that is, the square of *bd* is to the square of *fe*, as the axis *da* is to the portion *ae*.[22]

We can rather straightforwardly transcribe this geometrical argument into an elementary algebraic one as follows. Galileo set out to establish that

$$\frac{(bd)^2}{(fe)^2} = \frac{ad}{ae}$$

He begins by using a property of a circle (essentially that $y^2 = a^2 - x^2 = (a + x)(a - x)$, where a is the radius of the circle) to write

$$(bd)^2 = (id)(dk), \quad (fe)^2 = (ge)(eh)$$

Since equal segments are produced when parallel lines are cut by parallel lines, we have $eh = dk$ so that

$$\frac{(id)(dk)}{(ge)(eh)} = \frac{(id)(dk)}{(ge)(dk)} = \frac{(id)}{(ge)} = \frac{(ad)}{(ae)}$$

where the last equality follows from the similar triangles *ida* and *gea*. This establishes the desired result.

The proof that a ball projected horizontally (the case that Galileo actually considers) follows a parabolic path is now direct since

$$x = fe \propto t, \quad y = ae \propto t^2$$

so that y is proportional to x^2. This defines a parabola.

FURTHER READING

Allan Franklin's *The Principle of Inertia in the Middle Ages* and Marshall Clagett's *The Science of Mechanics in the Middle Ages* provide readable accounts of the emergence of mechanics as a science. Stillman Drake's 'Galileo: A Biographical Sketch' is a useful reference for the major events in Galileo's life. Colin Ronan's *Galileo* is a general biography of this central figure of modern science, while Lane Cooper's *Aristotle, Galileo, and the Tower of Pisa* and Michael Segre's 'Galileo, Viviani and the Tower of Pisa' investigate the actual historical record surrounding this famous 'experiment'. Ernan McMullin's essay, 'Introduction: Galileo, Man of Science,' gives a thoughtful summary of Galileo's scientific work.

PART III

The Newtonian universe

[T]hat which makes the earth move [in its orbit about the sun] is a thing similar to whatever moves Mars and Jupiter... . If [one] will advise me as to the motive power of one of these movable bodies, I promise I shall be able to tell him what makes the earth move. Moreover, I shall do the same if he can teach me what it is that moves earthly things downward.

Galileo Galilei, *Dialogue Concerning the Two Chief World Systems*

Every body continues in its state of rest, or of uniform motion in a right line, unless it is compelled to change that state by forces impressed upon it.

...

The change of motion is proportional to the motive force impressed; and is made in the direction of the right line in which that force is impressed.

...

Hitherto we have explained the phenomena of the heavens and of our sea by the power of gravity, but have not yet assigned the cause of this power. This is certain, that it must proceed from a cause that penetrates to the very centres of the sun and planets, without suffering the least diminution of its force; that operates not according to the quantity of the surfaces of the particles upon which it acts (as mechanical causes used to do), but according to the quantity of the solid matter which they contain, and propagates its virtue on all sides to immense distances, decreasing always as the inverse square of the distances. Gravitation towards the sun is made up out of the gravitations towards the several particles of which the body of the sun is composed; and in receding from the sun decreases accurately as the inverse square of the distances as far as the orbit of Saturn, as evidently appears from the quiescence of the aphelion of the planets; nay, and even to the remotest aphelion of the comets, if those aphelions are also quiescent. But hitherto I have not been able to discover the cause of those properties from phenomena, and I frame no hypotheses; for whatever is not deduced from the phenomena is to be called an hypothesis; and hypotheses, whether metaphysical or physical, whether of occult qualities or mechanical, have no place in experimental philosophy.

Isaac Newton, *Mathematical Principles of Natural Philosophy*

7

Newton's *Principia*

In this and the next two chapters we discuss the conceptual cornerstone of classical physics, namely Newtonian mechanics. We shall see how Newton, employing certain general principles of reasoning as guides, used the regularities of the solar system, in the form of Kepler's laws, to arrive at his own law of gravitation. He then reversed this reasoning process to derive not only Kepler's laws but also to account for the phenomena of the ocean tides. The stereotypical image of Newton represents him as the embodiment and culmination of the Age of Reason, as in the 1689 Godfrey Kneller portrait of him in the robes of a Cambridge don.[1] It may come as a surprise to learn that Newton had serious and long-standing interests in certain unorthodox theological subjects and in alchemy.

7.1 ISAAC NEWTON

Isaac Newton was born in the small English village of Woolsthorpe, in Lincolnshire, on Christmas day, 1642, eleven months after Galileo's death. (On the Gregorian calendar that was not adopted in England until 1752, Newton's birth date is January 4, 1643.) His father (1606–1642), also named Isaac Newton, died three months before his son's birth. Newton was a sickly baby and was not expected to live long. In 1645 his mother, Hannah Ayscough (*d.* 1679), married a wealthy minister, Barnabas Smith (1582–1653), and moved to a neighboring village, leaving young Isaac to be raised by his maternal grandmother. Even years later Newton harbored feelings of intense hatred toward his mother and stepfather for this desertion. It was also very likely a contributing factor to his psychological instabilities and to his sense of insecurity in later life. He received his education at local schools until at age twelve he was sent to the King's School in Grantham, where to this day initials allegedly his can be seen carved in the window ledge of the school house. Newton's mother was widowed again in 1653 and she decided that Isaac should return to their farm and manage it. Although there is some uncertainty about the exact date of Newton's return, by 1659 he was back at Woolsthorpe. It soon became evident that he had no talent for farming, so that in 1661 Newton matriculated at Trinity College, Cambridge. There he earned his keep by performing menial tasks for the fellows and wealthier students, much in the fashion of some college students today who wait tables in cafeterias to help finance their education.

Although this is the period usually thought of as the early part of the scientific revolution, since Copernicus, Kepler, Galileo and Descartes had already published

PART III   The Newtonian universe

their works, most universities in Europe and in England remained bastions of outmoded Aristotelian doctrine, and Cambridge was no exception. Moreover, Cambridge had endured hard times and had, with a few notable exceptions, a rather undistinguished faculty as a result of political intrigues and a recent civil war in England. Although Newton began with the usual study of the classical philosophers, he soon discovered the revolutionary works of more modern thinkers and wrote in his notebook '*Amicus Plato amicus Aristoteles magis amica veritas*' ('Plato is my friend, Aristotle is my friend, but my best friend is truth').[2] He attended lectures by the mathematician Isaac Barrow (1630–1677) and also became familiar with the *Optics* of Kepler and the *Geometry* of Descartes. The chemical writings of Robert Boyle (1627–1691) and the philosophical treatise of the Neoplatonist Henry More (1614–1687) greatly influenced his later thinking. ('Neoplatonism' is the term often used to refer to Platonic doctrine (especially the aspect of essential forms) fused with some type of mysticism or religious doctrine, such as Christianity in the teachings of St. Augustine (354–430).) Through More, Newton became aware of the Hermetic tradition that emphasized alchemy and magic in its explanation of the world. Newton combined both the mechanistic and the Hermetic philosophies into his own world view that, still in a Neoplatonic tradition, viewed matter as passive and sought a principle of activity outside of matter in space and in an aether, as we shall see in Chapter 11.

By the time he received his Bachelor's Degree in 1665 from Trinity College, Newton had already broken new mathematical ground by discovering the binomial theorem. He first began to develop his method of fluxions, or calculus, during the winter of 1664 and the spring of 1665. In his method of analysis, Newton conceived of lines as generated, not by the addition of ever smaller straight-line segments, but rather by continuous motion. The fluent, or independent, variable of this motion was the time. In 1665, the Great Plague spread from London to Cambridge and the University was closed. After receiving his degree, Newton returned to his home in Lincolnshire. It was in the period 1665–1667 – often characterized as the *annus mirabilis* of 1666 – that he conducted experiments in chemistry, made his early observations on light by using a prism to resolve the spectrum of sunlight, pursued his mathematical researches by inventing the calculus and laid the foundations of his work in mechanics and in gravitation by considering the forces necessary to keep the moon in its orbit (see Chapter 8). An October 1666 *Tract on Fluxions* summarizes his mathematical researches of the previous two years. These several labors would not be completed and made public for nearly a quarter of a century. Many years later Newton recalled this period in the following terms.

> In the beginning of the year 1665 I found the Method of approximating series & the Rule for reducing any dignity of any Binomial into such a series. The same year in May I found the method of Tangents..., & in November had the direct method of fluxions & the next year in January had the Theory of Colours & in May following I had entrance into the inverse method of fluxions. And the same year I began to think of gravity extending to the orb of the Moon & (having found out how to estimate the force with which [a] globe revolving within a sphere presses the surface of the sphere) from Kepler's rule... I deducted that the forces which keep the Planets in their Orbs

must [be] reciprocally as the squares of their distances from the centers about which they revolve: & thereby compared the force requisite to keep the Moon in her Orb with the force of gravity at the surface of the earth, & found them answer pretty nearly. All this was in the two plague years of 1665 & 1666. For in those days I was in the prime of my age for invention & minded Mathematicks & Philosophy more than at any time since.[3]

Few Newton scholars today take this statement at face value. It appears that, in these and other statements written after he had become famous, Newton was inclined to shade his recollections to strengthen his priority claims of discovery, especially as regards the calculus.

When Cambridge reopened in 1667, Newton was elected a Fellow of Trinity College. In June 1669, possibly in order to establish a claim of priority for his work on infinite series, he hurriedly assembled *De Analysi per Aequationes Numero Terminorum Infinitas* (*On Analysis by Equations Unlimited in the Number of their Terms*). The next month he communicated this to Barrow, still the Lucasian Professor at Trinity. In October of that year, just before his twenty-seventh birthday, Newton himself succeeded Barrow and became the second Lucasian Professor. He received his professorship upon the recommendation of his former teacher Barrow, who vacated the chair in order to pursue theological studies. This appointment freed Newton from all teaching responsibilities, except for an annual series of lectures, and allowed him to devote all of his efforts to research.

During 1670–1671 he compiled his *De Methodis Serierum et Fluxionum* (*On the Method of Series and Fluxions*) that combined his 1666 fluxion tract and the *De Analysi*. Newton toyed with the idea of having this *Method of Fluxions and Infinite Series* printed in the 1670s. In the severe depression that followed the 1666 Great Fire of London, there was no market for such mathematics books. However, more likely central to this decision not to publish was Newton's anxiety about possible criticism. This work, like the previous mathematics monographs mentioned here, went unpublished and was circulated only among a relatively select circle of mathematicians. It was not until 1736 that *Method* was finally published.[4] It is there that he gives the rules for what we would today term the differentiation of an algebraic function.

For the subject of his first annual professorial lectures in 1670, Newton chose his investigations in optics and in 1672 communicated his results to the Royal Society of London. This work was generally well received there, except most notably by Robert Hooke (1635–1703), a senior member of the Society, who questioned some of Newton's interpretations. Newton subsequently became involved in a series of protracted exchanges with Hooke and others. Then, as at later times in his life, Newton became enraged at any public criticism. As a result he withdrew into the isolation of his studies from about 1678 to 1684 and became an unremitting enemy of Hooke's. He refused to publish anything further. It appears as though Newton suffered his first nervous breakdown then and during this period of withdrawal became intensely interested in the Hermetic tradition and alchemy.

The modern reader may be shocked to learn that Newton wrote perhaps as many, or even more, words on alchemy and on his studies of ancient biblical texts as he did on the *Principia* and on similar 'respectable' topics. Historical evidence suggests that his ultimate goal in all of these diverse studies taken together was to reinstate a true religion as a defense against atheism that he saw as the root cause of the turmoil of his times. Newton was once offered the Mastership of Trinity College and, later, wanted to be Provost of King's College. He was prevented from doing either because he had not taken Holy Orders, as was usually required even to be a Fellow at Cambridge. He had received a special dispensation to become Lucasian Professor without Orders. His reason for refusing to take Holy Orders was that he was a Unitarian and could not subscribe to the doctrine of the Trinity, although this secret was closely guarded throughout his lifetime.

In 1679 Hooke posed the problem of determining the orbit of a planet given that the force acting on it varied inversely as the square of the distance of the planet from the sun. Hooke, Edmund Halley (1656–1742), the astronomer for whom the comet is named, and Sir Christopher Wren (1632–1723), the famous architect who supervised much of the rebuilding in London after the Great Fire of 1666, worked without success on this problem. In 1684 Halley happened to be visiting Newton at Cambridge and asked him if he knew the solution. Newton replied offhandedly that it was an ellipse. When pressed by Halley how he knew that, Newton answered that he had proved it sometime previously but had mislaid the proof. It was as a result of this encounter that Halley encouraged Newton to publish his results. During 1685 and 1686 Newton expanded this into his *Principia* that was published in 1687. There followed an unpleasant feud with Hooke over charges of plagiarism, that few could take seriously since the *Principia* was immediately recognized as a work of unrivaled scientific genius. Newton remained unrelenting toward Hooke, who was already in decline and ill health, until Hooke died in 1703. It was only then, with Hooke no longer a member of the Royal Society, that he would assume the presidency of that body.

As a result of the pressure of writing the *Principia* and of the controversy that followed, Newton sustained another nervous breakdown in 1692–1693. His mental state was made more precarious by the termination of his friendship with Nicholas Fatio de Duillier (1664–1753), a Swiss-born mathematician who lived in London, had many of the same interests as Newton and was the person whom Newton had been closest to in his adult life. The creative phase of his scientific career had ended by 1693 and his friends, feeling that a change of environment was essential for his mental well being, induced Newton to accept the post of Warden of the Mint in 1696. Within a few years, he was promoted to Master of the Mint. Newton resided henceforth in London and became a major social figure. He pursued his duties at the Mint with great enthusiasm, even taking an interest in apprehending counterfeiters, many of whom he sent to the gallows.

It was during this period that he became embroiled in a major controversy with Gottfried Leibniz (1646–1716) over the question of priority in inventing the calculus. De Duillier played a major role in bringing the matter to a boil and Newton occupied himself with this for the rest of his life, even after Leibniz died. It now seems clear that Newton formulated his version of the calculus first and did not publish it at the time, but waited until the 1704 edition of his *Optics*, while Leibniz independently discovered and published his in 1684. As part of this war of priority that pitted England against the Continent, Johann Bernoulli (1667–1748), a Swiss mathematician of great power and a co-worker with Leibniz on the calculus, formulated a problem[5] and gave the mathematicians of the world six months in which to solve it. In a single evening in 1697 Newton solved the problem and published the solution anonymously in the *Transactions of the Royal Society*. When Bernoulli saw the solution, he recognized Newton's hand at work and is claimed to have said '*Tanquam ex ungue leonem*' (loosely, 'The lion is known by his claw').[6]

In 1703, as we noted previously, Newton became President of the Royal Society and in 1705 he was knighted by Queen Anne to become the first scientist in England so honored. He mellowed considerably with age. Just two years before his death he remarked to his nephew:

> I do not know what I may appear to the world; but to myself I seem to have been only like a boy, playing on the sea shore, and diverting myself, in now and then finding a smooth pebble or a prettier shell than ordinary, whilst the great ocean of truth lay all undiscovered before me.[7]

Newton died on March 20, 1727, from gallstones. Shortly thereafter his body lay in state and was then buried in Westminster Abbey. Today a visitor to the Abbey can see the memorial statue to Newton, at the foot of which are buried not only Newton, but also Michael Faraday (1791–1867) and James Maxwell (1831–1879), whose researches laid the foundations for modern electromagnetic theory.

7.2 NEWTON'S PHILOSOPHY OF SCIENCE

As we saw in Chapter 3, a popular view of science represents the scientific method as consisting of successive steps – observation, hypothesis, prediction, confirmation. By this process science is seen to advance toward ever more general laws. This model of science is often traced back to Bacon in his *The New Organon*. There he criticizes the exclusive use of syllogistic or deductive reasoning and stresses the importance of induction in natural philosophy.

> But we can then only augur well for the sciences, when the ascent shall proceed by a true scale and successive steps, without interruption or breach, from particulars to the lesser axioms, thence to the intermediate (rising one above the other), and lastly, to the most general. For the lowest axioms differ but little from bare experiments; ....
>
> . . .
>
> In forming our axioms from induction, we must examine and try whether the axiom we derive be only fitted and calculated for the particular instances from which it is deduced, or whether it be more extensive and general.[8]

Newton, in the preface to the first edition of the *Principia*, reflects a philosophical outlook broadly consistent with this:

> [I] offer this work as the mathematical principles of philosophy [that is, exact mathematical science], for the whole burden of philosophy seems to consist in this – from the phenomena of

> motions to investigate the forces of nature, and then from these forces to demonstrate the other phenomena; and to this end the general propositions in the first and second Books are directed. In the third Book I give an example of this in the explication of the System of the World; for by the propositions mathematically demonstrated in the former Books, in the third I derive from the celestial phenomena the forces of gravity with which bodies tend to the sun and the several planets. Then from these forces, by other propositions which are also mathematical, I deduce the motions of the planets, the comets, the moon, and the sea.[9]

This approach is especially clear in Book III of the *Principia* where Newton uses the empirical fact that the orbits of the planets about the sun are ellipses and the fact that equal areas are swept out in equal times to deduce that the gravitational force is an attractive, inverse-square one. Using this generalization from celestial phenomena, he is then able to deduce (by reference to the relevant propositions of Book I) Kepler's three laws of planetary motion and to derive the tidal effects produced by the moon (and by the sun) on the seas of the earth (as we show in Chapter 9).

At the beginning of Book III of the *Principia* he states his Rules of Reasoning in Philosophy.

> RULE I
> *We are to admit no more causes of natural things than such as are both true and sufficient to explain their appearances.*
> RULE II
> *Therefore to the same natural effects we must, as far as possible, assign the same causes.*
> RULE III
> *The qualities of bodies, which admit neither intensification nor remission of degrees, and which are found to belong to all bodies within the reach of our experiments, are to be esteemed the universal qualities of all bodies whatsoever.*
> RULE IV
> *In experimental philosophy we are to look upon propositions inferred by general induction from phenomena as accurately or very nearly true, notwithstanding any contrary hypotheses that may be imagined, till such time as other phenomena occur, by which they may either be made more accurate, or liable to exceptions.*[10]

Basically, these urge us to seek explanations in terms of the smallest possible number of causes, to assume a common cause for similar effects, to take qualities common to all bodies we examine to be properties of all bodies in the universe and to treat careful inductions from empirical data as generally true until refuted by other phenomena. This is the essence of retroductive reasoning in which one uses the correctness of the consequences of an hypothesis as a warrant for the truth of the hypothesis itself.

Newton's approach to natural philosophy is often characterized by his *hypotheses non fingo* ('I do not frame (or feign) hypotheses') that appears in the General *Scholium* (or commentary) at the end of the *Principia*.

> [I] frame no hypotheses; for whatever is not deduced from the phenomena is to be called an hypothesis; and hypotheses, whether metaphysical or physical, whether of occult qualities or of mechanical, have no place in experimental philosophy. In this philosophy particular propositions are inferred from the phenomena, and afterwards rendered general by induction. Thus it was that the impenetrability, the mobility, and the impulsive force of bodies, and the laws of motion and of gravitation, were discovered.[11]

He returns to the same theme in one of the Queries to his *Optics*.

> [T]he main business of natural philosophy is to argue from phenomena without feigning hypotheses, and to deduce causes from effects....[12]

This is usually interpreted as meaning that Newton held the only legitimate activity of science to be the induction of laws from an examination of the phenomena of nature, without seeking more remote causes or explanations that are not directly observable by the senses. This is not so. To begin with, this quotation from the *Optics* appears in a passage where Newton is being highly speculative about the existence of an aether that can account for the propagation of light (and that we discuss in Chapter 13). Even in his *Principia*, that he was careful to keep at the mathematical and inductive level, he states:

> In mathematics we are to investigate the quantities of forces with their proportions consequent upon any conditions supposed; then, when we enter upon physics, we compare those proportions with the phenomena of Nature, that we may know what conditions of those forces answer to the several kinds of attractive bodies. And this preparation being made, we argue more safely concerning the physical species, causes, and proportions of the forces.[13]

That is, there are three different levels at which we must work: the mathematical (or deductive), where we analyze the implications of certain assumptions or axioms; the physical, where we use comparison with data to decide which of the many possible axioms or laws actually do correspond to nature; and, finally, the philosophical, where we seek the causes of these laws. In the *Principia* he attempted to do the first two as a preparation for the third that he also felt to be important. Newton sought the ultimate explanation for the physical phenomena we observe in the macrocosm (that is, the world we perceive with our senses) in terms of a microcosm that is not directly observable by the senses. He used an argument by analogy in which he assumed that the laws discovered in the macrocosm are also operative at the level of the microcosm. In his life Newton never did succeed in constructing an explanation of the causes behind his laws of mechanics and of gravitation.

With his *Principia*, Newton completed a revolution that had begun with Galileo. These creative scientists gave us a new way of looking at, or comprehending, our world. It is essential to appreciate that this was a truly creative (but certainly not an arbitrary) process, since their laws cannot be inductively reached simply by examining the data or phenomena of experience. Such constructions or conventions, while neither completely arbitrary nor uniquely dictated by nature, must be consonant with the world in order to be useful.

7.3 OUTLINE OF NEWTON'S ARGUMENT IN THE *PRINCIPIA*

We begin with an overview of the general structure of the *Principia* as it relates to mechanics and planetary motion. (In Chapters 8 and 9 we return to detailed discussions of several of the following topics.) It opens with a collection of definitions of such basic terms as 'mass', 'momentum' and 'force' (Section 7.5 below). The famous *scholium* on the nature of absolute space and time is also found here. There follow the three laws of motion and a series of corollaries to these, including what we would today term the vector addition law for forces (Section 7.4). Here, as elsewhere in the *Principia*, the second law of motion is, in modern notation, used in the form $F\Delta t = \Delta p$. Arguments and relations of this type in terms of increments are typical of Newton's presentation in this work. He often treated a continuously varying force acting on a body as the limit of a sequence either of successive impulses or of constant (finite) forces each acting for a small time interval.

Book I, *The Motion of Bodies*, begins with Section I containing mathematical lemmas on limits of areas, lines and arc lengths. All of the proofs are done under the guise of geometry, although with hindsight we can see that Newton introduced concepts of the calculus. Concerning the difference between the method (fluxions, or incremental and limit arguments – a 'new' analysis) by which he claimed to have discovered certain results and the style in which he presented them, Newton wrote the following around 1715.[14]

> By the help of the new *Analysis* [I] found out most of the Propositions in [my] *Principia Philosophiae*: but because the Ancients for making things certain admitted nothing into Geometry before it was demonstrated synthetically, [I] demonstrated the Propositions synthetically, that the Systeme of the Heavens might be founded upon good Geometry. And this makes it now difficult for unskillful Men to see the Analysis by which those Propositions were found out.[15]

More specifically, he also commented on his original proofs of Kepler's first and second laws.

> By the inverse Method of fluxions [calculus] I found in the year 1677 the demonstration of Kepler's Astronomical Proposition, viz. that the Planets move in Ellipses, which is the eleventh Proposition of the first book of Principles [*Principia*].[16]
>
> ...
>
> About... [1679] by the help of this method of Quadratures I found the Demonstration of Kepler's Propositions that the Planets revolve in Ellipses describing with a Radius drawn to the sun in the lower focus of the Ellipsis, areas proportional to the times.[17]

Because Newton did not publish his researches on the calculus until *Tractatus de Quatratura Curvarum* (*A Treatise on the Quadrature of Curves*)[18] appeared as a mathematical appendix to the 1704 edition of his *Optics*, some of his results on limits are included in Section I of Book I of the *Principia*.

Section II begins with two propositions establishing that, for the motion of a body about a fixed force center, the central character of that attractive force (toward the force center: a *central force*) is the necessary and sufficient condition for the orbit to lie in a plane and for equal areas to be swept out in equal times (Kepler's second law). (See Section 9.1 for the details of Newton's proof.) In other words, in modern terminology, the angular momentum (as a vector) must be constant in time. Section III discusses the motion of bodies in conic sections. Here Newton argues that motion in a plane conic section about a focus as center is the necessary and sufficient condition for an inverse-square central force law and that suitable initial data (the position and velocity) determine this conic section. He also proves

Kepler's third law. (We return in Section 9.2 to Newton's demonstrations for these claims.) Section VIII contains a proposition that is essentially an existence and uniqueness proof showing that, for any specified central force and any initial position r_0 and initial velocity v_0, the equation of motion ($F = ma$) determines the orbit $r = r(t)$. Finally, a series of propositions in Section XII treats, by means of a clever geometrical proof, the gravitational forces produced by a spherical shell and between two spherical shells.

In a sense, Newton's discussion of the central force problem in Book I of the *Principia* is purely hypothetical. That is, he presents a sequence of theorems and proofs, but neither the conditions under which the theorems are valid nor their results are related to the actual physical world. However, in Book III, *The System of the World*, Newton is concerned with applying his theorems to natural phenomena. It is here that he first states his Rules of Reasoning in Philosophy. Astronomical data are then listed that support Kepler's three laws of planetary motion for the moons of Jupiter and of Saturn, for the planets about the sun, and for the moon about the earth. The opening propositions use these data, via the propositions of Book I, to deduce that the force of gravity is an inverse-square one. Next, he shows that the inverse-square force that keeps the moon in its orbit is the same force that causes a body to fall at the surface of the earth with an acceleration g (32 ft/s^2 [9.80 m/s^2]). (See Sections 8.1 to 8.3.) His conclusion, in accord with his 'fewest causes' criterion of his Rules of Reasoning, is that it is the earth's gravity that keeps the moon in its orbit. He then argues, again from his astronomical data, that bodies gravitate toward each planet.

Finally, once more using his inductive rule of inference, Newton generalizes to a verbal statement of the law of universal gravitation between every pair of particles. As we discuss in Chapter 8, Newton's formulation of this law of gravitation was the culmination of a line of research that extended over a period of twenty years. From this law of gravitation, he arrives at Kepler's three laws of planetary motion. Newton also discusses the slight departures from these laws caused by the perturbation of the gravitational interactions of the planets among themselves.

7.4 NEWTON'S THREE LAWS OF MOTION

Let us now turn to the beginning of the *Principia* itself and discuss the laws of motion it contains. Even though Newton did not develop his laws of dynamics and of gravitation independently of each other, we present them separately for pedagogical reasons. We have come to the study of dynamics and must consider the causes of various states of motion. It seems fairly evident to us that forces cause the state of motion of a body to change, but precisely how to express this quantitatively is not a simple matter. Newton first had to define the quantity of matter (or mass) of a body. His definition in the *Principia* is not particularly satisfying.

> The quantity of matter is the measure of the same, arising from its density and bulk conjointly.[19]

Today we would express this as mass being the product of density times the volume of the body. Of course, we can now ask what density is. Since density is usually defined as mass per unit volume, this becomes circular. We simply take the mass m of a body to be an intrinsic property of the body and use the intuitive concept of its being the amount of 'stuff' or matter in the body. Later we consider an operational definition of mass.

Next, Newton introduced the quantity of motion as:

> The quantity of motion is the measure of the same, arising from the velocity and quantity of matter conjointly.[20]

In modern terminology we refer to this as the *momentum*, a vector quantity, written as $\boldsymbol{p} = m\mathbf{v}$, where m is the mass of the body and $\mathbf{v}$ its velocity.

Newton's first law of motion, often referred to as the *law of inertia*, states that a body will either remain at rest or continue its uniform motion in a straight line unless it is acted upon by external, unbalanced forces. Therefore, just as for Descartes, natural or unconstrained motion becomes uniform straight-line motion. Since forces arise because of the interaction of a body with its environment, and since a body never exists in complete isolation, we can only conjecture what would happen if no forces acted on a body. It is really a convention to say that no (net) force acts on a body when that body is unaccelerated. It is far from obvious from everyday experience that a body left to itself will continue in constant, straight-line motion (that is, that it will not slow down and eventually stop). In fact, Aristotle's view, as embodied in Eq. (2.1), appears much closer to common-sense intuition. The original wording of the first law in the *Principia* is:

> Every body continues in its state of rest, or of uniform motion in a right line, unless it is compelled to change that state by forces impressed upon it.[21]

Here 'right line' is a somewhat antiquated expression for 'straight line'.

Newton's second law of motion is the quantitative statement we need that relates the force acting on a body to the change in the state of motion produced in that body.

> The change of motion is proportional to the motive force impressed; and is made in the direction of the right line in which that force acts.[22]

In modern notation: if a force $\boldsymbol{F}$ acts for a time Δt, then the change in the momentum $\Delta \boldsymbol{p}$ is $\boldsymbol{F}\Delta t = \Delta \boldsymbol{p}$. (Actually, a literal transcription of Newton's Law II as stated in the *Principia* would be $\boldsymbol{F} \propto \Delta \boldsymbol{p}$, although later he uses the law in the form stated as we have expressed it.) In a sense, this contains nearly all of classical mechanics. When the mass m is a constant, as is very often the case in applications, then we can write the first law in the familiar form

$$\boldsymbol{F} = m\boldsymbol{a} \qquad (7.1)$$

Two comments about this equation are in order. First, one should appreciate that nowhere in the *Principia* does the equation $\boldsymbol{F} = m\boldsymbol{a}$ appear. The law is stated

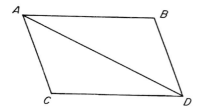

FIGURE 7.1 **The parallelogram law for forces**

verbally and used to set up proportions in applications made in the *Principia*. It was the Swiss mathematician Leonhard Euler (1707–1783) who, in 1752, appreciated the importance and general applicability of $\boldsymbol{F} = m\boldsymbol{a}$. He was the first to state this equation as the basic principle of mechanics.[23] Second, since Eq. (7.1) involves the acceleration $\boldsymbol{a}$, one needs a coordinate system (or reference frame) with respect to which to measure this. If two observers are accelerated with respect to each other and if they each measure the acceleration of a third object relative to themselves, then the two observers obtain different values for the acceleration of the object. However, if the observers are unaccelerated with respect to one another, they both observe the same value for the acceleration of the object. Therefore, in order not to introduce effects due to the acceleration of the observer, we must take care to apply Eq. (7.1) in a frame that is itself unaccelerated. We refer to these as *inertial frames*. In many situations, one can often effectively assume that an inertial frame of reference is one at rest with respect to the earth. Such a pragmatic move, though, does not address the question of principle about how one identifies an inertial frame generally.

Even though the modern vector notation we are using did not exist in Newton's time, he stated the parallelogram law of composition of two forces acting on a body. Two corollaries immediately follow the statement of his three laws of motion in the *Principia* and read:

> A body, acted on by two forces simultaneously, will describe the diagonal of a parallelogram in the same time as it would describe the sides by those forces separately.
>
> If a body in a given time, by the force M impressed apart in the place A [Figure 7.1], should with a uniform motion be carried from A to B, and by the force N impressed apart in the same place, should be carried from A to C, let the parallelogram ABCD be completed, and, by both forces acting together, it will in the same time be carried in the diagonal from A to D.
>
> . . .
>
> And hence is explained the composition of any one direct force AD, out of any two oblique forces AC and CD; and, on the contrary, the resolution of any one direct force AD into two oblique forces AC and CD: which composition and resolution are abundantly confirmed from mechanics.[24]

To complete the basic laws of classical mechanics, we need Newton's third law that, when one body exerts a force on another body, then the other body reacts on the first body with a force equal in magnitude but opposite in direction to the first force. This law states that forces always come in pairs. Notice that the action and reaction forces act on different bodies. Failure to appreciate this fact is the source of many famous paradoxes. Newton stated this law as:

> To every action there is always opposed an equal reaction; or, the mutual actions of two bodies upon each other are always equal, and directed to contrary parts.[25]

He then went on to explicate this as follows:

> Whatever draws or presses another is as much drawn or pressed by that other. If you press a stone with your finger, the finger is also pressed by the stone. If a horse draws a stone tied to a rope, the horse (if I may so say) will be equally drawn back towards the stone; for the distended rope, by the same endeavor to relax or unbend itself, will draw the horse as much towards the stone as it does the stone towards the horse, and will obstruct the progress of the one as much as it advances that of the other.[26]

Newton also offered an elaboration of the following type of argument in support of the third law.[27] Suppose a spherical earth were alone at rest in empty space and acted upon by no other bodies. Now consider the earth as two equal hemispheres in contact with each other. If the action (due to the gravitational attraction of one hemisphere for the other) and the reaction (by the second hemisphere) against it were not equal, then a net unbalanced force would act across the plane dividing these hemispheres. This would produce an acceleration of the sphere along a line perpendicular to this plane. The spherically symmetric body with no external forces applied would spontaneously accelerate off to infinity. However, from the symmetry of the problem, we see that there is no preferred direction so that the sphere must remain at rest. Hence, the gravitational force exerted on one hemisphere by the other must be precisely equal to the reaction force of this second hemisphere on the first.

7.5 THE LOGICAL STRUCTURE OF CLASSICAL MECHANICS

We present here a few comments on the logical structure and interconnectedness of Newton's laws of motion as he set them down in the *Principia*. Newton begins with a set of eight Definitions. Definition I concerns mass but is, as we saw earlier in this chapter, circular since it essentially defines mass in terms of density. However, this latter quantity is nowhere specified. Definition II introduces momentum, but this again involves the mass. Definitions III through VIII define various types of forces (inertial, impressed and centripetal) and state that a force is necessary to change the state of motion (or of rest) of a body. Definition VI states that the centripetal acceleration is proportional to the centripetal force.

Laws I (inertia) and II ($F = ma$) are then given, although they are really contained implicitly in the Definitions concerning forces. Law III (action and reaction) cannot have a precise meaning until the concept of force, and hence of mass, has been clarified (Definition I). These Laws are followed by a series of Corollaries, of which the first two state the parallelogram law for forces and the vector character of forces. The remaining four Corollaries expand upon the fact that each pair of bodies generates action–reaction forces between each other in a fashion that is independent of whether or not other bodies are present. Today we might paraphrase this by saying that all forces are two-body forces. That is, the net force acting on a given body is simply the (vector) sum of all the forces produced between that body and each other body separately. Although these last six propositions are termed corollaries by Newton, they do not all follow by pure deduction alone from the preceding Definitions and Laws, but also include as input some experimental information about the nature of forces in the real world.

Since the time of Newton there have been formal attempts to clarify the assumptions that are made in setting up the system of Newtonian mechanics. One example is given by the German physicist and philosopher Ernst Mach (1838–1916) in his *The Science of Mechanics*, a critical work that had great influence on Einstein in his own rethinking of the foundations of classical mechanics.

> Even if we adhere absolutely to the Newtonian points of view, and disregard the complications and indefinite features mentioned, which are not removed but merely concealed by the abbreviated designations 'Time' and 'Space,' it is possible to replace Newton's enunciations by much more simple, methodically better arranged, and more satisfactory propositions.[28]

Mach then went on to list a set of concise definitions and experimental facts that can be used to erect the system of classical mechanics. We outline below a similar scheme presented by the French mathematician and philosopher of science Henri Poincaré (1854–1912) in his *Science and Hypothesis*.

The acceleration a is defined operationally kinematically (in terms of length and time measurements relative to a chosen coordinate system) and is taken to be unproblematic. As part of our definition of the concept of force we accept the first and third laws. That is, we agree to say that no (net) force acts on a body that moves in uniform rectilinear motion. We also require that forces come in action–reaction pairs whenever bodies interact (which is the only way forces are ever generated). The first and third laws have now become conventions. For two bodies we can define the relative masses of the bodies by $(m_2/m_1) = (a_1/a_2)$. It is then an empirical question whether or not this ratio remains constant as the speeds and accelerations of the bodies change. Finally, force is defined quantitatively in terms of the second law, Eq. (7.1). If we are given an independent prescription for F (such as gravity), then we can use the second law to find a (once m has been assigned as previously). Poincaré concludes his discussion:

> Are the laws of acceleration and of composition of forces only arbitrary conventions? Conventions, yes; arbitrary, no – they would be so if we lost sight of the experiments which led the founders of the science to adopt them, and which, imperfect as they were, were sufficient to justify their adoption. It is well from time to time to let our attention dwell on the experimental origins of these conventions.[29]

None of these comments or attempts at clarifying the basis of classical mechanics should be taken as being critical of Newton's accomplishments. He creatively discovered these laws and correctly identified those qunatities (force, mass and acceleration) that determine the motion of a body. He stated more clearly and concisely than anyone before him the essential facts of mechanics as first vaguely perceived by Galileo: that each pair of bodies determines, mutually and independently of all other bodies, pairs of accelerations, the ratio of these accelerations being characteristic of each pair. This invariant characteristic we now call the ratio of the masses of this pair of bodies.

FURTHER READING

Louis More's *Isaac Newton* is a well-known, older biography of Newton, while Richard Westfall's magisterial *Never at Rest* is a masterpiece that incorporates the results of much modern scholarship on Newton. John Keynes' 'Newton the Man' was one of the earliest forays into Newton's 'unorthodox' investigations, Betty Dobbs' *The Foundations of Newton's Alchemy* develops this theme much more fully, and Frank Manuel's *A Portrait of Isaac Newton* borders on a psychoanalysis of the great scientist. Ernan McMullin's *Newton on Matter and Activity* analyzes Newton's ideas on the nature of matter and on the explanation for its attractive power. A much more technically demanding overview of theoretical reasoning in physics, from the time of Newton until the present, can be found in Malcolm Longair's *Theoretical Concepts in Physics*.

8

Newton's law of universal gravitation

A typical textbook summary of Newton's theory of gravitation presents his reasoning as a model in the application of scientific method. The scenario is roughly the following. Newton considered the rate at which the moon must 'fall' toward the earth in order to remain in its circular orbit and asked what centripetal acceleration was necessary to produce this motion. He concluded that the gravitational acceleration produced by the earth at the position of the moon decreased as the inverse square of the distance from the center of the earth compared to the known value of g at the surface of the earth. A bold generalization or induction from this yielded the law of universal gravitation between any two particles. Newton worried about the fact that the earth and moon were extended bodies, not point particles. By inventing the calculus he was able to prove that, as long as one was external to a uniform spherical mass, the sphere produced the same gravitational force as a point particle of equal mass situated at the center of the sphere. Then, using the law of gravitation plus his laws of motion, he deduced Kepler's three laws of planetary motion.

As we indicated in previous chapters and as we now show in more detail here and in the next chapter, the actual development, even as presented by Newton in his *Principia*, was a bit more circuitous and reflects his own philosophical view of the nature of science.[1] In the 1660s and beyond Newton was confronted with pieces of a puzzle concerning the motion of the planets. Kepler's three empirical laws of planetary motion were known to him, as were Galileo's laws of motion for bodies on the earth. Kepler believed that the planets moved in ellipses, while Galileo held they moved in circles. For Kepler the planets were driven along by 'spokes' of force radiating from a rotating sun, but Galileo's law of inertia stated that circular motion was self-perpetuating. To add to the confusion, Descartes had enunciated a law of inertia according to which bodies tend to persist in straight-line motion. For him the planets were held on their curved trajectories by vortices in an all-pervading cosmic aether. Newton was able to take this melange of fragmented facts and partial truths and ferret out a unified set of laws that correctly explained the motion of both heavenly and terrestrial bodies. In this chapter we examine some of the results of this stunning synthesis.

One of Newton's major advances in his study of the motion of heavenly bodies was to take laws discovered here on earth and apply them to the motions of the moon and of the planets. Galileo had a similar insight when he realized that if he understood why objects fell to the earth, then he would understand what kept the moon in its orbit. This is what Galileo had Salviati say in his *Dialogue Concerning the Two Chief World Systems* (see the first quotation for Part III (*The Newtonian universe*)).

PART III **The Newtonian universe**

8.1 NEWTON'S ASTRONOMICAL DATA AND DEDUCTIONS

In Book III of the *Principia*, Newton begins by listing the phenomena and data, or astronomical facts, about the bodies in the solar system.

> That the circumjovial planets, by radii drawn to Jupiter's centre, describe areas proportional to the times of description; and that their periodic times, the fixed stars being at rest, are as the $3/2th$ power of their distances from its centre.
>
> . . .
>
> That the circumsaturnal planets, by radii drawn to Saturn's centre, describe areas proportional to the times of description; and that their periodic times, the fixed stars being at rest, are as the $3/2th$ power of their distances from its centre.
>
> . . .
>
> That the five primary planets, Mercury, Venus, Mars, Jupiter, and Saturn, with their several orbits, encompass the sun.
>
> . . .
>
> That the fixed stars being at rest, the periodic times of the five primary planets, and (whether of the sun about the earth, or) of the earth about the sun, are as the $3/2th$ power of their mean distances from the sun.
>
> . . .
>
> Then the primary planets, by radii drawn to the earth, describe areas in no wise proportional to the times; but the areas which they describe by radii drawn to the sun are proportional to the times of description.
>
> . . .
>
> That the moon, by a radius drawn to the earth's centre, describes an area proportional to the time of description.[2]

From these six statements and the astronomical data that acompany them in the *Principia*, Newton concludes that the moons of Jupiter, the satellites of Saturn, the planets themselves and the earth's moon all obey Kepler's second law (equal areas in equal time) and third law (relating the mean radius and the period), and that the planets revolve about the sun (not about the earth) as center.

He then develops a series of propositions about the nature of the forces necessary to produce these phenomena.

> That the forces by which the circumjovial planets are continually drawn off from rectilinear motions, and retained in their proper orbits, tend to Jupiter's centre; and are inversely as the squares of the distances of the places of those planets from that centre.
>
> . . .
>
> That the forces by which the primary planets are continually drawn off from rectilinear motions, and retained in their proper orbits, tend to the sun; and are inversely as the squares of the distances of the places of those planets from the sun's centre.
>
> . . .
>
> That the force by which the moon is retained in its orbit tends to the earth; and is inversely as the square of the distance of its place from the earth's centre.[3]

In other words, by arguments of the type we give in the following sections and in the next chapter, Newton deduced that Kepler's second and third laws together require a central, inverse-square force.

That is how Newton presented matters in his *Principia*. However, early on in his investigations of the force that kept the planets in their orbits, Newton argued for the inverse-square nature of this force on the basis of Kepler's third law. In a manuscript written before 1669 he stated:

> Finally since in the primary planets the cubes of their distances from the sun are reciprocally as the squares of the numbers of revolutions in a given time: the endeavours of receding from the Sun will be reciprocally as the squares of the distances from the Sun.[4]

Notice that at this early date in his career Newton speaks in terms of the 'endeavours' of the planets to recede from the sun – that is, in terms of *centrifugal* (away from the center) rather than of *centripetal* (toward the center) force.

8.2 AN INVERSE-SQUARE LAW

Newton proved in Book I of his *Principia* that Kepler's second law of equal areas swept out by the planets in equal times required a central force between the planet and the sun. (In the next chapter we examine Newton's own geometrical argument for this result.) That is, the force necessary to divert the planet from its natural straight-line motion and keep it in its orbit has to point at each instant from the planet to the center of force located at the sun. In Book I he also established that Kepler's first law of elliptical orbits with the sun at a focus can be true only when this central force varies as the inverse of the square of the distance from the planet to the sun. (We consider his proof of this in the next chapter as well.)

Newton already knew the kinematical result that the centripetal acceleration acting on a planet (moving with a constant speed v in a circular orbit of radius R) can be written as $a_c = v^2/R$. In fact, this result was first given by the Dutch mathematician, astronomer and physicist Huygens. It was stated without proof in an appendix to his *Horologium Oscillatorium* (*On Pendulum Clocks*) in 1673.

> When two identical bodies move with the same velocity on unequal circumferences, their [centripetal] forces are in the inverse proportion to their diameters.
>
> ...
>
> When two identical bodies move on equal circumferences with unequal but constant velocities, ... the [centripetal] force of the more rapid is to that of the slower as the square of their velocities.[5]

The proofs were given in *De Vi Centrifuga* (*On Centrifugal Force*), published in 1703, after Huygens' death.[6] As early as 1669, in order to provide a priority claim, he sent an anagram to Henry Oldenburg (1618–1677), Secretary of the Royal Society of London, announcing these results,[7] although he had completed his study of circular motion around 1659.[8]

Newton independently discovered this relation for centripetal force.[9] In the *Principia* we find:

PART III   The Newtonian universe

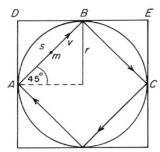

FIGURE 8.1   **Newton's argument for centripetal acceleration**

> The centripetal forces of bodies, which by equable motions describe different circles, tend to the centres of the same circles; and are to each other as the squares of the arcs described in equal times divided respectively by the radii of the circles.[10]

While we have quoted here Newton's statement from his *Principia* (1687), we should appreciate that he studied this question as long before as 1665.[11] In an early attempt (around 1665), Newton reduced the question to a problem of impact by considering two squares with a circle inscribed between them, as illustrated in Figure 8.1.[12] Suppose a mass m travels on the line segment AB with a uniform speed v. When this mass hits the cylindrical surface at point B, it will be 'reflected' upon impact just as it would if it were to collide with a rigid horizontal wall DBE. Upon collision, the component of the velocity perpendicular to this flat surface $(v_\perp = v/\sqrt{2})$ is simply reversed so that the total change in velocity upon impact is twice this value. If we denote by s the line segment AB, then the time between collisions is $\Delta t = s/v$. Again from the geometry of Figure 8.1, we see that $r = s/\sqrt{2}$ so that $\Delta t = \sqrt{2}r/v$. This implies that the rate of change of velocity (the centripetal acceleration) is $a_c = \Delta v/\Delta t = v^2/r$, just the result claimed. Even in 1665 Newton realized that the argument given above could be generalized to an N-sided polygon inscribed in a circle. Later, in a *scholium* in the *Principia*, Newton stated this argument as follows.

> In any circle suppose a polygon to be inscribed of any number of sides. And if a body, moved with a given velocity along the sides of the polygon, is reflected from the circle at the several angular points, the force, with which at every reflection it strikes the circle, will be as its velocity: and therefore the sum of the forces, in a given time, will be as the product of that velocity and the number of reflections; that is (if the species of the polygon be given), as the length described in that given time, and increased or diminished in the ratio of the same length to the radius of the circle; that is, as the square of that length divided by the radius; and therefore the polygon, by having its sides diminished *in infinitum*, coincides with the circle, as the square of the arc described in a given time divided by the radius. This is the centrifugal force....[13]

In the limit that $N \to \infty$, this polygon approaches a circle of radius r.

Armed with this result, we can today summarize Newton's argument for the law of attraction between the sun and a planet by combining his second law $\boldsymbol{F} = m\boldsymbol{a}$ (Eq. (7.1)) and Kepler's third law $R^3/\tau^2 = const$ (Eq. (5.2)) for the motion of a planet around the sun. (Here we do this only for a circular orbit, but in Chapter 9 we

Newton's law of universal gravitation

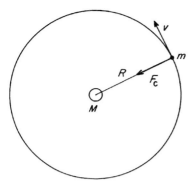

FIGURE 8.2 An inverse-square force law from Kepler's third law

outline Newton's more general argument for elliptical orbits. However, the logic of the argument is the same in both cases.) Newton's second law and the result for the centripetal acceleration we have just discussed together imply that the centripetal force F_c required to keep the planet in a circular orbit about the sun is $F_c = mv^2/r$. With the aid of Figure 8.2, we can express the speed v of the planet in its circular orbit of radius R as $v = 2\pi R/\tau$. If we combine these two relations, we obtain

$$F_c = \frac{mv^2}{R} = \frac{4\pi^2 m}{R^2}\left(\frac{R^3}{\tau^2}\right) \tag{8.1}$$

However, according to Kepler's third law, the quantity in parentheses in the last term of this equation is a constant (the same constant) for any planet orbiting the sun so that

$$F_c = \frac{4\pi^2 m}{R^2}(const) \tag{8.2}$$

The basic point of the above discussion is that Newton's second law and Kepler's third law together imply that the force exerted on a planet is an inverse-square one centered at the sun.

8.3 THE MOON'S CENTRIPETAL ACCELERATION

Then, in Book III, titled *The System of the World*, Newton asked what might be the source or nature of this inverse-square central force. This is answered with the following proposition.

> That the moon gravitates towards the earth, and by the force of gravity is continually drawn off from a rectilinear motion, and retained in its orbit.[14]

Let R_m be the radius of the moon's orbit about the earth and τ be the period for one complete revolution of the moon in its orbit. Then, as we just saw, we can find the speed of the moon about the earth as $v = 2\pi R_m/\tau$. Newton knew that the radius of the moon's orbit is about 60 times the radius of the earth itself. With the values

PART III The Newtonian universe

$r_e = 4{,}000$ miles $[6.44 \times 10^3$ km$]$ and $\tau = 27.3$ days, he obtained for the centripetal acceleration of the moon the value $a_c = 9 \times 10^{-3}$ ft/s^2 $[2.74 \times 10^{-3}$ m/s$^2]$. That is, at the location of the moon, the value of g has been reduced to $g_{\text{at moon}} = 9 \times 10^{-3}$ ft/s^2 $[2.74 \times 10^{-3}$ m/s$^2]$, as compared to its value at the earth's surface $g_{\text{at earth}} = 32$ ft/s^2 $[9.80$ m/s$^2]$.

But Newton already knew (recall the propositions of Book I of the *Principia* referred to above) that the force keeping the moon in its orbit is a $1/r^2$ one centered at the earth. Since

$$\frac{g_{\text{at earth}}}{g_{\text{at moon}}} = \frac{32}{9 \times 10^{-3}} = 3.56 \times 10^3 \cong \left(\frac{R_m}{r_e}\right)^2 = (60)^2 = 3600 \qquad (8.3)$$

Newton concluded that it was solely the earth's gravity that kept the moon in its orbit.

> And therefore the force by which the moon is retained in its orbit becomes, at the very surface of the earth, equal to the force of gravity which we observe in heavy bodies there. And therefore ... the force by which the moon is retained in its orbit is that very same force which we commonly call gravity; for, were gravity another force different from that, then bodies descending to the earth with the joint impulse of both forces would fall with a double velocity, and in the space of one second of time would describe [32] ... feet; altogether against experience.[15]

Newton used this centripetal force calculation, not (as is sometimes claimed) to learn that the force of gravity varies as $1/r^2$, but rather to conclude that the required $1/r^2$ force is due to gravity alone. Or, stated differently, this argument shows that the gravity of the earth is the only agent necessary to provide the proper rate of 'fall' of the moon. For example, no Cartesian vortices are needed. This assumes, of course, that earth exerts a gravitational force just as the sun does.

8.4   THE LAW OF GRAVITATION FOR POINT MASSES

Newton generalized the result of Eq. (8.2) to his *law of universal gravitation* that he stated as follows.

> That there is a power of gravity pertaining to all bodies, proportional to the several quantities of matter which they contain.
>
> ...
>
> The force of gravity towards the several equal particles of any body is inversely as the square of the distance of places from the particles....[16]

In modern notation we express this as

$$F = -\frac{Gm_1 m_2}{r^2} \qquad (8.4)$$

This is a central force since the line of action of the force is along the line joining the two bodies. Here G, the universal gravitational constant, is a proportionality constant. It is very difficult to measure G experimentally, because gravity is such a weak force, and nearly 100 years passed before Henry Cavendish (1731–1810) was able to do it. We cannot arbitrarily set $G = 1$ since, typically, units of mass, length and force have already been defined by other means. The numerical value of G is 6.67×10^{-11} m^3/kg s^2.

Newton stated the law governing gravity, but he did not attempt to explain it. The General *Scholium* at the end of the *Principia* contains Newton's famous disclaimer on this point.

> Hitherto we have explained the phenomena of the heavens and of our sea by the power of gravity, but have not yet assigned the cause of this power. This is certain, that it must proceed from a cause that penetrates to the very centres of the sun and planets, without suffering the least diminution of its force; that operates not according to the quantity of the surfaces of the particles upon which it acts (as mechanical causes used to do), but according to the quantity of solid matter which they contain, and propagates its virtue on all sides to immense distances, decreasing always as the inverse square of the distances.... But hitherto I have not been able to discover the cause of those properties of gravity from phenomena, and I frame no hypotheses....[17]

Since Eq. (8.4) gives the force of gravity in terms of the masses of the two bodies and of the distance between them, it is often referred to as action at a distance. The law makes no statement about how long it takes the force generated by one body to reach the other body. The force, or action, would seem to be propagated instantaneously over the distance between the two bodies and to be an inherent property of all matter. However, Newton himself did not accept these interpretations of his law, as he made clear in a letter he wrote on the subject.

> It is conceivable that inanimate brute matter should, without the mediation of something else which is not material, operate upon and affect other matter without mutual contact.... And this is one reason why I desired you would not ascribe innate gravity to me. ... [That] is to me so great an absurdity that I believe no man who has in philosophical matters a competent faculty of thinking can ever fall into it.[18]

Equation (8.4) states that two point particles m_1 and m_2, separated by a distance r, attract each other with a force directly proportional to the product of these masses and inversely proportional to the square of the distance between the particles. We have adopted the convention that the minus sign in Eq. (8.4) indicates a force of attraction. That is, m_1 exerts an attractive force $\boldsymbol{F}_{12}$ on m_2 and m_2 exerts an equal but opposite force $\boldsymbol{F}_{21}$ on m_1 (in accord with Newton's third law of action and reaction). Gravitational forces are always attractive. We can stress this point by expressing Newton's law of gravitation in vector form as

$$\boldsymbol{F}_{12} = -\frac{Gm_1 m_2}{r_{12}^2}\hat{r}_{12} \tag{8.5}$$

where r_{12} is the magnitude of the straight-line distance between m_1 and m_2 and $\hat{r}_{12}$ is a (dimensionless) unit vector that points from m_1 to m_2. Here $\boldsymbol{F}_{12}$ is the force exerted on m_2 by m_1, as indicated in Figure 8.3.

With his statement of the law of universal gravitation in Book III of the *Principia*, Newton finally presented a complete answer to a series of questions that he had been pondering over for two decades. Let us draw these threads together here. We have already seen that as early as the mid 1660s he applied Kepler's third law to the

PART III The Newtonian universe

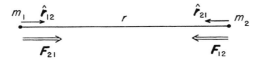

FIGURE 8.3 Mutually attractive gravitational forces

circular orbit of a planet to conclude that the central force keeping the planet in its orbit is an inverse-square one. At that time he also asked himself what the rate of acceleration or of 'fall' of the moon toward the earth was in order that the moon remain in its (nearly) circular orbit rather than going off in a straight line. With this alternative derivation of the expression for centripetal acceleration, Newton was able to treat a varying acceleration as the limit of constant accelerations, each applied for a sufficiently short time interval. This led him to the conclusion we discussed in Section 8.3 about the origin of the attractive force that kept the moon in its orbit. As we saw in Chapter 7, in 1679 Newton had an exchange with Robert Hooke about the orbit a body follows under the influence of an attractive inverse-square force. Newton proved to himself that the orbit is an ellipse but did not communicate this to Hooke.[19] In 1684 (probably August), Edmund Halley visited Newton at Cambridge. He made known to Newton Wren's open challenge (with a small prize promised for a solution) to derive the shape of the planets about the sun. In response, Newton began in the autumn of 1684 to write *De Motu Corporum* (*On the Motion of Bodies*). This work, with revisions, took until 1686 to complete. It was concerned with geometry and dynamics of motion and was to become essentially Book I of the *Principia*.[20]

8.5 GRAVITATION FOR EXTENDED BODIES

We emphasized that Eq. (8.4) is exact only for point masses. Newton worried for years about why this law should also hold for extended spheres that the earth and moon surely are.

> After I had found that the force of gravity towards a whole planet did arise from and was compounded of the forces of gravity towards all its parts, and towards every one part was in the inverse proportion of the squares of the distances from the part, I was yet in doubt whether that proportion inversely as the square of the distance did accurately hold, or but nearly so, in the total force compounded of so many partial ones; for it might be that the proportion which accurately enough took place in greater distances should be wide of the truth near the surface of the planet, where the distances of the particles are unequal, and their situation dissimilar. But by the help of Propositions 75 and 76, Book I, and their Corollaries, I was at last satisfied of the truth of the Proposition, as it now lies before us.[21]

However, there is no evidence that Newton tried to calculate the gravitational force produced by a sphere until 1685 (while writing the *Principia*).[22] This was long after he developed his calculus (*c.* 1665–1670). It is an involved story to give a detailed explanation for the long delay between Newton's first 1666 calculations on the moon's orbit and his proof, nearly twenty years later, that an inverse-square force produces elliptical orbits.[23] Inaccuracies in the known value of the earth–moon distance and the problem of the attractive force generated by a sphere were probably less serious difficulties than lack of a clear dynamical framework within which to do the necessary orbit calculations. Newton appears not to have fully and precisely formulated his dynamics until the 1680s.[24] Even the concept of universal gravitation was rather late in being finalized.

Today it is a fairly straightforward exercise in calculus (or even solid geometry) to prove that Eq. (8.4) is still valid between two homogeneous spheres, provided the spheres do not overlap. Newton developed this important result for spherical masses in a series of propositions.

> If to every point of a spherical surface there tend equal centripetal forces decreasing as the square of the distances from those points, I say, that a corpuscle placed within that surface will not be attracted by those forces any way.[25]

By 'centripetal force' here Newton means central force. This theorem states that inside a homogeneous hollow spherical shell a point mass experiences no net gravitational force. At any point interior to such a spherical shell, all of the gravitational forces from the shell exactly cancel each other.[26]

Next, if a point mass is placed outside a homogeneous spherical shell, then it is attracted toward the center of the sphere just as though all the mass of the shell were concentrated at a point at the center of the sphere.[27]

> The same things supposed as above, I say, that a corpuscle placed without the spherical surface is attracted towards the centre of the sphere with a force inversely proportional to the square of its distance from that centre.[28]

Once we have this result for a thin shell of mass, we can consider a solid sphere of uniform density to be made up of a series of concentric shells and hence conclude that a point mass placed external to a uniform spherical mass will be attracted by that sphere just as though all of its mass were concentrated at a point at the center of the sphere. Newton stated this result for two separated spherical masses as follows:

> In two spheres gravitating each towards the other, if the matter in places on all sides round about and equidistant from the centres is similar, the weight of either sphere towards the other will be inversely as the square of the distance between their centres.[29]

Finally, we can combine these theorems of Newton's to make an useful observation. Newton tells us that:

> If to the several points of a given sphere there tend equal centripetal forces decreasing as the square of the distances from the points, I say, that a corpuscle placed within the sphere is attracted by a force proportional to its distance from the centre.[30]

This means that if we have a homogeneous solid sphere (one of uniform density) and we place a particle inside this sphere, then the particle will experience a force of attraction toward the center of the sphere and this force will vary directly as the distance from the center of the sphere. The reason for this is that the part of the spherical mass in the shell beyond the location of the particle exerts no net force on the particle (as stated in the first proposition quoted above), while the mass in the spherical volume inside the location of the particle varies as the cube of this distance (since the volume of a sphere is $(4/3)\pi r^3$). But the force exerted by this inner sphere on the particle varies inversely as the square of the distance. Together, these facts establish Newton's claim made above. All of these simple results are true only for spheres (not, for example, for cubes or for other bodies of arbitrary shape).

8.6 INERTIAL AND GRAVITATIONAL MASSES

The concept of mass employed in Chapter 7 was actually that of the *inertial mass* m_i of a body as a measure of its resistance to acceleration when a given force is applied to it ($F = m_i a$). The gravitational mass m_g of a body determines the magnitude of the gravitational attraction between the body and any other body ($F = GMm_g/r^2$, or $w = m_g g$, where w is the *weight* of the body). A priori there is no reason in classical mechanics for these two masses of a given body to be related. However, experimental measurements from Newton's time on have shown that $m_g = m_i$. Newton does not use the terms 'inertial mass' and 'gravitational mass', but it is just these concepts he is concerned with in the following. In the *Principia* Newton discusses the exact proportionality between (inertial) mass and weight.

> [T]he [mass] is known by the weight of each body, for it is proportional to the weight, as I have found by experiments on pendulums, very accurately made, which shall be shown hereafter.[31]
>
> . . .
>
> The quantities of matter in pendulous bodies, whose centers... are equally distant from the centre of suspension, are in the ratio compounded of the ratio of the weights and the squared ratio of the times of the oscillations in a vacuum.[32]

As a proportion this latter proposition is just $m_i \propto \tau^2 w$. The period of a pendulum can then be written as[33]

$$\tau = 2\pi \sqrt{\frac{m_i \ell}{w}} = 2\pi \sqrt{\frac{\ell}{g} \frac{m_i}{m_g}} \qquad (8.6)$$

If m_i and m_g were different, then τ would vary from one pendulum to another made of a different substance even when both pendula had the same length ℓ. But, as Newton observes above, that is not the case.

Furthermore, if $m_i \neq m_g$, then $w = m_g g = m_i a$ would imply $a = (m_g/m_i)g$ and all bodies would not have equal acceleration at the surface of the earth. This is contrary to Galileo's great discovery, as Newton was well aware.

> It has been, now for a long time, observed by others, that all sorts of heavy bodies... descend to the earth *from equal heights* in equal times; and that equality of times we may distinguish to a great accuracy, by the help of pendulums. I tried experiments with [many substances].... By these experiments, in bodies of the same weight, I could manifestly have discovered a difference of matter less than the thousandth part of the whole, had any such been.[34]

Later in this same passage Newton points out that Kepler's third law of planetary motion would not be true if m_i and m_g were not equal. From Eqs. (8.1) and (8.4) we see that Kepler's law would become

$$\frac{R^3}{\tau^2} = const \frac{m_g}{m_i} \qquad (8.7)$$

The reason for the exact equality of inertial and gravitational masses can be understood on the basis of Einstein's general theory of relativity that we discuss in Chapter 18. However, in classical gravitational theory, this equality is simply a coincidence (or a brute fact of nature) with no fundamental explanation.

FURTHER READING

I. Bernard Cohen's long biographical sketch 'Isaac Newton' gives an accessible account of Newton's life and work. Cohen's 'Newton's Discovery of Gravity' is an essay intended for a general audience, while Leon Rosenfeld's 'Newton and the Law of Gravitation' is a more formal treatment of this great discovery. Of course, Richard Westfall's *Never at Rest* covers virtually all aspects of Newton's life and work. James Cushing's 'Kepler's Laws and Universal Gravitation in Newton's *Principia*' discusses in modern geometrical notation Newton's own argument for the gravitational attraction produced by a spherical mass.

9

Some old questions revisited

In the previous two chapters we discussed, largely in terms of modern mathematical notation, Newton's three laws of motion and his deduction of the law of universal gravitation. Here we attempt to give the reader some appreciation of the type of mathematical arguments that Newton actually employed in the *Principia*. Then we show how his laws of motion and of gravitation can be used to deduce Kepler's three laws as necessary logical consequences of this overarching theory. (Some of the material here is a bit more technically involved that that in previous chapters. The reader whose tastes such details do not suit can simply move onto Chapter 10.) As we have already seen in the last chapter, Newton accepted Kepler's laws and was able to argue from them to the inverse-square nature of the gravitational attraction between any two bodies. He then reversed the line of argument and, assuming the validity of the law of gravitation, derived Kepler's three laws of planetary motion. For example, we find in the *Principia*:

> The planets move in ellipses which have their common focus in the centre of the sun; and, by radii drawn to that centre, they describe areas proportional to the times of description.[1]

This contains Kepler's first and second laws together. Of course, we expect to be able to do more than simply recover Kepler's laws that were, after all, used to deduce the law of gravitation itself.

A major factor in our accepting the general structure of classical mechanics and gravitational theory (as for most theories in physics) is its empirical success in accounting for phenomena – both in detailed numerical predictions of specific cases and in uniting a broad range of hitherto apparently unrelated facts. Below we use Kepler's laws and the ocean tides as two illustrative examples of such success. First, though, we turn to the mathematics found in the *Principia*.

9.1 AN ILLUSTRATION OF NEWTON'S GEOMETRICAL PROOFS

In this section and the next we give a mathematical representation of the arguments that Newton used to derive Kepler's laws of planetary motion from his own laws of dynamics and of gravitation. Our presentation tries to remain faithful to the line of reasoning used by Newton and also to give some sense of the style and flavor of the *Principia*. It is, hopefully, a reasonable compromise between the original text and modern notation. By modern analytical standards, the imperfections of these basically geometric proofs of the *Principia* are apparent.

The logical structure of the line of argument we wish to pursue in Book I is the following. (The basic dynamical law is always taken to be $F = ma$.) Proposition 1 establishes that a central force implies both Kepler's second law of equal areas in equal times and the fact that the orbit is contained in a plane, while Proposition 2 proves the converse. Propositions 11, 12 and 13 show that, for a central force directed toward the focus, a plane conic section orbit requires an inverse-square force law. Proposition 17 proves that, if a body moves in a conic section about a given seat of central force, then a set of specified initial conditions uniquely determines that conic section. Corollary 1 of Proposition 13 concludes that, for an inverse-square force and a given set of initial conditions, one conic section always exists that satisfies both the initial conditions at $t = t_0$ and the equation of motion ($F = ma$) for all time. Proposition 42 argues that this solution, that has been explicitly exhibited, is the only possible orbit. This establishes Kepler's first law (and, in fact, allows for hyperbolas and parabolas as well as ellipses). Proposition 15 states Kepler's third law. (In this and the next section we discuss these eight propositions in this order.) Finally, Propositions 70, 71 and 75 treat the gravitational force produced by a uniform mass shell. (These last three propositions were discussed in Section 8.5.)

Now let us turn to Newton's argument for Kepler's second law (equal areas in equal times). In the *Principia* he states Kepler's second law of planetary motion as follows.

> *Proposition 1.* The areas which revolving bodies describe by radii drawn to an immovable centre of force do lie in the same immovable planes, and are proportional to the times in which they are described.[2]

Notice that a body must move in a fixed plane when under the influence of such a central force directed to a fixed force center. From a modern perspective, this follows from the observation that the initial velocity vector of the body and the (initial) instantaneous force vector from the body to the fixed force center (or, equivalently, the initial position vector from the force center to the body) define a plane. Because the acceleration vector (or force vector) lies in this plane, all changes in the velocity vector will lie in this plane too, as will all future velocity vectors. The radius vector then sweeps out a trajectory in this fixed plane.

We now give an elementary geometrical derivation (of the type employed by Newton) of Kepler's second law. A central force is always directed from the moving body to the fixed point S, as shown in Figure 9.1. If no force were acting, then in equal intervals of time a body initially at A would move to B and thence to C, where $AB = BC = v\Delta t$. If at B a large force, directed toward S, acts for an instant, then the path of the body will be diverted to the path BD. Since the change of velocity (or acceleration) points along CD and is directed parallel to SB, we see that CD is parallel to SB. It is a basic theorem of Euclidean geometry that the area of any triangle is equal to one-half the product of its base times its altitude perpendicular to that base. Since the triangles SAB and SBC have the same altitude h, they have equal areas because their bases, AB and BC respectively, are equal. However, since

PART III The Newtonian universe

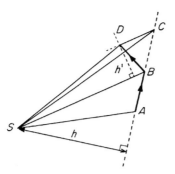

FIGURE 9.1 Newton's geometrical argument for Kepler's second law

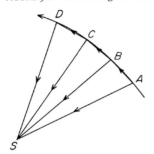

FIGURE 9.2 A continuous limit from incremental arguments

DC is parallel to SB, we see that triangles SBC and SBD, that share the base SB, also have the same altitude (h') so that the area of SBC equals that of SBD. This tells us that the area of SAB equals that of SBD. That is, in equal time intervals equal areas have been swept out.

If we take a smooth curve and approximate it by a series of straight-line segments as suggested in Figure 9.2 and allow instantaneous impulsive central forces to act successively at points B, C, D, etc., then the argument just given will show that the areas of all the triangles SAB, SBC, SCD, etc., swept out in equal times will be equal. As the equal time intervals become smaller and smaller, so that the points A, B, C, D, etc., on the curve become closer to one another, the central force will come to act continuously. In this limit we obtain the result that the radius vector from the force center to any body moving only under the action of a central force sweeps out equal areas in equal intervals of time.

We can also reverse the argument, now being given that a body sweeps out equal (planar) areas in equal times (so that the triangles SAB and SBD in Figure 9.1 have equal areas) and that the change in velocity (along DC) forms the closed triangle BDC lying in this common plane. Since the areas of triangles SAB and SBC are equal (because they share the common altitude h), so must triangles SBC and SBD be equal in area. This requires that they have a common altitude (h') for their common base SB. That means that DC is parallel to SB, or that the force is a central one. That is, if a body sweeps out equal areas in equal times, it is acted upon by a central force. This is essentially the proof that Newton gave for his claim:

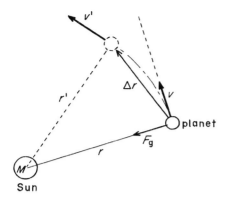

FIGURE 9.3   An incremental existence/uniqueness argument

Proposition 2. Every body that moves in any curved line described in a plane, and by a radius drawn to a point...immovable...describes about that point areas proportional to the times, is urged by a centripetal force directed to that point.[3]

Notice that this law of equal areas is true for any central force, not just for an inverse-square law.

9.2   KEPLER'S FIRST AND THIRD LAWS[4]

Let us begin by asking ourselves how one might determine the shape of a planetary orbit, given Newton's law of gravitation and his second law of motion. We first give a plausibility argument (not a strict mathematical proof) that these two laws together allow one in principle to calculate the possible orbits or trajectories of a planet or heavenly body about the sun. Since the mass of the sun ($M = 2.0 \times 10^{30}$ kg) is so much greater than the masses of the planets (even Jupiter, the most massive of the planets, has a mass of only $m = 1.9 \times 10^{27}$ kg), for simplicity we take the sun to be stationary in this discussion. Figure 9.3 shows a planet with an instantaneous velocity **v** (that is always, by its very definition, tangent to the orbit) and at a distance r from the sun. If, at some initial time, we are given the position (r) of the planet, then we can use

$$F(r) = -\frac{G\,Mm}{r^2} \qquad (9.1)$$

to compute the gravitational force acting on m. The law $\mathbf{F} = m\mathbf{a}$ will yield the instantaneous acceleration $\mathbf{a}$ ($= \Delta \mathbf{v}/\Delta t$) of the planet. Once we have been given the initial velocity **v**, we can find the velocity $\mathbf{v}'$ a short time Δt later as $\mathbf{v}' = \mathbf{v} + \Delta \mathbf{v} = \mathbf{v} + \mathbf{a}\Delta t$. The displacement $\Delta \mathbf{r}$ during this increment of time is $\Delta \mathbf{r} = \mathbf{v}\Delta t$. This gives us the new distance r' from the sun (since $\mathbf{r}' = \mathbf{r} + \Delta \mathbf{r}$) and we can then repeat the calculation for the next new position and so on to map out the entire trajectory. Both calculus and differential equations are usually employed today to carry out this program analytically. However, the important point for us to appreci-

PART III The Newtonian universe

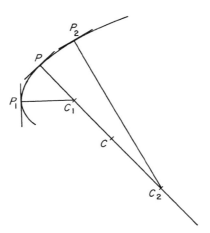

FIGURE 9.4 The radius of curvature

ate here is that, once the initial position and velocity of the planet have been specified, then Eq. (9.1) and $F = ma$ will uniquely grind out, or predict, all future positions and velocities of the planet. We can put this rather loosely by stating that once r_0 and v_0 are given initially, then 'nature' (or really Newton's law) determines $r(t)$ for all future times. We have not proved this statement here, but only attempted to make it appear reasonable.

Before we proceed to describe Newton's own derivations of Kepler's first and third laws, we must review some geometrical properties of conic sections. Central to this is the concept of curvature for a trajectory. We characterize the center of curvature[5] of a curve as the point of intersection of normal lines drawn through neighboring points on the curve. In Figure 9.4 the lines P_1C_1, PC and P_2C_2 represent lines perpendicular to the tangents to the curve at points P_1, P and P_2, respectively. As P_1 and P_2 approach P, the points C_1 and C_2 approach a point C, the center of curvature of the curve at P. The distance PC is the radius of curvature that we denote by ρ. This explicit characterization of the center of curvature was original with Huygens in his researches on envelopes to curves (1659)[6] and, independently, with Newton (1665).[7]

The classic definition of a conic section (recall Section 5.A) is the locus of all points such that the ratio of the distance from a fixed point (the focus F) to the perpendicular distance (d) from a fixed straight line (the directrix) is a constant (the eccentricity e). From the geometry of Figure 9.5, we see that this definition can be expressed analytically as (recall Eq. (5.5))

$$\frac{r}{d+x} = e \qquad (9.2)$$

(For $0 \leq e < 1$, we have an ellipse; for $e = 1$, a parabola; for $e > 1$, an hyperbola.) This allows us to deduce the important geometrical property for a conic section[8]

$$\rho \cos^3 \gamma = de \qquad (9.3)$$

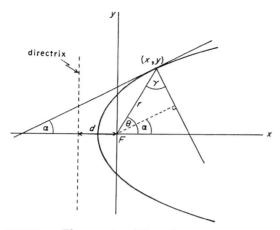

FIGURE 9.5 The geometry of Newton's argument

This relation is essential for our subsequent representation of Newton's argument for Kepler's first law. It should be evident that the result of Eq. (9.3) is a purely geometrical one that holds for any conic section. It is such properties that allowed Newton to obtain Kepler's first and third laws. For most modern readers Eq. (9.3) must seem a nonobvious, and in fact a bizarre, way to characterize a conic section.

We turn now to Newton's proof that, if a body moves in a conic section under the influence of a central force directed toward the focus, then that force must be an inverse-square one.

Proposition 11. If a body revolves in an ellipse; it is required to find the law of centripetal force tending to the focus of the ellipse.[9]

Propositions 12 and 13 are the same, except that they consider the cases of hyperbola and parabola, respectively.

In applying the dynamical equation of motion $F = ma$ for a central force $F(r) = f(r)\hat{r}$, Newton first used Kepler's second law that he had already established in Proposition 1. (Recall Section 9.1 above.) If we denote by v_θ the component of v in the θ-direction (that is, the component of v perpendicular to a or to r in Figure 9.5), then we see that the rate at which area is swept out (dA/dt) is equal to $\frac{1}{2}rv_\theta$. Since this is a constant by Kepler's second law, we can write

$$\frac{dA}{dt} = \frac{1}{2}r_0 v_{\theta_0} = \frac{1}{2}r_0 v_0 \cos\gamma_0 = \frac{1}{2}rv \cos\gamma = \frac{1}{2}r^2\omega = \text{const} \tag{9.4}$$

Here ω is the instantaneous angular speed of the body about the focus F and γ is the angle between r and the line to the center of curvature, as illustrated in Figure 9.5. Notice that Eq.(9.4) is of the form

$$v(\theta) \equiv \frac{ds}{dt} = \frac{2(dA/dt)}{r(\theta)\cos\gamma} = h(\theta) \tag{9.5}$$

where $h(\theta)$ is now some known function of θ. This quantitative relation between the speed v and the location of the planet is implied by Kepler's second law. It is essentially a recipe to guarantee motion under a central force (one such that the acceleration vector at each instant is directed toward a fixed point in space). Hence, for any $r = r(\theta)$, it is always possible to satisfy Eq. (9.4). This condition insures central-force motion and motion under the action of a central force implies Eq. (9.5). Therefore, we take this part of the central-force problem to have a solution. This allows a decomposition of the central-force problem into an essentially 'kinematical' part (Eq. (9.5)) involving the time and a purely geometrical part that determines the shape of the orbit. Let us see how this separation comes about.

Instead of treating the radial component of $\boldsymbol{a}$ directly, Newton effectively dealt with the projection of $\boldsymbol{a}$ along the normal direction (or along the direction of the radius of curvature ρ of Figure 9.5). That is, m can be considered as instantaneously moving on a circle of radius ρ and the necessary centripetal force is provided by the component of $\boldsymbol{F}$ along the normal direction as

$$\frac{mv^2}{\rho} = |f(r)| \cos\gamma \qquad (9.6)$$

If we use Eq. (9.4), we can rewrite this as

$$|f(r)| = \frac{4m(dA/dt)^2}{\rho \cos^3\gamma} \frac{1}{r^2} \qquad (9.7)$$

However, from Eq. (9.3) we see that the denominator of the first factor on the right side of this equation is a constant. This establishes that if the orbit is a conic section with the attraction toward a focus, then the central force must be an inverse-square one.

We turn next to the converse of this.

> *Proposition 17.* Supposing the centripetal force to be inversely proportional to the squares of the distances of places from the centre, and that the absolute value of that force is known; it is required to determine the line which a body will describe that is let go from a given place with a given velocity in the direction of a given right line.[10]

What Newton actually did in the proof of this proposition was to show that, once the focus F (toward which the central force is directed) is given, the values of the tangent and of the radius of curvature at one point on a conic uniquely specify the conic. This simply means that, for a given focus, only one conic section can pass through a given point, once the tangent and radius of curvature have been specified there. His geometrical argument is quite lengthy and cumbersome, so that we refer elsewhere for a paraphrase of the proof of it in the notation of this section.[11] Here we simply accept that result as established.

Suppose we then consider the following problem. A body of mass m orbits a fixed force center at F under the influence of an attractive inverse-square central force. We have already seen that the dynamical equation of motion $\boldsymbol{F} = m\boldsymbol{a}$ reduces to the two conditions of Eqs. (9.4) and (9.6) that must be satisfied for all time. Given arbitrary values of r_0, $\mathbf{v}_0$ (at $t = t_0$) and λ (the force constant in $f(r) = -\lambda/r^2$), we may determine the constant (dA/dt) of Eq. (9.4) because the value of the tangent yields α_0 (and, hence, γ_0) from the geometry of Figure 9.5. We may then choose a value of ρ_0 as (see Eq. 9.7))

$$\rho_0 = \frac{4m(dA/dt)^2}{\lambda \cos^3\gamma_0} \tag{9.8}$$

As we have just stated in the previous paragraph, this, along with r_0 and $\mathbf{v}_0$, will determine a unique conic section. Denote that unique conic section by $r(\theta)$. This conic is such that at $t = t_0$ it yields r_0 and $\mathbf{v}_0$ (that is, r_0, v_0 and α_0 for θ_0). It also satisfies the equation of motion (Eq. (9.6)) for all time (recall Proposition 11 and the following discussion, especially Eq. (9.7)). In modern terminology we have explicitly exhibited an $r(t)$ such that the initial conditions, $r(t_0) = r_0$, $\mathbf{v}(t_0) = \mathbf{v}_0$, are satisfied and that fulfills $\mathbf{F} = m\mathbf{a}$ for all subsequent times. This establishes that an inverse-square central force always has a conic section for an orbit. It is in Corollary 1 to Proposition 13 that Newton somewhat cryptically alludes to in this argument.[12]

> From the three last Propositions [11, 12 and 13] it follows, that if any body P goes from the place P with any velocity in the direction of any right line PR, and at the same time is urged by the action of a centripetal force that is inversely proportional to the square of the distance of the places from the centre, the body will move in one of the conic sections, having its focus in the centre of force; and conversely. For the focus, the point of contact, and the position of the tangent, being given, a conic section may be described, which at that point shall have a given curvature. But the curvature is given from the centripetal force and the velocity of the body given;....[13]

So, we have now addressed the question of the existence of the solution to the central-force problem and shown that there always exists one and only one conic section as a solution.[14] However, logically this still leaves open the question of whether or not there also exist other solutions that are not conic sections. While Newton was aware of this question and did attempt to address it,[15] his argument relied heavily on the intuition that an iterative (or incremental) argument of the type given at the beginning of the present section (see Figure 9.3) will lead to the only physical solution.

Perhaps it would be helpful for us to summarize these results in a form more familiar to the modern reader. There are just three possible types of orbits allowed for any heavenly body moving about the sun. One is an *ellipse* and this is the only type of closed orbit that Newton's law of gravitation allows. (Of course, a circle is a special case of an ellipse.) This motion is periodic and the planet never gets more than some maximum finite distance away from the sun. This is sometimes referred to as bounded motion or a reentry orbit. Two types of unbounded motion are also possible. In these cases the body (often a comet) can make just one pass by the sun and then never return to the solar system. For such unbounded motion the orbit is non-reentry. One of these orbits is an *hyperbola* and the other is a *parabola*. Sketches of these three curves are given in Figure 9.6. For our present purposes, there is really little difference between the hyperbola and the parabola. If the speed of a heavenly body is sufficiently great, then it will have an open orbit (hyperbola or

PART III The Newtonian universe

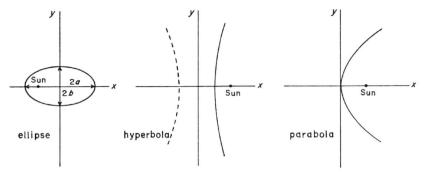

FIGURE 9.6 **Allowed types of inverse-square orbits**

parabola) and once it has passed by the sun, it will go off to infinity, escaping from the gravitational field of the sun. If the velocity is not so great, then the orbit will be bounded (an ellipse or circle) and periodic. Although we have seen that Kepler's first law follows from Newton's laws of motion and of gravitation, we get much more than just Kepler's first law in our result. Not only do ellipses turn out to be the only possible bounded orbits, but we also learn the exact nature of the two allowed types of open orbits.

It is worth mentioning that the result that the only possible closed orbits are ellipses depends upon the fact that Newton's law of gravitation describes a central force (since it always points directly from the planet to the sun) and that it varies exactly inversely as the square of the distance. If the gravitational attraction between two bodies varied, say, as $1/r^3$ rather than as $1/r^2$, the bounded orbits of the planets would not be ellipses (nor need they be closed). As we have just seen, one can prove mathematically that the only central force (directed to a focus) that can produce an ellipse is an inverse square law. In the *Principia* Newton states:

> That the forces by which the ... planets are continually drawn off from rectilinear motions, and retained in their proper orbits,... are inversely as the squares of the distances of the places of those planets from the sun's centre... is, with great accuracy, demonstrable from the quiescence of the aphelion points;....[16]

What Newton means by the 'quiescence of the aphelion points' is that the location of the point of greatest distance of a planet from the sun (r_2 of Figure 5.10) is fixed in space (relative to the frame of the sun) and does not advance as the planet makes successive revolutions in its orbit. He is claiming that if the force were not an inverse-square one alone, then the aphelion point would move in time. This is equivalent to saying that the orbit would not close smoothly upon itself (even if the motion remains bounded).

Finally, Newton addressed Kepler's third law.

> *Proposition* 15. The same things being supposed [that is, motion under the influence of a fixed inverse-square central force], I say, that the periodic times in ellipses are as the 3/2*th* power of their greater axes.[17]

This followed directly from Kepler's second law and from geometric properties of the ellipse that Newton already had at hand.[18] Kepler's third law depends on the $1/r^2$ form of the law of gravitation and upon the central nature of this force. (It is particularly easy to see that Kepler's third law follows from the inverse-square law of gravitation in the special case of circular orbits, as implied by Eq. (8.1).)

Therefore, as an exercise in logic, from Kepler's second law of equal areas we can conclude that the gravitational force of attraction must be a central one. Next, from Kepler's first law that the planetary orbits are ellipses we can deduce that the gravitational force follows an inverse-square law. From this knowledge we can then derive Kepler's third law relating the mean radius to the period of a planet. Kepler's third law is not logically independent of his first two laws. As we have seen, Newton began with Kepler's second and third laws and from these argued to Kepler's first law of elliptical orbits. We can prove this dependence, however, only within the framework of Newtonian mechanics. Without Newton's law of motion, $\boldsymbol{F} = m\boldsymbol{a}$, there is no way to see this connection.[19]

9.3 PERTURBATIONS

Actually, since all planets and heavenly bodies attract all other ones, the problem of a planet going around the sun is more complicated than we have presented it. If the sun and the planet were the only bodies in the entire universe, then the orbit would be a simple ellipse. However, the other planets in the solar system also exert forces on the planet and do slightly disturb the orbit from that of a perfect ellipse. These small disturbances are referred to as *perturbations*. Since all of these perturbative forces are governed by the law of gravity, Eq. (9.1), and since the gravitational effect of the sun is the major one for the planets, it is possible to calculate these small corrections to the elliptical orbits of the planets. The actual orbits of the planets do agree with the predictions made once these perturbations are included.

It is well known, however, that nature was kind (or perhaps cruel, given the ultimate 'deception' played on us, as we shall see in chapter 12) in that for a $1/r^2$ force the equation for the orbit can, as we have just seen, be solved exactly to yield conic sections. This much was known to Newton, but it was unclear to him that a many-body configuration, such as the solar system, would remain stable under the mutual perturbations caused by the planets on each other's orbits about the sun. Joseph-Louis Lagrange (1736–1813) and Pierre-Simon Laplace (1749–1827) are generally credited with taking care of the perturbations, resolving the issue in favor of the long-term stability of the solar system. But, the actual historical record is more complicated.[20] Of necessity, like all previous and later workers on the perturbations of the planets, Lagrange and Laplace had to employ approximate calculations in their analyses and truncated their equations in low orders of small expansion parameters. In 1766 Lagrange published a major paper on the theory of planetary perturbations and in 1773 Laplace obtained solutions in which all terms were periodic (or long-term recurrent, as opposed to being secular). Lagrange in 1776 and Siméon-Denis Poisson (1781–1840) in 1809 extended Laplace's results. By

1876 Simon Newcomb (1835–1909) demonstrated that, for the planetary perturbations, it is possible to represent the solutions by purely periodic functions, provided these series converge. It appeared that the stability of the solar system was very close to being established (a mere 'technicality' away). In an 1892 treatise, though, Poincaré showed that these series are in general divergent. Such results were, as we know today, the seeds of modern deterministic chaos theory (to be discussed in Chapter 12).

Observations of such perturbations led to the discovery of the planet Neptune. On March 13, 1781, William Herschel discovered the planet Uranus. At that time it was the farthest known planet from the sun. However, once the perturbation corrections due to the effects of Jupiter and of Saturn were made, the observed orbit of Uranus still did not agree with the predicted one. Both the Englishman John Adams (1819–1892) and the Frenchman Urbain Leverrier (1811–1877) independently calculated that there must be another as yet unobserved planet beyond Uranus that was causing further perturbations. On September 23, 1846, the German astronomer Johann Galle (1812–1910) found a new planet (Neptune) just where Leverrier had predicted it would be. We return to this topic in Section 11.2.

So we have seen, in outline at least, how Newton's mechanics and his theory of gravity are able in principle to account for the phenomena of planetary motion. Such problems were, of course, the origin of gravitational theory and of much of classical mechanics. We now show how this theoretical system is also able to account for another problem that had exercised mankind for ages.

9.4 THE OCEAN TIDES PRIOR TO NEWTON

That the level of the sea along the shore rose and fell with some regularity and that these variations were correlated with the position of the moon were observations already made in antiquity. Seleucus of Babylon (*c.* 150 B.C.) was able to demonstrate the effect of the moon on the tides.[21] The ancient Greeks remarked on the tides, even though tidal effects on the smaller Mediterranean sea are not nearly as pronounced as those associated with the Atlantic and Pacific oceans. Poseidonios of Rhodes (*c.* 135–*c.* 51 B.C.) first noted the greater-than-average tides (called spring tides) that occur around the times of new and full moons and the less-than-average tides (named the neap tides) that occur around the first and third quarters of the moon. He is also credited with suggesting that the tides are associated with the sun as well as with the moon.[22] The Roman scholar Pliny the Elder (23–79 A.D.) in his *Historia Naturalis* (*Natural History*) remarked on the tides and the relation between the rising and falling of the tides and the rising and falling of the water in a well.[23] The Islamic philosopher al-Kindi (*d. c.* 870 A.D.) wrote a treatise on the tides. The English Benedictine Bede the Venerable (672/673–735), arguably the outstanding scholar in Europe at the end of the seventh century, recorded his own observations of the tides.[24] There was no shortage of conjectures about the cause of these tides: expansion of the sea's waters due to heating by the rays of the moon, evaporation of sea water by these same rays, the existence of a large vortex or whirlpool that alternately sucked down and then expelled water, and several others.

Even in more modern times, a correct explanation of the tides was a major problem in the history of physics. In 1616 Galileo wrote a discourse on the tides in which he first discounted several specious explanations, such as the inclination of the sea bed and the agitation caused by the wind.[25] He also devoted the Fourth Day of his *Dialogue Concerning the Two Chief World Systems* to the tides. He believed that the ocean tides gave direct evidence for the motion of the earth.

> Among all sublunary things it is only in the element of water... that we may recognize some trace or indication of the earth's behavior in regard to motion and rest.[26]

Kepler held that the tides were due to the attraction of the ocean waters by the moon. In the preface to his *Astronomia Nova (New Astronomy)*, he wrote:

> Gravity is a mutual corporeal disposition among kindred bodies to unite or join together....
>
> ...
>
> If the earth should cease to attract its waters to itself, all the sea water would be lifted up, and would flow onto the body of the moon.
>
> The sphere of influence of the attractive power in the moon is extended all the way to the earth, and in the torrid zone calls the water forth, particularly when it comes to be overhead in one or another of its passages. This is imperceptible in enclosed seas, but noticeable where the beds of the ocean are widest and there is much free space for the water's reciprocation.
>
> ...
>
> [W]hen the moon departs, this congress of the waters, or army on the march toward the torrid zone, now abandoned by the traction that had called it forth, is dissolved.[27]

However, this would account for just one tide per day and Galileo rejected it.

> There are many who refer the tides to the moon, saying that this has particular dominion over the waters;... the moon, wandering through the sky, attracts and draws up toward itself a heap of water which goes along following it, so that the high sea is always in that part which lies under the moon. And since when the moon is below the horizon, this rising nevertheless returns, he tells us that he can say nothing to account for this....
>
> ...
>
> Please... spare us the rest [of this explanation]; I do not think there is any profit in spending the time to recount [it].[28]

In Galileo's opinion, what we today refer to as centrifugal force was the primary cause of the tides.

> Having established, then, that it is impossible to explain the movements perceived in the waters and at the same time maintain the immovability of the vessel which contains them, let us pass on to considering whether the mobility of the container could produce the required effect in the way in which it is observed to take place.[29]
>
> ...
>
> We have already said that there are two motions attributed to the terrestrial globe; the first is annual, made by its center along the circumference of its orbit about the ecliptic in the order of the signs of the zodiac (that is, from west to east), and the other is made by the globe itself revolving around its own center in twenty-four hours (likewise from west to east) around an axis which is somewhat tilted, and not parallel to that of its annual revolution.[30]

PART III The Newtonian universe

Here Galileo considers the motions of revolution of the earth about the sun and of the rotation of the earth about its own axis. During a twenty-four hour period, at any point on the surface of the earth (say, at the equator) these motions will once combine to yield a maximum value for the speed and once oppose each other to produce a minimum in the speed (relative to the fixed background of space). However, even if this were the correct mechanism, it would produce only one high- and one low tide per day:

> Now ... I shall resolve the question why, since there resides in the primary principle no cause of moving the waters except from one twelve-hour period to another (that is, once by the maximum speed of motion and once by its maximum slowness), the period of ebbing and flowing nevertheless commonly appears to be from one six-hour period to another.[31]

Galileo then attempted to account for the other tide in terms of the natural period of oscillation of large bodies of water contained in the sea basins.[32] This was qualitative in nature and did not really work.

Newton was able to use his law of gravitation to explain the phenomenon of the ocean tides caused by the moon.[33] He provided a correct understanding of the tides by considering the motion of a ring of water contained in a channel cut around the equator of the earth. This he did by first treating the perturbations caused by an orbiting body m on the motion of a small mass m' traveling around a large mass M. He then extended this to many small bodies in a common orbit and finally to a continuous fluid.[34] While Newton's theory of the tides did account for the major observed effects, it was essentially a 'static' theory because it did not take into consideration the dynamics of the water itself. As a result, effects such as the time lag between the passage of the moon over a given location and the actual appearance of the tide there remained a problem for his theory. A dynamical theory of the tides was due to Laplace. A complete quantitative tidal theory is extremely complex and remains an area of active research to this day. Rather than pursue the discussion of the tides as found in the *Principia*, we give a more intuitive description of the tides – but one still based directly on Newton's laws as found in the *Principia*.

9.5 THE EARTH–MOON SYSTEM AND TIDAL BULGES

We can understand in an elementary and semiquantitative fashion the origin of the forces that produce the tides by considering the acceleration a body would experience at various locations just above the surface of the earth. Two spheres attract each other with forces that act along the common line joining their centers. Therefore, two spheres, each traveling on circles, will orbit about a fixed point on this line. This is particularly easy to see in the case of two equal masses when this fixed center of their circular orbit would be exactly half way between the centers of the spheres. As one mass becomes larger, the center of rotation shifts toward the larger mass and the radius of its circular orbit decreases. Each mass travels on its own circle and both circles have their centers at the same fixed point. Figure 9.7

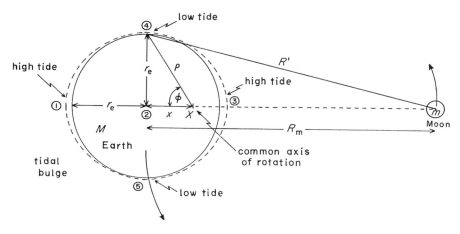

FIGURE 9.7 Tides in the earth–moon system

shows the earth–moon system orbiting about a common origin (X). (Appreciate that the north–south axis between the earth's poles is perpendicular to the plane of the page and passes through point 2 of Figure 9.7. The earth also executes its daily rotation about this NS axis.) The axis of rotation is perpendicular to the plane of the page and passes through the point X. The centripetal force required to keep this system orbiting about its common origin is provided by the mutual gravitational attraction between M and m. Since this mutual force of attraction between these two bodies always acts along a line joining them and since this line passes through the common center of rotation (point X of Figure 9.7), the angular speed ω of both the earth and moon must be the same. The condition for stable circular orbits is

$$F_c = \frac{GMm}{R_m^2} = M\omega^2 x = m\omega^2 (R_m - x) \qquad (9.9)$$

Here x is the distance from the center of the earth to X and R_m is the distance from the center of the earth to the center of the moon. If we solve this equation for x for the earth–moon system, we find $x = 0.726\, r_e$.[35] This common center of rotation is inside the earth (but not, of course, at its geometrical center). It is a little less than 3,000 miles [4.8×10^3 km] from the center of the earth to this point.

Let us now find the weight of a small mass m_0 placed above the earth at position 1 of Figure 9.7. The three forces indicated in Figure 9.8 must provide the required centripetal force directed toward X

$$F_c = m_0 g + \frac{Gmm_0}{(R_m + r_e)^2} - N_1 = m_0(x + r_e)\omega^2 \qquad (9.10)$$

Here N_1 is the reaction of a scale or balance pan, say, on m_0. N_1 would be the weight of m_0 as registered at position 1. We can solve this equation for N_1, use the fact that $(r_e/R_m) \cong 1/60$ to expand the inverse-square term to first order in (r_e/R_m), and finally employ Eq. (9.9) to obtain

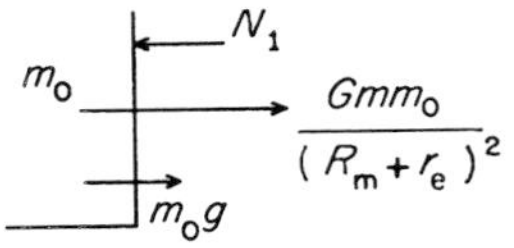

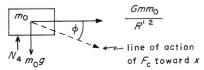

FIGURE 9.8 Collinear forces acting on a body at the earth's surface

FIGURE 9.9 Noncollinear forces acting on a body at the earth's surface

$$N_1 \cong m_0\left(g - \omega^2 r_e - \frac{2\,Gm}{R_m^3}r_e\right) \qquad (9.11)$$

A similar calculation at position 3 in Figure 9.7 yields the same result (to order (r_e/R_m)) so that $N_1 \cong N_3$. At point 4 in Figure 9.7 the gravitational attraction due to the moon (acting along R') will be practically parallel to R_m since $R' \cong R_m$. In Figure 9.9 we show the forces acting at m_0 there. Since the centripetal force F_c must point directly at X, we deduce from the geometry of Figures 9.7 and 9.9 that[36]

$$N_4 \cong m_0(g - \omega^2 r_e). \qquad (9.12)$$

We also find, as we would expect from the symmetry of points 4 and 5 about point 2, that $N_4 \cong N_5$.

What do we conclude from all of this? The apparent weights of an object at points 1 and 3 of Figure 9.7 are equal, but each less than the apparent weight of that same object at 4 or 5. If we think of a solid spherical earth covered with a layer of water, then a kilogram of water at 4 (or at 5) would weigh more than a kilogram of water at 1 (or at 3). This means that the water pressure (essentially the force that a fluid exerts on its surroundings) will be greater at 4 than at 1. Water will flow from regions 4 and 5 (where the pressure is higher) to regions 1 and 3 until a high enough level of water accumulates to produce pressure equilibrium. So, we can understand that tidal bulges (of nearly equal heights) will be produced on that region of the earth directly under the moon and on the opposite side of the earth. If the earth were not spinning about its own north–south axis (that is perpendicular to the plane of the page in Figure 9.7), but simply revolving about the common center X, then these two tidal bulges would remain fixed over the same places on the earth at all times. In fact, though, the earth spins through these bulges or, equivalently, these bulges move relative to the surface of the earth. When one of these bulges comes to a shore line, the water from the bulge is spread out on the land as a high tide. In a sense, the water 'hangs' in space while the earth rotates 'through' it.

In Figure 9.7 we drew the earth–moon configuration as though the plane of the moon's orbit were in the plane of the equator of the earth. Actually, the orbit of the moon is inclined at nearly 30° to the equator. Therefore, the situation is much more

Some old questions revisited

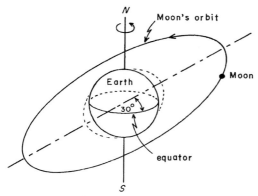

FIGURE 9.10 **Inclination of the moon's orbit**

like that depicted in Figure 9.10. Since the tidal bulge on either side of the earth remains nearly directly 'under' the moon, the two high tidal bulges that any fixed point on the surface of the earth experiences within a twenty-four hour period will be of unequal heights. The exception occurs when the moon is over the equator. The equatorial tides are then produced.[37]

9.A NEWTON AND YOUNG ON WAVE INTERFERENCE

In his *Principia* Newton discussed the effects of tides in certain bays in terms of interference effects.

> Further, it may happen that the tide may be propagated from the ocean through different channels towards the same port, and may pass quicker through some channels than through others; in which case the same tide, divided into two or more succeeding one another, may compound new motions of different kinds. Let us suppose two equal tides flowing towards the same port from different places, one preceeding the other by six hours;... [E]very six hours alternately there would arise equal floods, which, meeting with as many equal ebbs, would so balance each other that for that day the water would stagnate and be quiet. If the moon then declined from the equator [recall Figure 9.10 and the discussion following it], the tides in the ocean would be alternately greater and less;... and... two greater and two less tides would be alternately propagated towards that port. But the two greater floods would make the greatest height of the waters to fall out in the middle time between both; and the greater and less floods would make the waters to rise to a mean height in the middle time between them, and in the middle time between the two less floods the waters would rise to their least height. Thus in the space of twenty-four hours the waters would come, not twice, as commonly, but once only to their greatest, and once only to their least height;... An example of this Dr. *Halley* has given us, from the observations of seamen in the port of *Batshaw*, in the kingdom of *Tunquin* [Gulf of Tonkin], in the latitude of 20°50′ north. In that port... [the waters] flow and ebb, not twice, as in other ports, but once only every day.... There are two inlets to this port and the neighboring channels, one from the sea of *China*, between the continent and the island of *Leuconia* [Philippines],[38] the other from the *Indian* sea, between the continent and the island of *Borneo* [Figure 9.11]. But whether there be really two tides propagated through the said channels, one from the *Indian* sea in the space of twelve hours, and one from the sea of *China* in the space of six hours, which therefore happening at the third and

129

PART III The Newtonian universe

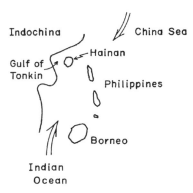

FIGURE 9.11 **Tides in the Gulf of Tonkin**

ninth lunar hours, by being compounded together, produce those motions; or whether there by any other circumstances in the state of those seas, I leave to be determined by observations on the neighboring shores.[39]

In his less mathematical *The System of the World* Newton gave essentially the same discussion of the phenomenon except that the last two sentences of the quotation above were replaced by the following.

> There are two inlets from the ocean to this port: one more direct and short between the island *Hainan* and the coast of *Quantung*, a province of *China*; the other round about between the same island and the coast of *Cochim*; and through the shorter passage the tide is propagated more rapidly to *Batshaw*.[40]

The first explanation by Newton seems to indicate the effect is due to two tides coming from different oceans (the China Sea and the Indian Ocean) and the second description attributes the effect to a tide that must go around the island of Hainan, once by a direct route and once by a longer route, in order to reach the Gulf of Tonkin. In either case, though, Newton is discussing constructive and destructive interference produced when two different waves meet.

In order to understand why it is that these two tides that are six hours apart do not always completely cancel each other to produce no tidal effect, it is necessary to see that the high- and low tides for a single tide are not of equal amplitude (refer to Figure 9.12). Except when the moon is directly over the equator, the single-tide profile will be like that indicated in Figure 9.12. If two of these patterns six hours apart are superimposed as shown in Figure 9.13, the combined tide will have one high point and one low point per 24 hour period.

In 1802, Thomas Young (1773–1829) discovered that certain interference effects of light could be explained by an analogy with water waves. We have just seen that this interference effect of water waves was used by Newton in his *Principia* to explain why some bays experienced only one tide, not two, in a 24-hour period. Young himself used an analogy that was quite similar to Newton's original idea.

> Suppose a number of equal waves of water to move upon the surface of a stagnant lake, with a certain constant velocity, and to enter a narrow channel leading out of the lake. Suppose then another similar cause to have excited another equal series of waves, which arrive at the same chan-

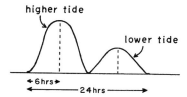

FIGURE 9.12 **Tidal profiles for the two channels**

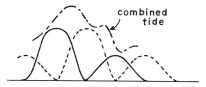

FIGURE 9.13 **Superposition of tidal profiles**

nel, with the same velocity, and at the same time with the first. Neither series of waves will destroy the other, but their effects will be combined: if they enter the channel in such a manner that the elevations of one series coincide with those of the other, they must together produce a series of greater joint elevations; but if the elevations of one series are so situated as to correspond to the depressions of the other, they must exactly fill up those depressions, and the surface of the water must remain smooth.

Now I maintain that similar effects take place whenever two portions of light are thus mixed; and this I call the general law of the interference of light.

. . .

[W]henever two portions of the same light arrive to the eye by different routes, either exactly or very nearly in the same direction, the light becomes most intense when the difference of the routes is any multiple of a certain length, and least intense in the intermediate state of the interfering portions; and this length is different for light of different colours.[41]

FURTHER READING

James Cushing's 'Kepler's Laws and Universal Gravitation in Newton's *Principia*' presents in elementary modern notation Newton's geometrical arguments for Kepler's laws (in the style of Sections 9.1 and 9.2 of this chapter), while J. Bruce Brackenridge's *The Key to Newton's Dynamics* and Dana Densmore's *Newton's Principia* give detailed expositions and analyses of these propositions in the *Principia*. Popular accounts of various aspects of the fascinating topic of the tides and of their history can be found in Robert Ball's *Time and Tide*, in George Darwin's, *The Tides and Kindred Phenomena in the Solar System*, in Albert Defant's *Ebb and Flow, and Water*, in Peter Goldreich's 'Tides and the Earth–Moon System', in Paul Melchior's *The Earth Tides*, and in Francis Wylie's *Tides and the Pull of the Moon*.

PART IV

A perspective

That Man is the product of causes which had no prevision of the end they were achieving; that his origin, his growth, his hopes and fears, his loves and his beliefs, are but the outcome of accidental collocations of atoms; that no fire, no heroism, no intensity of thought and feeling, can preserve an individual life beyond the grave; that all the labours of the ages, all the devotion, all the inspiration, all the noonday brightness of human genius, are destined to extinction in the vast death of the solar system, and that the whole temple of Man's achievement must inevitably be buried beneath the debris of a universe in ruins – all these things, if not quite beyond dispute, are yet so nearly certain, that no philosophy which rejects them can hope to stand. Only within the scaffolding of these truths, only on the firm foundation of unyielding despair, can the soul's habitation henceforth be safely built.

How, in such an alien and inhuman world, can so powerless a creature as Man preserve his aspirations untarnished? A strange mystery it is that Nature, omnipotent but blind, in the revolutions of her secular hurryings through the abysses of space, has brought forth at last a child, subject still to her power, but gifted with sight, with knowledge of good and evil, with the capacity of judging all the works of his unthinking Mother. In spite of Death, the mark and seal of the parental control, Man is yet free, during his brief years, to examine, to criticise, to know, and in imagination to create. To him alone, in the world with which he is acquainted, this freedom belongs; and in this lies his superiority to the resistless forces that control his outward life.

Bertrand Russell, *A Free Man's Worship*

10

Galileo's *Letter to the Grand Duchess*

In this chapter we study one of the dramatic clashes that took place during the overthrow of the Aristotelian world view that had given man a unified picture of himself and of his surroundings. In the Galileo episode this became hopelessly entangled with the issue of intellectual freedom versus institutional authority. However, there were other relevant factors as well: the personalities of Galileo and of those who opposed him and, possibly, the social structure of the patronage system of support for artists and scientists in those times.

10.1 THE BACKGROUND

Perhaps it will help the reader if we begin by listing several key events that preceded the direct confrontation between Galileo and the Church authorities.[1]

1. March 1610 – Galileo published the *Sidereus Nuncius* (*The Starry Messenger*) in which he discussed his telescopic observations and discoveries. He showed that there were many more stars in the heavens than had been thought previously, that the surface of the moon was rough, containing mountains, plains and valleys, much like the surface of the earth itself, and that Jupiter had a set of moons orbiting the central planet, very much as the planets orbit the sun in the Copernican system. This work had great popular appeal and wide impact.
2. December 1613 – Galileo wrote a letter to one of his scientific pupils, the Benedictine monk Benedetto Castelli (1578–1643), concerning his own views on the relation between science and religion.
3. December 1614 – Father Thomas Caccini (1574–1648), a Dominican friar, publicly attacked Galileo for his views on the Copernican system not being antithetical to the teachings of the Bible.
4. February 1615 – A copy of Galileo's letter to Castelli was sent to the Holy Office in Rome, along with a document denouncing the views of Galileo's supporters. Later the same year Galileo completed his *Letter to the Grand Duchess*. While this was not published then, it was eventually widely circulated.
5. December 1615 – Galileo went to Rome to defend his views on the Copernican system.
6. February 1616 – In an audience with the Jesuit Robert Cardinal Bellarmine

(1542–1621), Galileo was officially forbidden to hold the Copernican theory that was also condemned at that time.

7 March 1616 – The Congregation of the Index prohibited publication of Copernicus' *De Revolutionibus*, pending suitable corrections.

8 October 1623 – Galileo published *Il Saggiatore* (*The Assayer*) in which he responded to Jesuit astronomers who were critical of his views on the origin of comets. Although Galileo's hypothesis on comets would later be proven incorrect, this work presented some of Galileo's most important ideas on the philosophy of science.

9 February 1632 – His *Dialogue Concerning the Two Chief World Systems* was published. In it Galileo presented both the Ptolemaic and Copernican models of the universe. Although he never explicitly supported the Copernican theory, the arguments given are such that any intelligent reader could readily see that Galileo took it to be the correct theory.

10 August 1632 – The Holy Office ordered sales of the *Dialogue* suspended.

11 October 1632 – Galileo was summoned to Rome to stand trial.

12 June 1633 – Galileo was convicted, read his statement of abjuration and was sentenced to indefinite imprisonment in his villa.

Symptoms of the trouble to come can already be seen, at least in retrospect, in 1611 or 1612 when certain defenders of the Aristotelian world view, alarmed by the success of Galileo's teachings, began to publish attacks against him. These were not so much concrete refutations of his work as they were largely personal attacks on him, claiming that his conclusions contradicted (in their opinion) long-held views on Aristotle and on the Bible. Some of these enemies were educated and influential people whose work Galileo felt was not of sufficient merit to warrant a direct response from him, but rather from one of his own students to demean those men even more. Galileo typically responded sharply and without mercy to critics who were educated enough to know better than to defend such positions. His sharp tongue and venomous pen gained him fame, but also often exacerbated an already bad situation.

The *Letter* referred to in the title of this chapter was occasioned by an exchange that took place in December of 1613, at a meal presided over by the Grand Duke Cosimo II in Pisa, where the Tuscan Court was spending the winter. In attendance was Benedetto Castelli, a cleric and brilliant pupil of Galileo's. There Castelli, newly appointed to the chair of mathematics at the University of Pisa, discussed the satellites of Jupiter and Galileo's views on the correctness of the Copernican theory of the universe as opposed to the Ptolemaic one. One of the other guests was an ardent Aristotelian, a professor of philosophy at the University of Pisa. This academic argued that Galileo's teachings about the motion of the earth contradicted Holy Scripture. The Dowager Grand Duchess, Christina of Lorraine (1565–1636), was also present and she appeared quite upset at this possibility, even though Castelli defended Galileo's views well. A few days after that, Castelli wrote to Galileo about this encounter. Later in December of the same year, presumably in part to allay the fears of the Grand Duchess, Galileo responded to Castelli with a long letter on the question of scientific inquiry versus religious belief.

> [I] think it would be the better part of wisdom not to allow anyone to apply passages of Scripture in such a way as to force them to support as true conclusions concerning nature, the contrary of which may afterward be revealed by the evidence of our senses or by necessary demonstration. Who will set bounds to man's understanding? Who can assure us that everything that can be known in the world is known already?[2]

Galileo subsequently elaborated on this theme and eventually wrote the final version of his *Letter to the Grand Duchess Christina* in 1615. In December of 1614, Father Caccini denounced Galileo and the Copernican system as heretical and dangerous to the state. Since this had been done in public, Galileo now felt that the charges had to be answered. Against the advice of friends, Galileo now decided to fight his critics.

It was in 1616 that Galileo's opponents took official actions that would ultimately prove of pivotal importance for his condemnation by the Inquisition in 1633. Among these was a decree issued by the Congregation of the Index suspending temporarily publication of Copernicus' *De Revolutionibus* and also stating that any teaching about the earth's motion was false, contrary to the Bible and a threat to Catholicism. Ostensibly the battle concerned the Copernican system that Galileo adopted but, as Galileo appreciated, a central issue was who had the authority to define the criteria for deciding the meaning of Scripture and, broadly, intellectual freedom versus authority (such as the freedom of scientific inquiry).

Let us now look at some of the details of a few key encounters that took place leading up to the famous trial.

10.2 A BASIC ISSUE

In April of 1615, Bellarmine wrote a letter in which he set forth his own views on the Copernican system and on Galileo's teachings. The letter was addressed to a Carmelite priest Paolo Foscarini (1580–1616), who had written a book defending the Copernican system against charges of conflict with Scripture.

> I say that it appears to me that Your Reverence and Sig. Galileo did prudently to content yourselves with speaking hypothetically and not positively, as I have always believed Copernicus did. For to say that assuming the earth moves and the sun stands still saves all the appearances better than eccentrics and epicycles is to speak well. This has no danger in it, and it suffices for mathematicians. But to wish to affirm that the sun is really fixed in the center of the heavens and merely turns upon itself without traveling from east to west, and that the earth is situated in the third sphere and revolves very swiftly around the sun, is a very dangerous thing, not only by irritating all the theologians and scholastic philosophers, but also by injuring our holy faith and making sacred Scripture false.
>
> . . .
>
> I say that if there were a true demonstration that the sun was in the center of the universe and the earth in the third sphere, and that the sun did not go around the earth but the earth went

around the sun, then it would be necessary to use careful consideration in explaining the Scriptures that seemed contrary, and we should rather have to say that we do not understand them than to say that something is false which had been proven.

. . .

And if you tell me that Solomon spoke according to the appearances, and that it seems to us that the Sun goes round when the earth turns, as it seems to one aboard ship that the beach moves away, I shall answer thus. Anyone who departs from the beach moves away, yet knows that this is an error and corrects it, seeing clearly that the ship moves and not the beach; but as to the sun and earth, no sage has needed to correct the error, since he clearly experiences that the earth stands still and that his eye is not deceived when it judges the sun to move, just as he is likewise not deceived when it judges that the moon and the stars move. And that is enough for the present.[3]

In May of the same year, Galileo drafted a letter to Bellarmine.

To me the surest and swiftest way to prove that the position of Copernicus is not contrary to Scripture would be to give a host of proofs that it is true and that the contrary cannot be maintained at all; thus, since no two truths can contradict one another, this and the Bible must be perfectly harmonious. But how can I do this, and not be merely wasting my time, when those Peripatetics who must be convinced show themselves incapable of following even the simplest and easiest of arguments, while on the other hand they are seen to set great store in worthless propositions?

. . .

I should not like to have great men think that I endorse the position of Copernicus only as an astronomical hypothesis which is not really true. Taking me as one of those most addicted to his doctrine, they would believe all its other followers must agree, and that it is more likely erroneous than physically true. That, if I am not mistaken, would be an error.[4]

A major issue is at stake here, one that had a long history of development. Bellarmine in his letter says that the Copernican model 'saves all the appearances better than' the Ptolemaic one and Galileo says that he does not want men to think that he endorses 'the position of Copernicus only as an astronomical hypothesis that is not really true.' There is a basic tension between the formalistic and the realistic approaches to a description of the physical world. Do the concepts in a theory represent entities actually existing in physical reality or are they merely mathematical constructs whose only function is to allow us to give a concise description and to provide us a means of making calculations? In Section 4.7 we saw that one representation of Plato's conception of the proper goal of astronomy was that circular motions were to be used to save the appearances presented by the movement of the planets through the heavens. Once astronomy found a suitable geometrical construction, its work was done. It was not appropriate to ask whether or not the circles and epicycles really were the orbits of the heavenly bodies.

Ptolemy and many astronomers until Copernicus worked within this Platonic tradition. In *The Almagest*, after he outlined his system of cycles and epicycles, Ptolemy stated:

Let no one, seeing the difficulty of our devices, find troublesome such hypotheses. For it is not proper to apply human things to divine things nor to get beliefs concerning such great things from such dissimilar examples. For what is more unlike than those which are always alike with respect to those which never are, and those which are impeded by anything with those which are not even impeded by themselves? But it is proper to try and fit as far as possible the simpler hypotheses to the movements in the heavens; and if this does not succeed, then any hypotheses possible. Once all the appearances are saved by the consequences of the hypotheses, why should it seem strange that such complications can come about in the movements of heavenly things?[5]

Ptolemy makes two central points here. First, in the last two sentences of this quotation, we find that astronomy must use the simplest set of hypotheses to save the appearances. As required by observation, the model can be made more and more complex, but only as far as is necessary to fit the available data. Second, and very important for the intellectual climate that prevailed at the time, he tells us that we ought not be concerned with the fact that the resulting model appears very complicated, since we are attempting to use human intelligence and laws discovered on a corruptible earth to explain something in another realm, namely the changing motions of the heavenly bodies. This was a widely accepted belief, even up through the Middle Ages: one set of laws governs the heavens and another set the earth. It would be foolish to expect to understand the former in terms of the latter.

One of the first thinkers to hold that the same physical laws are operative in the heavens as on the earth was Nicholas of Cusa (1401–1464), a scholar and churchman who denied strict circular motion for the planets. The realist school has origins going as far back as Aristotle and its later evolution can be traced through the writings of Simplicius, Cusa, Copernicus, Brahe, Kepler and Galileo. They believed that not only must the models or hypotheses employed explain the observations, but they must also be in accord with generally accepted philosophical principles and correspond to reality. In an attempt to reconcile astronomical observations with the Aristotelian and (apparently) Scriptural doctrine of an earth at rest, Tycho Brahe formulated his own model of the universe. It was a compromise between the Ptolemaic and Copernican ones. The moon and the sun still circle a stationary earth at the center of the celestial sphere, but the five remaining planetary orbits are centered about the sun. It can be seen as an attempt to harmonize a realistic commitment to a model with an acceptance of a literal interpretation of Scripture.

In contrast to this realist view, many Christian astronomers of medieval Europe required only that the hypotheses used should be as simple as possible and save the appearances as well as possible. This attitude of indifference regarding the reality of such hypotheses continued in some quarters even after Copernicus' *De Revolutionibus* made its impact upon astronomers. Luther was among the first to attack the Copernican model in the name of Holy Scripture. Astronomy must hereafter conform to the views of philosophy and theology. The Jesuit astronomer Christopher Clavius (1538–1612), an influential scientist at the Collegio Romano in Rome, held that, to be probable, astronomical hypotheses must not only save the appearances, but must also conform to the laws of physics and not contradict Church teaching. So, without a viable alternative, one might accept the Copernican model as highly probable – except that it appeared to conflict with Scripture. Clavius opposed the Copernican system and remained a supporter of the Ptolemaic one.

So, in addition to disagreeing on the proper method to resolve apparent conflicts

PART IV **A perspective**

between Scripture and science, Bellarmine and Galileo also differed on a fundamental philosophical issue: that of realism versus instrumentalism for the interpretation of astronomical theories. The stage was set for battle.

10.3 THE *LETTER TO THE GRAND DUCHESS*

In relation to the problem of interpreting the Bible, Galileo argued that the Scriptures used common-sense language to treat matters of faith and morals, and that the Bible was not a scientific treatise. He also believed that an important precedent was being set for the relation between religion and the sciences of observation and experiment. His letter to the Grand Duchess appeared in 1636, years after it had been written (1615) and it was then suppressed in Catholic countries. In full, the letter runs to over forty pages. Several excerpts from it are given at the end of this chapter. Here we simply summarize some of the major points Galileo makes in the letter.

1. Academic philosophers (referring to Aristotelians of his day) value opinions more than truth. Galileo's work refuted the arguments of Aristotle and of Ptolemy. His persecutors did not really understand his arguments. The main thrust of their objection was their belief that science was against the Bible, rather than demonstrably wrong.
2. Science makes no claim about religion. The sun does seem to move and the earth to stand still. This is the language of the common people for whom the Bible was intended. There exists a question of interpretation.
3. Experience and the reason God gave us ought to be used to decide the meaning of Scripture, rather than a blind acceptance of authority. In a later work (1623) titled *The Assayer*, Galileo very tellingly made a case for the superiority of reason and observation over mere belief in accounting for matters of fact.

   > If Sarsi [a pen name of one of Galileo's critics] wants me to believe with Suidas [a tenth-century Greek lexicographer] that the Babylonians cooked their eggs by whirling them in slings, I shall do so; but I must say that the cause of this effect was very different from what he suggests. To discover the true cause I reason as follows: 'If we do not achieve an effect which others formerly achieved, then it must be that in our operations we lack something that produced their success. And if there is just one single thing we lack, then that alone can be the true cause. Now we do not lack eggs, nor slings, nor sturdy fellows to whirl them; yet our eggs do not cook, but merely cool down faster if they happen to be hot. And since nothing is lacking to us except being Babylonians, then being Babylonians is the cause of the hardening of eggs, not friction of the air.'[6]

4. The purpose of the Bible is not to teach science. Galileo again stressed the unity of truth.
5. He was concerned with the danger of allowing dogma to silence intellectual inquiry.

6 Finally, he urged scholars to seek the sense of the Bible with the aid of the sciences.

Galileo's basic position is that one must strike a proper balance between simply accepting a literal interpretation and taking sufficient cognizance of the historical context of the language used in the Bible. This is similar to the stance adopted today by many Biblical scholars. He effectively, even if sarcastically, makes this case in the *Letter* with his discussion of the famous biblical passage from the Book of Joshua (10:10–15) on which much of the controversy focused. There Joshua commanded the sun to stand still so that the Israelites could slay the Amorites and hence save the town of Gibeon (northwest of Jerusalem in ancient Palestine).

> Yahweh drove them [the Amorites] headlong before Israel, defeating them completely at Gibeon.... And as they fled from Israel..., Yahweh hurled huge hailstones from heaven on them.... Then Joshua spoke to Yahweh, the same day that Yahweh delivered the Amorites to the Israelites. Joshua declaimed:
>
> 'Sun, stand still over Gibeon,
> and, moon, you also, over the Vale of Aijalon'.
> And the sun stood still, and the moon halted,
> till the people had vengeance on their enemies.
>
> Is this not written in the Book of the Just? The sun stood still in the middle of the sky and delayed its setting for almost a whole day.[7]

In the *Letter*, Galileo, employing his best rhetorical form, attempted to use the position of his adversaries (that one should take Scriptures in their literal meaning) to argue against the Ptolemaic model and for the Copernican one.[8] He claimed that (according to the Ptolemaic model) the day could not be greatly lengthened by the sun's standing still. That is, since the motion of the sun along the ecliptic is in the opposite direction to the rotation of the sphere of the fixed stars (recall Figure 4.1), the sun's stopping its own proper motion (relative to the rotating celestial sphere) would simply shorten the day (not lengthen it). The way to lengthen the day, he argued, would be to make the sun speed up its own proper motion until that motion was equal to that of the celestial sphere. But that would certainly be at variance with a literal reading of Joshua. If one attempts to save the reading by saying – in accord with 'and the moon halted' – that the entire system of celestial rotations ceased (that is, the celestial sphere and all of the planets that it drove), then he must still depart from a literal interpretation of Scripture. Galileo then went on to argue, not wholly convincingly, that the Copernican model (as interpreted by him) is actually more nearly consistent with a literal interpretation of Joshua's command. His basic picture of the 'Copernican' (actually, very much a Keplerian – recall the *anima motrix* of Section 5.4) universe was that the sun, 'as the chief minister of Nature and in a certain sense the heart and soul of the universe, infuses by its own rotation not only light but also motion into other bodies which surround it.'[9] Hence, if the rotation of the sun (about its own axis) were to cease, then, Galileo tells us, so would the motion of the planets and that would account for Joshua's description of events as found in Scripture. (Of course, today we do not accept the rotation of the sun as the cause of the motions of the planets.) Galileo presented his opponents with a dilemma: either accept a literal interpretation of the Bible and along with it the ('heretical') Copernican model or give up a literal interpretation to save the ('orthodox') Ptolemaic model. Galileo's real aim here was to move the debate away from a literal interpretation. We have met this ploy of Galileo's before (toward the end of Section 6.5) in which he accepts the position of an opponent (there, Aristotle), only to draw a conclusion that (apparently, at least) refutes the position itself.

In 1616, after an interview conducted by Bellarmine, Galileo was admonished not to hold or defend the Copernican theory that was said to be contrary to the Bible. This instruction from Bellarmine did not forbid teaching (in the sense of relating hypothetically) the theory. However, an unsigned, unwitnessed document was entered into the files of the Inquisition in which Galileo was 'commanded and enjoined, in the name of His Holiness the Pope and the whole Congregation of the Holy Office, to relinquish altogether the said opinion that the Sun is the center of the world and immovable and that the Earth moves; nor further to hold, teach, or defend it in any way whatsoever, verbally or in writing; otherwise proceedings would be taken against him by the Holy Office; which injection they said Galileo acquiesced in and promised to obey.'[10] Galileo appeared unaware of the injunction in this note that was not made public at the time. Following his initial audience with Bellarmine came many years of (voluntary) seclusion in which Galileo wrote some of his most famous works. In 1624 *The Assayer*, that was a vitriolic attack on his enemies, many of whom were Jesuits, was published and in 1632 the *Dialogue Concerning the Two Chief World Systems*. The *Dialogue* made some pretense at being technically neutral in not explicitly championing the Copernican theory and in not stating outright that the earth could not be at rest. Nevertheless, the arguments and evidence presented were so overwhelmingly in favor of the Copernican model that the Ptolemaic model was effectively demolished. However, it was not this support of the heliocentric theory alone that resulted in Galileo's being brought before the Inquisition in 1632. The personality of Pope Urban VIII played an important part as well.

10.4 GALILEO AND URBAN VIII

After the great success of his early telescopic observations, Galileo traveled to Rome in the spring of 1611 to discuss his work with the Jesuit astronomers, especially Clavius, at the Roman College. The visit was a triumphant one. He was fêted by the Jesuits and also made a member of the Lincean Academy, a forerunner of modern scientific societies. This Academy would play a significant role in the publication of some of Galileo's future works. Later in 1611, at a dinner given in Florence by the Grand Duke, Galileo met Cardinal Maffeo Barberini (1568–1644). This influential ecclesiastical official belonged to a wealthy Florentine family and took a lively interest in the arts and in science. At this time Barberini showed himself partial to Galileo's work and remained friendly for many years.

Given his 1616 interview with Bellarmine, the resulting decree prohibiting work on the Copernican system and his subsequent disputes with the powerful Jesuits (most especially Father Horatio Grassi (1590–1654) against whom *The Assayer* was directed), Galileo was heartened to see his admirer and patron, Maffeo Barberini, become Pope Urban VIII in August of 1623. *The Assayer* was dedicated to the new Pope and published under the auspices of the Lincean Academy. In the spring of 1624 Galileo journeyed to Rome to see the Pontiff and he was well received there. Without written records of these exchanges between the two men, we can never know with certainty just what was said. However, there are references to these exchanges that allow some inferences to be made.[11] They do seem to have discussed the relation between scientific theories (such as Copernicus') and matters of belief as contained in revealed writings. In these audiences with the Pope, Galileo could reasonably have reminded him that many years previously he, Barberini, agreed that arguments based on the ocean tides provided evidence for the motion of the earth and hence for the Copernican theory. (Today we know that Galileo's explanation of the tides was incorrect, although this fact is of no consequence here.) A likely reply (one that he was known to favor) to that on Urban's part would be that we ought not rule out the possibility of God accomplishing the observed effect (here, the tides) by some other means. In any event, the Pope did grant Galileo permission to discuss the rival Ptolemaic and Copernican theories, but only 'hypothetically' and without taking sides on a choice between them. While this would lead to Galileo's writing the *Dialogue*, it must nevertheless have become evident to Galileo that Urban felt a greater responsibility to preserving the authority of Scripture than to advancing knowledge by scientific enquiry.

When Galileo wrote the *Dialogue*, he concluded it by having Simplicio (Galileo's caricature of Simplicius, the sixth-century commentator on Aristotle's work) summarize the argument based on the tides as:

> As to the discourses we have held, and especially this last one concerning the reasons for the ebbing and flowing of the ocean, I am really not entirely convinced; but from such feeble ideas of the matter as I have formed, I admit that your thoughts seem to me more ingenious than many others I have heard. I do not therefore consider them true and conclusive; indeed, keeping always before my mind's eye a most solid doctrine that I once heard from a most eminent and learned person, and before which one must fall silent, I know that if asked whether God in His infinite power and wisdom could have conferred upon the watery element its observed reciprocating motion using some other means than moving its containing vessels, both of you would reply that He could have, and that He would have known how to do this in many ways which are unthinkable to our minds. From this I forthwith conclude that, this being so, it would be excessive boldness for anyone to limit and restrict the Divine power and wisdom to some particular fancy of his own.

To which Salviati, Galileo's spokesman, replies:

> An admirable and angelic doctrine, and well in accord with another one, also Divine, which, while it grants to us the right to argue about the constitution of the universe (perhaps in order that the working of the human mind shall not be curtailed or made lazy) adds that we cannot discover the work of His hands. Let us, then, exercise these activities permitted to us and ordained by God, that we may recognize and thereby so much the more admire His greatness, however much less fit we may find ourselves to penetrate the profound depths of His infinite wisdom.[12]

PART IV A perspective

Some have suggested that a proud Urban came to believe (possibly in part due to goading by Galileo's enemies) that Galileo was ridiculing him in the person of Simplicio.[13] Or, it could have been that the Pope was enraged at what he saw as Galileo's betrayal of the trust implied by the permission granted to publish the *Dialogue* – given that this work did effectively make a compelling case for the Copernican model. This appeared to challenge the authority of Scripture. Whatever the actual factors were, Urban was prompted to appoint his nephew, Cardinal Francesco Barberini (1597–1679), as chairman of a commission to assess the charges against Galileo. As a result of this committee's report, the matter was formally brought before the Inquisition in 1632. In June of 1633, Galileo was convicted by the Inquisition and made to recant. After sentence was passed, Galileo had to recite a humiliating formula of abjuration.

So far we have presented this conflict mainly in terms of intellectual and religious ones: freedom of inquiry versus institutional authority and realism versus instrumentalism (as related to a literal interpretation of Scripture). We now mention the more controversial thesis that another important dimension to this episode was the patronage by which the wealthy and powerful supported the pioneers of science.[14] The universities of the seventeenth century were not generally receptive to new ideas that threatened the established view of the world and most scientists, like Galileo, were unable to support themselves in their work and had to seek a sponsor. In the patronage system, major discoveries, or high ground held, increased the prestige of the scientist, reflected honor on his patron and increased the chances of continued support by the patron (or, perhaps even better, support by another, more powerful, patron). On the other hand, failure or loss of face could mean loss of livelihood for the scientist. Several of Galileo's career moves can be best understood as motivated by a desire and need to obtain or improve his position in this system. One could not coast for long on past accomplishments, but had continually to reassert dominance. The fame he attained made him one of the most sought-after of clients. His career ambitions crossed those of another influential body of the day – the Jesuits. (Recall that Bellarmine, Clavius and Grassi were Jesuits.) Both were dependent upon patronage and, thus, the stakes were high when they came into conflict. In this light, some attempt to understand the alleged role of the Jesuits in the intrigues against Galileo as an effort to gain the intellectual high ground by discrediting a major competitor. Even if such facts about patronage do give additional insight into the events leading to the trial of Galileo, they can constitute just one of many threads in a complex story. Intrigues and patronage or not, if the *Dialogue* had not appeared, there would likely have been no trial.[15]

10.5 RELIGION *VIS-À-VIS* NATURAL PHILOSOPHY

So, in spite of all of the complexities of the trial of Galileo, it is perhaps still fair to say that a key issue remains institutional authority versus the freedom of the individual (at least as regards intellectual inquiry). This is, of course, a recurrent theme throughout modern history and is with us even to this day. In the Galileo case, this took on the guise of the Church versus Science (at least retrospectively), but that was an historical contingency. The dynamics peculiar to that time and place – the personalities of Bellarmine, of Urban and of Galileo himself, possibly also the system of patronage for scientists and artists – complicated the issue and came to play significant roles in the final outcome of this episode.

It is evident that there was not yet a separation of religious authority and science in Galileo's Italy. An English contemporary of Galileo's, Thomas Hobbes (1588–1679), argued for the rigorous separation of theology from philosophy. Hobbes' belief was that, while science would reveal more about God's creation and hence increase man's appreciation of God, organized religion was in fact a source of civil discord (as argued in his *Leviathan*, where he also defended the absolute sovereignty of kings). Hobbes' philosophy was influenced by Galileo's views on the laws of motion and he did meet Galileo in Italy around 1636. We mention Hobbes as one figure in the struggle to separate formal religious influences from science (or natural philosophy), but this is not meant to imply that such a transition was either sudden or complete. After all, Newton, as we saw previously, had strong (if somewhat unorthodox) religious beliefs and his understanding of God influenced his broader scientific views (even as expressed in the *Principia* and still more so in the Queries to the *Optics*).

Finally, even though the Galileo episode is often used to show the adverse effect religion (actually religious authority) can have on science, let us consider the following opinion of the philosopher Alfred Whitehead (1861–1947) on the role of medieval theology in producing a climate conducive to the rise of science.

> [I] have... [not] yet brought out the greatest contribution of medievalism to the formation of the scientific movement. I mean the inexpungable belief that every detailed occurrence can be correlated with its antecedents in a perfectly definite manner, exemplifying general principles. Without this belief the incredible labours of scientists would be without hope. It is this instinctive conviction, vividly poised before the imagination, which is the motive power of research: – that there is a secret, a secret which can be unveiled.
>
> . . .
>
> When we compare this tone of thought in Europe with the attitude of other civilizations when left to themselves, there seems but one source for its origin. It must come from the medieval insistence on the rationality of God, conceived as with the personal energy of Jehovah and with the rationality of a Greek philosopher. Every detail was supervised and ordered: the search into nature could only result in the vindication of the faith in rationality.... [I] mean... the impress on the European mind arising from the unquestioned faith of centuries.
>
> . . .
>
> In Asia, the conceptions of God were of a being who was either too arbitrary or too impersonal for such ideas to have much effect on instinctive habits of mind. Any definite occurrence might be due to that fiat of an irrational despot, or might issue from some impersonal, inscrutable origin of things. There was not the same confidence as in the intelligible rationality of a personal being. I am not arguing that the European trust in the scrutability of nature was logically justified even by its own theology. My only point is to understand how it arose. My explanation is that the faith in the possibility of science, generated antecedently to the development of modern scientific theory, is an unconscious derivative from medieval theology.[16]

PART IV   **A perspective**

10.A GALILEO'S *LETTER TO THE GRAND DUCHESS*

The following pages contain several excerpts from Galileo's long *Letter*.

Some years ago, as Your Serene Highness well knows, I discovered in the heavens many things that had not been seen before our own age. The novelty of these things, as well as some consequences which followed from them in contradiction to the physical notions commonly held among academic philosophers, stirred up against me no small number of professors – as if I had placed these things in the sky with my own hands in order to upset nature and overturn the sciences. They seemed to forget that the increase of known truths stimulates the investigation, establishment, and growth of the arts; not their diminution or destruction.

Showing a greater fondness for their own opinions than for truth, they sought to deny and disprove the new things which, if they had cared to look for themselves, their own senses would have demonstrated to them. To this end they hurled various charges and published numerous writings filled with vain arguments, and they made the grave mistake of sprinkling these with passages taken from places in the Bible which they had failed to understand properly, and which were ill suited to their purposes.

...

Possibly because they are disturbed by the known truth of other propositions of mine which differ from those commonly held, and therefore mistrusting their defense so long as they confine themselves to the field of philosophy, these men have resolved to fabricate a shield for their fallacies out of the mantle of pretended religion and the authority of the Bible. These they apply, with little judgment, to the refutation of arguments that they do not understand and have not even listened to.

...

For Copernicus never discusses matters of religion or faith, nor does he use arguments that depend in any way upon the authority of sacred writings which he might have interpreted erroneously. He stands always upon physical conclusions pertaining to the celestial motions, and deals with them by astronomical and geometrical demonstrations, founded primarily upon sense experiences and very exact observations. He did not ignore the Bible, but he knew very well that if his doctrine were proved, then it could not contradict the Scriptures when they were rightly understood. And thus at the end of his letter of dedication, addressing the pope, he said:

'If there should chance to be any exegetes ignorant of mathematics who pretend to skill in that discipline, and dare to condemn and censure this hypothesis of mine upon the authority of some scriptural passage twisted to their purpose, I value them not, but disdain their unconsidered judgment. For it is known that Lactantius – a poor mathematician though in other respects a worthy author – writes very childishly about the shape of the earth when he scoffs at those who affirm it to be a globe. Hence it should not seem strange to the ingenious if people of that sort should in turn deride me. But mathematics is written for mathematicians, by whom, if I am not deceived, these labors of mine will be recognized as contributing something to their domain, as also to that of the Church over which Your Holiness now reigns.'

...

The reason produced for condemning the opinion that the earth moves and the sun stands still is that in many places in the Bible one may read that the sun moves and the earth stands still. Since the Bible cannot err, it follows as a necessary consequence that anyone takes an erroneous and heretical position who maintains that the sun is inherently motionless and the earth movable.

With regard to this argument, I think in the first place that it is very pious to say and prudent to affirm that the holy Bible can never speak untruth – whenever its true meaning is understood. But I believe nobody will deny that it is often very abstruse, and may say things which are quite different from what its bare words signify.

. . .

But I do not feel obliged to believe that that same God who has endowed us with senses, reason, and intellect has intended to forgo their use and by some other means to give us knowledge which we can attain by them. He would not require us to deny sense and reason in physical matters which are set before our eyes and minds by direct experience or necessary demonstrations.

. . .

I would say here something that was heard from an ecclesiastic of the most eminent degree: 'That the intention of the Holy Ghost is to teach us how one goes to heaven, not how heaven goes.'

. . .

Hence it would probably be wise and useful counsel if, beyond articles which concern salvation and the establishment of our Faith, against the stability of which there is no danger whatever that any valid and effective doctrine can ever arise, men would not aggregate further articles unnecessarily.

. . .

I entreat those wise and prudent Fathers to consider with great care the difference that exists between doctrines subject to proof and those subject to opinion. Considering the force exerted by logical deductions, they may ascertain that it is not in the power of the professors of demonstrative sciences to change their opinions at will and apply themselves first to one side and then to the other. There is a great difference between commanding a mathematician or a philosopher and influencing a lawyer or a merchant, for demonstrated conclusions about things in nature or in the heavens cannot be changed with the same facility as opinions about what is or is not lawful in a contract, bargain, or bill of exchange.

. . .

From this it is seen that the interpretation which we impose upon passages of Scripture would be false whenever it disagreed with demonstrated truths. And therefore we should seek the incontrovertible sense of the Bible with the assistance of demonstrated truth, and not in any way to try to force the hand of Nature or deny experiences and rigorous proofs in accordance with the mere sound of words that may appeal to our frailty.[17]

FURTHER READING

Stillman Drake's *Discoveries and Opinions of Galileo* is a convenient source for English translations of and commentaries on several of Galileo's shorter works, including the *Letter to the Grand Duchess Christina*. Giorgio de Santillana's *The Crime of Galileo* is a standard reference for an account of Galileo's encounter with the Church of Rome, while Colin Ronan's *Galileo* is a readable general biography of Galileo. In *The Galileo Affair* Maurice Finocchiaro gives a carefully documented history of this famous clash. On the system of patronage and how it affected scientists such as Galileo, see Richard Westfall's *Essays on the Trial of Galileo* and Mario Biagioli's *Galileo Courtier*.

11

An overarching Newtonian framework

In modern science we judge a theory by its ability to make detailed quantitative predictions within some domain (vertical coherence) and we also look for applications across various fields outside its original domain (horizontal coherence).[1] In this way we generate a coherent network of concepts, laws and theories that is usually taken to be a sign of a correct representation of the phenomena of nature. The present chapter recapitulates our exposition of the discovery and development of the law of universal gravitation as a useful illustration of this process and discusses how this became interwoven with our ideas about space and time. Here we tell a retrospective story in which facts are selectively arranged in a sequence to produce a hopefully coherent narrative. This, of course, does not imply that there was any inherent need that events had to develop as they did.

11.1 A REVOLUTION

As we have seen (Sections 4.5 and 4.7), in the Aristotelian view of the cosmos the earth was at the center of the celestial sphere and all the space between these was divided into concentric regions (the homocentric spheres of Eudoxus), each region being the domain of one of the planets. This system of spherical regions turned one within the other to account for the observed motions of the stars and planets. The entire system was driven by the motion of the outermost shell, the celestial sphere. Nothing existed beyond the celestial sphere. In this model, the motions of the heavens were responsible for all change and variety in the sublunary world. The earth was naturally at rest in the center and all terrestrial motion (on or just above the surface of the earth) was (at least partially) indirectly driven from the heavens.

This Aristotelian theory of heavenly motion, like the two-sphere universe, was a good first step toward an understanding of the physical world. Since this model required a central, stationary earth with natural circular motion for the heavenly bodies, any new cosmology having a planetary earth could well require new laws of motion. These two problems were inextricably connected. The entire pattern of Aristotelian thought was a complex one. The *horror vacui* that we mentioned in Chapter 6 not only explained the terrestrial phenomena of water rising in a suction pump and the apparent impossibility to the ancient Greeks of creating a vacuum, but it was also related to the contiguous heavenly spheres that transferred rotation to each other. Each spherical shell was filled with its own element since the transmission of a rotation through a vacuum would seem implausible. The concepts of matter and space were inseparably linked in this scheme, so that one could not exist without the other. There would be no 'place' for a vacuum to be located. Hence, a vacuum was logically impossible. The central position of the earth gave it a unique status, as well as man, the most important creature on earth. An entire fabric of thought was woven together and an attack on any part was perceived as threatening to unravel the whole.

In Chapter 5 we saw that the heliocentric model, in the form first proposed by Copernicus, was not able to gain a decisive victory over the geocentric model of Ptolemy. There was no revolution in world view during Copernicus' lifetime. In fact, the Copernican revolution is not found in his *De Revolutionibus*. This work laid the foundations for that revolution by posing certain questions and Newton's *Principia* terminated the revolution by codifying a set of laws and principles that explained the motion of the planets. The real appeal of Copernicus' *De Revolutionibus* was for professional astronomers because of the technical and computational advantages that resulted for a moving earth. Chapter 5 also discussed the qualitative and semiquantitative aesthetic advantages of the Copernican model. The Gregorian calendar, first adopted in 1582, was based on computations made using this model. The real resistance and debate came from outside astronomical circles. As we indicated in the previous chapter, Copernicanism was seen by some as potentially destructive to the whole of Christian thought, especially to the Scriptures.

The English poet and divine John Donne (1572–1631) was aware of Copernicanism and feared that it might well be true. He saw this as an evil that would infect a man's mind and rend the fabric of the traditional Aristotelian and Ptolemaic cosmology. His poem, *An Anatomy of the World*, communicates this sense of apprehension and of sadness over a loss: '[The] new philosophy calls all in doubt,... 'Tis all in pieces, all coherence gone.'[2] This is the same author who penned 'No man is an island, entire of itself; every man is a piece of the continent, part of the main.'[3] Here the theme is again that of the coherence of reality, as in the traditional metaphor of the great chain of being.[4] This was rooted in the Platonic concept of a plurality of actually existing beings necessitated by a type of principle of 'plenitude' of the Good, linking the lowest forms of being continuously in an ascending order with the highest. Each had its assigned place in a grand scheme. Unraveling any one thread of this tapestry was seen as endangering the entire world view.

Much of the technical difficulty that Copernicus, and other astronomers of his time, had in fitting a heliocentric model to the motion of the planets was that some of the old observational data they relied on were of poor quality or contradictory. The empirical basis for implementing the Copernican revolution was the work of the Danish astronomer Brahe. In Section 5.3 we saw that he gathered an incredible amount of accurate data on the stars and planets. His were the most consistent and accurate naked-eye observations, being reliable to about four minutes of arc. This was more than twice as good as that of the best observers of antiquity. As with so many other advances in science, this one required a significant improvement in experimental technique or accuracy. Brahe's colleague Kepler used this data to discover the three laws that today bear his name. Kepler was able to discover that the planetary orbits are ellipses, not compound circles, because of Brahe's more accurate data. However, if Brahe's data had been much more accurate (say, as good as that obtainable with a modern telescope), then Kepler would have seen that the orbits are not precisely ellipses (recall the perturbations of Chapter 9) and might not have discovered his first law. It is unclear whether Newton could have formulated his law of gravitation without Kepler's laws.

Beginning in 1609, Galileo studied the heavens with a telescope. The effect on the thinking of the times was dramatic. He found many new stars so that the heavens were more populated and larger than had been generally believed previously. The moon was seen to have a rugged surface, appearing very much like a mountainous region on earth. There did not seem to be any difference between the composition of bodies on the earth and in the heavens. Galileo observed the moons of Jupiter. This provided a visible model of the Copernican solar system. Galileo appreciated that the language of nature is mathematics. In his essay *The Assayer* he states:

> Philosophy is written in this grand book, the universe, which stands continually open to our gaze. But the book cannot be understood unless one first learns to comprehend the language and read the letters in which it is composed. It is written in the language of mathematics, and its characters are triangles, circles, and other geometric figures without which it is humanly impossible to understand a single word of it; without these, one wanders about in a dark labyrinth.[5]

That is, the patterns of nature become apparent only upon proper analysis. This is similar in spirit to Aristotle's views on the difference between the confused mass of phenomena of which we first become aware, versus the underlying first principles that become clearer to the intellect after proper study (see Section 1.1).

We can now appreciate that the basic laws of motion and of gravity have a certain simplicity. In principle, those two mathematical laws allow one to say everything that can be said about the problem of planetary motion, as we discussed in Chapter 9. Although the motions of the planets are complicated and extremely difficult to calculate, the basic laws themselves are simple. However, this simplicity can be seen only in the language of mathematics and after considerable analysis. In spite of this great achievement, it is sobering to remind ourselves that to date we know of no law or theory that encompasses or explains all physical phenomena in the universe, but only various laws that each explain different classes of phenomena.

11.2 A BROAD COHERENCE

The method by which gravitation was discovered is thoroughly modern and in fact is a paradigm for scientific research and advancement as we know it today. Based on Brahe's improved data, Kepler empirically discovered his three laws. Galileo studied idealized situations, both theoretically and experimentally, and made an extrapolation to his (perhaps not wholly correct) concept of inertia. Newton unified these with his own laws of inertia and gravitation. In his theory, these laws have no known origin; they are quantitative and precise laws that tell us how things behave, but not why. In his original argument that the sun attracts the planets with an inverse-square force, Newton basically restated what Kepler said in his own laws. However, Newton made a great generalization to the mutual attraction between all bodies in the universe. From his more general laws he was able to derive what had gone before – namely Kepler's laws of planetary motion – and to explain the effects of the ocean tides, as well as why the earth is almost a sphere (due to gravitational attraction) and why it is slightly bulged at the equator (due to the centrifugal force arising from rotation about its axis). All this was in the best Baconian tradition of careful advance from one level of generalization to another. As we stressed previously, in addition to induction Newton was also in fact using the hypothetico-deductive method. At times he seemed to hint at this process, as in his Rule IV for reasoning (see Section 7.2), but in other places he put more emphasis on the inductive aspects of his method. In a letter to his colleague Roger Cotes (1682–1716), Newton stated:

> [First] principles are deduced from phenomena and made general by induction, which is the highest evidence that a proposition can have in this philosophy. And the word 'hypothesis' is here used by me to signify only such a proposition as is not a phenomenon nor deduced from any phenomena, but assumed or supposed – without any experimental proof.[6]

Similarly, in a letter to Henry Oldenburg, the Secretary of the Royal Society who maintained an international correspondence with the leading scientists of the world, Newton discussed how he arrived at his law of gravitation and how he felt it should be tested.

> And I told you that the theory which I propounded was evinced to me... by deriving it from experiments concluding positively and directly.
>
> ...
>
> [I]f the experiments which I urge be defective, it cannot be difficult to show the defects; but if valid, then by proving the theory, they must render all objections invalid.[7]

Once the Copernican model, Kepler's laws, or Newton's laws of motion and of gravitation, are accepted, there are many tests to which they can be subjected and other facts with which they must be consistent. For example, in 1675 the Danish astronomer Olaus Roemer (1644–1710) observed that sometimes the moons of Jupiter emerged from behind that planet eight minutes sooner than predicted and sometimes eight minutes too late. They appeared early when the earth was near Jupiter and late when the earth was far from Jupiter. Rather than doubting the regularity of the motion of the planets (that one would expect from Kepler's laws), Roemer attributed the time difference to the finite speed of propagation of light. Realize that you 'see' an object only after light from that object reaches your eye. The object need no longer be where you 'see' it when the light actually reaches your eye. As Figure 11.1 makes clear, the time difference Roemer observed is that taken by light to travel across the diameter of the earth's orbit. This was the first determination of the finite speed of light. Here one good law was used to discover another one (an example of horizontal coherence). Recall that in Chapter 5 we saw how the absolute sizes of the planetary orbits could be related, by triangulation, to

PART IV **A perspective**

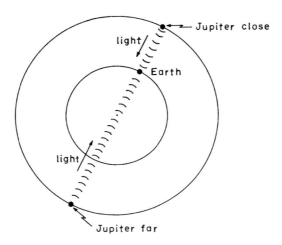

FIGURE 11.1 **Roemer's determination of the speed of light**

the size of the earth itself. That is, Roemer already knew the (numerical) value of the size of the earth's orbit about the sun (see Section 4.A). We can easily summarize his calculations as follows. The radius (R_e) of the earth's orbit about the sun is given in terms of the radius (r_e) of the earth itself as $R_e = 24{,}400 r_e$ (with $r_e = 4{,}000$ miles [6.44×10^3 km]). Since the measured delay time t was 22 minutes, Roemer was able to calculate c (the speed of light) as $c = (2R_e/t)$, or about 1.48×10^5 miles/s [2.38×10^5 km/s]. This is to be compared to the currently accepted value of 1.86×10^5 miles/s [3.00×10^5 km/s]. Even though Roemer's original value is about 20% low by present standards, it is reasonably good for a first determination and, more important, it shows c to be finite.[8]

As we mentioned in passing in Section 9.3, a faith in Newton's laws of motion and of gravity allowed astronomers to use the unexplained perturbations in the orbit of Uranus to predict the existence and location of the planet Neptune. In 1781, during a star survey of the heavens, William Herschel found a bright object that moved relative to the fixed stars from night to night. This was a new planet, Uranus. It soon became evident that the orbit of Uranus was not quite an ellipse. By 1820 the elaborate and accurate perturbation theory of the great French mathematician Laplace was applied to the motion of Uranus taking into account the gravitational effects of the two closest and largest planets then known, Jupiter and Saturn. The predictions of these calculations did not agree with the observed orbit of Uranus. The German mathematician Bessel attempted to produce agreement by increasing the mass of Saturn in the calculation, but this then varied Saturn's orbit too much to agree with other observations. Because of this, he told John Herschel that he felt the residual perturbations of Uranus's orbit were due to an as yet undiscovered planet beyond Uranus. In 1843 John Adams, still an undergraduate at St. John's College in Cambridge, began to calculate the expected mass, orbit and location of the new planet. In September of 1845, he presented his results to the astronomer James Challis (1803–1882), director of the Cambridge Observatory and then in November of that year to the English mathematician Sir George Airy (1801–1892), who was Astronomer Royal at the Greenwich Observatory. As a result of this and of a communication from the French mathematician Leverrier, Airy suggested in 1846 that a systematic search of the skies be made by the Cambridge observatory. Adams had difficulty in persuading those in charge of Cambridge Observatory to look for this planet. Challis finally began his search in late July of 1846. But, because he was not thorough and persistent enough in his observations, he did not find the new planet. Meanwhile, Leverrier had made similar but more extensive calculations and in November of 1845 published his predictions. In 1846 he sent his results to Johann Galle of the Berlin Observatory. In a matter of hours in September of 1846, the same day he received Leverrier's letter, Galle found the new planet, Neptune.

Similarly, in the early part of the twentieth century, the American astronomer Percival Lowell (1855–1916) made a detailed study of the irregularities of Neptune's orbit. He attributed these to the influence of an as yet unseen planet beyond Neptune. His *Memoir on a Trans-Neptunian Planet* was not published until 1915. Nevertheless, he earlier predicted the probable location of the new planet. In 1905 he began a systematic search for it at the observatory he founded in Flagstaff, Arizona. It was not until early 1930 that Clyde Tombaugh (1906–1997), also working at the Lowell Observatory, actually observed the long sought-for planet. It was named Pluto and denoted by a symbol made from Percival Lowell's initials. Subsequent reexamination of Lowell's dynamical calculations showed them to be much in error. The actual values of the parameters of Pluto accepted today differ greatly from those predicted by Lowell. Thus, the discovery of Pluto does not fit the scenario of a triumphant quantitative victory of gravitation in which the exact parameters of the orbit and mass were theoretically predicted and subsequently observed. Nevertheless, Pluto was discovered as the result of a systematic search initiated by the belief in its existence because of irregularities that could be accounted for by Newtonian theory.

The discoveries both of Neptune and of Pluto are examples of the cumulative effects of knowledge. This is a major factor in the scientific enterprise. This intertwining of the threads of nature's fabric is reminiscent of the complex pattern of Aristotelian thought, much of which has been replaced by modern science. All of the tests of Newton's theory of gravitation that we have discussed so far are based on astronomical observations rather than on laboratory experiments. Newton's theory was also given direct experimental verification by Cavendish in his determination of the universal gravitational constant G (see Section 8.4).

However, not everything about the planetary orbits is correctly explained by Newton's laws. There is a small precession, or rotation, of Mercury's elliptical orbit in its plane that is not predicted by classical perturbation theory. We can picture the elliptical orbit of Mercury as rotating slowly, as indicated in Figure 11.2. During each revolution of Mercury about the sun the perihelion (or distance of closest approach of Mercury to the sun) advances through an angle $\Delta\theta$. As a result, the orbit of the planet never quite closes, as shown in Figure 11.3. Now according to classical Newtonian theory, if only a central $1/r^2$ force acts between the sun and

PART IV A perspective

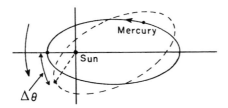

FIGURE 11.2 A precessing ellipse

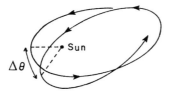

FIGURE 11.3 The shift of the perihelion

Mercury, then the orbit of Mercury should be exactly an ellipse and the perihelion of the orbit should remain fixed at precisely the same point in space (as shown in Section 9.2). But the perihelion of Mercury is observed to advance by 5,601 seconds of arc per century. When the perturbative effects due to other bodies in the solar system are calculated in Newton's theory (see Section 18.2), 5,558 seconds of arc per century are accounted for. However, there remain 43 seconds of arc per century unaccounted for classically. (Appreciate how small an advance of 43" per century is, since there are 3,600" in 1°.) Astronomical observations and perturbation calculations were sufficiently accurate that this was an outstanding problem even in the nineteenth century.

There were several attempts made to save Newtonian theory. We mention only two. In 1859 Leverrier, who had previously successfully predicted the existence of Neptune, postulated another planet, Vulcan (the consort of Venus in Roman mythology), as the cause of these residual perturbations. This time, though, no such planet was ever found. It was also suggested that Newton's law of gravitation might not be exactly correct, but should be modified to include an additional small $1/r^3$ term. One can prove mathematically that a small additional term of this form will produce a precessing ellipse. However, such an explanation is ad hoc and not very appealing since it is 'cooked up' just to explain this one case. Einstein's theory of general relativity does quantitatively and naturally explain this discrepancy with Newton's law of gravitation. His theory predicts much else besides. As we shall see in Chapter 18, Einstein's theory contains Newton's law as a limiting case.

Interestingly enough, each of these predictively successful theories of gravitation – Newton's and later Einstein's – was closely intermeshed with a certain conception of space.

11.3 VIEWS ON SPACE PRIOR TO NEWTON

From antiquity to the present, beliefs about the nature of space have undergone drastic changes. Today the proverbial 'person in the street', if he or she chose to think about it at all, would probably characterize the underlying empty space of the physical world as being infinitely divisible (continuous), having no intrinsically preferred direction (isotropic), being basically the same everywhere (homogeneous) and having infinite extent. It is also likely that such a person would conceive of space as a passive receptacle or background in which matter can exist, although the existence of space would not be seen to depend upon matter being present. However, even though these characteristics of space may appear reasonable enough to many people now, we should realize that each is an abstract concept, not immediately evident to sense perception. Very likely the primitive notion of 'place' came first and was quite local in its connotation. In this section we sketch the evolution of our abstract conceptions of space.

For Democritus, who is today famous for his 'atomic' picture of matter, space (or the void) was an infinite empty extension that simply contained the motion of matter (the atoms) but that exerted no influence on the matter. In a later age, the Roman poet and philosopher Lucretius argued as follows in his treatise *The Nature of the Universe* that the universe was of infinite extent.

> Again, if for the moment all existing space be held to be bounded, supposing a man runs forward to its outside borders, and stands on the utmost verge and then throws a winged javelin, do you choose that when hurled with vigorous force it shall advance to the point to which it has been sent and fly to a distance, or do you decide that something can get in its way and stop it? For you must admit and adopt one of the two suppositions; either of which shuts you out from all escape and compels you to grant that the universe stretches without end.[9]

The basic idea here is that a boundary or end to space would require a 'wall' that would be a place beyond the assumed edge of space. This argument can then be repeated indefinitely to prove that the universe (either a void or one with matter in it) must be infinite in extent. But an infinite universe with bodies existing in space implies an infinite space. With Lucretius space became an infinite receptacle into which matter could be placed. This is similar to our 'person in the street' view stated previously.

However, earlier with Plato, matter and space became inextricably united. He identified matter with empty space because he equated the world of physical bodies to that of geometric forms. The four elements that made up the world were given the following regular spatial structures: water the icosahedron (a twenty-sided solid), air the octahedron (eight-sided solid), fire the pyramid and earth the cube. For him physics became geometry. (Recall a similar geometrical influence in Kepler's five regular solids and his third law as discussed in Section 5.4.)

Subsequently, in Aristotle's view space had an active influence on matter since it determined the natural motions of bodies:

> Further, the typical locomotions of the elementary natural bodies – namely, fire, earth, and the like – show not only that place is something, but also that it exerts a certain influence. Each is carried to its own place, if it is not hindered, one up, the other down.[10]

The different conditions in various regions of space determined the natural motion of a body towards its suitable place. Since a void would have no properties of direction or change of condition from one region to another, Aristotle felt there would be no natural motion in such a void. From this he concluded the impossibility of the existence of a void.

These three major theories of space in antiquity can be summarized as the atomistic view (stressing the physical character of space itself), the Platonic one (stressing its mathematical aspects and the connection between space and matter) and the Aristotelian one (stressing the causal nature of space). With Plato, classical Greek philosophy and science represented space as something inhomogeneous due to local geometric changes and then with Aristotle as anisotropic due to the apparently intrinsic distinction between up and down.

In Chapter 6 we saw that the sixth-century (A.D.) philosopher John Philoponus criticized Aristotle's theory of motion. He also broke with the Aristotelian tradition when he held that space was separate from matter, being pure dimensionality and possessing no qualitative differences from one region to another. For him, space exerted no influence over matter. Later, under the press of the new physics of Copernicus, Galileo and Kepler, the reality of the void became accepted, as did the existence of an independent, infinite, structureless space. Newton would incorporate these properties into his conception of space.

In his *Principles of Philosophy*, Descartes took a very different view when he identified matter with pure geometrical extension: '[T]he nature of body consists not in weight, nor in hardness, nor colour and so on, but in extension alone.'[11] For him there was nothing in the universe but matter and motion. Or, since he equated matter with geometrical extension, the basic entities of the universe were extension and motion. From this nearly Platonic identification of matter and space, Descartes concluded, much as Aristotle had before him, that a void (in the sense of space devoid of matter) was a logical impossibility.[12] Once he deduced that a material void could not exist, he used an argument similar to Lucretius' above to show that, since space had to be infinite, so must the material universe.

We pick up another thread of our story by digressing to ancient Judaism where we find the origin of an idea that subsequently became influential. Here there was a connection between space and God. In fact, the Hebrew word for 'place' (*makom*) is used as a name for God. The idea of the Divine Omnipresence is evident in the well-known Psalm 139:

> Where could I go to escape your spirit?
> Where could I flee from your presence?
> If I climb the heavens, you are there,
> there too, if I lie in Sheol [Hell].
> If I flew to the point of sunrise,
> or westward across the sea,

> your hand would still be guiding me,
> your right hand holding me.[13]

This identification of space with God would eventually influence Newton through the writings of the Cambridge philosopher Henry More (see Section 7.1) who sought a proof for the existence of God by identifying God with space that permeates and acts on all matter. Although he strongly disagreed with Descartes on key philosophical issues, including the identification of space with matter, More believed that space was necessarily infinite. For him space was immaterial and hence a spirit, sharing many of the attributes of God. More identified this infinite, necessary, spiritual entity with God.

> There are not less than twenty titles by which the Divine Numen [power or spirit] is wont to be designated, and which perfectly fit this infinite internal place (*locus*); ... the very Divine Numen is called, by the Cabalists, MAKOM, that is Place (*locus*).[14]

The material (and contingent) world was finite, being situated in an infinite (necessarily existing) space.

11.4 NEWTON'S ABSOLUTE SPACE

We now turn to the view of space that became dominant after the success of the *Principia*. Newton kept his physics and his metaphysics separate, except in his theory of space. His famous '*hypotheses non fingo*' ('I do not frame hypotheses') was his guide for excluding occult, metaphysical or religious entities in his other scientific endeavors. In addition to More's philosophical influence on Newton, that of Isaac Barrow was also important. Barrow developed his own system of geometry in which space was identified with the divine omnipresence. Newton felt that absolute space was a logical and ontological necessity, required even for the validity of his first law of motion (the law of inertia). In his preface to the first edition of the *Principia*, he stated geometry to be a part of mechanics:

> Therefore geometry is founded in mechanical practice, and is nothing but that part of universal mechanics which accurately proposes and demonstrates the art of measuring.[15]

That is, Newton conceived of space and its structure as being a part of the empirical science of mechanics. He opened Book I of the *Principia* with a lengthy *scholium* on the natures of relative and absolute spaces.[16] We can summarize it by saying that absolute or mathematical space exists independently of material objects, while relative space is what we determine by the position of material bodies. Newton realized that absolute motion (relative to this absolute space) could not be detected kinematically (by studying just the motions of bodies without examining the forces that act on them). Only relative motions can be detected by such observations.

However, he did offer a dynamical argument, based on centrifugal forces, that would allow one to demonstrate the existence of absolute motion. Here is the essence of his famous thought experiment intended to establish this.[17] Suppose one

PART IV   **A perspective**

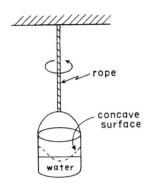

FIGURE 11.4   Newton's bucket 'experiment'

hangs a bucket of water from a rope, as shown in Figure 11.4. The surface of water will be flat. Next, wind the rope so that it is tightly twisted and will spin the bucket, once released, about the vertical axis as it unwinds. Both the bucket and the water are initially at rest, with the rope wound. Suddenly, by an impulsive twist applied to the bucket, release the bucket and set it into rapid rotation in the same direction that the rope will drive the bucket as it uncoils. At first the bucket is spinning rapidly relative to the water, but the water remains essentially at rest and its upper surface flat. As the motion of the spinning bucket is transferred to the water inside it, the surface of the water will become increasingly concave upward, until the water is at rest relative to the wall of the bucket. Newton argued that the nonflatness of the water's surface, caused by the centrifugal forces of rotation, showed that the reference frame of the water and bucket was now a noninertial one. In the beginning (when the bucket is moving relative to the water with its surface still flat) there was relative motion between the water and the bucket, and at the end (when the water with its concave surface is no longer moving relative to the bucket) there was no such relative motion. Newton claimed that relative motion produces no such effects as these, so that this observed difference was due to and also demonstrated absolute (not relative) motion.[18] He argued that a body has only one true circular motion. The basic idea here for Newton is that true (or absolute) motions are those caused by forces, whereas relative motions are generated without any force being impressed upon a body. It would seem that, even in empty space, it should be possible to use such an 'experiment' to detect absolute motion, simply by observing the behavior of the surface of the liquid in a container as described above.

Unfortunately, the experiment can only be performed in the real, or actually existing, world so that it remains a matter of conjecture what the outcome would be in an empty world. As the German physicist and philosopher Ernst Mach pointed out in his *The Science of Mechanics*:

> Newton's experiment with the rotating vessel of water simply informs us, that the relative rotation of the water with respect to the sides of the vessel produces *no* noticeable centrifugal forces, but that such forces *are* produced by its relative rotation with respect to the mass of the earth and the other celestial bodies. No one is competent to say how the experiment would turn out if the sides of the vessel increased in thickness and mass till they are ultimately several leagues thick. The one

experiment only lies before us, and our business is, to bring it into accord with the other facts known to us, and not with the arbitrary fictions of our imagination.[19]

Mach's essential point is that we cannot in fact empty the universe of all matter (except for the rotating bucket and its water) to see how the surface of the water would really behave under such circumstances. Therefore, we can only conjecture what shape the surface of the water would be. Newton's argument then loses its force and is not conclusive. In Mach's opinion, Newton had proved nothing.

Newton felt that an absolute space (in which the center of the universe is forever at rest[20]) was essential to give the law of inertia operational content. Furthermore, for Newton absolute space showed the divine omnipresence. In the Queries to his *Optics* we find:

> [D]oes it not appear from phenomena that there is a Being incorporeal, living, intelligent, omnipresent, who in infinite space (as it were in his sensory) sees things themselves intimately, and thoroughly perceives them, and comprehends them wholly by their immediate presence to himself?[21]
>
> . . .
>
> [T]he instinct of brutes and insects can be the effect of nothing else than the wisdom and skill of a powerful, ever-living agent, who being in all places, is more able by His will to move the bodies within His boundless uniform sensorium, and thereby to form and reform the parts of the Universe, than we are by our will to move the parts of our own bodies.[22]

Several of Newton's more religiously inclined colleagues welcomed this interpretation of space as the sensorium of God and claimed that he had rescued science from atheism.[23]

However, not all of Newton's contemporaries accepted his ideas about the nature of space. The British philosopher George Berkeley (1685–1753), the Dutch scientist Huygens, and the German mathematician–philosopher Leibniz held absolute space and absolute motion to be fictions. For Berkeley it was absurd that space could be anything but relative, on the theological ground that, if space were absolute, then there would be something beside God that would be infinite, unchangeable and, hence, existing forever. Leibniz objected strenuously both to the existence of absolute space and to the identification of it with God. In an exchange of a long series of letters between himself and Dr. Samuel Clarke (1675–1729), a supporter of Newton, Leibniz held that the relationship between material objects was itself sufficient for the concept of space and that absolute space was not needed. He had an analogy with a tree of genealogy that shows the kinships among persons by placing them in definite positions in the scheme. Why would one hypothesize this system and give it an absolute existence independent of the persons it relates? Space without matter in it was, for him, meaningless.

The difference between this view and Newton's is basically the following. For Newton, space was the container of all material objects and could exist without matter. It also served as a means of individuating objects (by their different positions in space).[24] For Leibniz, space was the positional quality of material objects and was inconceivable without matter.[25]

PART IV **A perspective**

11.5 PHYSICAL VERSUS MATHEMATICAL SPACES

Although the great mathematician Euler believed in the necessity of absolute space and attempted to prove its existence a priori by demonstrating the logical necessity of the law of inertia, the outstanding French writers on mechanics – Lagrange, Laplace and Poisson – showed no real interest in absolute space. For them, it was merely a working hypothesis needing no theoretical justification. Mach held that all metaphysical posits must be eliminated from science. For him, the concept of an absolute space that acts but cannot be acted upon was inimical to scientific thought, a claim later echoed by Einstein. Mach related the unaccelerated motion of a mass to the center of mass of the entire universe, rather than to space itself. This objection to an unobservable absolute inertial reference frame was reiterated by Einstein in speaking of Galileo's work.

> Let me interpolate here that a close analogy exists between Galileo's rejection of the hypothesis of a center of the universe for the explanation of the fall of heavy bodies, and the rejection of the hypothesis of an inertial system for the explanation of the inertial behavior of matter. (The latter is the basis of the theory of general relativity.) Common to both hypotheses is the introduction of a conceptual object with the following properties:
>
> (1) It is not assumed to be real, like ponderable matter (or a 'field').
> (2) It determines the behavior of real objects, but it is in no way affected by them.
>
> The introduction of such conceptual elements, though not exactly inadmissible from a purely logical point of view, is repugnant to the scientific instinct.[26]

In similar spirit, an insistence that a theory should contain no entities that are in principle unobservable or for which a procedure for measurement cannot be specified was part of the background out of which the philosophical movement of positivism or logical empiricism grew. In the 1920s a group of philosophers known as the Vienna Circle (Moritz Schlick (1882–1936), Rudolf Carnap (1891–1970), Philipp Frank (1884–1966) and Kurt Gödel (1906–1978), among others) held that any proposition was meaningful only insofar as it was based on experience and observation. By this criterion, ethical, metaphysical and religious statements might be taken to have no objective meaning. This is an example of the influence science has upon culture or philosophy, just as we have previously seen cases of influence in the other direction (for example, the 'eternal' nature of circular motion in Aristotelian physics requiring that the planetary orbits be circles).

Once it became clear that absolute space and absolute motion were not physically detectable, more and more scientists felt that these concepts should be eliminated from exact science. The last attempt made to save an absolute reference frame was to identify it with the aether at the end of the nineteenth century. As we shall see in Chapters 13 and 16, this proved futile and led to Einstein's special theory of relativity. However, even after the notion of absolute space had been abandoned, physical space was for some time still conceived of as Euclidean (or as flat, in the sense of ordinary plane or solid geometry). The mathematician Carl Gauss

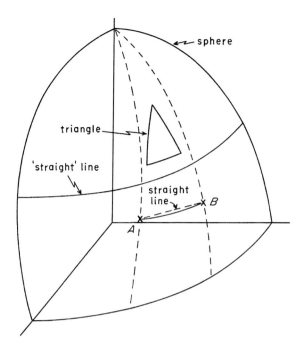

FIGURE 11.5 Geometry on the surface of a sphere

(1777–1855) realized that non-Euclidean geometry was a logically consistent possibility and that the nature of the geometry of the physical world could only be decided by experiment. He attempted to detect by direct observation a deviation from Euclidean geometry, but was unable to do so. Bernhard Riemann (1826–1866) intimated, and Einstein gave convincing evidence for the position, that the structure of space (or of geometry) in the physical world is determined by the distribution of matter. Here is a situation in which simplicity as a criterion for the truth of a conjecture (that of Euclidean space) proved to be misleading. Of course, as we have seen previously in Section 11.1, what appears to be simple is often dependent upon one's background knowledge and philosophical predilections. Generally agreed upon, objective criteria of simplicity do not seem to exist. Hence, simplicity as a guide to truth (as opposed merely to economy of effort) in formulating and developing a theory or hypothesis can be quite subjective.

The basic issue between Newton and Leibniz was the distinction of physical space from mathematical space. Quite aside from any physical reality, one can construct mentally and logically an entity referred to as Euclidean space. This is the ordinary space familiar to all from plane geometry. As Gauss and Riemann realized in the nineteenth century, one can also logically and consistently construct other types of mathematical spaces. The relevant question for physics is whether or not physical space is like mathematical space in having an existence of its own independent of the matter in it. For Newton, this was the case and physical space was identified with Euclidean space. Einstein, as we shall see in Chapter 18, postulated that physical space and matter are inextricably related. (Recall that Plato had a similar belief, although for very different reasons.)

To get some idea of what a non-Euclidean space is, let us consider life on a two-dimensional surface other than a plane. Perhaps the simplest case is a sphere. If we were two-dimensional creatures living on the surface of a sphere (no up or down away from the surface), then many of the 'laws' or theorems valid for plane Euclidean geometry would no longer be true. As we see from Figure 11.5, if we were to travel along a 'straight' line long enough we would eventually return to our starting point, unlike what would happen on a plane. For a triangle (a three-sided figure bounded by 'straight' lines) drawn on the surface of a sphere, the sum of its three angles would be more than 180°. (In fact, the departure of this sum from 180° is a measure of the curvature of the surface.) Pythagoras' theorem is no longer true either. The shortest distance between two points A and B on the surface of the sphere could not be a true straight line, since this would require tunneling through the volume of the sphere. Rather, the arc AB on the surface itself would be the shortest path from A to B on the sphere. Finally, this spherical surface has no boundaries, but does have a finite surface area $4\pi R^2$, where R is the radius of the sphere. Lucretius' argument for the infinity of space would certainly not be valid in this case. The javelin would simply travel along the surface back to its starting point.

These examples of logically possible mathematical and physical spaces provide counter evidence for our 'intuition'. The naive expectations we have, based on our everyday experiences, may not be satisfied exactly by physical reality. Even such 'obvious' matters as the nature of space ultimately become (at least in part) empirical questions to be settled by direct observation and experiment. The underlying structure of space does influence the geometrical or physical operations that are possible in that space.

Still, many may come to feel uneasy about taking such modern views of space as literally true representations of our world. On this point, the theoretical physicist Felix Bloch (1905–1983) had a charming recollection of a conversation he, as a young man, had with the already legendary Werner Heisenberg (1901–1976).

> We were on a walk and somehow began to talk about space. I had just read Weyl's book *Space, Time and Matter*, and under its influence was proud to declare that space was simply the field of linear operations.
> 'Nonsense,' said Heisenberg, 'space is blue and birds fly through it.'[27]

FURTHER READING

In *The Relevance of Physics* Stanley Jaki weaves a unified picture that encompasses the history of Western thought and shows the generality and yet the limitations of this science. David Park tells a marvelous story about the origins and development of physical theory in his *The How and the Why*. Max Jammer's *Concepts of Space* is a history of theories of space in physics and contains a really insightful Foreword by Einstein on the two conceptions of space mentioned in this chapter. Alexandre Koyré's *From the Closed World to the Infinite Universe* has become a classic in the history of the development, mainly in the sixteenth and seventeenth centuries, of ideas about the nature of our universe.

12

A view of the world based on science: determinism

As we indicated previously in Chapter 3, there is an element of belief in the very foundation of what is usually seen as the objective scientific enterprise. Traditionally, science has assumed that basically simple laws exist that explain the myriad phenomena of nature. This is a belief and science cannot prove it is a correct one. Throughout the history of Western thought there has been a tendency to reduce the phenomena of nature to a few simple laws or principles. One obvious motivation for this could well be a desire or felt need to make the world seem understandable to us. Today it is not uncommon to see the claim made that modern man has a view of the world based on science and that science has replaced religion. In addition to its acceptance of a basic simplicity in the fundamental laws of nature, this world view is often characterized as being philosophically materialistic, in the sense that matter and its interactions are regarded as comprising the entire universe, with the mind (or spirit) assigned a dependent reality (if any at all). Such a description of the world is typically taken, especially in older writing, to be completely deterministic, although quantum mechanics is widely believed to have changed this aspect of the philosophical materialism of modern science. (We return to determinism versus indeterminism in the quantum world in Chapters 21–24.)

In the present chapter we begin with an impressionistic, as opposed to a systematic, review of how science arrived at such a view of the world. Some of this will touch on material we have seen before. We then turn to a discussion of a fundamental change in outlook about the status of determinism – even in classical mechanics – that has been thrust upon us by developments in the study of deterministic chaos in recent decades.

12.1 THE BELIEF IN SIMPLE LAWS

We begin our story with several philosophers who lived (in the sixth and fifth centuries B.C.) in Ionia, an ancient region on the west coast of Asia Minor and on the adjacent islands of the Aegean Sea. The city most often associated with them is Miletus. Although their teachings varied considerably, theirs was a materialistic philosophy holding that all substances in the cosmos were derived from a single substrate (or, at most, from a few basic elements). They attempted to explain the phenomena of reality in terms of matter or physical forces, rather than in terms of mythology as had been typical before their time. The first of the Ionian philosophers was Thales. He introduced abstract geometry into Greek thinking, is claimed to have predicted the eclipse of the sun that occurred in 585 B.C. and took water to be the basic element of which all matter was composed. For Anaximander the universe was infinite in extension and duration. He constructed a spherical model for the heavens (planets and stars) with the earth at the center and had as his basic element a rather indefinite substance termed *the unlimited*. Anaximenes assumed the earth to be flat and held that the stars were fixed on a type of celestial hemisphere (this latter idea surviving in the form of the celestial sphere until the time of Galileo) and that air was the basic element of the universe. Heraclitus (*c.* 540–*c.* 480 B.C.) considered fire to be the essential element of the cosmos and, related to the transformative nature of fire, he stressed continual change as the most important characteristic of phenomena. Anaxagoras (*c.* 500–*c.* 428 B.C.) taught that Mind was the sole initiator or organizer of an undifferentiated mixture into the cosmos as we know it. Many of these concepts fit nicely into the atomism of Leucippus (*fl. c.* 450 B.C.) and that of his most famous pupil, Democritus.[1]

A different tradition – one that had a profound influence on the work of Plato – traces its origins to Pythagoras (*c.* 580–*c.* 500 B.C.). Although born at Samos in Greece, he migrated to southern Italy around 532 B.C. to be allowed greater freedom of thought and of teaching. His philosophy attempted to give a unified view of nature based on the prime importance of number as the fundamental entity for understanding the universe. This philosophy is sometimes characterized as 'all is number', meaning that all existing objects were composed of form, not of matter. In his cosmology a central role was assigned to the tetractys of the decade (the progression 1, 2, 3, 4 since $1 + 2 + 3 + 4 = 10$, the 'perfect' number). Examples of important tetractys were:

point, line, surface, solid;

earth, water, air, fire.

In education, the classical quadrivium (geometry, arithmetic, astronomy and music) had its origin here. Pythagoras is credited with discovering that consonant musical intervals are expressed by simple ratios involving 1, 2, 3 and 4. Harmony was important in his philosophy and the master was supposed to be able to hear the harmony of the heavenly spheres. He founded the Pythagorean brotherhood. This school recognized the importance of mathematics and began to formalize the Western tradition of rational thought. Theirs was an explanation of nature based on a priori mathematical–mystical postulates. Harmony and whole numbers (or integers) were so important for the Pythagoreans that it is said they attempted to keep to themselves the discovery of irrational numbers and put to death one of their members when he let the secret out. (An irrational number (for example $\sqrt{2}$) is one that cannot be expressed as the ratio of two integers.) This aspect of a simple unifying principle in Pythagoras' teachings left its mark on the works of Plato and of his students. For example, in Aristotle's opinion the basic natural motions were uniform rectilinear and uniform circular, since each of these was simplest for its purpose: limited motion of finite duration and eternal, unvarying motion, respectively.

PART IV **A perspective**

We now move forward considerably in time. William of Ockham was an influential fourteenth-century Scholastic philosopher who founded nominalism that denied the independent existence of universals (in contradiction to Plato's theory of forms). Although the law of economy or of parsimony was not original with him, he employed it frequently in his work. Today we characterize this concept with his famous razor, '*non sunt multiplicanda entia praeter necessitatem*' ('entities are not to be multiplied beyond necessity'), even though it seems not to be found in precisely this form in his writings. Put more succinctly, the basic idea is that the best explanation is the simplest one that works. In this same spirit, Copernicus objected to the equant of Ptolemy largely on aesthetic grounds and felt that uniform circular motion for the planets (but now with the sun, rather than the earth, at rest in the center) had a perfection of simplicity. His description, in terms of circular orbits concentric about the sun, was simpler than Ptolemy's at a qualitative level. On the other hand, Kepler had largely mystical reasons for preferring the theory of Copernicus to that of Ptolemy. Copernicus' theory in its original simple form was not supported quantitatively by observation. Even though Kepler had certain philosophical predilections, he did use agreement with experiment or with observation as a decisive test. It was precisely this criterion that eventually led him to reject circular motion and discover the ellipse as the shape of the planetary orbits. Once this was done, a great simplicity and economy became apparent. Bacon, in his *The New Organon*, warned people to resist the tendency of human nature to impose our wishes and expectations upon our description and explanation of empirical facts. He spoke of anticipation versus observation. By his admonition, we are to eschew the a priori and examine nature to see what it has to tell us. For him, sense data should have primacy over theoretical constructs.

Galileo helped formulate the method of analysis and induction in physics. He assumed the existence of a simple, ordered world possessing regularity. Recall his statement about the book of nature being written in the language of mathematics (see Section 11.1). Newton, in his *Rules of Reasoning in Philosophy* that he gave in the *Principia*, stated what is essentially Ockham's razor as a criterion of simplicity:

> We are to admit no more causes of natural things than such as are both true and sufficient to explain their appearances.[2]

Central to the philosophy of Leibniz was the principle of sufficient reason according to which nothing exists that is unnecessary. This was the best of all possible worlds. (The French satirist François-Marie Voltaire (1694–1778) wrote *Candide* in caustic response to this belief.) For Leibniz there was a certain economy in nature. This same theme was central to the work of Pierre-Louis de Maupertuis (1698–1759). He believed that nature always acted in such a way as to minimize something. For mechanics, he postulated that this something, that he termed the *action*, was the product of mass, speed and distance. He attempted to furnish a theological foundation for mechanics. Maupertuis claimed to obtain several experimentally verifiable results from his principle, but often imprecisely and with a certain amount of 'fudging'. However, Euler and Lagrange gave precise, mathematical formulations of

Maupertuis' vague idea. For example, if a body is constrained to the surface of a sphere and an impulse is imparted to that body, it will move from its initial location to its final position along that path (on the surface of the sphere) that requires the least transit time. Euler maintained the theological view of Maupertuis and held that phenomena could be explained not only in terms of causes but also in terms of purpose. He believed that, since the universe was the creation of a perfect God, nothing could happen in nature that did not exhibit this maximum or minimum property. In Euler's program all the laws of nature should be derivable from this principle of maximum or minimum. The fact that Newton's second law of motion was deduced from such a principle lent great support to this claim. This was the beginning of the use of variational principles that are common in physics today (but without the theological trappings). Basically, some quantity, such as Maupertuis' action, is assumed to be kept a minimum during the motion of a system and from this the laws of motion for the system are deduced. Mach saw the purpose of science to be to represent the phenomena of nature in the simplest and most economical way. All a priori and metaphysical propositions were to be eliminated from science.

These repeated attempts, over so long a period of time, to discover or construct basic and simple laws of nature may say more about us than it does about the external nature that we represent with them. We seem to have felt need to find or construct ultimate explanations and these may arguably have some survival value (a theme developed in the early years of the twentieth century by the French philosopher of science Émile Meyerson (1859–1933)[3]).

Now that we have indicated how a belief in the existence of simple laws was employed as one criterion (along with empirical adequacy) for judging correct or true laws of nature, let us examine the origin of the deterministic view of the world typically associated with classical mechanics.

12.2 THE MEANING OF DETERMINISM

The pre-Newtonian world was a very anthropocentric one with much mystery about nature. After Newton, the material universe came to be seen as completely deterministic in principle. That is, given the exact initial positions and velocities of all particles, the law $\mathbf{F} = m\mathbf{a}$ determines the future trajectories exactly forever in the future, assuming the forces $\mathbf{F}$ are known. We discussed this in Section 9.2. There we argued that, with the initial conditions $\mathbf{r}_0$ and $\mathbf{v}_0$ specified, the trajectory $\mathbf{r}(t)$ of an object is completely determined for the future. We might expect to be able to generalize this to a hoped-for result, for a collection of N particles, as

$$\left.\begin{array}{l} \mathbf{r}_j(t_0) \\ \mathbf{v}_j(t_0) \\ \mathbf{F}_j = m_j \mathbf{a}_j \end{array}\right\} \Rightarrow \mathbf{r}_j(t), j = 1, 2, \ldots, N \qquad (12.1)$$

PART IV **A perspective**

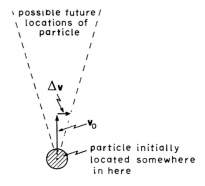

FIGURE 12.1 **The propagation and growth of initial uncertainties**

This does seem quite plausible and it can be established mathematically for a wide class of well-behaved forces. However, there are singular situations that are important in physics (such as the mutual gravitational interactions in a many-body system) and for which this result cannot be established. Below we return to some of these difficulties. Nevertheless, for the moment let us cling to the (overly) optimistic view that was generally accepted (at least for classical mechanics) until fairly recent times.

Of course, there may be an uncertainty arising from our ignorance of the precise initial conditions. For example, suppose we know that at $t = 0$ a particle is located somewhere within the shaded region of Figure 12.1 and that its initial velocity is v_0 with some uncertainty Δv. Even if no forces act on it, we cannot be certain of its exact location at some future time. We can only say that it will be somewhere within the cone containing those possible future positions that are consistent (according to Newton's laws of motion) with our imprecise initial data. As time goes on, we have less and less idea of just where the particle is. This classical statistical uncertainty is an uncertainty in the practical sphere (that is, in our knowledge about the world, rather than in the physical world itself), but not one in principle. This is different from the inherent uncertainty usually associated with quantum mechanics, as we discuss in Chapter 20.

Such mechanical determinism in the seventeenth century reinforced views of divine foreknowledge. Absolute determinism (in principle) in the equations of mathematical physics was consonant with the generally accepted theological belief in an omnipresent, omniscient God. Such a belief preceded Newton's formulation of his mathematical laws of motion. Although Newton certainly believed in the existence and action of such a God, he was less certain that the laws of mechanics could alone represent the deterministic evolution of a stable universe. In Newton's opinion, as expressed in the Queries to his *Optics*, the mechanical universe required the active intervention of God, not just to create and order it, but also to maintain it.

> To what end are comets, and whence is it that planets move all one and the same way in orbs concentric, while comets move all manner of ways in orbs very eccentric; and *what hinders the fix'd stars from falling upon one another?*[4]

A view of the world based on science: determinism

> . . .
>
> For it became Him who created [all material things] to set them in order. And if He did so, it's unphilosophical to seek for any other origin of the world, or to pretend that it might arise out of a chaos by the mere laws of Nature; though being once formed, it may continue by the laws for many ages. For while comets move in very eccentric orbs in all manner of positions, blind fate could never make all the planets move one and the same way in orbs concentric, some inconsiderable irregularities excepted, which may have risen from the mutual actions of comets and planets upon one another, and which will be apt to increase, *till this system wants a reformation*.[5]

Here, in the sentences we have set off in italics, Newton touches upon the stability of the solar system, a question that we return to in Chapter 18.

By the latter part of the eighteenth century, the absolute determinism and self-sufficiency of the mechanical universe became an accepted article of faith for many. This classical determinism was boldly stated by Laplace, the great mathematician and theoretical astronomer who perfected pertubation-theory calculations and used them to argue for the stability of the solar system. In his *Philosophical Essays on Probabilities*, Laplace expressed his views as follows:

> All events, even those which on account of their insignificance do not seem to follow the great laws of nature, are a result of it just as necessarily as the revolutions of the sun. In ignorance of the ties which unite such events to the entire system of the universe, they have been made to depend upon final causes or upon hazard, according as they occur and are repeated with regularity, or appear without regard to order; but these imaginary causes have gradually receded with the widening bounds of knowledge and disappear entirely before sound philosophy, which sees in them only the expression of our ignorance of the true causes.
>
> Present events are connected with preceding ones by a tie based upon the evident principle that a thing cannot occur without a cause which produces it. This axiom, known by the name of *the principle of sufficient reason*, extends even to actions which are considered indifferent.
>
> . . .
>
> We ought then to regard the present state of the universe as the effect of its anterior state and as the cause of the one which is to follow.[6]

Similarly, in the introduction to the 1814 edition of his *Analytic Theory of Probability*, we find:

> If an intelligence, for one given instant, recognizes all the forces which animate Nature, and the respective positions of the things which compose it, and if that intelligence is also sufficiently vast to subject these data to analysis, it will comprehend in one formula the movements of the largest bodies of the universe as well as those of the minutest atom: nothing will be uncertain to it, and the future as well as the past will be present to its vision. The human mind offers in the perfection which it has been able to give to astronomy, a modest example of such an intelligence.[7]

When the Emperor Napoléon Bonaparte (1769–1821) questioned him concerning whether or not he had mentioned God in his treatise *Celestial Mechanics*, Laplace is said to have responded, 'Je n'avais pas besoin de cette hypothèse-là.'[8] ('I did not need that hypothesis.') Laplace found no role for God in keeping the universe running.

PART IV A perspective

12.3 WHY THE CLOCKWORK UNIVERSE?

We tend to take classical mechanics as the paradigmatic warrant for a belief in a completely deterministic ('clockwork') universe. For the present, let us accept as a rough and ready definition of *determinism* the requirement that the present state of the universe (or system) plus the laws of mechanics uniquely determine the future state of the universe. Of course, we must spell out in more detail just what we mean by 'state' and 'laws'. But, more generally, let us first ask what the basis is for this belief in determinism. As we indicated in the previous section, prior to Newton's time there was already a theological underpinning in terms of a God seen as a lawgiver. If one accepts a God who runs the universe in an orderly and law-like fashion, then it makes sense for such a person to seek to discover these laws as represented in the workings of His creation. (Recall Whitehead's observation about this at the end of Section 10.5.) The point here is that a belief in or an inclination toward the acceptance of a law-like evolution of the universe predated any specific set of analytical or mathematical laws of physics. Furthermore, Newton was a transition figure. While he definitely did believe in the existence of a God who was responsible for the orderly evolution of the physical universe, he did not believe that the mathematical laws as represented in his *Principia* were in themselves sufficient to explain or to predict the long-term stability and future evolution of the physical universe. It was only after Newton that the determinism and complete predictive accuracy of his laws of mechanics became accepted.

Let us return to the type of argument by which one might use Newton's laws to justify a belief in absolute determinism. Consider the motion of a planet around a fixed force center under the influence of the gravitational attraction between the force center and the planet. The relevant laws are Newton's second law of motion and his law of universal gravitation. The basic question now is whether classical systems governed by these laws evolve deterministically and whether they are stable against small perturbations, so that we have effective long-range predictive power. We have previously outlined an intuitively appealing argument (represented by Eq. (12.1)) that specified initial conditions and the law that governs the time evolution of the system together uniquely determine the trajectory $r(t)$ for any arbitrary time t in the future. However, it is essential to appreciate that, in general, this is a very difficult mathematical problem since it takes the form of a differential equation

$$\ddot{x} \equiv \frac{d^2 x}{dt^2} = f(x) \tag{12.2}$$

for the function $x = x(t)$. Because $f(x)$ can be an arbitrary and complicated function of x, relatively little can be said without rather stringent restrictions being placed on $f(x)$.

Now it so happens that, for the central-force problem of a planet orbiting the sun, the equation corresponding to Eq. (12.2) can be solved exactly to obtain conic sections as the orbits. (Recall the discussion of Section 9.2.) The question then is whether or not this admittedly important example is representative of the general situation in classical mechanics. The answer is no, as we shall see in more detail later when we discuss the topic of chaos in classical systems. It typically turns out that one loses effective long-time predictive power for nonlinear systems governed by laws of the form of Eq. (12.2). (In the $1/r^2$ central-force case just discussed, the resulting differential equation can be transformed into a linear one, but this is exceptional and not generally the rule for such differential equations.)

However, let us press on a bit with the $1/r^2$ central-force problem we were able to handle. We can ask whether or not these orbits are stable against small perturbations. In order to keep the mathematics as simple as possible so as not to obfuscate the central point of the discussion, we can consider only the question of the stability of circular orbits. We are then able to prove the stability of a circular orbit for a $1/r^2$ attractive central force, when the system is displaced slightly from its circular orbit (say, by an impulsive force that acts just once) and then allowed to move again under the action of this one central force alone.[9] For a more general force field, the question is much more complicated and stability cannot be guaranteed. (Recall the discussion of Section 9.3.) This illustrates that the question of the long-term stability of a mechanical system is a rather sensitive one. Again, though, until quite recently this stability result for the simplest case just mentioned was taken as being representative of the general situation, rather than as being an exception to the type of behavior to be expected.

One of our strong reasons for believing in science, and in physics in particular, is the ability that its laws give us to make precise predictions. For example, the laws of mechanics and of gravity accounted for the locations of the planets and of comets over long periods of time. There are, however, other physical phenomena, such as the weather, for which we obviously have only short-term predictive power (say, two or three days at best), rather than the long-term predictive power we seem to have for the solar system. The traditional view was that systems consisting of only a few parts (for example, the planets moving about the sun) are amenable to the precise calculations necessary for meaningful prediction, whereas complex systems, such as a collection of gas molecules (or the atmosphere) are simply beyond our calculational abilities. However, no in-principle difference was seen between these two types of systems as regards determinism. One was just too complicated to handle computationally – a mere practical limitation.

This traditional view, or intuition, was based on an examination of the (relatively few) physical problems that we could solve analytically (as in the case of the two-body central-force problem above). Such integrable systems are, essentially by definition, the ones that can be treated by the methods of (classical) mathematical analysis. The exactly soluble problems of classical mechanics usually turn out to be *separable*. This means, roughly, that the equations describing their behavior separate into sets of individual one-body problems. That might already have given us some hint about how special, and not surprisingly, atypical, these cases are. Nevertheless, we took our 'old' insights, formed on the basis of those systems we were able to handle analytically, and assumed them to be typical of all physical systems. (Of course, what else could one reasonably do but form a picture of the physical universe based on a theory that appeared to be enormously successful?) But, as we have just indicated, such integrable systems turn out to be very special.[10] Our intuition, or general picture of the world, was based on a poor induction from too narrow a range of systems. For nearly 300 years we thought we understood classical mechanics, but we didn't. As we shall see later, there are many simple mechanical systems that exhibit chaotic behavior and for which we have no effective predictive power (even though these systems are governed by completely deterministic laws).

So, we are brought full circle back to the question with which we began this discussion: What is the warrant for our belief in the completely deterministic evolution of classical systems? Historically, an important element supporting that belief was the predictive accuracy of the deterministic laws of classical physics. For, suppose the situation were such that a set of proposed deterministic laws of classical physics, in the overwhelming majority of the cases to which they were applied, proved incapable of yielding useful, reliable predictions. A defender of 'in-principle' determinism could still claim that, at the most fundamental level, the universe is governed by deterministic laws, but that we are simply unable to specify initial data precisely enough to do the requisite calculations that would allow us to 'see' this determinism reveal itself. Who would be impressed by such an argument or theory, since its effective 'predictions' would be empirically indistinguishable from those that would obtain in a universe that was at base completely indeterministic? Today we know that 'most' classical physical systems can (and often do) exhibit chaotic behavior. So, is the world fundamentally deterministic and is this determinism masked in a few atypical situations (the traditional, standard view), or is it fundamentally indeterministic but exhibiting apparent or effective deterministic behavior in a few atypical situations? Is there any way to decide this issue? (We return to this question, from the perspective of quantum mechanics, in Section 24.2.)

This should make us appreciate that there is both an *epistemological* and an *ontological* dimension to the basic issue in question. The former term refers, in the present context, to what we know (about the world) and the latter to actual existence claims about that world. We discussed determinism both as a property of the theories or equations of physics and as a property of the actual world itself. We began with the observed phenomena of the world (say, astronomical data as embodied in Kepler's three empirical laws) and then constructed a theoretical framework, or a system of equations (here, Newton's laws of motion and of gravitation), to account for these phenomena and to make new quantitative predictions. It was in these laws or equations, that represent our world, that we discovered and explored the property of determinism. Once we satisfied ourselves of the accuracy and reliability of these mathematical laws, we then were willing to transfer (perhaps unconsciously) the generic features of these laws (or representations of our world) back to the physical world itself. In other words, we went beyond a mere instrumentalist view of the laws of science (in which we demand only empirical adequacy of our laws and nothing more) and took them realistically as literally true representations of the world. What is our warrant for this transfer? Let us now examine some difficulties that chaotic systems present for the argument we have just sketched.

12.4 AN UNWARRANTED OPTIMISM

By the term *chaos* we mean the complicated, unpredictable, seemingly random behavior of a system. The weather provided a paradigm case of this type of time evolution. We contrasted such behavior with the apparently orderly, highly predictable motion of the planets and comets in the heavens. (Recall in this regard, however, comments made in Section 9.3 on the instability of perturbation calculations with respect to the long-term behavior of the solar system, as pointed out by Poincaré.) Furthermore, there are common systems that, under different conditions, exhibit now one, then the other, of these two patterns of evolution. A fluid flowing at moderate rates shows a smooth, orderly behavior (termed laminar flow). However, when the fluid moves rapidly, the flow becomes highly disordered or turbulent. A single billiard ball moving on a smooth square table and suffering only elastic collisions with the rigid sides of the table has a very (and, in fact, simply) predictable behavior. On the other hand, one billiard ball moving among many other (even initially at rest) billiard balls (as in a typical game of billiards) is unstably susceptible to arbitrarily small external influences (such as random impacts from air molecules or virtually imperceptible vibrations of the table). After a relatively short time (that is, after several collisions), we have no effective predictive power as to its whereabouts and velocity on the table, even if we had perfect initial information (data) about its location and velocity.

The subject of deterministic chaos has become so important in recent years because it is now apparent that chaotic (classical) dynamical systems occur in many different fields of science. Examples of chaotic behavior can be illustrated in (among many others) such disparate physical systems as: the classical many-body problem (even just the three-body problem will do), turbulence and heat convection in fluids, chemical reactions (for example, the Belousov–Zhabatinsky reaction), cardiac dynamics, electrical circuits (the van der Pol equation), ecology and population dynamics, mechanical vibrations and buckling phenomena (Duffing's equation), information processing in the brain, the solar system (the orbit of Pluto on a long enough time scale) and, possibly, the stock market.[11] The generic source of this classical dynamical chaos is, as we indicate below, the extraordinarily rapid separation[12] of system trajectories during dynamical evolution. There is extreme sensitivity to initial conditions: trajectories that begin close together at some initial time will in a very short time be separated widely in space, leading to loss of effective predictive ability for the long-term behavior of the system. (This is sometimes referred to as the 'butterfly effect', since, fancifully, a butterfly flapping its wings in Brazil may, eventually, set off a tornado in Texas.)

The general mathematical form of the equations that exhibit chaotic behavior is

PART IV A perspective

$$\dot{x} \equiv \frac{dx}{dt} = f(x, \lambda; t) \tag{12.3}$$

(Appreciate that $x(t)$ is a vector that may have many components, $x = (x_1, x_2, x_3, \ldots)$, so that Eq. (12.3) can represent a coupled set of differential equations.) Here λ is an external parameter (such as the amplitude of a disturbance or the temperature). The function f need not (and often does not) depend explicitly upon the time. Dynamical chaos arises only when nonlinearity is present in these equations. (By 'nonlinearity' we mean that $x(t)$ enters into the right-hand side of Eq. (12.3) other than linearly (the first power). For example, a quadratic term is nonlinear.) Nonlinearity is a necessary, but not a sufficient, condition for chaotic behavior. We often make linear approximations to these equations, or sometimes find that the dominant forces present give rise to linear dynamical equations (for which there can be no chaotic behavior), and a more detailed treatment then produces nonlinear terms in the (more nearly exact) equations of motion. We saw an example of this in our outline of the two-body central-force problem. The resulting equation could be transformed into a linear one (and had conic sections as its exact, closed-form solutions), as long as we neglected the perturbing effects due to the gravitational influence of other planets.

The field of nonlinear differential equations is an enormously difficult one so that often one cannot decide rigorously whether or not a given system of such equations exhibits truly chaotic behavior. Fortunately, certain important features of the long-term behavior of the solutions to these continuous, or differential, equations can frequently be obtained by studying related (discrete) mapping problems.[13] This correspondence makes it possible to study certain universal features of chaos for complicated nonlinear differential equations by analyzing the generic properties of these much simpler discrete maps. Since this allows us to gain some understanding of chaotic behavior, we discuss two examples of these maps in the next section.

12.5 TWO MAPS AS EXAMPLES

We begin with one particularly simple but rich and important example of a nonlinear system of equations that produces chaotic behavior, namely the *logistic map*, that arises from a model of population dynamics.[14] The logistic map consists of an iterative procedure for generating the value of a variable x (say, the fraction of some species in a population) at the $(n + 1)$ step (or time interval) in terms of the preceding step (n)

$$x_{n+1} = ax_n(1 - x_n) \tag{12.4}$$

For small values of x, this equation is approximately linear and the slope of the curve, or the rate of growth, depends only upon the parameter a. However, as the parameter a is increased, the negative term, $-ax^2$, can retard further growth. Successive iterations of this equation can be given a very simple geometrical

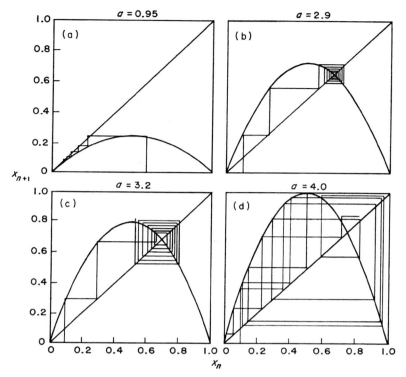

FIGURE 12.2 The logistic map for various values of the parameter a in Eq. (12.4)

interpretation in terms of Figure 12.2. Putting x_n into the right-hand side of this iterative equation, obtaining x_{n+1} on the left and then inserting this value into the right to obtain the next iteration is equivalent to a series of reflections, along horizontal and vertical lines, between the parabola

$$y = ax - ax^2 \qquad (12.5)$$

and the straight line $y = x$, as indicated in Figure 12.2. (For example, if the parabola is above the line $y = x$, then this just amounts to taking the first value, x_n, going vertically up to the parabola, reflecting along the horizontal to the straight line $y = x$ and then moving again along the vertical back to the parabola. This is illustrated in the left-hand portions of Figures 12.2(b) and (c).)

The long-term, or asymptotic, behavior of this iterative map depends upon the value of a. We restrict a to the range $0 < a < 4$, since we want x to remain in the range between 0 and 1 here.[15] We define a *fixed point*, denoted by x^*, as a value of x for which the logistic map of Eq. (12.5) has the property that $x_{n+1} = x_n$. It is relatively straightforward[16] to show that $x^* = 0$ is a stable fixed point[17] when $0 < a < 1$ and that it is the only one for a in this range of values. In the range $1 < a < 3$, $x^* = 0$ becomes unstable and another value of x^* then becomes stable. For $a > 3$, both fixed points are unstable. Then the behavior of the limiting value of

PART IV A perspective

x becomes more complicated and requires extensive numerical iteration as illustrated in Figures 12.2(c) and (d).

Let us summarize several important subintervals of the range of a.

i $a < 1 \Rightarrow$ extinction ($x_{n+1} \to 0$) always. That is, $x = 0$ is a stable fixed point.
ii $a > 1 \Rightarrow x = 0$ becomes an unstable fixed point and a new stable one arises located at the intersection of the parabola and the 45° line at $x = (a-1)/a$.
iii $1 < a < 3 \Rightarrow$ the population moves to this new fixed point.
iv $a > 3$ (say, 3.2) $\Rightarrow$ the population cycles between two limit points.
v a still larger $\Rightarrow$ the number of limit points for these cycles doubles and redoubles until it goes to infinity at $a \approx 3.57$. These doublings are termed 'bifurcations'.
vi $a \sim 4 \Rightarrow$ population is chaotic and jumps all over the range of values between 0 and 1.

Notice that, for $x_0 = 0$ or $x_0 = 1$, all of the x_n remain zero forever for all a, so that there is extreme sensitivity to initial conditions.

Another important illustration is provided by the standard map

$$x_{n+1} = x_n + y_{n+1}$$
$$y_{n+1} = y_n + k\sin x_n \qquad (12.6)$$

If we take x_n to be the angular displacement of a rotator and y_n to be its angular velocity, then these equations (the standard map) represent the equations of motion of a kicked rotator. Such a mechanical system is illustrated in Figure 12.3. Only for $k = 0$ is a closed-form solution known (in terms of standard functions). For $k \neq 0$, solutions must be obtained numerically. The transition from ordered to chaotic behavior occurs around $k \approx 1$. This transition from ordered to disordered (or chaotic) behavior as k increases is easily seen in Figure 12.4. This is a *phase-space* portrait (that is, a plot of position (x) versus momentum (or, sometimes, the velocity) (y)) for various values of k. ('Phase space' is the term for a mathematical space whose coordinate axes are labeled with the position, say x, and the velocity (or the momentum p) of a system. A point in this space gives the (classical) state of the system.) The case for $k = 0$ corresponds to regular (or integrable) dynamics (here, a simple rotator executing uniform circular motion) and the system point at successive times falls on the smooth curves indicated. Even for $k = 0.5$, the motion, while now less regular, still has the same general qualitative features as that for an integrable system (that is, the motion is still fairly smooth and predictable). By the time $k = 1.0$, the motion is a combination of regular and chaotic. In the latter, the location of the system point jumps around erratically and essentially fills an area. It does not trace out a smooth curve. The motion for values of k beyond unity is indistinguishable from a random walk problem,[18] since large regions of chaotic behavior dominate.

We have just outlined the behavior of two simple systems that exhibit extremely complex deterministic chaos. These are only two examples and we might expect that, even though the results are interesting in their own right, they may be peculiar

A view of the world based on science: determinism

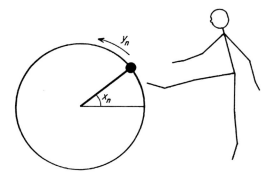

FIGURE 12.3 **A kicked rotator**

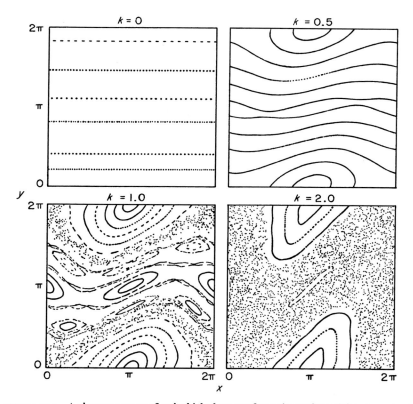

FIGURE 12.4 **A phase-space map for the kicked rotator for various values of the parameter k in Eq. (12.6)**

to these particular cases. However, the importance of these examples for our purposes is that such features turn out to be generic ones of most mechanical systems under appropriate circumstances. This is not meant to be obvious and we certainly have not here either proved this claim or even made it particularly plausible. It is the universality of these features that makes an appreciation of maps central to an understanding of deterministic chaos. The case sketched in Figure

12.2(d) illustrates an essential characteristic of chaotic behavior: the loss of any useful predictive power for the long-time behavior of the system. Specifically, in terms of the system (say, x represents a population) governed by Eq. (12.4) and graphed in Figure 12.2(d), we see that after a few iterations of the map (that is, after a few 'generations') we know little more than that $0 \leq x \leq 1$. But, we would have known that much even in the absence of our 'theory'.

Another way to put this is that two 'populations' beginning very close together (that is, having values of x that are at nearly the same point on the horizontal axis of Figure 12.2(d)) can soon be widely separated after several iterations (or 'generations'). Our expectation might have been that by beginning with two initial points fairly close together we would surely be able to have the later points also remain very close together. Chaotic systems rapidly magnify small differences into large ones so that we become unable to control the separation of the iterated points. It is this extreme sensitivity to initial conditions (so that initially neighboring trajectories in phase space diverge very rapidly in time) that renders meaningful prediction of future behavior effectively impossible for chaotic systems.

There are precise, technical mathematical criteria for chaotic behavior, but we do not enter into these here. However, the behavior (or 'output') of deterministic chaotic systems can be so irregular (and unpredictable) as to be indistinguishable from that of truly random systems. For a random sequence, the length of a set of instructions (say, a computer program) to compute the sequence is roughly as long as the sequence itself. There is no simple rule that will allow one to compute or predict the nth term in the sequence short of writing out the entire sequence itself. No finite algorithm exists for computing a chaotic orbit. This is why the amount of time (or computational effort) increases without limit when we attempt to calculate the future values of the state variables (say, the position and velocity). Hence, for a chaotic system there is no faster way to find out how such a system will evolve than to watch the system itself actually evolve. (That is, there is no usable 'rule' for passing directly from the initial conditions or state to the final one.) Even though an individual system's behavior cannot be predicted, the average behavior of a large set of them can often be (in terms of probabilities).

It is for these reasons that we previously stated that belief in determinism as being more fundamental than probabilities in the actual physical universe is an act of faith, rather than a position demanded, or even particularly well warranted, by the laws of classical physics and by the observed behavior of physical systems. In a sense, determinism reigns almost nowhere and chaos nearly everywhere, so that Newtonian (Laplacian) determinism remains only a theorist's unattainable dream.[19] We return to this randomness–determinateness conundrum when we discuss quantum mechanics later.

While deterministic chaos is an area of intense popular interest and research activity at present, it remains unclear just how significant or truly revolutionary it is for the conceptual foundations of physics, or science in general. Although it certainly does have great practical implications for science, the question of any revolution in the fundamental laws of nature themselves is still to be resolved.[20]

FURTHER READING

In Chapter 8 of *The Relevance of Physics* Stanley Jaki examines the often uneasy relation between physics and metaphysics and the function that regulative principles, such as the notion of simplicity, have played in the history of science. Ian Hacking's *The Emergence of Probability* and Lorraine Daston's *Classical Probability and the Enlightenment* recount how the concept of probability came to play so important a role in modern times. Florin Diacu and Philip Holmes' *Celestial Encounters* traces the origins of chaos theory, from the work of Poincaré on the many-body problem in classical mechanics to recent technical developments. Of the huge number of books and articles on deterministic chaos and related topics, very readable introductory overviews of the field are contained in Joseph Ford's 'How Random Is a Coin Toss?', James Crutchfield *et al.*'s 'Chaos,' Roderick Jensen's 'Classical Chaos,' Keith Briggs' 'Simple Experiments in Chaotic Dynamics' and Max Dresden's 'Chaos: A New Scientific Paradigm – or Science by Public Relations?'. James Gleick's *Chaos* is a popularization of the origin and growth of chaos theory in recent times. In *Order Out of Chaos* Ilya Prigogine and Isabelle Stengers argue that randomness and irreversible processes are the proper starting point for science, rather than order and reversible processes as traditionally assumed. Stephen Kellert's *In the Wake of Chaos* examines the philosophical implications of modern chaos theory and John Earman's *A Primer on Determinism* does this in much more technical detail. Ian Stewart's *Does God Play Dice?* discusses the mathematics of chaos. A sampling of the many technical works on deterministic chaos is: Heinz Schuster *Deterministic Chaos*, Michael Thompson and Bruce Stewart *Nonlinear Dynamics and Chaos*, S. Neil Rasband *Chaotic Dynamics of Nonlinear Systems*, John Guckenheimer and Philip Holmes *Nonlinear Oscillations, Dynamical Systems, and Bifurcations of Vector Fields* and Joseph McCauley *Chaos, Dynamics, and Fractals*.

PART V

Mechanical versus electrodynamical world views

Science, to the ordinary reader of newspapers, is represented by a varying selection of sensational triumphs, such as wireless telegraphy and aeroplanes, radioactivity and the marvels of modern alchemy. It is not of this aspect of science that I wish to speak. Science, in this aspect, consists of detached up-to-date fragments, interesting only until they are replaced by something newer and more up-to-date, displaying nothing of the systems of patiently constructed knowledge out of which, almost as a casual incident, have come the practically useful results which interest the man in the street. The increased command over the forces of nature which is derived from science is undoubtedly an amply sufficient reason for encouraging scientific research, but this reason has been so often urged and is so easily appreciated that other reasons, to my mind quite as important, are apt to be overlooked. It is with these other reasons, especially with the intrinsic value of a scientific habit of mind in forming our outlook on the world, that I shall be concerned in what follows.

The instance of wireless telegraphy will serve to illustrate the difference between the two points of view. Almost all the serious intellectual labour required for the possibility of this invention is due to three men – Faraday, Maxwell, and Hertz. In alternating layers of experiment and theory these three men built up the modern theory of electromagnetism, and demonstrated the identity of light with electromagnetic waves. The system which they discovered is one of profound intellectual interest, bringing together and unifying an endless variety of apparently detached phenomena, and displaying a cumulative mental power which cannot but afford delight to every generous spirit. The mechanical details which remained to be adjusted in order to utilise their discoveries for a practical system of telegraphy demanded, no doubt, very considerable ingenuity, but had not that broad sweep and that universality which could give them intrinsic interest as an object of disinterested contemplation.

Bertrand Russell, *The Place of Science in a Liberal Education*

13

Models of the aether

We shall see in Chapter 14 that, with Maxwell's great work, *A Treatise on Electricity and Magnetism* (1873), our conception of an electromagnetic wave (of which ordinary light is but one example) became that of a wave consisting of electric (E) and magnetic (B) fields propagating along at the speed of light (c). However, most waves of which we have some immediate experience, such as water waves, sound waves in air and waves in a vibrating string, are transmitted through some material medium. The obvious question, then, is what is the nature of the medium that transmits optical and other electromagnetic effects. At first sight this may not appear to pose much of a problem when these effects are transmitted through a material medium such as air, water or a solid. But electromagnetic waves do propagate through what we usually term a vacuum, as between the sun and the earth. In this chapter we review the history of certain ideas concerning light and electromagnetism.

13.1 EMERGENCE OF THE OPTICAL AETHER

As we saw previously, Descartes believed that all space was a *plenum*, everywhere filled with matter so that there were no voids and could exist no vacuum. An aether permeated all of space. For him, interactions could take place only via pressure and impact; that is, through the tangible action of some intermediary agent or matter. He, and other members of what became the Cartesian school, felt that instantaneous action at a distance was senseless. For example, Newton's law of gravitation ($F = -GMm/r^2$) (for masses M and m) and Coulomb's law for electrostatics ($F = -\text{const } Qq/r^2$) (for charges Q and q) simply state the force laws between two bodies, but they do not indicate how that force is transmitted from one body to another. No speed of propagation of disturbance enters either equation. It is for this reason that they are referred to as *action-at-a-distance* theories. Descartes favored a projectile or corpuscular theory for light. One of the requirements of that theory was that light travel more rapidly in a dense medium than in a rarified one or in vacuum. On the other hand, Pierre de Fermat (1601–1665) postulated a principle of least action (related to the minimization, or variational, principles mentioned in Section 12.1), from which it followed that light traveled more slowly in a denser medium.

Hooke, a contemporary and antagonist of Newton's, suggested a theory in which light was a vibratory motion transmitted through a medium as a series of wave fronts. The effect of polarization[1] could not be explained by this theory so that

Newton and others had an argument against the wave nature of light. In the years before Newton wrote the *Principia*, he held that all space was filled with an aether of variable density. However, in the General *Scholium* to Book III of that work, Newton argued that Kepler's first and third laws were incompatible with the existence of a dense aether. This does not mean, though, that Newton gave up altogether on the notion of an aether. We saw in Section 8.4 that Newton (in a letter to Richard Bentley (1662–1742) in the early 1690s) most emphatically dissociated himself from any belief in action at a distance through a void and spoke in terms that supported transmission of action by direct contact through an intermediary. In his later years he thought that there was some type of active aether, somehow associated with his notion of God's omnipresence, that was responsible for gravitational action. Although Newton was never definite about the nature of this active agency, space for him was not simply a passive void through which matter moved.[2] Light was distinct from this aether but interacted with it. Newton was not specific about just what light was, any more than he was about gravity. While corpuscles were only one of the possibilities for light left open by Newton, the fact that he was against Hooke's wave theory was taken by many as indicating that Newton supported a corpuscular theory similar to Descartes'.

Huygens favored the wave theory of light. Since two crossed beams of light pass through one another without any evident effect on either beam, Huygens argued that light could not consist of corpuscles because these would collide with one another and produce scattering of light in the crossed beams. He proposed that light waves were propagated as disturbances through a tenuous, highly elastic medium. In these early wave theories, light was considered to be a longitudinal wave,[3] much like a sound wave. However, Huygens' theory did not gain acceptance, in part due to the belief that Newton had unreservedly supported a corpuscular theory of light. Newton's law of gravitation was so immensely successful that action at a distance became an accepted, even if mysterious and not wholly understood, fact. If the propagation of gravity were instantaneous, then there was no need for an aether. Without an aether, there was no wave theory and the corpuscular theory appeared more reasonable. As a matter of fact, in the late eighteenth century George-Louis Lesage (1724–1803) proposed a corpuscular explanation of gravity by assuming that all space was filled with a great number of tiny particles traveling at high speeds in all directions. A single body alone and at rest in space would be bombarded equally from all sides and hence receive no net impulse in any direction. However, two bodies near each other would partially shield one another from these streams of bombarding particles on the sides facing each other and hence receive a net impulse (or 'force' of attraction) directed toward each other. It is fairly simple to show geometrically (for two spheres whose radii are small compared to the distance between them) that this effect would vary inversely as the square of the distance between the bodies. This theory has several flaws among which are the facts that a moving planet would receive more vigorous blows on its front side than on its back one and hence slow down and that the effect of attraction would be proportional to (the product of) the volumes of the two bodies (rather than to their masses, as experience shows).

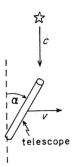

FIGURE 13.1   **Stellar aberration due to a moving earth**

In 1728 James Bradley discovered stellar aberration. Basically this means that if a star is located directly overhead, as shown in Figure 13.1, then one must incline a telescope (located on a moving earth) at a small angle α to the vertical in order to see the star. This was taken as evidence in favor of the corpuscular theory of light, in analogy with vertically falling rain passing down through an inclined stove pipe (without the rain touching the inside of walls of the pipe) when that pipe moves along in the rain. On the other hand, Euler favored the wave theory of light since bodies do not lose any detectable mass when they emit light. If corpuscles were being emitted, this mass would have to come from the light source itself, whereas if light is a wave sent through a medium, no mass would be lost by the body. He also advanced the idea that the source of all electrical phenomena, and of gravity too, was the same aether responsible for the propagation of light.

Young was the first to use the concepts of constructive and destructive interference for light waves. He did this by analogy with water waves on the surface of an otherwise calm lake. As we saw in Section 9.A, Newton applied the idea of destructive interference in water waves to solve the problem of the anomalous tides at Batshaw on the Gulf of Tonkin. Young was able to explain the effect of Newton's interference rings that appear when light is reflected from two different interfaces as shown in Figure 13.2. The rays reflected from the upper surface and from the lower one can interfere constructively or destructively depending on the path difference between the two glass surfaces. (See the last quotation from Young in Section 9.A.) Since Dominique Arago (1786–1853) and Augustin-Jean Fresnel (1788–1827) demonstrated experimentally that two beams of light polarized at right angles cannot be made to produce interference effects, Young proposed that light consists of transverse vibrations[4] so that there would be two linearly independent directions of polarization at right angles to the direction of propagation. It was assumed that the restoring forces in the aether were proportional to the lateral displacement of the aether. The wave theory of light was by then firmly established. In 1850 Jean-Bernard Foucault (1819–1868) and Armand-Hippolyte Fizeau (1819–1896) made a direct measurement of the speed of light in water and found it to be less than the speed of light in air. This was a major victory for the wave theory.

PART V Mechanical versus electrodynamical world views

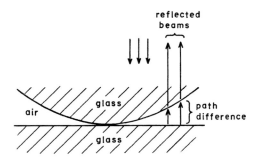

FIGURE 13.2 Newton's interference rings

13.2 THE ELASTIC SOLID AETHER

Generations of great mathematicians on the Continent and in England now worked on the elastic solid theory of aether. Claude-Louis Navier (1785–1836) developed precise mathematical equations of motion for an elastic medium. Augustin-Louis Cauchy (1789–1857) worked on the same equations and Poisson found a solution to these equations. Poisson showed that both transverse and longitudinal waves had to be present for any elastic solid that can be compressed and distorted.

In 1839 James MacCullagh (1809–1847) proposed a new type of elastic solid that, rather than resisting compression and distortion as is usually the case, had a potential energy that depended only upon the rotation of its elements. In such a medium, transverse waves alone were transmitted and with a speed of propagation $v = \sqrt{\mu/\rho}$, where ρ is the density of the medium and μ a constant that measures the stiffness or resistance of the elastic medium to rotation. His mathematical equations were similar in form to those that Maxwell would later propose for the electromagnetic field. MacCullagh had solved, in a mathematically consistent fashion, the problem of an elastic medium that would transmit transverse waves only. However, his work was never really taken very seriously for over forty years because there existed no plausible mechanical model for such a peculiar solid medium that would resist only rotations of its elements of volume and would be incompressible.

In 1853 the great German mathematician Riemann suggested an aether resisting compression and rotation in which the compressional properties were responsible for gravity and electrostatic effects and the rotational ones for optical and magnetic phenomena. He began the unification of optics and electromagnetism. Unfortunately, Riemann never really pursued this theory. Probably the last serious attempt at an elastic solid model of an aether that treated just optical phenomena alone was made in 1867 by Joseph Boussinesq (1842–1929). In his theory there was just one aether that was the same everywhere, whether inside material bodies or not. It was the interaction between the particles of the aether and those of ordinary matter that accounted for the optical properties of matter. However, such restricted theories, that accounted for optics but not for electric and magnetic effects, soon fell out of vogue once the electromagnetic theory of light was formulated. We now turn to that development.

13.3 THE ELECTROMAGNETIC AETHER

Out of the physical reality that Faraday attached to magnetic lines of force grew the concept of a field propagating its effects contiguously from one point in space to another. In an essay titled 'On the Physical Lines of Magnetic Force,' Faraday came to see the line of force as a physical entity.

> On a former occasion certain lines about a bar-magnet were described and defined (being those which are depicted to the eye by the use of iron filings sprinkled in the neighbourhood of the magnet), and were recommended as expressing accurately the nature, condition, direction, and amount of the force in any given region either within or outside of the bar. At that time the lines were considered in the abstract. Without departing from or unsettling anything then said, the inquiry is now entered upon of the possible and probable *physical existence* of such lines.[5]

After reviewing the experimental evidence from gravity, static electricity, electric currents and induction, and magnetic phenomena, he concluded:

> Now all these facts, and many more, point to the existence of physical lines of force external to the magnets as well as within. They exist in curved as well as in straight lines; for if we conceive of an isolated straight bar-magnet, or more especially of a round disc of steel magnetized regularly, so that its magnetic axis shall be in one diameter, it is evident that the polarities must be related to each other externally by curved lines of force; for no straight line can at the same time touch two points having northness and southness. Curved lines of force can, as I think, only consist with physical lines of force.[6]

Faraday saw these physical lines of force as exerting influence upon each other as a mechanism for transmitting effects across space.

> It appears to me possible, therefore, even probable, that magnetic action may be communicated to a distance by the action of the intervening particles, in a manner having a relation to the way in which the inductive forces of static electricity are transferred to a distance [by transferring its action from one contiguous particle to the next]; the intervening particles assuming for the time more or less of a peculiar condition, which (though with a very imperfect idea) I have several times expressed....[7]

As a possible mechanism for the transmission of these magnetic effects, Faraday suggested the aether of light so that an electromagnetic theory of light was hinted at.

> Such an action may be a function of the ether; for it is not at all unlikely that, if there be an ether, it should have other uses than simply the conveyance of radiations.[8]

In such a field-theory view, the fundamental mechanism is one of continuous transmission at a finite speed as opposed to instantaneous action at a distance. Faraday was keenly aware of the break he was making with the preponderance of mathematical thought of his day.

> [I] was led to suspect that common induction itself was in all cases an *action of contiguous particles*, and that electrical action at a distance (i.e., ordinary inductive action) never occurred except through the influence of the intervening matter.

PART V   **Mechanical versus electrodynamical world views**

> The respect which I entertain towards the names of Epinus, Cavendish, Poisson, and other most eminent men, all of whose theories I believe consider induction as an action at a distance and in straight lines, long indisposed me to the view I have just stated; and though I always watched for opportunities to prove the opposite opinion, and made such experiments occasionally as seemed to bear directly on the point,... it is only of late, and by degrees, that the extreme generality of the subject has urged me still further to extend my experiments and publish my view. At present I believe ordinary induction in all cases to be an action of contiguous particles consisting in a species of polarity, instead of being an action of either particles or masses at sensible distances; and if this be true, the distinction and establishment of such a truth must be of the greatest consequence to our further progress in the investigation of the nature of electric forces.[9]

13.4 THOMSON'S AND MAXWELL'S MODELS

It was William Thomson (1824–1907), later to become Lord Kelvin, who first gave some degree of mathematical respectability to Faraday's concept of an electric medium. In 1841, as a seventeen-year-old first-year student at Cambridge, Thomson wrote a paper in which he showed the mathematical equivalence of the lines of force in certain electrostatic problems with the lines of heat flow in analogous thermal problems in an infinite solid. For example, in Figure 13.3 we show the electrostatic field pattern for a positive charge Q placed opposite a positively charged flat surface and in Figure 13.4 the lines of heat flow from a heat source located near a thermally insulated barrier. The mathematical equations for these two problems are identical, with charge playing the role of the heat source and lines of force that of lines of heat flow. The importance of this precise analogy was that it related a theory (electrostatics) formulated in terms of action at a distance with a theory (steady heat flow) formulated in terms of transfer from one contiguous particle to another in a material medium. Faraday was glad for such support for his views.

> Professor W. Thomson, in referring to a like view of lines of force applied to static electricity, and to Fourier's law of motion for heat, says that the lines of force give the same mathematical results of Coulomb's theory, and by more simple processes of analysis (if possible) than the latter; and afterwards refers to the 'strict foundation for an analogy on which the *conducting power of a magnetic medium for lines of force* may be spoken of.'[10]

Years later Maxwell evaluated this insight of Thomson's as follows:

> This paper first introduced into mathematical science that idea of electrical action carried on by means of a continuous medium, which, though it had been announced by Faraday, and used by him as the guiding idea of his researches, had never been appreciated by other men of science, and was supposed by mathematicians to be inconsistent with the law of electrical action, as established by Coulomb, and built on by Poisson.[11]

Thomson was one of the first members of the 'Cambridge school' of natural philosophers founded by George Green (1793–1841). Between the time of Newton and Green – nearly a hundred years – Cambridge lapsed far behind the great mathematics of the Continent. Green's work began a new tradition at Cambridge.

Models of the aether

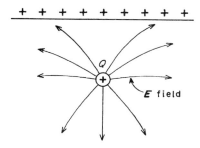

FIGURE 13.3   Electric lines of force near a conducting plane

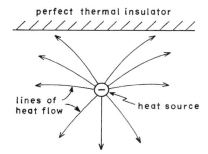

FIGURE 13.4   Lines of heat flow near an insulating plane

William Strutt (Lord Rayleigh) (1842–1919), Maxwell, Horace Lamb (1849–1934), Joseph Thomson (1856–1940), Joseph Larmor (1857–1942) and Augustus Love (1863–1940) would follow.

In 1846 Thomson studied an analogy between the electric field and the displacements of the elements of an elastic solid under strain. He associated the *E* field with these displacements from equilibrium. Ten years later he proposed a rotary interpretation of the *B* field since a magnetic field produces a rotation of the plane of polarization of a light wave. In 1889 Thomson devised a mechanical model for an aether of the type earlier proposed by MacCullagh in which the medium resists only rotation of its elements and remains incompressible. He conceived the solid as made up of spheres arranged on a tetrahedron (a pyramid) with neighboring spheres at the vertices, as depicted in Figure 13.5. Each sphere was connected to its neighbors by rigid rods with spherical caps so that the rods would slip on the spheres. Such a configuration was rigid against compression. On each rigid rod was mounted a pair of oppositely rotating flywheels (see Figure 13.6) that would, by Newton's first law of inertia, resist any change in orientation of these connecting rods. Thomson's early attempts at a mechanical interpretation of electromagnetism attracted the attention of the young Maxwell. Much later, in the preface to his great *Treatise*, Maxwell recorded his conceptual debt to both Faraday and Thomson.

> [B]efore I began the study of electricity I resolved to read no mathematics on the subject till I had first read through Faraday's *Experimental Researches in Electricity*. I was aware that there was supposed to be a difference between Faraday's way of conceiving phenomena and that of the mathematicians, so that neither he nor they were satisfied with each other's language. I had

PART V Mechanical versus electrodynamical world views

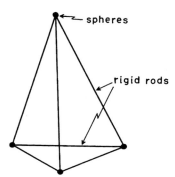

FIGURE 13.5 **An element of the 'aether'**

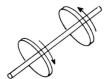

FIGURE 13.6 **Counter-rotating flywheels**

also the conviction that this discrepancy did not arise from either party being wrong. I was first convinced of this by Sir William Thomson, to whose advice and assistance, as well as to his published papers, I owe most of what I have learned on the subject.

As I proceeded with the study of Faraday, I perceived that his method of conceiving the phenomena was also a mathematical one, though not exhibited in the conventional form of mathematical symbols. I also found that these methods were capable of being expressed in the ordinary mathematical forms, and thus compared with those of the professed mathematicians.

For instance, Faraday, in his mind's eye, saw lines of force traversing all space where the mathematicians saw centres of force attracting at a distance: Faraday saw a medium where they saw nothing but distance: Faraday sought the seat of the phenomena in real actions going on in the medium, they were satisfied that they had found it in a power of action at a distance impressed on the electric fluids.

When I had translated what I considered to be Faraday's ideas into a mathematical form, I found that in general the results of the two methods coincided, so that the same phenomena were accounted for, the same laws of action deduced by both methods, but that Faraday's methods resembled those in which we begin with the whole and arrive at the parts by analysis, while the ordinary mathematical methods were founded on the principle of beginning with the parts and building up the whole by synthesis.

I also found that several of the most fertile methods of research discovered by the mathematicians could be expressed much better in terms of ideas derived from Faraday than in their original form.[12]

Maxwell eventually accepted Thomson's association of the B field with the rotations of the aether and he further saw the E field as the velocity of an incompressible fluid that had sources and sinks. The type of mechanical model by which Maxwell attempted to interpret his equations in the 1860s is illustrated in Figure 13.7.[13] In this picture an elastic aether has distributed throughout it layers of small particles that play the role of electric charges and of 'idle' wheels. The electric current flowing

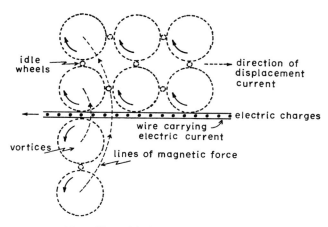

FIGURE 13.7 Maxwell's model of the electromagnetic aether

in the wire sets the aether in its immediate neighborhood rotating to form vortices. The loops of closed vortex filaments concentric with the wire are the lines of the magnetic field. The layer of rotating vortices just next to the wire sets the particles in the aether in motion (to the right in the figure). However, since the aether is to be elastic, it allows only a small lateral displacement of these particles and, after this very limited initial displacement, these particles, always having to remain in contact with the vortices and not being allowed to slip, therefore rotate in the opposite sense of the first layer of vortices. These spinning idle wheels communicate their rotation to the next layer of the aether and form a second layer of vortex rings that also rotate in the same sense as the first layer (clockwise above the wire and counter-clockwise below the wire in Figure 13.7). The initial lateral motion of the idle particles was associated with the electric field (also termed the electric displacement by Maxwell). This effect was a transient one and existed only for an instant after the current in the wire began to flow. Similarly, once the current in the wire ceased, there would be a transient electric displacement, this time in the opposite direction, as the aether returned to its initial equilibrium position. As long as the current in the wire remained steady, there would be no further electric displacement, but only a constant magnetic field consisting of the vortex filaments. This gave an explanation of Faraday's law of induction. The model also made it clear how a set of orthogonal transverse E and B fields would be propagated outward from a wire that carried a varying current. In a similar vein, the electrostatic field was conceived of as a stress due to the displacement of the medium from equilibrium. This electric tension will remain as long as the medium remains under stress and will disappear once the medium is allowed to return to equilibrium.

In an 1861 letter to Thomson, Maxwell described this medium.

> I suppose that the 'magnetic medium' is divided into small portions or cells, the divisions or cell walls being composed of a single stratum of spherical particles, these particles being 'electricity.' The substance of the cells I suppose to be highly elastic, both with respect to compression and distortion; and I suppose the connection between the cells and the particles in the cell walls to

PART V   Mechanical versus electrodynamical world views

be such that there is perfect rolling without slipping between them and that they act on each other tangentially.[14]

He then went on to detail to Thomson the mechanics of electromagnetic propagation in such a medium. A more extensive description of this is given by Maxwell in the first excerpt in Section 13.A at the end of the chapter.

13.5 MAXWELL'S ARGUMENTS FOR THE AETHER

In a long article on the aether for the ninth edition of the *Encyclopaedia Britannica*, Maxwell concluded by discussing the vortex lines that would have to exist for an indefinitely long period of time in the neighborhood of a permanent magnet. Since any known fluid that can have vortices must also be viscous so that the vortices must ultimately dissipate as heat, he had to admit that the aether possessed another peculiar property. Nevertheless, he was adamant about its existence and even hinted at a function for the aether beyond the realm of mere physics.

> No theory of the constitution of the aether has yet been invented which will account for such a system of molecular vortices being maintained for an indefinite time without their energy being gradually dissipated into that irregular agitation of the medium which, in ordinary media, is called heat.
>
> Whatever difficulties we may have in forming a consistent idea of the constitution of the aether, there can be no doubt that the interplanetary and interstellar spaces are not empty, but are occupied by a material substance or body, which is certainly the largest, and probably the most uniform body of which we have any knowledge.
>
> Whether this vast homogeneous expanse of isotropic matter is fitted not only to be a medium of physical interaction between distant bodies, and to fulfill other physical functions of which, perhaps, we have as yet no conception, but also... to constitute the material organism of beings exercising functions of life and mind as high or higher than ours are at present, is a question far transcending the limits of physical speculation.[15]

The great virtue of the aether and of Faraday's lines of force in Maxwell's opinion was that these allowed one to do away with the mystical concept of action at a distance. In a talk before the Royal Institution in London he expressed his views on action at a distance versus the continuous field concept. An excerpt of this is given in the second, long quotation in the following section. This and other writings of his make it clear how seriously Maxwell took the aether as a really existing physical medium of transmission for electromagnetic phenomena.

13.A MAXWELL ON THE AETHER VERSUS ACTION AT A DISTANCE

In describing the aether, Maxwell wrote (see Figure 13.7 again):

> In the first part of this paper I have shewn how the forces acting between magnets, electric currents, and matter capable of magnetic induction may be accounted for on the hypothesis of the magnetic field being occupied with innumerable vortices of revolving matter, their axes coinciding with the direction of the magnetic force at every point of the field.

The centrifugal force of these vortices produces pressures distributed in such a way that the final effect is a force identical in direction and magnitude with that which we observe.

In the second part I described the mechanism by which these rotations may be made to coexist, and to be distributed according to the known laws of magnetic lines of force.

I conceived the rotating matter to be the substance of certain cells, divided from each other by cell-walls composed of particles which are very small compared with the cells, and that it is by the motions of these particles, and their tangential action on the substance in the cells, that the rotation is communicated from one cell to another.

I have not attempted to explain this tangential action, but it is necessary to suppose, in order to account for the transmission of rotation from the exterior to the interior parts of each cell, that the substance in the cells possesses elasticity of figure, similar in kind, though different in degree, to that observed in solid bodies. The undulatory theory of light requires us to admit this kind of elasticity in the luminiferous medium, in order to account for transverse vibrations. We need not then be surprised if the magneto-electric medium possesses the same property.

According to our theory, the particles which form the partitions between the cells constitute the matter of electricity. The motion of these particles constitutes an electric current; the tangential force with which the particles are pressed by the matter of the cells is electromotive force, and the pressure of the particles on each other corresponds to the tension or potential of the electricity.[16]

On the aether versus action at a distance, Maxwell argued:

I have no new discovery to bring before you this evening. I must ask you to go over very old ground, and to turn your attention to a question which has been raised again and again ever since men began to think.

The question is that of the transmission of force. We see that two bodies at a distance from each other exert a mutual influence on each other's motion. Does this mutual action depend on the existence of some third thing, some medium of communication, occupying the space between the bodies, or do the bodies act on each other immediately, without the intervention of anything else?

The mode in which Faraday was accustomed to look at phenomena of this kind differs from that adopted by many other modern inquirers, and my special aim will be to enable you to place yourselves at Faraday's point of view, and to point out the scientific value of that conception of *lines of force* which, in his hands, became the key to the science of electricity.

When we observe one body acting on another at a distance, before we assume that this action is direct and immediate, we generally inquire whether there is any material connection between the two bodies; and if we find strings, or rods, or mechanism of any kind, capable of accounting for the observed action between the bodies, we prefer to explain the action by means of these intermediate connections, rather than to admit the notion of direct action at a distance.

. . .

[I]n many cases the action between bodies at a distance may be accounted for by a series of actions between each successive pair of a series of bodies which occupy the intermediate space; and it is asked, by the advocates of mediate action, whether, in those cases in which we cannot perceive the intermediate agency, it is not more philosophical to admit the existence of a medium which we cannot at present perceive, than to assert that a body can act at a place where it is not.

To a person ignorant of the properties of air, the transmission of force by means of that invisible medium would appear as unaccountable as any other example of action at a distance, and yet in this case we can explain the whole process, and determine the rate at which the action is passed on from one portion to another of the medium.

. . .

The force is therefore a force of the old school – a case of *vis a tergo* – a shove from behind.

PART V   **Mechanical versus electrodynamical world views**

...

I have no time to describe the methods by which every question relating to the forces acting on magnets or on currents, or to the induction of currents in conducting circuits, may be solved by the consideration of Faraday's lines of force. In this place they can never be forgotten. By means of this new symbolism, Faraday defined with mathematical precision the whole theory of electro-magnetism, in language free from mathematical technicalities, and applicable to the most complicated as well as the simplest cases. But Faraday did not stop here. He went on from the conception of geometrical lines of force to that of physical lines of force. He observed that the motion which the magnetic or electric force tends to produce is invariably such as to shorten the lines of force and to allow them to spread out laterally from each other. He thus perceived in the medium a state of stress, consisting of a tension, like that of a rope, in the direction of the lines of force, combined with a pressure in all directions at right angles to them.

This is quite a new conception of action at a distance, reducing it to a phenomenon of the same kind as that action at a distance which is exerted by means of the tension of ropes and the pressure of rods. When the muscles of our bodies are excited by that stimulus which we are able in some unknown way to apply to them, the fibres tend to shorten themselves and at the same time to expand laterally. A state of stress is produced in the muscle, and the limb moves. This explanation of muscular action is by no means complete. It gives no account of the cause of the excitement of the state of stress, nor does it even investigate those forces of cohesion which enable the muscles to support this stress. Nevertheless, the simple fact, that it substitutes a kind of action which extends continuously along a material substance for one of which we know only a cause and an effect at a distance from each other, induces us to accept it as a real addition to our knowledge of animal mechanics.

For similar reasons we may regard Faraday's conception of a state of stress in the electromagnetic field as a method of explaining action at a distance by means of the continuous transmission of force, even though we do not know how the state of stress is produced.[17]

FURTHER READING

Chapters IV, V and VIII of the first volume of Sir Edmund Whittaker's classic *A History of the Theories of Aether and Electricity* gives a technically detailed history of the evolution of mathematical theories of the aether from the time of Bradley through that of Maxwell. Peter Harman's *Energy, Force and Matter* presents a broad overview of the conceptual development of nineteenth-century physics and of the role the aether played in this. Jed Buchwald's *The Rise of the Wave Theory of Light* treats both optical theory and experiment in the early nineteenth century.

14

Maxwell's theory

Both empirical considerations and mechanical models of the aether provided the foundation upon which Maxwell built his classical theory of electromagnetism. In this chapter we first discuss some formal aspects of this theory and then turn to observational consequences of it that led to a profound conflict between the principles of classical mechanics and those of electromagnetic theory. This will set the stage for our treatment of relativity in later chapters.

14.1 MAXWELL'S EQUATIONS

Prior to Maxwell's great synthesis, the basic laws of the separate fields of electricity and of magnetism were, respectively, Coulomb's law for the electric field E produced by a static point charge q and the Biot–Savart law for the magnetic field B produced by a wire carrying a current i. (See Section 14.A for mathematical statements of these laws and for the mathematical details that support many of the claims made in the present section.) Each of these two laws involves a proportionality constant (say, k_1 and k_2, respectively) that must be determined by experiment. (These are analogous to the constant G in Newton's law of gravitation, Eq. (8.4).) That is, the constants k_1 and k_2 are fixed independently and by different types of phenomena (electrostatics and magnetostatics, respectively). It turns out that the ratio k_1/k_2 has the dimensions of a velocity squared that we denote by c^2 for reasons that will become evident shortly.

The content of these dynamical laws, that in a sense are for electromagnetic phenomena what Newton's laws were for classical gravitational effects, is that they describe in quantitative detail how charges at rest and in motion (currents) produce, respectively, electric and magnetic fields. The laws of electromagnetism as they stood after Maxwell's *Treatise* implied that, in empty space (in a region where there are no charges or currents), these electric and magnetic fields satisfy a wave equation. This means that electric and magnetic fields can propagate through space at the speed c. Hence, propagating E and B fields are solutions to Maxwell's equations. The equations further require that these E and B fields be orthogonal to each other. Figure 14.1 shows the relative orientations of these fields and the direction of propagation (c), while Figure 14.2 illustrates a propagating set of sinusoidal fields.

One of the great achievements of Maxwell's theory is that it predicts that the speed of propagation of these electromagnetic waves must be numerically equal to the square root of the ratio of the proportionality constant (k_1) that appears in

PART V   Mechanical versus electrodynamical world views

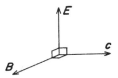

FIGURE 14.1   *E* and *B* fields mutually orthogonal to the propagation direction *c*

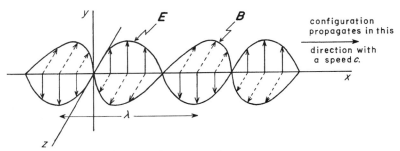

FIGURE 14.2   A propagating plane electromagnetic wave

Coulomb's law (see Eq. (14.9) in Section 14.A) of (electrostatic) attraction (or repulsion) between point charges and the corresponding proportionality constant (k_2) that occurs in the basic law (Eq. (14.10) of Section 14.A) determining the (magnetic) force acting between two current-carrying wires. The essential point is that these constants had each been numerically determined by independent experiments on the basis of electrostatic and magnetic measurements, respectively, and the speed of light *c* had also been measured experimentally prior to Maxwell's work. The value of this predicted speed turns out to be numerically equal to the speed of light (*c*) in vacuum. While this could be mere coincidence, it seems more reasonable to consider this as an indication that light is an electromagnetic wave. In his *Treatise* Maxwell observed:

> The quantity *v*... which expresses the velocity of propagation of electromagnetic disturbances... is... equal to $1/\sqrt{\mu_0 \varepsilon_0}$.
>
> ...
>
> On the theory that light is an electromagnetic disturbance,... *v* must be the velocity of light [*c*], a quantity the value of which has been estimated by several methods. On the other hand, $1/\mu_0 \varepsilon_0$ is [determined by methods that]... are quite independent of the methods of finding the velocity of light. Hence the agreement or disagreement of the values of *c* and of $1/\sqrt{\mu_0 \varepsilon_0}$ furnishes a test of the electromagnetic theory of light.[1]

Furthermore, this velocity (*c*) that enters into Maxwell's equations appeared to offer the hope of defining an absolute reference frame, since *c* had to be the speed relative to some particular frame. The aether was to serve as that frame with respect to which the speed of light had the numerical value *c* (3×10^8 m/s). That is, Maxwell's equations (Eqs. (14.13a)–(14.13d)) contain a velocity explicitly, while Newton's second law of motion (**F** = m**a**) involves only the acceleration, but not a velocity. Hence, while Newton's dynamics could not single out a unique inertial frame,

Maxwell's seemed to do so. We return to this question when we discuss the special theory of relativity in Chapter 16.

14.2 THE DISPLACEMENT CURRENT

Although we tend to lose sight of the fact today, the concept of an actual electromagnetic aether was important for Maxwell's own development of the equations (Eqs. (14.13) in Section 14.A) that now bear his name. One of his major insights in arriving at this final set of equations was the hypothesis of the so-called *displacement current* (the term $\mu_0\varepsilon_0\partial E/\partial t$ on the right side of Eq. (14.13d)), without which propagating electromagnetic waves would not be possible.

How did Maxwell arrive at the displacement current? It is commonly argued that he did this by a purely logical argument to remove a mathematical contradiction. The laws of electricity and magnetism Maxwell inherited from his predecessors included Ampère's law (Eq. (14.18)) that gave the magnetic field in terms of the currents producing it, and the continuity equation (Eq. (14.13e)) that expressed the conservation of charge. When these equations are combined, we find that they are consistent only for steady-state situations – that is, for ones in which the density of electric charges does not change with time. (See Section 14.A for the mathematical details of this argument.) Now, so the standard story goes, Maxwell saw that this conflict could be resolved by modifying Ampère's law (Eq. (14.18)) to include a new term in the current responsible for generating the magnetic field (to obtain Eq. (14.13d)), so that everything became consistent.

Certainly, this is a very neat and logically compelling reconstruction of how Maxwell's great discovery was surely made. But, does it correspond to the sense of what we find in the historical record?[2] It seems not quite, since Maxwell did not focus on any mathematical inconsistency. He did, however, reason that, for a time-dependent electric field ($\partial E/\partial t \neq 0$), a material medium would become polarized, just as we know a dielectric does. Figure 14.3 illustrates the effect of an electric field E applied to a medium consisting of positive and negative charges bound to each other. Under the influence of this imposed field, a 'molecule' becomes polarized (or 'stretched') into the configuration indicated in the figure. For a constant E field, the effect is a transient one and each pair of charges soon equilibrates as shown. However, if the E field varies with time, then the charge configurations are constantly in motion. These moving charges constitute a current – the displacement current. In his *Treatise*, Maxwell tells us that:

> Any increase of this displacement is equivalent, during the time of increase, to a current of positive electricity from within outwards, and any diminution of the displacement is equivalent to a current in the opposite direction.
>
> The whole quantity of electricity displaced through any area of a surface fixed in the dielectric is measured by the quantity which we have already investigated as the surface-integral of induction through that area....[3]

This would make sense even for an electric field in a 'vacuum' that was not empty,

PART V   Mechanical versus electrodynamical world views

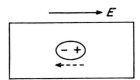

FIGURE 14.3   Electric polarization in a medium

but filled with the aether (recall his model of this aether in Figure 13.7 and the accompanying discussion).

Even though the aether was not used explicitly in many of the formal advances that were made in the development of electromagnetic theory, such a medium was seen as being necessary for the conceptual coherence of the underlying physical processes. For instance, something had to support the stresses and vortices that constituted the field itself. Whether or not we today find it convenient to recall this, Maxwell arrived at his displacement current and his account of the propagation of electromagnetic waves because he took seriously the existence of an electromagnetic aether as a genuine physical substance.

14.3 THE FINAL CLASSICAL THEORY

In his paper of 1865 on 'A Dynamical Theory of the Electromagnetic Field,' Maxwell presented only the mathematical equations that described the electromagnetic field and did not discuss vortices and idle wheels (recall Figure 13.7 again). The equations have proven to be correct and have survived. Eventually, every mechanical model proposed for the aether made some incorrect prediction and they all were finally abandoned. Although this had not yet happened by the end of the nineteenth century, even then it became clear that the aether was an 'immaterial' medium unique in its own right. Thereafter, continued, unsuccessful attempts to detect effects produced by this rarefied aether eventually led to the special theory of relativity, as we shall see in later chapters.

This sequence of developments in the field of electromagnetism is typical of many areas of modern science. The basic phenomena were first observed experimentally and then laws governing these were formulated. In time, the laws were given a concise mathematical form that, in a sense, summarized all of the experimental observations and also allowed new relations to be perceived readily. A physical model, in this case the aether, was also constructed as an aid to conceptualizing the laws and as an attempt to understand electromagnetic phenomena in terms of what was believed to be a more basic and less complicated subject, namely classical mechanics. It was only after consistent defeat that mechanical models of electromagnetism were abandoned and the electromagnetic theory was recognized as a fundamentally distinct branch of physics. We now discuss an experiment that led to a result that was unanticipated by the theory of the electromagnetic aether.

14.4 THE MICHELSON–MORLEY EXPERIMENT

The Michelson–Morley experiment has often been referred to as the most famous negative experiment in the history of physics. Its chief designer was Albert Michelson (1852–1931), an American experimental physicist who devoted the major portion of his professional life to making extremely accurate measurements of the speed of light. In 1907 these efforts brought him the Nobel Prize in physics, thus making him the first American scientist to be so honored. Although Michelson was born in Strelno, Prussia, his family emigrated to the United States when Albert was just two years old. He grew up in a small and rugged mining town in California. At age seventeen Michelson entered the U.S. Naval Academy at Annapolis. His early frontier years in California plus his naval training produced a ramrod-straight personality that characterized him throughout his life. He was often cantankerous and certainly was not a warm person. While still an instructor at the Academy and using his own simple equipment, he made his first determination of the speed of light in 1878. In order to pursue his scientific career in earnest, Michelson resigned from the Navy in 1881 and became a professor of physics in 1883 at the newly founded Case School of Applied Science in Cleveland, Ohio. It was there in 1887 that he collaborated with another Case faculty member, Edward Morley (1838–1923), to perform an experiment sensitive enough to detect the earth's motion through the aether. From 1889 to 1892 Michelson was a professor of physics at Clark University in Worcester, Massachusetts, but found his relationship with its president so uncongenial that he accepted a position as professor and the first chairman in the Department of Physics at the recently established University of Chicago. He remained there until his retirement in 1929.

The discussion we now give of this landmark experiment is predicated upon Newtonian concepts (in particular, the existence of a universal time for all observers). The assumption is that the aether is the rest frame with respect to which $c = 3 \times 10^8$ m/s and that our apparatus is moving through this privileged frame with a speed v, as indicated in Figure 14.4.[4] The ratio v/c is quite small since the speed of the earth (in its orbit around the sun) through space is about 18 miles/s [29 km/s] while the speed of light is 1.86×10^5 miles/s [3.00×10^5 km/s], so that $v/c \approx 10^{-4}$. The basic idea of the experiment is to send light down to a mirror and back (a distance ℓ), once along the line of motion of the earth and once at right angles to that motion and to measure the time difference for these two round-trip journeys.

We begin by considering the path from the light source to the mirror as parallel to **v** (as illustrated in Figure 14.5) and compute the time for a light beam to travel from the source to the reflector and back again. Notice that, if the apparatus were at rest, then we would obtain simply $2\ell/c$ for the total time of flight. In all cases the mirror and light source are held rigidly at a fixed distance ℓ from each other by the arm of the apparatus. As seen from the 'aether' frame (through which this apparatus moves with a speed v), the time t_D taken for the light signal (traveling at a speed c relative to the aether) to travel from the light source (S) down to the mirror (M') is obtained

PART V Mechanical versus electrodynamical world views

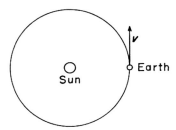

FIGURE 14.4 Motion of the earth through the aether

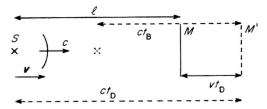

FIGURE 14.5 In a stationary aether, a beam of light (speed c) moving parallel to an apparatus (speed v)

from Figure 14.5 as $ct_D = (\ell + vt_D)$ so that

$$t_D = \frac{\ell}{c - v} \tag{14.1}$$

and for the return flight the time back is obtained from $ct_B = (\ell - vt_B)$ so that

$$t_B = \frac{\ell}{c + v} \tag{14.2}$$

Therefore, the total time of flight down and back, parallel to **v**, is $T_\parallel = (t_D + t_B)$ or

$$T_\parallel = \frac{2\ell}{c}\left(\frac{1}{1 - (v/c)^2}\right) \tag{14.3}$$

The results of Eqs. (14.1) and (14.2) can also be obtained immediately by using the classical law for the addition of velocities and realizing that the speed of the light relative to the apparatus arm is $(c - v)$ in the first case and $(c + v)$ in the second.

Next, we turn to the case in which the apparatus is aligned perpendicular to **v**, as indicated in Figure 14.6. Again we have shown the path of the light as seen from the rest frame of the aether. Here $t_D = t_B$ and from Pythagoras' theorem on right triangles we can write $c^2 t_D^2 = (\ell^2 + v^2 t_D^2)$ so that

$$T_\perp = 2t_D = \frac{2\ell}{\sqrt{c^2 - v^2}} = \frac{2\ell}{c}\frac{1}{\sqrt{1 - (v/c)^2}} \tag{14.4}$$

The overall experimental arrangement is shown in Figure 14.7. We expect from Eqs.

200

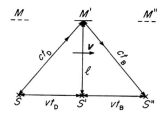

FIGURE 14.6 **In a stationary aether, a beam of light (speed c) moving perpendicular to an apparatus (speed v)**

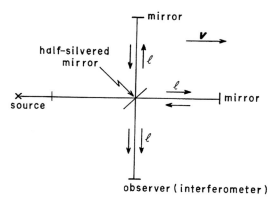

FIGURE 14.7 **An idealized version of the Michelson–Morley experiment**

(14.3) and (14.4) that there will be a time difference of

$$T_\| - T_\perp = \frac{2\ell}{c}\left[\frac{1}{1-(v/c)^2} - \frac{1}{\sqrt{1-(v/c)^2}}\right] \tag{14.5}$$

This time difference is not measured directly in the experiment, but an interference effect is looked for. Since the wave trains are both in phase when they begin their trips along these different paths and since the times of flight of the light signals along these two paths are different, the two light beams will be slightly out of phase when they are recombined. This phase difference can produce an interference fringe pattern. If we now begin to rotate the apparatus (about an axis through the intersection of the two arms and perpendicular to the plane of the page in Figure 14.7), the path difference decreases until it is zero when both arms are inclined at 45° to the direction of **v**. The fringe pattern then shows constructive interference. Therefore, as the apparatus is rotated through 90° from its original position (effectively the parallel and perpendicular arms are interchanged), the change in time difference is twice that given by Eq. (14.5) so that

$$\Delta T_{\text{total}} = \frac{4\ell}{c}\left(\frac{1}{1-(v/c)^2} - \frac{1}{\sqrt{1-(v/c)^2}}\right) \cong \frac{4\ell}{c}\left(\frac{1}{2}\frac{v^2}{c^2}\right) = \frac{2\ell}{c}\frac{v^2}{c^2} \tag{14.6}$$

We have approximated the denominators here by an expansion in $(v/c)^2$ since (v/c) is

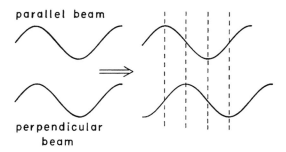

FIGURE 14.8   A fringe shift produced by interference

very small. The effect to be detected is second order in (v/c) (about 10^{-8}), so that the experiment is a difficult one requiring great precision.

Let us look in a bit more quantitative detail at these results. Because $T_\parallel$ is expected to be greater than $T_\perp$ from Eq. (14.5), the wave train traveling parallel to **v** will fall somewhat behind the wave train going via the other route, as indicated in Figure 14.8. We can compute the number of fringe shifts expected if we recall that for a periodic wave $\lambda v = \lambda/\tau = c$, where λ is the wavelength and v the frequency. For each amount of time τ (the period of the wave) that one wave train falls behind the other, there will be one complete fringe shift in the interference pattern. For a complete fringe shift, the pattern would go from one of constructive interference back to one of constructive interference, while for half a fringe shift it would pass from one of constructive interference to one of total destructive interference. From Eq. (14.6) we see that, for $\ell \cong 11$ m as in one version of the experiment, the total number n of fringes shifted as the apparatus is rotated through 90° is $n \approx 0.4$ for a typical wavelength (say, $\lambda \cong 5.5 \times 10^{-7}$ m) in the visible part of the spectrum used by Michelson. However, the Michelson–Morely experiment gave a null result. No fringe shift was detected.

14.5 PRECURSORS TO RELATIVITY

In 1892 George FitzGerald (1851–1901), a professor of natural and experimental philosophy at Trinity College in Dublin, found a way to avoid the implications of the negative result of the Michelson–Morley experiment. He showed that if the measuring apparatus contracted slightly in the direction of its motion through the aether, then a null result would be expected. In order to see this quantitatively, let us consider a Michelson–Morley apparatus as pictured in Figure 14.7, but this time with an arm of length $\ell_\parallel$ along **v** and another $\ell_\perp$ at right angles to **v**. If we correspondingly modify Eqs. (14.3) and (14.4) and then demand that $T_\parallel = T_\perp$ in order to agree with experiment, we find that

$$\ell_\parallel = \ell_\perp \sqrt{1 - (v/c)^2} \tag{14.7}$$

That is, if ℓ_0 is the length of a rod at rest (or traveling perpendicular to the direction of motion) while ℓ is its length when moving and aligned parallel to **v**, then the FitzGerald conjecture becomes

$$\ell = \ell_0\sqrt{1 - (v/c)^2} \qquad (14.8)$$

Notice that, once again, this is a second-order effect depending on $(v/c)^2$. However, it was very difficult to understand why the tenuous aether should be able to shorten a body moving through it. In this explanation FitzGerald still adhered to Newtonian concepts of space and time.

The contraction hypothesis gained acceptance among physicists, but not simply because it explained the negative results of the Michelson–Morley experiment in the manner we have just indicated. The same contraction effect was also arrived at independently later in 1892 by the Dutch theoretical physicist Hendrik Lorentz (1853–1928), a professor of mathematical physics at the University of Leiden and later a recipient of the 1902 Nobel Prize in physics for his theory of electromagnetic radiation. Lorentz was one of the outstanding theoretical physicists of his day and he constructed a highly successful theory of electrons interacting with each other through the medium of the aether. The original 1892 version of Lorentz's theory was unable to account for the negative result of the Michelson–Morley experiment, unless a contraction hypothesis like Eq. (14.8) was adopted. By employing a purely mathematical 'trick', Lorentz was led to a line of reasoning, first expressed in 1892 and later more fully developed in 1895, that accounted for the contraction of a body moving through the aether. In studying Maxwell's equations in a moving frame, he found that the mathematical analysis and solution of this problem was much simpler in terms of a new set of variables, x', y', z', t', instead of the physical variables x, y, z, t. At this point we need not concern ourselves with the precise mathematical relation between x', y', z', t' and x, y, z, t. (These equations, today known as the Lorentz transformations and published by him in their modern form in 1904, will be developed in Chapter 16 in a different context.) When Lorentz wrote out the expression for the electric force in this new set of variables (or frame of reference), he found that it was not the same as in the initial frame. Since Lorentz believed that all molecular forces and interactions were basically electromagnetic in nature, he made a *molecular force hypothesis*[5] that assumed that molecular forces transform (or appear with respect to different reference frames) just as do electromagnetic forces. From this hypothesis he was able to derive what is now referred to as the Lorentz–FitzGerald contraction hypothesis of Eq. (14.8). In an 1895 paper in which he discussed the Michelson–Morley experiment, Lorentz, after having argued that the contraction hypothesis would account for the null result, attempted to make this contraction plausible as follows.

> Surprising as this hypothesis may appear at first sight, yet we shall have to admit that it is by no means far-fetched, as soon as we assume that molecular forces are also transmitted through the aether, like the electric and magnetic forces of which we are able at the present time to make this assertion definitely. If they are so transmitted, the translation will very probably affect the action between two molecules or atoms in a manner resembling the attraction or repulsion between charged particles. Now, since the form and dimensions of a solid body are ultimately conditioned

PART V Mechanical versus electrodynamical world views

> by the intensity of molecular actions, there cannot fail to be a change of dimensions as well.[6]

The result of this theory, and of all experimental attempts to measure the motion of the earth relative to the aether, seemed to be that nature conspired to prevent detection of the privileged reference frame at absolute rest. This was elevated to a new principle of physics by Poincaré, the incisive mathematician, theoretical physicist and philosopher who had been a student at the École Polytechnique, like so many of France's leading scientists, and later became a professor at the University of Paris. At an 1899 Sorbonne lecture, Poincaré summarized his views:

> I regard it as very probable that optical phenomena depend only on the *relative* motions of the material bodies, luminous sources, and optical apparatus concerned and that this is true...*rigorously*.[7]

Poincaré questioned the very existence of the aether. In 1904 at a Congress of Arts and Science at St. Louis, he termed this the *principle of relativity* according to which only relative velocities between inertial observers could be detected, but not their individual absolute velocities. He concluded from this principle not only that the laws of physics must be the same for all inertial observers, but also that no velocity can exceed that of light. In that same year Lorentz produced a theory of electrons that was consistent with this principle, yet he still clung to his belief in the existence of an aether. In Chapter 16 we present Einstein's formulation of special relativity that had no place in it for an aether and that was eventually accepted universally. Well, almost universally, as we see from the following conclusion to the 1916 edition of Lorentz's *The Theory of Electrons*:

> I cannot speak here of the many highly interesting applications which Einstein has made of this principle [of relativity]. His results concerning electromagnetic and optical phenomena...agree in the main with those which we have obtained in the preceding pages, the chief difference being that Einstein simply postulates what we have deduced, with some difficulty and not altogether satisfactorily, from the fundamental equations of the electromagnetic field. By doing so, he may certainly take credit for making us see in the negative result of experiments like those of Michelson, Rayleigh and Brace, not a fortuitous compensation of opposing effects, but the manifestation of a general and fundamental principle.
>
> Yet, I think, something may also be claimed in favour of the form in which I have presented the theory. I cannot but regard the aether, which can be the seat of an electromagnetic field with its energy and its vibrations, as endowed with a certain degree of substantiality, however different it may be from all ordinary matter. In this line of thought, it seems natural not to assume at starting that it can never make any difference whether a body moves through the aether or not, and to measure distances and lengths of time by means of rods and clocks having a fixed position relatively to the aether.
>
> It would be unjust not to add that, besides the fascinating boldness of its starting point, Einstein's theory has another marked advantage over mine. Whereas I have not been able to obtain for the equations referred to moving axes *exactly* the same form as for those which apply to a stationary system, Einstein has accomplished this by means of a system of new variables slightly different from those which I have introduced. I have not availed myself of his substitutions, only because the formulae are rather complicated and look somewhat artificial, unless one deduces them from the principle of relativity itself.[8]

In the next chapter we discuss a series of experiments that attempted to adjudicate between the world view of classical mechanics and that of relativity.

14.A MAXWELL'S EQUATIONS IN MATHEMATICAL FORM

Coulomb's law takes the form[9]

$$E = k_1 \frac{q}{r^2}\hat{r} \qquad \left\{ E = \frac{1}{4\pi\varepsilon_0}\frac{q}{r^2}\hat{r} \right\} \qquad (14.9)$$

for the electric field E produced by a static point charge q and the Biot–Savart law is

$$B = k_2 \frac{i\ell \times r}{r^3} \qquad \left\{ B = \frac{\mu_0}{4\pi}\frac{i\ell \times r}{r^3} \right\} \qquad (14.10)$$

for the magnetic field B produced by a (small) wire (of directed length ℓ) carrying a current i. Here k_1 and k_2 are two constants to be determined by experiment. For two long, parallel straight wires carrying currents i and i' and separated by a distance d, the force dF on a segment of wire of length $d\ell$ is

$$\frac{dF}{d\ell} = 2k_2 \frac{ii'}{d} \qquad \left\{ \frac{dF}{d\ell} = \frac{\mu_0}{2\pi}\frac{ii'}{d} \right\} \qquad (14.11)$$

As stated in the text, the ratio k_1/k_2 has the dimensions of a velocity squared (that we have denoted by c^2).

$$\frac{k_1}{k_2} = c^2 \qquad \left\{ \frac{1}{\mu_0 \varepsilon_0} = c^2 \right\} \qquad (14.12)$$

In terms of the velocity c of Eq. (14.12), the laws of electromagnetism (in modern notation) as they stood after Maxwell's *Treatise* are

$$\nabla \cdot E = 4\pi\rho \qquad \left\{ \nabla \cdot E = \frac{\rho}{\varepsilon_0} \right\} \qquad \text{(Coulomb's law)} \qquad (14.13a)$$

$$\nabla \cdot B = 0 \qquad \{\nabla \cdot B = 0\} \qquad \text{(no magnetic charges)} \qquad (14.13b)$$

$$\nabla \times E = -\frac{1}{c}\frac{\partial B}{\partial t} \qquad \left\{ \nabla \times E = -\frac{\partial B}{\partial t} \right\} \qquad \text{(Faraday's law of induction)} \qquad (14.13c)$$

$$\nabla \times B = \frac{1}{c}\frac{\partial E}{\partial t} + \frac{4\pi}{c}j \qquad \left\{ \nabla \times B = \mu_0\varepsilon_0\frac{\partial E}{\partial t} + \mu_0 j \right\} \qquad \begin{array}{l}\text{(Ampère's law with}\\ \text{Maxwell's displacement}\\ \text{current)}\end{array} \qquad (14.13d)$$

$$\nabla \cdot j + \frac{\partial \rho}{\partial t} = 0 \qquad \left\{ \nabla \cdot j + \frac{\partial \rho}{\partial t} = 0 \right\} \qquad \text{(conservation of charge)} \qquad (14.13e)$$

$$F = q\left(E + \frac{v}{c} \times B\right) \qquad \{F = q(E + v \times B)\} \qquad \text{(Lorentz's force law)}[10] \qquad (14.13f)$$

In empty space (where $j = \rho = 0$), the first four equations imply that each component of the E and B fields satisfies the wave equation[11]

$$\left(\nabla^2 - \frac{1}{c^2}\frac{\partial^2}{\partial t^2}\right)E = 0 \qquad \left\{\left(\nabla^2 - \mu_0\varepsilon_0\frac{\partial^2}{\partial t^2}\right)E = 0\right\} \qquad (14.14)$$

That is, if we let $f(r, t)$ represent any of the components E_x, E_y, E_z or B_x, B_y, B_z, then f satisfies the equation

$$\left(\nabla^2 - \frac{1}{c^2}\frac{\partial^2}{\partial t^2}\right)f(r, t) = 0 \qquad (14.15)$$

As an illustration, if we consider a one-dimensional problem, so that f depends only upon x and t, then Eq. (14.15) becomes

$$\left(\frac{\partial^2}{\partial x^2} - \frac{1}{c^2}\frac{\partial^2}{\partial t^2}\right)f(x, t) = 0 \qquad (14.16)$$

It is straightforward to show that the most general solution to Eq. (14.16) is[12]

$$f(x, t) = g(x - ct) + h(x + ct) \qquad (14.17)$$

where g and h are arbitrary functions. But Eq. (14.17) represents waves of fixed shape propagating (to the left or to the right) with a speed c. Hence, propagating E and B fields are solutions to Maxwell's equations. The conditions of Eqs. (14.13c) and (14.13d) then require that these E and B fields be orthogonal to each other. These results are illustrated in Figures 14.1 and 14.2 of Section 14.1.

However, there remains an important point to be emphasized. Even though the laws of electricity and magnetism that Maxwell inherited from his predecessors were essentially Eqs. (14.13a)–(14.13e), there was one major difference. Rather than Eq. (14.13d), Maxwell had to begin with Ampère's law

$$\nabla \times B = \frac{4\pi}{c}j \qquad (14.18)$$

A difficulty is readily seen when we take the divergence of Eq. (14.18) to obtain[13]

$$\nabla \cdot (\nabla \times B) \equiv 0 = \frac{4\pi}{c}\nabla \cdot j \qquad (14.19)$$

or

$$\nabla \cdot j = 0 \qquad (14.20)$$

If we now combine this with the continuity equation (Eq. (14.13e)), we find that

$$\frac{\partial \rho}{\partial t} = 0 \qquad (14.21)$$

This implies that Ampère's law is valid only for steady-state situations. But we often have physical charge distributions that are time-dependent. This is the 'inconsistency' referred to in Section 14.2.

FURTHER READING

Ivan Tolstoy's *James Clerk Maxwell* is a well-known biography of the founder of modern electromagnetic theory. Dorothy Livingston's *The Master of Light* is an engaging and informative story of a great scientist as written by his daughter. Jed Buchwald's *From Maxwell to Microphysics* is a scholarly study of electromagnetic theory in the last quarter of the nineteenth century. An intermediate modern technical treatment of basic electromagnetic theory can be found in *Foundations of Electromagnetic Theory* by John Reitz and Frederick Milford, while John Jackson's *Classical Electrodynamics* remains the standard advanced reference for that subject.

15

The Kaufmann experiments

In the early years of the twentieth century, Walter Kaufmann (1871–1947) performed a series of experiments to determine the variation of the electron's mass with velocity. Today physics texts often cite these experiments as confirming the formula

$$m = \frac{m_0}{\sqrt{1 - (v/c)^2}} \qquad (15.1)$$

derived on the basis of the special theory of relativity (see Section 17.2). The solid curve in Figure 15.1 shows a plot of m/m_0 versus $\beta = v/c$ and the circles represent data points.[1] In fact though, as we show below, Kaufmann's data were originally taken as evidence against the special theory of relativity. This chapter is the story of how that verdict was arrived at and how, finally, reversed.

In a fifteen-year span approximately centered around 1900, a question of considerable interest to physicists was the possible electromagnetic origin for part or all of the mass of the electron. Although detailed, quantitative arguments can be given, the basic idea is simply that the mass of a body can be considered as a measure either of its resistance to acceleration or, equivalently, of the work necessary to set the body into motion. Classical electrodynamics predicts for a charged body an electromagnetic energy that increases rapidly with speed. Even early experiments to determine the charge to mass ratio e/m for the electron indicated that the experimentally determined mass increased sharply as the speed v of the electron approached the speed of light c. It was then a reasonable question to ask whether or not this empirically observed variation of mass could be accounted for wholly or in part by a model of the electron based on classical electrodynamics.

15.1 RIVAL THEORIES OF ELECTROMAGNETIC MASS

The main characters in this episode were Kaufmann, Max Abraham (1875–1922), Alfred Bucherer (1863–1927), Lorentz and Planck. While Planck and Lorentz are familiar enough physicists, Kaufmann, Abraham and Bucherer are not generally as well known. Kaufmann studied at the Universities of Berlin and Munich, receiving his Ph.D. from Munich in 1894. While an assistant in the Physics Institute at the University of Berlin (1896–1899), in 1897 he performed experiments on cathode rays.[2] He confirmed that they were negatively charged particles, but did not feel that he was warranted in identifying them as electrons.[3] The credit for that discovery goes

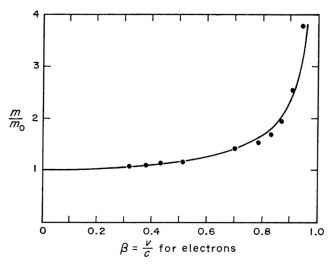

FIGURE 15.1 **Electron mass variation with velocity**

to the British physicist J. Thomson of Cambridge University in 1897. Kaufmann was subsequently on the faculties at the Universities of Berlin, Göttingen, Bonn and Königsberg. It was while at Göttingen and Bonn that Kaufmann performed the experiments (1901–1906) we discuss. There Kaufmann became a close associate of Abraham, who in 1902 proposed his own model of the electron with its charge uniformly distributed over the surface of a rigid sphere. Abraham had been a doctoral student of Planck's at Berlin and was later on the faculties of the Universities of Göttingen, Illinois, Milan and Aachen. He was often embroiled in controversy as a result of his openly critical nature. His death was also a tragic and protracted affair. Although well known among theorists in his own time, Abraham's name is recognized today mainly by those physicists aware of his *Theorie der Elektrizität* in one of its many revised editions – usually the famous (often translated) version by Abraham and Becker that graduate students in physics still consult.

A rival model of the electron was proposed in 1904 by Bucherer, who studied at Johns Hopkins and Cornell Universities before completing his graduate work at Strasbourg in 1895 and subsequently joining the faculty at Bonn. In Bucherer's model, the electron was deformed as it moved through the aether, but in such a fashion that its volume remained constant. Kaufmann's experiments could not be used to make a definite decision between the models of Abraham and of Bucherer.

Beginning in 1892 Lorentz developed his theory of electrons. By 1904 that theory was, in its predictions, indistinguishable from Einstein's special theory of relativity (the subject of Chapters 16 and 17). Lorentz's model of the electron at rest pictured it as a uniform spherical surface charge. As the electron was set into motion through the aether, its transverse dimensions remained unaffected, but its length in the direction of motion was contracted (recall the Lorentz–FitzGerald contraction of Eq. (14.8)). The variation of mass with velocity in this model turns out to be precisely that given by Eq. (15.1). It was during 1906–1907 that Planck subjected the

PART V **Mechanical versus electrodynamical world views**

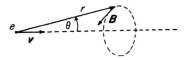

FIGURE 15.2 **The magnetic field produced by a moving charge**

reduction of Kaufmann's data to a careful logical analysis and showed that they actually favored the Einstein–Lorentz predictions, rather than those of Abraham. Finally, by 1908 Bucherer became sufficiently suspicious of the reliability of Kaufmann's work that he performed a different and more accurate set of measurements as a result of which he agreed with the Einstein–Lorentz theory. By that time he had already abandoned his own model of the electron.

Although some technical details are sketched in Section 15.A at the end of the chapter, here we give a simple discussion of the central idea motivating all of these models. For a charge q moving at a velocity v small compared to the speed of light, one can use the Biot–Savart law to calculate the magnetic field B produced by that moving charge.[4] Figure 15.2 illustrates this situation.[5] According to classical theory, the energy carried by an electromagnetic field can be pictured as distributed (or 'stored') throughout space in terms of electric and magnetic energy densities, where, for example, the magnetic energy density per unit volume is $U_{mag} = \frac{1}{2}B^2$. If we add up (or integrate) this magnetic energy over all space and equate the resulting total magnetic energy W_{mag} to the work done on the electron to bring it up to a speed v (that is, to the kinetic energy), we find[6]

$$m = \frac{e^2}{6\pi c^2 a} \tag{15.2}$$

where a is the radius of the electron. This expression is valid only for small velocities (really, in the limit $v \to 0$) and represents the electromagnetic mass of the electron under these conditions. When v becomes appreciable compared to c, then m itself becomes highly velocity dependent.

Abraham's basic idea was to provide an electromagnetic foundation for all mechanics.[7] This was essentially a complete reversal from an earlier tendency on the part of theorists, such as Maxwell, to provide a mechanical basis for electromagnetic phenomena via mechanical models of the aether (as we saw in Chapter 13). The beauty of Abraham's arguments is that they are very general and avoid detailed (and messy) calculations.

15.2 KAUFMANN'S EXPERIMENTS

During the period 1901–1906 Kaufmann published the results of a series of experiments designed to measure the variation of the charge to mass ratio of the electron. Within the framework of classical electrodynamics there was no reason to expect

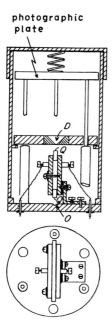

FIGURE 15.3   A sketch of Kaufmann's actual apparatus

the charge e of the electron to vary with v, whereas we have just seen that the electromagnetic mass of the electron should be velocity dependent. Therefore, the observed decrease in the measured ratio e/m was interpreted as an increase in m.

The diagram of Figure 15.3 (based on one of his own publications) shows the basic design of Kaufmann's apparatus. The exterior cylindrical container (placed in a vacuum chamber) has a pair of vertical condenser plates separated by quartz insulators (Q) (this conductor–insulator–conductor arrangement sometimes being termed a 'capacitor'). At O a few grains of radium chloride (supplied by Pierre (1859–1906) and Marie Curie (1867–1934)) provided a source of high-speed β rays. (In previous experiments the identity of Becquerel (or β) rays and cathode rays (or electrons) had been established.) A horizontal electric field was maintained between the plates (by means of a potential difference V across the condenser). The entire apparatus was surrounded by a stack of permanent magnets that provided a uniform horizontal magnetic field B (parallel to the E field) everywhere in the interior of the chamber. Those electrons that had the proper velocity to pass from the source O up between the plates emerged through a tiny (0.2 mm) diaphragm D into an evacuated region having only the magnetic field. They eventually hit the horizontal photographic plate at the top of the chamber and produced a series of exposures at points on the plate. Figure 15.4 gives a schematic of this apparatus. Although Kaufmann did a series of experiments, all were quite similar so that we tell here a 'composite' story that typifies the entire series.[8]

If there were no E field, then the B field alone would produce the circular trajectory indicated in the x–z plane of Figure 15.4 and the electron would impinge

PART V **Mechanical versus electrodynamical world views**

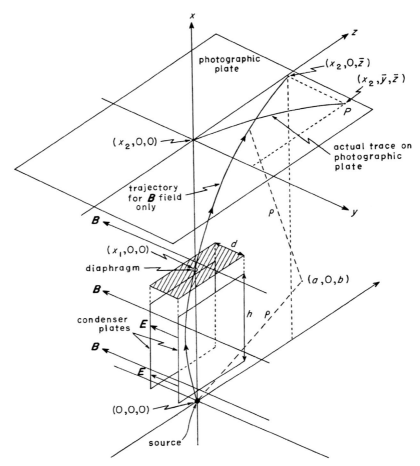

FIGURE 15.4 **A diagram of Kaufmann's apparatus**

on the photographic plate at point $(x_2, 0, \bar{z})$. When both the *E* and *B* fields are on, the electron also receives a horizontal acceleration during the time it travels between the condenser plates so that it finally hits the photographic plate at a point $(x_2, \bar{y}, \bar{z})$. That is, the actual trajectory shown in Figure 15.4 is the composition of two (orthogonal) motions. Since the initial speed of an electron as it left the source was already high (an appreciable fraction of the speed of light *c*) and since the acceleration it received from the electric field in the region between the plates was relatively small, Kaufmann was justified in making the simplifying assumption that v, the speed of the electron, remained essentially constant throughout its motion. In that case, the motion projected into the *x*–*z* plane of Figure 15.4 is uniform circular (see Figure 15.5). This allows us to express the radius (ρ) of this circle in terms of the speed v of the electron, its e/m ratio and the known strength *B* of the magnetic field.[9] Since three points in a plane uniquely determine the circle that passes through them $((0, 0, 0), (x_1, 0, 0)$ and $(x_2, 0, \bar{z})$ in Figure 15.5), we see that, from the known dimensions of his apparatus (x_1, x_2) and the measured value for $\bar{z}$, Kaufmann could

The Kaufmann experiments

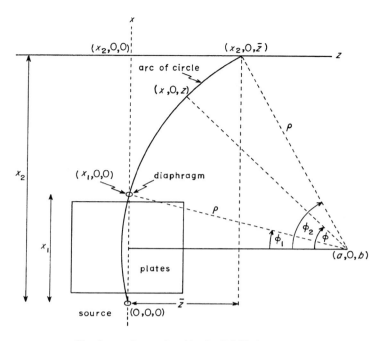

FIGURE 15.5 Circular motion produced by the B field alone

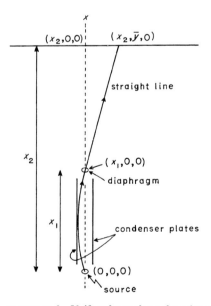

FIGURE 15.6 Uniformly accelerated motion produced by the E field

determine the radius ρ of the trajectory of the electron. This, plus the theoretical relation among the velocity, the strength of the magnetic field and this radius of curvature, allowed Kaufmann to obtain the result $\bar{z} = (e/m)(A/v)$, where A is a constant characterizing the apparatus.[10]

Similarly, in the x–y plane we have (i) uniformly accelerated motion[11] in the region between the condenser plates of Figure 15.4 and then (ii) uniform rectilinear motion after leaving the condenser (and having passed through the hole $(x_1, 0, 0)$ in the diaphragm of Figure 15.4, as illustrated in Figure 15.6). Once again, the (approximate) constancy of the speed v along the trajectory allows us to integrate the equation of motion for the y-coordinate to find that the second experimentally measured coordinate $\bar{y}$ is given as $\bar{y} = (e/m)/(A'/v^2)$, where A' is another constant characteristic of the apparatus.[12] The point is that we have expressions for e/m in terms of the measured coordinates $(\bar{y}, \bar{z})$. If we characterize any of the theories of the electromagnetic mass of the electron by a function $\psi(\beta)$ (where the various forms of $m(v) = m_0 \psi(\beta)$ are given explicitly in Eqs. (15.12) and (15.15) of Section 15.A), then we can write the equations needed to analyze the Kaufmann data as

$$\bar{z} = \left(\frac{e}{m_0}\right)\left(\frac{A}{c}\right)\left[\frac{1}{\beta\psi(\beta)}\right] \tag{15.3}$$

$$\bar{y} = \left(\frac{e}{m_0}\right)\left(\frac{A'}{c^2}\right)\left[\frac{1}{\beta^2\psi(\beta)}\right] \tag{15.4}$$

Here (e/m_0) is the low-velocity limit value of the charge to mass ratio for the electron (obtained either from a least-squares fit to the data or from different experiments done by others (see Section 15.1)).

With this as background, let us now summarize the results of Kaufmann's experiments. Kaufmann's first experimental results on the e/m values for high-speed β rays ($0.787 \leq \beta$ (see note[13]) ≤ 0.945) were published in a 1901 paper. It is a curious paper, to be studied carefully.[14] However, we pass over its false starts here and discuss his 1902 paper. In that year Kaufmann published more data and analyzed them in terms of Abraham's mass. He discovered an algebraic error in a formula of his 1901 paper and corrected that in subsequent analysis. This gave the major contribution to the difference between the results stated in his 1901 and 1902 papers. Kaufmann also made geometrical corrections for the dimensions of his apparatus. Here, for simplicity of presentation, we continue in the following discussion to use the notations $\bar{y}$ and $\bar{z}$ for the experimentally observed quantities. These data are shown in Table 15.1.[15] In spite of the fact that he realized that a small experimental error in determining β would cause a large uncertainty in m because of the rapid variation in $\psi(\beta)$ (Eq. (15.12)) for β close to unity, Kaufmann nevertheless declared:

> The mass of the electrons that constitute the Becquerel rays is dependent on velocity. The dependency can be demonstrated exactly by the formula of Abraham. Therefore the mass of electrons is purely of an electromagnetic nature.[16]

In 1903 Kaufmann published further data and in 1905 attempted to decide definitively among the theories of Abraham, Bucherer and Lorentz. This work was finally summarized in a massive review article of 1906. We discuss the 1905 paper before moving onto alternative interpretations of Kaufmann's data. There appeared the

Table 15.1. *Kaufmann's 1902 data*

$\bar{z}$	$\bar{y}$	β	$\psi(\beta)$
0.348	0.0839	0.957	3.08
0.461	0.1175	0.907	2.49
0.576	0.1565	0.847	2.13
0.688	0.198	0.799	1.96

Table 15.2. *Kaufmann's 1905 data*

$\bar{z}$	$\bar{y}_{exp}$	$\bar{y}_{theo}$		$\delta = (\bar{y} - y_t) \times 10^4$		β	
		Abraham	Lorentz	Abraham	Lorentz	Abraham	Lorentz
0.1350	0.0246	0.0251	0.0246	−5	0	0.974	0.924
0.1919	0.0376	0.0377	0.0375	−1	+1	0.922	0.875
0.2400	0.0502	0.0502	0.0502	0	0	0.867	0.823
0.2890	0.0545	0.0649	0.0651	−4	−6	0.807	0.765
0.3359	0.0811	0.0811	0.0813	0	−2	0.752	0.713
0.3832	0.1001	0.0995	0.0997	+6	+4	0.697	0.661
0.4305	0.1205	0.1201	0.1202	+4	+3	0.649	0.616
0.4735	0.1404	0.1408	0.1405	−4	−1	0.610	0.579
0.5252	0.1666	0.1682	0.1678	−16	−12	0.566	0.527

famous set of nine data points (Table 15.2) that Planck would later scrutinize.[17] All these data were new and independent of Kaufmann's previous measurements. On the basis of the best fit to his data, Kaufmann claimed to be able to distinguish the best theory. At the end of his 1905 paper are some far-reaching conclusions:

> The prevalent results decidedly speak against the correctness of Lorentz's assumption as well as Einstein's. If on account of that one considers this basic assumption refuted, then one would be forced to consider it a failure to attempt to base the entire field of physics, including electrodynamics and optics, upon the principle of relative movement. A choice between the theory of Abraham and Bucherer for the time being is impossible and does not seem to be attainable by observations of the type described above due to the largely numerical identity of the values of $\psi(\beta)$. Whether Bucherer's formula for the optics of moving bodies in the realm of possible observation can yield the same results as Lorentz's, still has to be proven.[18]

15.3 PLANCK'S ANALYSIS OF KAUFMANN'S WORK

As early as 1904, Lorentz questioned whether or not Kaufmann's analysis of his 1902 data really gave conclusive support to Abraham's model over Lorentz's own model. What Lorentz did was to use Kaufmann's method of analysis (basically Eqs. (15.3) and (15.4) above), but substituting for the function $\psi(\beta)$ that corresponding to his own theory (Eq. (15.15)), rather than to Abraham's (Eq. (15.12)). He found a fit nearly as good as Kaufmann's. Lorentz concluded that either theory provided an acceptable fit to the data.

However, it was Planck who plausibly reversed the interpretation of Kaufmann's data from disconfirmation of relativity to support for that theory. Planck's was a classically beautiful application of strict logic to a rather confused situation.[19] In an important 1906 paper he used what are essentially form-invariance arguments (of the type to be discussed in Section 17.2) to obtain a modification of Newton's second law of motion for a charged particle in external E and B fields.[20] This is the relativistically correct form of the law still used today and implied that Eq. (15.1) is the form of the mass to be used in Newton's second law ($d\boldsymbol{p}/dt = \boldsymbol{F}$, where the momentum is $\boldsymbol{p} = m(v)\boldsymbol{v}$).

While emphasizing the great value and beauty of the work of Lorentz and of Einstein on the relativity principle and the importance of investigating its consequences, Planck acknowledged Kaufmann's (1905) experiments.

> To be sure, this question [of the acceptability of the relativity principle] appears to be already answered through the recent and important measurements of W. Kaufmann, that is, however in the negative sense so that every further investigation seems to be unnecessary. In the meantime I would still like to consider it possible, in view of the extremely complex theory of these experiments, that the principle of relativity could be reconciled with these observations if one would more carefully elaborate them.[21]

Later in 1906 Planck returned to the problem posed by Kaufmann's results.[22] Using a method of analysis different from Kaufmann's, Planck computed the expected deflections on the basis of the Abraham and of the Lorentz theories. He was forced to conclude, as had Kaufmann, that Abraham's theory did provide a better fit than the Einstein–Lorentz one. However, he pointed out the discomforting fact that, if one used the empirical numbers as provided by Kaufmann (Table 15.2 above) and Planck's method of analysis, then for either form of $m(v)$, one of the data points gave $\beta > 1$. That was inconsistent for both theories. (Incidentally, Kaufmann might have had suspicions himself since one of his 1903 data points yielded $\beta = 1.04$ as listed in his own table.)

Although the mathematical details of Planck's analysis are rather involved,[23] the logic of his argument is easy to outline. Kaufmann had a set of nine data points (Table 15.2). From the value of $\bar{z}$ alone Planck was able to find the value of a quantity u that he defined as $u = m_0 c/p$.[24] Notice that this value of u is given directly from the data and does not depend upon the specific form of $m(v)$. It is essential for Planck's subsequent argument that only the value of $\bar{z}$, but not of $\bar{y}$, was required to get u from the experimental data.[25] Planck used the same value for e/m_0 as Kaufmann had.[26] On the basis of $m(v)$ as specified by each theory, Planck was able to obtain a theoretical expression for p. From these and his explicit expression for u, he extracted values of β for each theory. All of this was done without yet using the experimental value of $\bar{y}$. With β known, he was able to predict[27] theoretical values of $\bar{y}$ for each theory. The results are given in Table 15.3. Furthermore, Planck's equations for $\bar{z}$ and $\bar{y}$ were such that they could also be used, given the experimental values ($\bar{y}, \bar{z}$), to find the numerical value of β – independently of a choice between Abraham's and Lorentz's theory.[28] It is for the

Table 15.3. *The results of Planck's 1906 analysis of Kaufmann's data*

$\bar{z}$	u	$\bar{y}$	Abraham		Lorenz	
			β	$\bar{y}_{theo}$	β	$\bar{y}_{theo}$
0.1354	0.3871	0.0247	0.9747	0.0262	0.9326	0.0273
0.1930	0.5502	0.0378	0.9238	0.0394	0.8762	0.0415
0.2423	0.6883	0.0506	0.8689	0.0526	0.8237	0.0555
0.2930	0.8290	0.0653	0.8096	0.0682	0.7699	0.0717
0.3423	0.9634	0.0825	0.7542	0.0853	0.7202	0.0893
0.3930	1.100	0.1025	0.7013	0.1054	0.6728	0.1099
0.4446	1.236	0.1242	0.6526	0.1280	0.6289	0.1328
0.4926	1.360	0.1457	0.6124	0.1511	0.5924	0.1562
0.5522	1.510	0.1746	0.5685	0.1823	0.5521	0.1878

first of Kaufmann's data points (Table 15.2) that the troublesome value of $\beta = 1.033$ mentioned above arises.

Of course, Planck realized that there were auxiliary assumptions made in his analysis. In a 1907 paper he focused on one of these. There was some question about how well Kaufmann was able to determine the value of the electric field strength in his apparatus and how constant he could keep it from run to run in his experiment. (For example, if a good vacuum were not maintained, then the β rays could ionize residual air molecules thus reducing the electric field strength in the region between the condenser plates. Leakage of charge from the condenser plates would have had a similar effect.) Planck suggested reanalyzing Kaufmann's data, this time replacing the allegedly known value of the electric field strength E in the region between the condenser plates by α, where α was a free parameter to be determined to produce agreement for each data point $(\bar{y}, \bar{z})$.[29] In his previous analysis, the value of α was taken to be fixed at $E_{max} = 2.013 \times 10^6$ V/m (as given by Kaufmann). He also used a more recently determined value of e/m_0.[30] The results are given in Table 15.4. It is apparent that for either theory $\alpha < E_{max}$. Since α represents the electric field strength between the plates, its value should be a characteristic of the apparatus and hence the same value for all of the data taken. Therefore, the fact that α for the Lorentz theory has less variation (2%) than for the Abraham theory (5%) allows us to conclude that Kaufmann's data favor the Einstein–Lorentz relativity theory. This was Planck's line of reasoning. In either case, though, it is a close call (2% versus 5%) and scarcely overwhelming evidence in favor of relativity. The important outcome of Planck's analysis was that Kaufmann's experiments no longer presented a stumbling block to the acceptance of relativity theory.

In a 1907 review article on relativity, Einstein presented a graph (see Figure 15.7) of $\bar{y}$ versus $\bar{z}$ for Kaufmann's data and compared it to those for relativity. In this graph the circles represent Kaufmann's nine data points. The solid curve appears to consist of straight-line segments joining these points. The crosses are Einstein's

PART V Mechanical versus electrodynamical world views

Table 15.4. *The results of Planck's 1907 reanalysis of Kaufmann's data*

			Abraham		Lorenz	
$\bar{z}$	u	$\bar{y}$	β	α	β	α
0.1354	0.4226	0.0247	0.9655	18,840	0.9211	17,970
0.1930	0.6006	0.0378	0.9045	18,920	0.8572	17,930
0.2423	0.7515	0.0506	0.8424	18,770	0.7994	17,810
0.2930	0.9050	0.0653	0.7779	18,520	0.7414	17,650
0.3423	1.052	0.0825	0.7194	18,580	0.6891	17,800
0.3930	1.200	0.1025	0.6650	18,560	0.6400	17,860
0.4446	1.350	0.1242	0.6157	18,430	0.5953	17,820
0.4926	1.485	0.1457	0.5756	18,240	0.5586	17,710
0.5522	1.649	0.1746	0.5321	18,040	0.5186	17,590

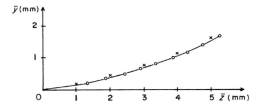

FIGURE 15.7 Einstein's representation of Kaufmann's data

predictions for $\bar{y}$ when $\bar{z} = 1, 2, \ldots, 5$ mm. Einstein's evaluation of this comparison was:

> Considering this difficulty of investigation one might view the identity as being sufficient. The present deviations however are systematic and quite outside of the margin of error concerning Kaufmann's examination. The fact that Mr. Kaufmann's calculations are without mistakes can be deduced from the fact that Mr. Planck by using another method of calculation was led to results that entirely concur with those of Mr. Kaufmann.[31]

He then acknowledged that the theories of Abraham and of Bucherer provided a better fit, but offered his opinion that they had a small probability of being correct since, in their construction, they were ad hoc and constrained in nature. That is, they were more restricted in the range of phenomena to which they applied than was (his own) relativity theory (that applied to all physical phenomena).

15.4 SUBSEQUENT DETERMINATIONS OF e/m_0

We saw that even after Planck's reanalysis of Kaufmann's data, the issue was still not clearly decided in favor of either the Abraham or the Lorentz theory. For completeness, we now indicate the results of e/m_0 determinations in the years following Kaufmann's experiments. In 1906 Adolf Bestelmeyer (1875–1954) used

Table 15.5. *Bestelmeyer's 1907 measured values for* e/m

β	(e/m)	Abraham $e/m_0 = 1.720$	Lorentz $e/m_0 = 1.733$
0.195	1.697	1.694	1.700
0.247	1.678	1.678	1.679
0.322	1.643	1.647	1.640

Table 15.6. *Bucherer's 1909 extracted values for* e/m

β	Lorentz e/m_0	Abraham e/m_0
0.3173	1.752	1.726
0.3787	1.761	1.733
0.4281	1.760	1.723
0.5154	1.763	1.706
0.6870	1.767	1.642

the secondary cathode rays ejected from a metal by incident X rays and measured e/m for these electrons that were first sent through crossed E and B fields acting as a velocity selector and then deflected by the magnetic field alone.[32] His results are given in Table 15.5. (In this and subsequent tables all values of e/m have been multiplied by 10^{-7} for convenience.) The value of e/m_0 was adjusted to give the best fit for each of the theories. It is clear that for these rather small values of β neither theory is definitively favored over the other.

In 1908 Bucherer used β rays from a radium fluoride source and subjected them to crossed electric and magnetic fields, much as Bestelmeyer had done.[33] In reducing this data, Bucherer extracted the value of e/m_0 for each measurement and listed these. That theory was to be preferred that yielded nearly constant values for e/m_0 as β varied. This is the criterion that subsequent experiments also used to judge one theory over another. At the beginning of his paper Bucherer stated that while both classical electron theory (Abraham's) and relativity (Lorentz's) had other empirical successes, his own theory of a constant-volume electron was contradicted by dispersion phenomena. It remained only to decide between Abraham's theory and Lorentz's. Table 15.6 shows that e/m_0 is more nearly constant for Lorentz's theory than for Abraham's. Still, there was some controversy in the literature about the validity of Bucherer's conclusions because of certain technical details of his experiment.

In 1914 the question of which theory yielded the most nearly constant values of e/m_0 was definitively settled.[34] A refinement of Bucherer's method was employed to obtain 26 data points for $0.39152 \leq \beta \leq 0.80730$. The superiority of Lorentz's theory over Abraham's is manifest from Figure 15.8.[35] These results were subsequently corroborated independently in 1915.[36]

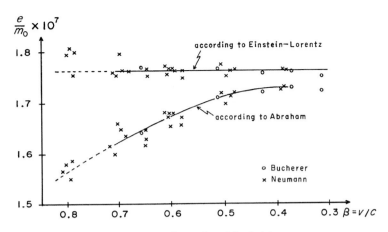

FIGURE 15.8 Conclusive evidence in favor of special relativity

15.5 CONCLUSIONS

Aside from the rather obvious conclusion that Kaufmann's 1905 data were not at first taken as providing support for the special theory of relativity as opposed to classical theories, this case study illustrates several interesting features of the development and acceptance of a scientific theory. The Kaufmann experiment, even though it initially appeared to refute special relativity, did not act as a definitive and deciding ('crucial') experiment in the sense that a strict falsificationist view of science would lead one to expect.[37] The essential idea behind a *crucial experiment* is that a single set of observations can unambiguously refute a theory or decide among competing theories. However, as the Kaufmann experiments nicely illustrate, a single major hypothesis (here Abraham's model versus Lorentz's) is not tested in isolation, but only in conjunction with certain background or auxiliary assumptions (here a theory or model of Kaufmann's apparatus).[38] When disagreement between prediction and observation obtains (as it did in this case), one is not certain whether the primary hypothesis to be tested or one of the auxiliary assumptions is at fault. (Recall the discussion of hypothetical propositions in Section 1.2.) It is unclear that genuine crucial experiments exist in interesting physical situations. We return to this general issue in Section 24.1 when we consider the underdetermination of a scientific theory by its empirical base.

In summary we can say that science does not always operate according to the simple scheme

$$\text{hypothesis} \to \text{prediction} \to \text{refutation} \to \text{rejection of hypothesis}$$

In fact, Planck, aware of Kaufmann's results, still developed a covariant formulation of Newton's second law because of the great potential and generality of relativity. Even after he himself had verified Kaufmann's calculations, he did not reject relativity. Einstein went still further when he took Kaufmann's data as agreeing with relativity. On the grounds that they were ad hoc and contrived in nature, he discounted the likelihood of the two classical models being correct in spite of their admittedly better agreement with that experiment. By the time that new observations conclusively settled the e/m question in 1914, the special theory of relativity was already firmly established through its many other successes. The lack of conclusive support for relativity in this one area no longer had any aspect of a crucial experiment. This question had been reduced to the status of a minor skirmish or niggling detail that would be, and in fact was, taken care of by subsequent measurements.

15.A SOME TECHNICAL DETAILS

The essentials of Abraham's method for calculating the mass variation of the electron based on various models is the following. In advanced formulations of classical mechanics, one uses a formalism in which the equations of motion (essentially Newton's second law of motion, $\boldsymbol{F} = m\boldsymbol{a}$) are obtained from a function termed the Lagrangian L ($L = T - V$, where T is the kinetic energy and V the potential energy of the system being studied). For our purposes here, it will be sufficient to know that, once one has L, the momentum $\boldsymbol{p}$ of the system is obtained as[39]

$$\boldsymbol{p} = \frac{\partial L}{\partial \boldsymbol{v}} \tag{15.5}$$

where $\boldsymbol{v}$ is the velocity of the particle. The mass m of the particle is then identified (really, defined) as[40]

$$\boldsymbol{p} = m\boldsymbol{v} \tag{15.6}$$

and Newton's second law has the form

$$\frac{d\boldsymbol{p}}{dt} = \boldsymbol{F} \tag{15.7}$$

So, once we have $L = L(v)$, we can straightforwardly obtain $m = m(v)$.

Abraham's model for an electron was that of a rigid, uniformly charged sphere of radius a.[41] He showed that the Lagrangian for this model is

$$L_A = -\frac{e^2}{16\pi a} \frac{(1-\beta^2)}{\beta} \ln\left(\frac{1+\beta}{1-\beta}\right) \tag{15.8}$$

where $\beta = v/c$ (and we have used the subscript A to denote Abraham). Equation (15.5) then implies that

$$p_A = \frac{\partial L}{\partial v} = \frac{e^2}{8\pi ac} \frac{1}{\beta} \left[\left(\frac{1+\beta^2}{2\beta}\right) \ln\left(\frac{1+\beta}{1-\beta}\right) - 1\right] \tag{15.9}$$

In the limit $\beta \ll 1$, this reduces to

$$p_A \xrightarrow[\beta \to 0]{} \frac{e^2}{6\pi ac^2} v \equiv m_0 v \tag{15.10}$$

Notice that this value of m_0 agrees with Eq. (15.2), as it must. Therefore, we can express p as

$$p_A = \frac{3}{4}m_0\frac{c}{\beta}\left[\left(\frac{1+\beta^2}{2\beta}\right)\ln\left(\frac{1+\beta}{1-\beta}\right) - 1\right] \tag{15.11}$$

From Eq. (15.6), we find

$$m_A = \frac{3}{4}m_0\frac{1}{\beta^2}\left[\left(\frac{1+\beta^2}{2\beta}\right)\ln\left(\frac{1+\beta}{1-\beta}\right) - 1\right] \equiv m_0\psi_A(\beta)$$

$$= m_0\left(1 + \frac{2}{5}\beta^2 + \frac{9}{35}\beta^4 + \cdots\right) \tag{15.12}$$

where the power series represents an expansion for $\beta \ll 1$.

For the Lorentz electron we have a spherical electron (when at rest) that becomes flattened into an ellipsoid. By similar calculations one then finds

$$L_L = -m_0c^2(1-\beta^2)^{1/2} \tag{15.13}$$

$$p_L = m_0c\beta(1-\beta^2)^{-1/2} \tag{15.14}$$

$$m_L = m_0(1-\beta^2)^{-1/2} \equiv m_0\psi_L(\beta)$$

$$= m_0\left(1 + \frac{1}{2}\beta^2 + \frac{3}{8}\beta^4 + \cdots\right) \tag{15.15}$$

These two theories differ from each other only in second order (that is, in terms of order β^2). For this reason, high speeds were necessary in order to be able to discriminate between them experimentally.

FUTHER READING

Stanley Goldberg's 'The Abraham Theory of the Electron: The Symbiosis of Experiment and Theory' gives a good overall discussion of the issues involved in this attempt at a classical model of the electron. Technical details of the analysis of the Kaufmann experiments and some consideration of the methodological questions involved can be found in James Cushing's 'Electromagnetic Mass, Relativity, and the Kaufmann Experiments.' Much historical background and a wealth of technical detail are given in Arthur Miller's thorough study *Albert Einstein's Special Theory of Relativity*.

PART VI

The theory of relativity

The beauty and clearness of the dynamical theory, which asserts heat and light to be modes of motion, is at present obscured by two clouds. I. The first came into existence with the undulatory theory of light, and was dealt with by Fresnel and Dr. Thomas Young; it involved the question, How could the earth move through an elastic solid, such as essentially is the luminiferous ether? II. The second is the Maxwell–Boltzmann doctrine regarding the partition of energy.

Lord Kelvin (William Thomson), *Nineteenth Century Clouds Over the Dynamical Theory of Heat and Light*

Newton, forgive me; you found the only way which, in your age, was just about possible for a man of highest thought- and creative power. The concepts, which you created, are even today still guiding our thinking in physics, although we now know that they will have to be replaced by others farther removed from the sphere of immediate experience, if we aim at a profounder understanding of relationships.

Albert Einstein, *Autobiographical Notes*

16

The background to and essentials of special relativity

In the popular mind Albert Einstein is identified as the architect of the theory of relativity and as the embodiment of modern scientific genius. Also, among a somewhat older set, he is probably remembered as a man with great humanitarian concerns, as well as the person whose theoretical work was the foundation for the atomic bomb. For one who was to become a singular legend even in his own lifetime, Einstein's background and early years were inauspicious enough. In fact, in later life his own assessment of his ability was that 'I have no particular talent, I am merely extremely inquisitive.'[1] Like Newton, he seems to have had an exceptional power of concentration. We begin with a sketch of Einstein the man and then turn to his arguments for the relativity principle.

16.1 ALBERT EINSTEIN

Einstein was born in Ulm, Germany, into a family none of whose members had shown any brilliance. One year after Albert's birth, the family moved to Munich. His father ran a series of small businesses, most of which ended in failure. Young Albert was not a particularly good pupil, largely because he disliked the rigid German school system that placed great emphasis on learning by rote. He attended a Catholic elementary school, where he was the only Jewish student in the class, and then entered the Luitpold Gymnasium in Munich when he was ten. In Germany at that time the gymnasium was the equivalent (in age groups of the students) of American junior high and high schools combined. Students usually attended these from ages ten through eighteen. Einstein found the intellectual atmosphere almost uniformly oppressive. Throughout his life he thoroughly disliked coercion and yet at the same time he had a devotion to the laws of nature. A similar paradox of his personality manifested itself in his humanitarian concern for mankind in general and his absence of close, personal friendships. In 1930 he expressed himself on this as follows.

> My passionate interest in social justice and social responsibility has always stood in curious contrast to a marked lack of desire for direct association with men and women. I am a horse for single harness, not cut out for tandem or teamwork. I have never belonged wholeheartedly to any country or state, to my circle of friends, or even to my own family. These ties have always been accompanied by a vague aloofness, and the wish to withdraw into myself increases with

the years. Such isolation is sometimes bitter, but I do not regret being cut off from the understanding and sympathy of other men. I lose something by it, to be sure, but I am compensated for it in being rendered independent of the customs, opinions, and prejudices of others, and am not tempted to rest my peace of mind upon such shifting foundations.[2]

His early interest in science was not due to anything that happened in his formal schooling, but to two events that left an impression on him for the rest of his life.

> A wonder... I experienced as a child of 4 or 5 years, when my father showed me a compass. That this needle behaved in such a determined way did not at all fit into the nature of events, which could find a place in the unconscious world of concepts (effect connected with direct 'touch'). I can still remember – or at least I believe I can remember – that this experience made a deep and lasting impression upon me. Something deeply hidden had to be behind things. What man sees before him from infancy causes reaction of this kind; he is not surprised over the falling of bodies, concerning wind and rain, nor concerning the moon or about the fact that the moon does not fall down, nor concerning the differences between living and non-living matter.
>
> At the age of 12 I experienced a second wonder of a totally different nature: in a little book dealing with Euclidean plane geometry, which came into my hands at the beginning of a school year. Here were assertions, as for example the intersection of the three altitudes of a triangle in one point, which – though by no means evident – could nevertheless be proved with such certainty that any doubt appeared to be out of the question. This lucidity and certainty made an indescribable impression upon me. That the axiom had to be accepted unproved did not disturb me. In any case it was quite sufficient for me if I could peg proofs upon propositions the validity of which did not seem to me dubious. For example I remember that an uncle told me the Pythagorean theorem before the holy geometry booklet had come into my hands. After much effort I succeeded in 'proving' this theorem on the basis of the similarity of triangles; in doing so it seemed to me 'evident' that the relations of the sides of the right-angled triangles would have to be completely determined by one of the acute angles. Only something which did not in similar fashion seem to be 'evident' appeared to me to be in need of any proof at all. Also, the objects with which geometry deals seemed to be of no different type than the objects of sensory perception, 'which can be seen and touched.' This primitive idea, which probably also lies at the bottom of the well known Kantian problematic concerning the possibility of 'synthetic judgments a priori,' rests obviously upon the fact that the relation of geometrical concepts to objects of direct experience (rigid rod, finite interval, etc.) was unconsciously present.[3]

When Albert was fifteen his family moved to Milan, Italy, where his father sought new fortunes in business. Einstein remained at the gymnasium in Munich for a year but then dropped out of school with no diploma and traveled for a year or so. At the same time, he renounced his German citizenship. Eventually he completed his secondary education in Aarau, Switzerland, and then entered the Swiss Federal Polytechnic School (Eidgenössische Technische Hochschule, or ETH) in Zurich. The greater intellectual freedom of the Swiss system appealed to an independent thinker like Einstein. As a university student he rarely attended lectures, using instead the notes of his friend Marcel Grossmann (1878–1936), and spent most of his time reading the scientific classics of Hermann von Helmholtz (1821–1894), Gustav Kirchhoff (1824–1887), Ludwig Boltzmann (1844–1906), Maxwell and Heinrich Hertz (1857–1894). One of Einstein's teachers at Zurich was Hermann Minkowski (1864–1909), an outstanding European mathematician of his day. At this period in his life Einstein was fairly indifferent to highly advanced mathematics. It was only much later, when he was formulating the general theory of relativity, that he came to believe in the importance of the more abstract fields of mathematics as being central to the advance of physical theory. Einstein received his university degree in 1900 but was unable to find a position at any university. He obtained his Swiss citizenship in 1901 and later, with the assistance of Grossmann, secured a job as a patent examiner at the government patent office in Bern, Switzerland. In 1903 he married Mileva Maric (1875–1948) who had also been a student at Zurich. It was while in Bern that Einstein did some of his most creative work, even though he was afforded no contact with the leading physicists of his day. He discussed his ideas with an Italian engineer and friend, Michelangelo Besso (1873–1955). In 1905 Einstein published his dissertation research on determining molecular dimensions and was awarded a Ph.D. from the University of Zurich. That year, four other papers of his also appeared in the German journal *Annalen der Physik*. The first, 'On a Heuristic Viewpoint Concerning the Production and Transformation of Light,' provided an understanding of the photoelectric effect on the basis of light quanta. His 'On the Movement of Small Particles Suspended in a Stationary Liquid Demanded by the Molecular Kinetic Theory of Heat' explained Brownian motion and provided quite direct evidence for the existence of atoms. 'On the Electrodynamics of Moving Bodies' contained the special theory of relativity and 'Does the Inertia of a Body Depend Upon Its Energy Content?' established the equivalence of mass and energy, a subject discussed in Chapter 17. Incidentally, the citation of Einstein's 1921 Nobel Prize in physics refers not to the theory of relativity, but to his work on the photoelectric effect and in mathematical physics. Surprisingly enough, it is unclear whether or not Einstein even knew specifically of the null result of the Michelson–Morley experiment at the time he formulated the special theory of relativity. In any event, it appears that the experiment played no role in the line of reasoning by which he arrived at relativity, although there has been recent controversy on this point.[4] There were, however, other unsuccessful attempts to detect motion relative to the aether and Einstein was not unaware of these.

Before we turn to the special theory of relativity, let us complete this brief biographical sketch of Einstein. His 1905 papers soon attracted wide attention and in 1909 he was appointed an *extraordinarius* professor (associate professor) of physics at the University of Zurich. It is interesting to note that, for political reasons, the position was first offered to Friedrick Adler (1879–1960), formerly a fellow student with Einstein at the ETH. When Adler learned that Einstein was also being considered for the post, he wrote to the university officials that, were it possible to get a man of Einstein's ability on the faculty, it would be foolish to appoint Adler.[5] In fairly rapid succession Einstein went from his position at Zurich (1909) to a full professorship at Prague (1910), back to Zurich (1912) at the ETH, and finally in 1913 jointly to the University of Berlin and the Kaiser Wilhelm Institute as a member of the Prussian Academy of Sciences. While in Berlin Einstein completed his general theory of relativity that he published in 1916. There he remained until 1933 when he moved to the Institute for Advanced Study at Princeton where he stayed until his death.

Einstein became a confirmed pacifist during and after the First World War and he also supported the cause of Zionism as a means of gaining independence for the Jewish people. With the great experimental confirmation of his general theory of relativity in 1919, Einstein became a very public figure and he attempted to use this prestige to further the causes of world peace and Zionism. From the years 1921 to 1923 he traveled and lectured in Europe, England, the United States, the Orient and Palestine, and in 1925 in South America. During the unrest of the postwar years in Germany, an organization was formed whose purpose was to discredit Einstein and his views, attacking even his physics as non-German and as detrimental to the true progress of science. One of the enthusiastic members of this group was Philipp Lenard (1862–1947), winner of the 1905 Nobel Prize in physics for his experimental work on the photoelectric effect. Lenard backed the Hitler movement and was an early member of the National Socialist Party. Later in 1933 Lenard published an attack on Einstein. There he held Einstein up as an example of the insidious Jewish influence on Germany and charged that Einstein's work on relativity was based on false premises and was leading scientists into error. For Lenard, Einstein, the Jew, could not be considered a good German. Again in a 1935 inaugural address, Lenard inveighed against Einstein's influence on German science and charged that no good German could be the intellectual follower of a Jew. He concluded these remarks with a 'Heil Hitler.'[6]

The last decade or so that Einstein spent in Berlin, from 1924 to 1933, was marked with increasing political upheaval. Throughout this period, and for the rest of his life, he attempted to formulate a unified field theory of gravitation and electromagnetism, a preliminary version of which he published in 1929. Even his last version in 1950 proved unacceptable so that Einstein spent the last thirty-five years of his life on this unsuccessful quest. When the Hitler regime began its purging of the university faculties in the early 1930s in order to have only 'suitable' German teachers – meaning no faculty members who were Jewish or who had Jewish wives – Einstein left Berlin and assumed his position at Princeton. He also resigned from the Prussian Academy of Sciences in order to spare his old friend Planck and others who remained behind any embarassment. His works were burned publicly in Berlin before the State Opera House. Einstein was no longer an unqualified pacifist and in 1939 he wrote President Franklin Roosevelt (1882–1945) a letter urging the United States to begin development of an atomic bomb lest Germany be the first to obtain such a weapon.

16.2 EINSTEIN'S SKEPTICISM ABOUT CLASSICAL PHYSICS

Let us now turn to Einstein's special theory of relativity that completely restructured our concepts of space and time and that occasioned a thorough revision of the Newtonian world view. Philosophically Einstein is usually identified with logical positivism since he did not accept metaphysical presuppositions as conclusive, but based the elements of his theory upon observed facts. However, he believed that general theories could not be arrived at by a direct examination of the empirical facts alone, but that a free invention of the human mind was also required, after which the theory must be tested to see whether or not it agreed with the phenomena of nature. By his own later recollections, those philosophers who had the greatest influence on him were Hume, Mach and Poincaré. Ever since his friend Besso drew his attention to Mach's *The Science of Mechanics* (see Sections 7.5 and 11.4), that work greatly influenced Einstein's reexamination of the foundations of classical mechanics. He saw 'Mach's greatness in his incorruptible skepticism and independence.'[7] He credited his study of Hume and of Mach with helping him develop the critical attitude necessary to revise his concepts of space and time. His immediate scientific predecessors of greatest importance to his scientific development were Maxwell and Lorentz. Maxwell's theory unified optics and electromagnetism and Einstein considered that work greater than his own. Lorentz based his theory not upon action at a distance but upon the fields that exist only in empty space (or in the aether) and upon charged particles that create these fields so that space, not matter, bore the electromagnetic field. Even as a student, Einstein appreciated that the classical mechanical equivalence of all inertial frames and the form invariance of Maxwell's equations were incompatible. He asked himself what one would see if one could travel at the speed of light and hence ride along with a light wave. In such a frame of reference the electromagntic wave would appear 'frozen' in space and the E and B fields would not oscillate in a plane transverse to the direction of propagation. The young Einstein felt that such an asymmetry between this reference frame and all others traveling at less than the speed of light was not realistic. At age sixty-seven Einstein recalled this as follows.

> [I] had already hit [upon a paradox] at the age of sixteen: If I pursue a beam of light with the velocity c (velocity of light in a vacuum), I should observe such a beam of light as a spatially oscillatory electromagnetic field at rest. However, there seems to be no such thing, whether on the basis of experience or according to Maxwell's equations. From the very beginning it appeared to me intuitively clear that, judged from the standpoint of such an observer, everything would have to happen according to the same laws for an observer who, relative to the earth, was at rest. For how, otherwise, should the first observer know, i.e., be able to determine, that he is in a state of fast uniform motion?[8]

Here Einstein points out that this 'frozen' field that has only a spatial, but not a temporal, variation is not a solution to Maxwell's equations (or, equivalently, to the wave equation, Eq. (14.16), that describes the propagation of a wave through space).[9]

At the beginning of his 1905 relativity paper, he returned to the vexing problem of certain asymmetries. Here we have an example of the type of *Gedankenexperiment* (thought experiment) that Einstein made famous.

> It is known that Maxwell's electrodynamics – as usually understood at the present time – when applied to moving bodies, leads to asymmetries which do not appear to be inherent in the phenomena. Take, for example, the reciprocal electrodynamic action of a magnet and a conductor. The observable phenomenon here depends only on the relative motion of the conductor and the magnet, whereas the customary view draws a sharp distinction between the two cases in which either the one or the other of these bodies is in motion. For if the magnet is in motion

PART VI **The theory of relativity**

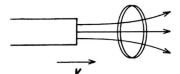

FIGURE 16.1 **A magnet and its field move through a stationary metal ring**

and the conductor at rest, there arises in the neighbourhood of the magnet an electric field with a certain definite energy, producing a current at the places where parts of the conductor are situated. But if the magnet is stationary and the conductor in motion, no electric field arises in the neighbourhood of the magnet. In the conductor, however, we find an electromotive force, to which in itself there is no corresponding energy, but which gives rise – assuming equality of relative motion in the two cases discussed – to electric currents of the same path and intensity as those produced by the electric forces in the former case.

Examples of this sort, together with the unsuccessful attempts to discover any motion of the earth relatively to the 'light medium,' suggests that the phenomena of electrodynamics as well as of mechanics possess no properties corresponding to the idea of absolute rest. They suggest rather that, as has already been shown to the first order of small quantities, the same laws of electrodynamics and optics will be valid for all frames of reference for which the equations of mechanics hold good. We will raise this conjecture (the purport of which will hereafter be called the 'Principle of Relativity') to the status of a postulate, and also introduce another postulate, which is only apparently irreconcilable with the former, namely, that light is always propagated in empty space with a definite velocity c which is independent of the state of motion of the emitting body. These two postulates suffice for the attainment of a simple and consistent theory of the electrodynamics of moving bodies based on Maxwell's theory for stationary bodies. The introduction of a 'luminiferous ether' will prove to be superfluous inasmuch as the view here to be developed will not require an 'absolutely stationary space' provided with special properties, nor assign a velocity-vector to a point of the empty space in which electromagnetic processes take place.[10]

The effect that Einstein is discussing in the introduction to this paper is illustrated in Figures 16.1 and 16.2. If the magnet is in motion (so that the B field is changing with time at a fixed point in space) as in Figure 16.1, then Faraday's law of induction implies that an electric field is generated there in space due to the changing magnetic flux. That is, Faraday's law (Eq. (14.13c)) requires that a changing magnetic field $B(r, t)$ at a point in space gives rise to an electric field E at that same point. Therefore, a current flows in the conducting ring situated at rest in this changing field. On the other hand, when the magnet is at rest (so that the B field does not change with time at a fixed point in space) and the ring moves in the direction indicated in Figure 16.2, no electric field is generated throughout space, but, because of the Lorentz force (Eq. (14.13f)), the conduction charges in the ring are set into motion. In either case, the same physical current flows in the ring. This can be proven from the equations of classical electrodynamics (see Section 16.A). The actual physically observable effect, namely the flow of current in the ring, depends only upon the relative velocity of the magnet and ring, whereas the classical electromagnetic explanation is in principle very different depending upon whether it is the magnet or the ring that is in motion. In both cases the currents are exactly the same, provided only that the magnitude v is the same in both situations. Einstein felt that a correct theory should be for-

The background to and essentials of special relativity

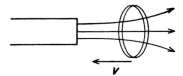

FIGURE 16.2 A metal ring moves through a stationary magnetic field

mulated in such a way that only the relative velocity entered, so that just one account would cover both cases discussed above. We shall see that the E and B fields get transformed into each other as we pass from one inertial frame to another and in such a fashion that there exists a unified description, valid in any inertial frame, of these induction phenomena.

16.3 THE POSTULATES

A careful analysis of the meaning of simultaneity was a starting point of Einstein's 1905 paper on relativity.

> If at the point A of space there is a clock, an observer at A can determine the time values of events in the immediate proximity of A by finding the positions of the hands which are simultaneous with these events. If there is at the point B of space another clock in all respects resembling the one at A, it is possible for an observer at B to determine the time values of events in the immediate neighbourhood of B. But it is not possible without further assumption to compare, in respect of time, an event at A with an event at B. We have so far defined only an 'A time' and a 'B time.' We have not defined a common 'time' for A and B, for the latter cannot be defined at all unless we establish *by definition* that the 'time' required by light to travel from A to B equals the 'time' it requires to travel from B to A. Let a ray of light start at the 'A time' t_A from A towards B, let it at the 'B time' t_B be reflected at B in the direction of A, and arrive again at A at the 'A time' t'_A.
>
> In accordance with definition the two clocks synchronize if
> $$t_B - t_A = t'_A - t_B.$$
> We assume that this definition of synchronization is free from contradiction, and possible for any number of points.[11]

This is the basis for Einstein's criterion of simultaneity and it (along with the assumed constancy of c) then leads to the usual Lorentz transformations (to be discussed in Section 16.5). Here the speed of light is defined as being the same for a trip from one point to another as it is on the return trip (or, equivalently, that the speed of light is independent of the direction of propagation of the light). This criterion for simultaneity is, however, a convention on Einstein's part and cannot be checked experimentally, unless it is possible to measure the one-way speed of light.[12] For example, if, rather than accepting the relativity principle, one assumes only the constancy of the round-trip speed of light and the experimentally verified effect of time dilation (see Section 16.4), one does not obtain the Lorentz transformations, but a more general set.[13] Among these other possible conventions is one that leads to an absolute simultaneity among inertial frames, but to no observable differences from the predictions of special relativity as given by the usual Lorentz transformations.[14] This is true in spite of the fact that these other choices result in a preferred frame of reference. Although we do not pursue those alternatives here and do, instead, adopt Einstein's convention for simultaneity and the Lorentz transformations for the following discussion, we want the reader to be aware of the element of convention in this program of special relativity.

The types of considerations about simultaneity we have just discussed led Einstein to make two assumptions from which follow the special theory of relativity:

1. The form of the laws of physics is the same in all inertial frames. (This implies that the laws of physics cannot be used to measure an absolute velocity.)
2. The speed of light has the same constant value for all inertial observers.

The exact wording of Einstein's statements are:

1. The laws by which the states of physical systems undergo change are not affected, whether these changes of state be referred to the one or the other of two systems of co-ordinates in uniform translatory motion.
2. Any ray of light moves in the 'stationary' system of co-ordinates with the determined velocity c, whether the ray be emitted by a stationary or by a moving body.[15]

Of course, postulate (2) 'explains' the null result of the Michelson–Morley experiment, defining it as an ill-posed problem.

It is also clear, however, that something has to 'give' in classical mechanics to make (2) compatible with (1). The logical essence of the dilemma prior to Einstein was that (i) Newton's laws of motion plus (ii) Maxwell's equations plus (iii) classical concepts of space–time together implied that it must be possible to determine the velocity of an observer relative to the light medium. That is, at least one of (i)–(iii) must be incorrect. It turns out that Maxwell's equations are exactly correct, while (i) and (iii) are at fault when v approaches c.

Einstein's choice of Maxwell's equations over Newton's law was not an arbitrary one or merely a lucky guess. He was aware that all attempts to account for electromagnetic phenomena in terms of mechanical models had failed. (Recall the discussion in Chapter 13 of mechanical pictures of the electromagnetic aether.) It was for this same reason that Lorentz tried to reduce mechanics to electrodynamics.[16] The null result of, for example, the Michelson–Morley experiment (or, perhaps in Einstein's case, of other aether-drift experiments) implied that the earth does not move with respect to the aether. Therefore, Einstein could conclude that there was no reason to assume the existence of the aether and no means of determining an absolute velocity. It follows that our 'common sense' law for adding velocities cannot be correct. We develop this and other consequences of postulates (1) and (2) in this and in the next chapter.

As an interesting aside, we point out that Einstein's 1905 relativity paper – probably the most profoundly revolutionary single paper in the history of physics –

contains no explicit references to the contemporary theoretical or experimental research of his time nor to any exchanges of ideas with the leading physicists of his day. The only acknowledgment is in the last sentence of the paper.

> In conclusion I wish to say that in working at the problem here dealt with I have had the loyal assistance of my friend and colleague M. Besso, and that I am indebted to him for several valuable suggestions.[17]

Einstein once commented that he had never met a real physicist until he was thirty.

A notable contrast in conceptual approaches between Lorentz's theory of electrons (that we discussed in Section 14.5) and Einstein's special theory of relativity gives us an opportunity to make an important philosophical point to which we return in later chapters (especially in Section 23.5). Einstein spoke of theories of principle (such as thermodynamics or special relativity) and constructive theories (such as the kinetic theory of gases or the Lorentz aether-based theory of the electron). In his opinion, the virtues of the former were epistemic security (of the deduced consequences once the basic principles were accepted) and generality of applicability, while the latter had the advantage of providing clarity of comprehension (or of understanding). Einstein characterized these two types of theories as follows:

> We can distinguish various kinds of theories in physics. Most of them are constructive. They attempt to build up a picture of the more complex phenomena out of the materials of a relatively simple formal scheme from which they start out.... When we say we have succeeded in understanding a group of natural processes, we invariably mean that a constructive theory has been found which covers the processes in question.
> Along with this most important class of theories there exists a second, which I will call 'principle-theories.' These employ the analytic, not the synthetic, method.
>
> ...
>
> The advantages of the constructive theory are completeness, adaptability, and clearness, those of the principle theory are logical perfection and security of the foundations.[18]

It is this distinction that formed the basis of the methodological criticism that Einstein leveled against the models of Abraham and of Bucherer – that were useful only for a restricted class of phenomena – versus his own (broadly applicable) theory when he discussed the import of the Kaufmann experiments (see the end of Section 15.3).

16.4 TIME DILATION AND LENGTH CONTRACTION

To begin to tease out a wide range of predictions from Einstein's two concise assumptions, let us turn to a physical argument based on a simple thought experiment. We define a *proper time* interval as the time between two events measured at a fixed point in space in a given frame. That is, it is the time interval between two events in the (common) rest frame of the two events when these events occur at the same location in that frame.

PART VI The theory of relativity

FIGURE 16.3 A round trip for a light beam as seen in the rest frame of the source and mirror

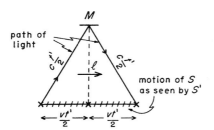

FIGURE 16.4 A round trip for a light beam as seen by a moving observer

Now consider sending a light signal a distance ℓ from a source (L) to a mirror (M) and back to the source, as indicated in Figure 16.3. The apparatus is so constructed that L and M are rigidly held a distance ℓ apart, although we assume that the apparatus itself is in uniform motion (to the right in Figure 16.3) with a speed v. In this figure we take the point of view of an observer in the rest frame (S) of the apparatus. For this observer in S we have very simply for the total time of the round trip of the light signal $\Delta t = 2\ell/c$. This is a proper elapsed time in S. However, this process observed in S' appears as shown in Figure 16.4, since S is in motion (with a speed v) relative to S'. From Pythagoras' theorem applied to the right triangle of the figure, we deduce that the round-trip time $\Delta t'$ as seen by an observer in S' is related to the Δt in S as[19]

$$\Delta t' = \frac{\Delta t}{\sqrt{1 - (v/c)^2}} \qquad (16.1)$$

Therefore, proper times are dilated. This is equivalent to saying that a moving clock runs slowly.

A proper length is one measured by an observer at rest with respect to that length. Let ℓ_0 be the length of a rod (here the distance from A to B) in its rest frame (S') and consider the process illustrated in Figure 16.5. As B passes the origin of S, let $t = t_1$ and, when A passes the origin of S, let $t = t_2$, these times being measured in S. Therefore, $\Delta t = (t_2 - t_1)$ is a proper time interval. In S (by definition) we have, for the distance between A and B, $\ell = v\Delta t$. Since ℓ_0 is the corresponding distance (from A to B) in S' and the corresponding time interval is $\Delta t' = (t'_2 - t'_1)$, then $\ell_0 = v\Delta t'$, where $\Delta t'$ is not a proper time. With use of Eq. (16.1) we then obtain

$$\ell = \ell_0 \frac{\Delta t}{\Delta t'} = \ell_0 \sqrt{1 - (v/c)^2} \qquad (16.2)$$

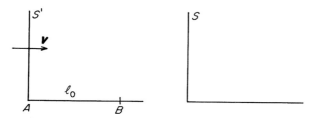

FIGURE 16.5 **A thought-experiment argument for length contraction**

We saw in Section 14.5 (Eq. (14.8)) that, after the null result of the attempt by Michelson and Morley to detect motion relative to the aether, FitzGerald showed that such a result would be expected provided a measuring apparatus contracted slightly in the direction of its motion through the aether according to the relation of Eq. (16.2). A similar result followed from Lorentz's theory of electromagnetic forces. Although Eq. (16.2) is the same formula as the Lorentz–FitzGerald contraction hypothesis, its physical significance is very different. Equation (16.2) is a relation between measured lengths in the two frames, not a statement about an actual contraction of a length.[20]

We now discuss a direct experimental confirmation of the time dilation effect. High-energy particles, known as cosmic rays, arrive at the earth from all directions in outer space. Through nuclear interactions in the upper atmosphere, these cosmic rays produce showers of unstable particles called μ mesons (or muons). For our present purposes, it is sufficient to know that these muons are particles charged either positively or negatively and very similar to electrons (only considerably more massive since $m_\mu = 207\, m_e$) and that they decay into other particles after a characteristic time known as the half-life. That is, if we could place a large collection of newly produced muons at rest on a table and observe them then, on the average, after about 1.5×10^{-6} s half of them would each have decayed into two neutrinos (v) – actually, into a neutrino v and an antineutrino $\bar{v}$ (but this distinction is not important for us here) – and an electron (e^-) as

$$\mu^- \rightarrow e^- + v + \bar{v} \tag{16.3}$$

In the present case we can forget about the neutrinos that are uncharged, massless particles that carry off energy and momentum. From the meaning of the half-life, it follows that, if initially we had 1000 muons, then after 1.5×10^{-6} s we would have just 500; after another 1.5×10^{-6} s, 250; then 125, and so on. This half-life is unaffected by forces external to the muon, so that these decaying particles act as autonomous clocks whose rate of decay in their own rest frames is an invariant quantity.

The basic idea of the experiment is to determine the length of time that a moving muon appears to live as seen by an observer at rest on the earth. If the moving muons on the average survive much longer than 1.5×10^{-6} s, then there is an indication of time dilation. As we can see from Eq. (16.1), the effect will be large only when the speed v of the muons is close to the speed of light. The muons actually

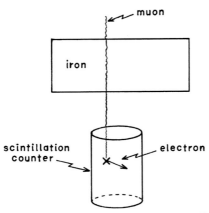

FIGURE 16.6 An experiment to detect time dilation

studied had speeds between $0.9950c$ and $0.9954c$. The first experimental confirmation of this time dilation effect for high-speed radioactive particles was made by Bruno Rossi (1905–1993) in 1941. The version we are presenting was carried out as a demonstration at the Massachusetts Institute of Technology in 1962.[21]

The first set of data was taken atop a mountain 6,300 ft [1.92 km] above sea level. A $2\frac{1}{2}$ ft [0.76 m] layer of iron was used to slow down and stop some of the muons showering to earth through the atmosphere. The experimental arrangement is indicated in outline in Figure 16.6. The amount of iron that a muon can penetrate depends upon the speed of the muon. Only those having speeds in a certain narrow range (here around 0.995 the speed of light) are stopped and brought to rest in the scintillation counter. Those traveling too slowly are stopped in the iron and those traveling too rapidly pass right through the counter without being brought to rest. The characteristic of a scintillation counter is that when a high-speed charged particle enters or passes through it, a single flash of light is given off by the counter. Actually, very many individual blips of light are produced as the particle moves through the counter, but they come so close together that they are detected as a single tiny flash. This light signal is so weak that it must be amplified electronically before we can see it. The details of this amplification process need not concern us here. The output of this counter–amplifier system consists of one flash made by the initial muon entering the counter and another, later flash by the electron generated in the decay process of Eq. (16.3). The upshot is that one obtains a record of how many muons per hour decay in the counter and how long each lasts there before the decay. For example, in one of the hour-long runs on the mountain top, 568 decays were recorded, as was the distribution of their survival times before decay.

We now ask the following question. Given this distribution of survival times of the muons at rest and assuming it were to remain valid for muons moving at $0.995c$ past us, how many of the 568 muons in this sample would survive long enough to reach sea level, 6,300 ft [1.92 km] below the point of these observations? The amount of time it would take a muon traveling at $0.995c$ to cover 6,300 ft [1.92 km] is (for $c = 3.0 \times 10^5$ km/s) 6.4×10^{-6} s, or 4.27 half-lives. The data from the actual

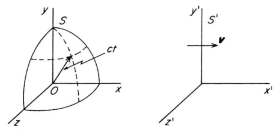

FIGURE 16.7 A spherically propagating light wave

experiment show that just 27 muons should survive long enough to make the trip to sea level where they could then decay in a scintillator. However, when the experiment was actually repeated at sea level, it was found that 412 muons decayed per hour in the scintillator.

Since many more muons survive than would have been the case had their lifetimes been the same in their own rest frame as in the frame of the earth past which they were moving, we can conclude that time dilation does occur. Furthermore, since the half-life of a muon in its rest frame is 1.5×10^{-6} s and since during the 6,300-ft [1.92-km] trip the sample was reduced from 568 to 412, we can see that the number of half-lives n elapsed in the rest frame of the muons is given as $(1/2)^n = (412)/(568) = 0.725$, so that n is 0.464. This means that only 0.464 of a half-life elapsed in the rest frame of the muons rather than the 4.27 half-lives in the laboratory. From Eq. (16.1) we see that the time-dilation factor for $(v/c) = 0.995$ would yield a value in reasonable agreement with this experiment. So, we do have some direct experimental support for the effect of time dilation.

16.5 THE LORENTZ TRANSFORMATIONS

We can use the postulates given in Section 16.3 and the result stated in Eq. (16.1) to obtain the general set of transformations that connect descriptions of events in two inertial frames. Figure 16.7 shows two inertial frames S and S' in uniform relative motion with a velocity **v** directed along the common x-, x'-axes. For convenience we take the origins O and O' to coincide at the instant $t = t' = 0$. One can use the assumptions that space is homogeneous (no preferred origin) and isotropic (no preferred direction) to argue that the equations of transformation must be linear in the variables and reduce to the following simple form for motion along the common x-, x'-axes.[22]

$$x' = ax + bt$$
$$y' = y \tag{16.4}$$
$$z' = z$$
$$t' = dx + ft$$

Here $a(v)$, $b(v)$, $d(v)$ and $f(v)$ are unknown functions of the velocity v. In order to determine them, we need only consider a few special situations. Since we have chosen S and S' to coincide initially ($x = x' = 0$ when $t = t' = 0$), then for an observer in S' watching the origin O of S (the point $x = 0$), we have $x' = -vt'$ for the location of O as seen in S'. Putting this into Eqs. (16.4), we deduce that $b = -vf$. Similarly, if an observer in S watches the origin O' of S' ($x' = 0$), then $x = vt$ for that point so that Eqs. (16.4) then imply that $a = f$. Next, since the (one-way) speed of light in all inertial frames is to be c (postulate 2 of Section 16.3), observers in the frames S and S' of Figure 16.7 will give as their descriptions of the coordinates of the advancing wave front $x = ct$ and $x' = ct'$, respectively. It then follows from Eqs. (16.4) that $d = -av/c^2$. Finally, suppose that at $x = y = z = 0$ in frame S a light is turned on at time t_1 and then off later at time t_2 so that the proper time interval $\Delta t = t_2 - t_1$ has elapsed. From the last of Eqs. (16.4) we can calculate the corresponding times t'_1 and t'_2 to find $\Delta t'$ as their difference. If this result is equated to that of Eq. (16.1), we have the value of $a(v)$. Putting all of these results together, we arrive at the *Lorentz transformations*

$$x' = \gamma(x - vt)$$

$$y' = y \quad (16.5)$$

$$z' = z$$

$$t' = \gamma\left(t - \frac{vx}{c^2}\right)$$

where

$$\gamma = \frac{1}{\sqrt{1 - (v/c)^2}}$$

These relations constitute, in essence, a 'dictionary' or rule for taking the description (x, y, z, t) of an *event* (by which we simply mean the space–time coordinates themselves) in one inertial frame (S) and transcribing it into the corresponding description (x', y', z', t') of the same event in another inertial frame (S'). The important logical point not to be lost sight of in the mathematical manipulations is that Eqs. (16.5) follow directly from Einstein's two postulates (and his convention on simultaneity). Notice that, when $(v/c) \ll 1$ (or, at least formally, $c \to \infty$), then the Lorentz transformations of Eqs. (16.5) reduce to the familiar Newtonian transformations between inertial frames: $x' = (x - vt)$, $t' = t$ (so that we recover 'absolute' time).

In order to be able in the next chapter to complete our discussion of the formal implications of Einstein's two postulates, we must have at hand the relativistic law for the addition of velocities. We can readily obtain this directly from the Lorentz transformations by considering two observers, one in S and one in S', and a third

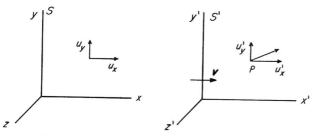

FIGURE 16.8 The relativistic composition of velocities

object (P) moving through space. Let u be the velocity of that object relative to S and u' be its velocity relative to S', as indicated in Figure 16.8. For simplicity we restrict the motion of P to the x–y (or x'–y') plane so that there is no z-component of the velocity. Since, by definition, the x'-component of u' is just the rate of change of x' with respect to t' (and similarly for the other components of u), one can use Eqs. (16.5) to deduce[23]

$$u'_x = \frac{u_x - v}{1 - \frac{vu_x}{c^2}}$$

$$u'_y = \frac{u_y}{\gamma\left(1 - \frac{vu_x}{c^2}\right)} \tag{16.6}$$

Notice that when $(v/c) \ll 1$ (or, once again, $c \to \infty$), Eqs. (16.6) reduce to the familiar Newtonian forms $u'_x \to (u_x - v)$ and $u'_y \to u_y$.

16.A A TECHNICAL DETAIL ON EINSTEIN'S *GEDANKENEXPERIMENT*

We can see directly the equivalence of the physical currents produced by the configurations illustrated in Figure 16.1 and 16.2 if we rewrite Faraday's law of induction (Eq. (14.13c)) in its integral form

$$\mathscr{E} \text{ (electromotive force)} = \int_{\text{circuit}} \mathbf{E} \cdot d\boldsymbol{\ell} = -\frac{1}{c}\frac{d}{dt}\int_{\text{surface}} \mathbf{B} \cdot d\mathbf{A} \tag{16.7}$$

In the first case, when the magnet, and hence its $\mathbf{B}$ field, is in motion (Figure 16.1), the integral on the right side of Eq. (16.7) becomes[24]

$$\lim_{\Delta t \to 0} \frac{1}{\Delta t} \int_{\text{surface}} \left[\mathbf{B}(\mathbf{r} - \mathbf{v}\Delta t) - \mathbf{B}(\mathbf{r})\right] \cdot d\mathbf{A} \tag{16.8}$$

That is, the $\mathbf{B}$ field that at time t is at $\mathbf{r} - \mathbf{v}\Delta t$ will at time $t + \Delta t$ be at $\mathbf{r}$. Similarly, in the second case, when the ring moves with a velocity $-\mathbf{v}$ through a static $\mathbf{B}$ field (Figure 16.2), the time rate of change of the integral in Eq. (16.7) arises due to the motion of the moving surface area enclosed by the ring as

PART VI **The theory of relativity**

$$\lim_{\Delta t \to 0} \frac{1}{\Delta t}\left[\int_{\Sigma(t+\Delta t)} \boldsymbol{B} \cdot \mathrm{d}\boldsymbol{A} - \int_{\Sigma(t)} \boldsymbol{B} \cdot \mathrm{d}\boldsymbol{A}\right] = \lim_{\Delta t \to 0} \frac{1}{\Delta t}\int_{\Sigma}\left[\boldsymbol{B}(\boldsymbol{r}-\boldsymbol{v}\Delta t) - \boldsymbol{B}(\boldsymbol{r})\right]\cdot \mathrm{d}\boldsymbol{A}$$

that is identical to the right side of Eq. (16.8).

FURTHER READING

Philipp Frank's *Einstein* is an excellent biography written by a longtime friend and colleague of Einstein's. Ronald Clark's *Einstein* is a popularized biography written largely in the style of a novel. Jeremy Bernstein's *Einstein* is a brief nontechnical work that concentrates on certain conceptual issues related to Einstein's achievements in physics. Abraham Pais' *Subtle is the Lord* is an exhaustive and authoritative account of Einstein's life and of his work on relativity written by an accomplished theoretical physicist with a style for biography. Paul Schilpp's *Albert Einstein* begins with Einstein's own autobiographical notes, then follows with a collection of essays on Einstein's ideas by the who's who of theoretical physics and mathematics of the day and finally concludes with Einstein's responses to those essays. Arthur Miller's *Albert Einstein's Special Theory of Relativity* discusses in comprehensible technical detail Einstein's thought experiments in formulating relativity. Edwin Taylor and John Wheeler's *Spacetime Physics* is a lively, readable and thought-provoking elementary technical presentation of special relativity.

17

Further logical consequences of Einstein's postulates

In this chapter we consider several of the empirical implications that follow from Einstein's two postulates and relate these to experimental tests of special relativity. The emphasis is on how many diverse, quantitative predictions result from two verbal statements that appear so disarmingly simple and qualitative in nature. Just as Newton's system of classical mechanics and theory of gravity gained universal acceptance because of its many quantitative successes and its wide range of applicability to seemingly unrelated phenomena (recall Chapter 11), so Einstein's special theory of relativity soon became the dominant theory because of the great scope of its achievements.

17.1 RELATIVISTIC DOPPLER EFFECT

We begin by considering the effect that relative motion has on the measured frequency of light. In a classical, or nonrelativistic, context this was originally discussed by Christian Doppler (1803–1853) in 1842. The relativistic version was first treated by Einstein in his 1905 'relativity' paper. In Figure 17.1 let observer B be moving parallel to and in the same direction as the propagation of the wave front, all as seen by observer A. From the point of view of A, for whom the wavelength (or distance between successive wave fronts) is λ, the time t that it takes for two successive wave fronts to pass the moving observer B is $t = (\lambda + vt)/c$.[1] Denote by t' (a proper time) the time B measures between the passage of these two wave fronts (past B). By definition the frequency v' is just $v' = 1/t'$. Since the times t' and t are related by the time dilation equation (Eq. (16.1)), we find[2] for the relation between the observed frequencies v' and v

$$v'_{\text{red}} = \sqrt{\frac{1-(v/c)}{1+(v/c)}}\, v \qquad (17.1)$$

This is the relativistic Doppler-shift formula. We have placed the subscript 'red' on this because, when one moves away from a light source (really, in the same direction as the wave train), one observes the light downshifted to lower frequency (that is, shifted toward the red end of the spectrum for visible light).

Furthermore, we see that when the observer B's velocity $\mathbf{v}$ makes an angle α with the direction of propagation of the wave front (see Figure 17.2), then Eq. (17.1) becomes[3]

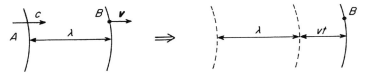

FIGURE 17.1 The origin of the relativistic Doppler-shift effect

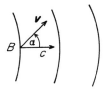

FIGURE 17.2 The situation when the motion of the observer is not along the direction of propagation of the light

$$v' = \frac{(1 - \frac{v}{c}\cos\alpha)}{\sqrt{(1 - (v/c)^2}} v \qquad (17.2)$$

For the observer B moving directly toward A ($\alpha = 180°$ in Eq. (17.2)), the expression for the Doppler shift becomes

$$v'_{\text{blue}} = \sqrt{\frac{1 + (v/c)}{1 - (v/c)}} v \qquad (17.3)$$

For obvious reasons, this is often referred to as the *blueshift*.

Since there is no time dilation classically, the expected expression for the Doppler shift based on an aether model is

$$v' = (1 - \frac{v}{c}\cos\alpha)v \qquad (17.4)$$

We see from Eqs. (17.2) and (17.4) that the relativistic correction to the Doppler shift is of order $(v/c)^2$, just as was the effect sought in the Michelson–Morley experiment. The difference between the relativistic and nonrelativistic expressions for the Doppler effect is in essence due to time dilation.[4]

17.2 MASS–ENERGY EQUIVALENCE

We now recapitulate the logic of the argument by which Einstein obtained one of the most famous and important formulas of special relativity.

1. From his two basic postulates (Section 16.3) for the special theory of relativity, Einstein was able to derive the Lorentz transformations (Eqs. (16.5)) that connect

(or 'translate') a description or characterization of an event (x, y, z, t) in a frame S with that (x', y', z', t') in a frame S' moving with a uniform (or constant) velocity v (here taken to be along the x-axis). From these follows the time-dilation formula of Eq. (16.1).[5] In turn, from this one obtains the Doppler-shift formulas of Eqs. (17.1) and (17.3).

2. If we consider the special case (that will be sufficient for our purposes here) of a plane-polarized electromagnetic wave propagating along the positive x-axis (recall Figures 14.1 and 14.2), then the requirement of the form invariance of Maxwell's equations under these Lorentz transformations implies that the **E** and **B** fields transform as[6]

$$\mathbf{E}' = \gamma\left(1 - \frac{v}{c}\right)\mathbf{E} \tag{17.5a}$$

$$\mathbf{B}' = \gamma\left(1 - \frac{v}{c}\right)\mathbf{B} \tag{17.5b}$$

(Here we have also taken the relative velocity **v** to be parallel to the x-axis.) Since the energy carried by a wave is proportional to the square of the amplitude of oscillation, it is not surprising that the energy U per unit volume carried by an electromagnetic wave is proportional to $(E^2 + B^2)$.[7] This was known to be true in classical electromagnetic theory.[8] Therefore, the relation between the energy density U as seen by S and U' as seen by S' is just

$$U' = \gamma^2\left(1 - \frac{v}{c}\right)^2 U = \left(\frac{1 - (v/c)}{1 + (v/c)}\right)U \tag{17.6}$$

If we now consider a region bounded by two successive wave fronts (by definition a distance λ (the wavelength) apart) and of cross-sectional area A (as indicated in Figure 17.3), then the volume of this region as seen in S is just $V = \lambda A$, while that of the same volume as seen in S' is $V' = \lambda' A$. (Notice that the area A is not Lorentz-contracted because it is aligned at right angles to the relative velocity **v** relating the frames S and S'.) From the Doppler-shift formula of Eq. (17.1) (relating v and v', or equivalently, λ and λ'), we therefore find that $V' = \sqrt{(1 + (v/c))/(1 - (v/c))}\,V$. This implies that the energy content $(E = UV)$ of this light pulse is related in the two frames as

$$E' = E\gamma\left(1 - \frac{v}{c}\right) \tag{17.7}$$

(Realize that here E stands for the energy, not for the electric field strength.) Similarly, if we consider the frame S' moving to the left along the x-axis (by letting $v \to -v$), then we find

$$E' = E\gamma\left(1 + \frac{v}{c}\right) \tag{17.8}$$

243

PART VI The theory of relativity

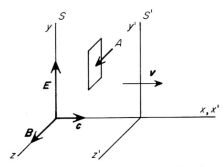

FIGURE 17.3 The geometry for computing the energy content of a light wave in different inertial frames

In his 1905 paper Einstein observes:

> It is remarkable that the energy and the frequency of a light complex vary with the state of motion of the observer in accordance with the same law.[9]

What he is referring to here is that, if we take Planck's expression for the energy of a photon $\varepsilon = h\nu$ (see Section 19.2) and use the Doppler-shift formula for ν (Eqs. (17.1) and (17.3)), then we obtain at once the relation between E' and E derived above from (essentially classical) electrodynamics.

3. We again demand form invariance of the combined laws of classical mechanics and of electrodynamics for a particle of charge e moving in an electromagnetic field under the Lorentz force (Eq. (14.13f))[10]

$$F = \frac{dp}{dt} = e(E + v \times B) \tag{17.9}$$

so that in the frame S' this law has the form

$$\frac{dp'}{dt'} = e(E' + v' \times B') \tag{17.10}$$

Realize that the relation (Eqs. (17.5)) among E', B' and E, B was already known from the form invariance of Maxwell's equations under Lorentz transformations. Then we are led (as Planck pointed out in 1907; recall Section 15.3) to modify the (relativistic) expression for the momentum p to

$$p = \frac{m_0 v}{\sqrt{1 - (v/c)^2}} \tag{17.11}$$

Here m_0 is the rest mass of the particle (the mass of the particle as determined in its own rest frame).

4. It then follows from the work–energy theorem (see Section 17.A for details) that the relativistic expression for the kinetic energy K becomes (with $K = 0$ for $v = 0$)

Further logical consequences of Einstein's postulates

$$K = m_0 c^2 \left[\frac{1}{\sqrt{1-(v/c)^2}} - 1 \right] = m_0 c^2 (\gamma - 1) \tag{17.12}$$

5. By considering a source at rest in frame S that emits two light pulses, each of energy content $\frac{1}{2}L$, one traveling along the positive x-direction and one along the negative x-direction, Einstein argued as follows.[11] If E_1 is the energy content of the source before the two pulses are emitted and E_2 its energy content after the pulses have been emitted, then conservation of energy requires that:

In frame S: $\quad E_1 = E_2 + \frac{1}{2}L + \frac{1}{2}L = E_2 + L$

In frame S'
$$E_1' = E_2' + \frac{1}{2}L\gamma\left(1 - \frac{v}{c}\right) + \frac{1}{2}L\gamma\left(1 + \frac{v}{c}\right) = E_2' + L\gamma$$

Here we have used Eqs. (17.7) and (17.8) in writing down the relations for frame S'. If we subtract these two equations, we find

$$(E_1' - E_1) = (E_2' - E_2) + L(\gamma - 1)$$

But, Einstein observes, $E' - E$ is simply the energy of the same body as observed from two different frames in relative motion, so that this must be the kinetic energy K. Therefore we see that

$$K_1 = K_2 + L(\gamma - 1) \tag{17.13}$$

so that

$$\Delta K = L(\gamma - 1) \approx \frac{1}{2}\left(\frac{L}{c^2}\right)v^2$$

where the last form is valid when $(v/c)^2 \ll 1$ (in the classical regime). Einstein's conclusion is:

If a body gives off the energy L in the form of radiation, its mass diminishes by L/c^2.[12]

Thus we arrive at one of the most famous formulas in modern physics

$$E = mc^2 \tag{17.14}$$

Notice that Einstein did not use the actual relativistic expression for the kinetic energy in order to obtain this dramatic result. However, if we do use Eq. (17.12) in Eq. (17.13), then we find

$$m_1 c^2 (\gamma - 1) = m_2 c^2 (\gamma - 1) + L(\gamma - 1)$$

so that

$$m_1 c^2 = m_2 c^2 + L$$

If we rewrite this as

PART VI   The theory of relativity

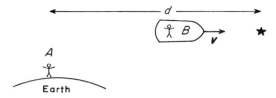

FIGURE 17.4   **A trip to illustrate the twin paradox**

$$(m_1 - m_2) = \frac{L}{c^2}$$

then we see particularly clearly the equivalence between mass and energy.

17.3   THE TWIN PARADOX

The postulates and many of the implications of the special theory of relativity appear so counterintuitive (relative to our every-day, classically based 'common sense') that there is often an uneasy, almost visceral feeling that it cannot really be correct or self-consistent. There have been many attempts to argue that the theory leads to contradictions. One of the most famous historically, and one that still captures the imagination of a wide audience, is the so-called twin paradox. In this section we state the problem and discuss its resolution.[13]

In the 'twin paradox' one of two twins rockets away from the earth for a certain distance, then turns around and rockets back, as indicated in Figure 17.4. To her it appears that the earth and the other twin made a round trip in the opposite direction. Will they both age the same amount? Will each claim that the other aged less because of time dilation? Or will one age less than the other?

The paradox arises because at first sight it appears as though either twin can use the time-dilation formula (Eq. (16.1))

$$\Delta t' = \frac{\Delta t_{proper}}{\sqrt{1 - (v/c)^2}} \qquad (17.15)$$

to show that the other twin after the journey is younger than she. In such an argument Δt_{proper} is taken to be the amount that a twin ages in her own frame (all at a fixed position in that frame). However, each cannot after the trip be younger than the other. It is logically possible that one or the other could be less old (relative to the other) than she would have been had the trip not been taken, even if our 'common sense' makes us feel uncomfortable with that result. The source of the paradox lies in the apparent symmetry of the problem. Can we not equally well consider the one twin to remain on earth while the other rockets off and back or take the second twin to remain at rest in her rocket ship while the earth speeds off and back? One inertial frame is as good as another.

The resolution of the paradox lies in the fact that both frames are not inertial frames, since the rocket ship does undergo an acceleration when it stops, turns around and then returns to earth. An observer in this frame could detect this acceleration (say, by watching a plumb bob swing away from the vertical during the periods of deceleration and acceleration). We know that the observer at rest on the earth remains in an inertial frame at all times. Therefore, if Δt_{proper} is the time elapsed during the journey for the twin in the rocket ship (actually, $\Delta t_{proper}/2$ on the way out and $\Delta t_{proper}/2$ on the way back, each being proper times for twin B), Eq. (17.15) tells us that when the two twins meet at the end of the trip, the one in the rocket ship will have aged less than the one who remained on earth. That is, we could (as a thought experiment) place along the path of B's trip a large number of observers with clocks, all synchronized with each other and all at rest with respect to A in A's frame.[14] These observers could take readings (all in A's frame) as B passes them, record these times and (later) communicate them to A. In this fashion the $\Delta t'$ on the left side of Eq. (17.15) could be measured. (Appreciate that the differences in elapsed times of B's successive locations as measured in A's frame are not proper elapsed times.) Because this argument is made in one inertial frame (A's), even though B changes inertial frames at the turnaround point, the use of Eq. (17.15) is valid. We therefore conclude that the twin who takes the trip ages less than the one who remains on the earth.[15] Neither one has gotten any younger. One has simply aged less than the other.

An equivalent way to see this is to consider three Lorentz frames to demonstrate this effect of asymmetrical aging. Let observer A remain on the earth as previously, B rocket off to the distant star and C, traveling with a velocity $-v$, pass B just as B reaches the star. As they pass, C could note the elapsed time ('age') as recorded on B's clock and then keep track of the additional time that elapses (for C) on the trip from the star back to A on the earth. The total elapsed time as reported by C would be the sum of two proper times and would be related to A's elapsed time ($\Delta t'$) by Eq. (17.15).

However, if we are careful in our reasoning, we must be able to view the trip from either frame and obtain consistent results. As outlined in Figure 17.4, assume that twin A stays on the earth while twin B leaves the earth at speed v and travels a distance d as measured in the earth's frame. Then B comes to a sudden stop and returns to earth, again with speed v. During the whole trip they broadcast their heartbeats to each other. Because they are identical twins, their hearts beat at the same frequency v when on the earth. In what follows we neglect any effects of the starting, turning and stopping accelerations on the heartbeat of B. As early as 1913 the German physicist Max von Laue (1879–1960) explained that any effect on the clock (or twin) during the period of acceleration could not account for the asymmetric aging of the twins, simply because we can make the stretches of uniform motion as long as we please compared to the turnaround time during which acceleration takes place.[16] Therefore, we assume that almost no time is spent accelerating, so that B is essentially moving half the time away from the earth and half the time toward the earth, all with speed v.

In A's frame, the outward and inward journeys each last a time $t = d/v$. For B the distance is Lorentz-contracted and appears to be $d\sqrt{1 - (v/c)^2}$. Hence each journey takes her (as measured in her own frame) only a time $t' = (d/v)\sqrt{1 - (v/c)^2}$. The total journey lasts a time $2\,d/v$ for A, during which she counts $v(2\,d/v)$ heartbeats for herself. For B, the total journey lasts a time $(2\,d/v)\sqrt{1 - (v/c)^2}$ during which she counts $v(2\,d/v)\sqrt{1 - (v/c)^2}$ heartbeats for herself. Thus, twin A counts more heartbeats for A during the trip than B does for B, so that B has aged less than A at the end of the trip. (This is treated in even more detail in Section 17.A at the end of the chapter, where we calculate how many heartbeats each records for the other twin during the complete trip.) There is really no paradox, but only a result that may at first sight strike us as surprising.

We emphasized that, within the framework of the special theory of relativity, we are unable to calculate the effect that the periods of deceleration and acceleration (during the turnaround at the 'star') has on twin B. However, within the general theory of relativity one can calculate the effect on the measured time of twin B. If B experiences a uniform acceleration g caused by a gravitational field during the turnaround, then the elapsed times, or 'ages', Δt_A and Δt_B for the twins are related as[17]

$$\sqrt{1 - (v/c)^2}\,\Delta t_A = \Delta t_B + \left[1 - \frac{1}{\sqrt{1 - (v/c)^2}}\right]\frac{4}{3}t^* \qquad (17.16)$$

This implies, for $(v/c) \ll 1$

$$\Delta t_A - \Delta t_B \approx \left(\frac{v}{c}\right)^2\left(\frac{1}{2}\Delta t_B - \frac{2}{3}t^*\right) \qquad (17.17)$$

Here $t^* = 2v/g$ is the turnaround time. For $t^* = 0$, Eq. (17.16) reduces to Eq. (17.15). Therefore, depending upon the length of the trip (Δt_B) compared to the turnaround time (t^*), B could return younger or older than A. However, for a long enough trip (Δt_B) and fixed turnaround time (t^*), B is always younger than A upon her return. That is the essence of von Laue's argument mentioned earlier.

This question of differential time rates in various inertial frames has also been subjected to experimental test. In October of 1971, atomic clocks (that keep very accurate time) were flown around the world (once on a westward journey and once on an eastward journey) in commercial jets. The elapsed times were compared with those for other atomic clocks kept on the earth. The predicted and observed times were in general agreement.[18]

17.4 SIMULTANEITY AND COEXISTENCE

The great interest in the twin paradox probably stems from the counterintuitive nature of what may seem to be a natural reading of it: that something as fundamen-

Further logical consequences of Einstein's postulates

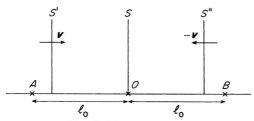

FIGURE 17.5   **The relativity of simultaneity**

tal (and, we feel, as absolute) as the rate of an aging process can apparently be influenced by the state of relative motion of an observer. This is not the only example of conflicts that arise when our intuitions, based on experience in one regime of phenomena, are confronted by predictions of a theory designed to cover phenomena from a very different regime (here velocities v small compared to the speed of light c versus velocities comparable to c). This can force us to reexamine some of our basic concepts. Here we consider two of these: simultaneity and coexistence.

It is instructive to examine this simultaneity question within the context of a specific example as a prelude to our discussion of the meaning of coexistence. The very symmetrical situation illustrated in Figure 17.5 will do for our purposes. Let O be the origin of the inertial frame S and consider two other inertial frames S' and S'' in motion with velocities $\mathbf{v}$ and $-\mathbf{v}$, respectively, relative to S. Two stations A and B are each located a distance ℓ_0 from O as shown in the figure. At time $t=0$ (in frame S) light signals leave A and B for O. That is, event 1 occurs at position A at time $t=0$ so that we denote it as $(-\ell_0, 0)$, while event 2 at B is $(\ell_0, 0)$. It is evident that, by arrangement, events 1 and 2 are simultaneous in frame S so that $\Delta t = 0$. From the Lorentz transformation for the time coordinates (the last of Eqs. (16.5)), we find for the time difference between these two events as seen by an observer in frame S'

$$\Delta t' \equiv t'_2 - t'_1 = \gamma\left(-\frac{v\ell_0}{c^2} - \frac{v\ell_0}{c^2}\right) = -2\gamma\frac{v\ell_0}{c^2} < 0$$

That is, for an observer in frame S', event 1 (at A) takes place after event 2 (at B). A similar calculation for frame S'' (that is, $v \to -v$) yields

$$\Delta t'' \equiv t''_2 - t''_1 = 2\gamma\frac{v\ell_0}{c^2} > 0$$

so that in this frame event 1 (at A) occurs before event 2 (at B). A more picturesque, but potentially misleading, way to state this is that the future for S'' is the present for S and the past for S' (and vice versa).

Still using Figure 17.5, we now turn to a situation in which there are three events. In frame S these are: event 1 (at B), $(\ell_0, 0)$; event 2 (at O), $(0, \tau)$; and event 3 (again at B, $(\ell_0, 2\tau)$. That is, in frame S the order of events (say, emitting a flash of light) is 1,

249

then 2, then 3, each separated by a time interval τ. For frames S' and S'', respectively, we find (again from the last of Eqs. (16.5))

$$\Delta t' \equiv t'_3 - t'_2 = \gamma\left(\tau - \frac{v\ell_0}{c^2}\right)$$

and

$$\Delta t'' \equiv t''_2 - t''_1 = \gamma\left(\tau - \frac{v\ell_0}{c^2}\right)$$

This means that if we were to choose a speed v such that

$$v = \frac{\tau c^2}{\ell_0} = \left(\frac{\tau}{\ell_0/c}\right)c$$

then we would have $\Delta t' = \Delta t'' = 0$. This implies that events 2 and 3 are simultaneous in frame S', while events 1 and 2 are simultaneous in frame S''. If we were now to say that 2 and 3 *coexist* (in S') and 1 and 2 coexist (in S''), then we might be tempted to say that 1 and 3 coexist. That would be peculiar indeed, since there is no frame in which events 1 and 3 are simultaneous. This follows simply from the fact that 1 and 3 are separated by the proper time 2τ (since they both occur at B). We tend to think of existence and coexistence as something absolute, in the sense of either being or not being, independently of any particular observer or frame. The difficulty we created here arose because we effectively equated coexistence with simultaneity. The latter is a concisely defined mathematical condition, while the former implicitly carries much wider connotations for us. Of course, in light of the comments we made earlier about the conventional aspect of the criterion for simultaneity in the special theory of relativity, we could also see this tension between simultaneity and coexistence as resulting from that convention.[19]

If, however, one takes the message of special relativity to be that there is no absolute distinction between past and future, a person can (but need not) arrive at the so-called 'block-universe' model in which existence is seen as deterministic.[20] Such a view would not seem to be essentially different from the Laplacian determinism of classical physics (recall Chapter 12). Whether or not such an in-principle determinism can be consistent with quantum mechanics we discuss in Chapters 23 and 24.

17.A SOME CALCULATIONAL DETAILS

The work–energy theorem of mechanics states that the net work (W) done on a system is exactly equal to the change in the kinetic energy of that system. We can apply this theorem to obtain the relativistic expression for the kinetic energy (Eq. (17.12)) as

$$\Delta K = W = \int F \cdot dr = \int F \cdot v dt = m_0 \int_0^v v \cdot d\left(\frac{v}{\sqrt{1-(v/c)^2}}\right) = m_0 c^2 \int_0^v d\left(\frac{1}{\sqrt{1-(v/c)^2}}\right)$$

$$= m_0 c^2 \left[\frac{1}{\sqrt{1-(v/c)^2}} - 1\right] = m_0 c^2 (\gamma - 1)$$

In this appendix we also pursue in more detail the resolution of the twin paradox via the method of each twin sending and receiving information throughout the journey. Let us begin by asking how many heartbeats each counts for the other from the broadcast of the heartbeats. We must recognize that the frequency received differs from the frequency emitted, owing to the Doppler effect. If source and observer are separating, the apparent (received) frequency is lower than the actual (emitted) frequency; if they are approaching each other, the apparent (received) frequency is higher (see Eqs. (17.1) and (17.3)). During the outward journey, B receives a redshifted signal from A for a time $t'_{red} = (d/v)\sqrt{1-(v/c)^2}$. The instant she reverses direction, the redshift turns to a blueshift. During the inward journey, B receives a blueshifted signal for a time $t'_{blue} = (d/v)\sqrt{1-(v/c)^2}$

Twin A also receives a redshifted signal from B during the outward journey. When B reverses direction, A continues to receive the redshifted signal for an additional time d/c, the time it takes for signals to reach her when B is farthest away. Thus A will receive a redshifted signal for a total time $t_{red} = (d/v + d/c) = (d/v)(1 + v/c)$ and a blueshifted signal for the remaining time $t_{blue} = (d/v - d/c) = (d/v)(1 - v/c)$. It is this time delay that is the essential asymmetry of the problem. The total number of B's heartbeats as counted by A is $(v_{red} t_{red} + v_{blue} t_{blue}) = v(2d/v)\sqrt{1-(v/c)^2}$. That is exactly what B counted for herself (see section 17.3). The total number of A's heartbeats counted by B is $(v_{red} t'_{red} + v_{blue} t'_{blue}) = v(2d/v)$. That is exactly what A counted for herself. Again, we see that there is no paradox.

FURTHER READING

Edwin Taylor and John Wheeler's *Spacetime Physics* can be consulted for more details about the technical material covered in this chapter. Hasok Chang's 'A Misunderstood Rebellion: The Twin-Paradox Controversy and Herbert Dingle's Vision of Science' tells an engrossing story of the history of the twin paradox and of some of the reactions to it. Talal Debs and Michael Redhead's 'The Twin "Paradox" and the Conventionality of Simultaneity' presents an exceptionally clear and comprehensive discussion of this perennial riddle.

18

General relativity and the expanding universe

The special theory of relativity was developed mainly to deal with electromagnetic phenomena and with the space–time background in which they take place. The general theory of relativity was formulated by Einstein to give a description of gravitational phenomena. He was bothered by the fact that the gravitational and inertial masses of a body are always precisely equal. We pointed out previously in Chapter 8 that there is in Newtonian physics a conceptual distinction that can be made between inertial mass (m_i) and gravitational mass (m_g)[1]

$$F = m_i a = \frac{GM}{r^2} m_g$$

Although they are in fact found to be equal experimentally, we may ask why this should be the case. As we shall now see, the general theory of relativity is based on the assumed exact equality of gravitational and inertial masses. What was considered an accident or mere coincidence in classical physics, Einstein recognized as a fundamental law of nature. This was similar to what he and Poincaré each did in interpreting the impossibility of detecting any motion relative to the aether as meaning that there is no aether and that only relative motions are physically meaningful.

18.1 THE BASIC PRINCIPLES

We begin with the *equivalence principle* that Einstein first put forward (but not under that name) in a 1907 review article.[2] Basically the principle states that, in a small region of space–time (*locally*), it is not possible operationally to distinguish between a frame 'at rest' in a uniform gravitational field and a frame being uniformly accelerated through empty space. A famous thought experiment concerns two observers in elevators, as illustrated in Figure 18.1. In (a) an observer is at rest in the uniform gravitational field of the earth.[3] When a ball is released, it falls toward the floor of the elevator with an acceleration $a = g$, where g is just the acceleration due to gravity. In (b) the elevator is located in empty space, but now accelerated upward at a rate $a = g$. Here, too, if a person releases a ball, one will see it 'fall' to the floor of the elevator. If both elevators were windowless, an observer inside could not distinguish situation (a) from (b) by measuring the acceleration of objects inside the elevator.

General relativity and the expanding universe

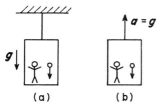

FIGURE 18.1   **The equivalence principle**

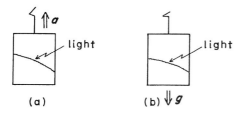

FIGURE 18.2   **The 'fall' of light in a gravitational field**

If we now assume, as Einstein did, that no experiment performed inside the elevator can distinguish between a uniform gravitational field, as in (a), and a uniform acceleration in empty space, as in (b), then we can draw two important conclusions. First, if one common acceleration g in (b) of Figure 18.1 is physically indistinguishable from case (a) for all bodies, then m_i and m_g must be equal. Otherwise, all bodies would accelerate with a common value g relative to the elevator in situation (b), but not in situation (a). We can see this explicitly as follows. The description of the motion in case (a) of Figure 18.1 is just $m_i a = m_g g$, or $a = (m_g/m_i)g$. On the other hand, in case (b) the observed acceleration of all bodies would necessarily be $a = g$. From this it follows that $m_i = m_g$, as claimed.

Second, we can also obtain a surprising prediction for the behavior of light in a gravitational field. In Figure 18.2(a) the reference frame of the elevator is shown as being uniformly accelerated upward through empty space. If an observer at rest in space outside the elevator were to watch a beam of light travel from one side of the elevator to the other, she would see the light move along a straight horizontal line, while at the same time, due to the upward acceleration of the elevator, the floor of the elevator would move upward during the transit of the light beam. An observer standing on the floor of the elevator would therefore see the light beam 'drop', as indicated in Figure 18.2(a). If the equivalence principle is correct, then a beam of light should also 'drop', or be bent, as it passes through a uniform gravitational field, as shown in Figure 18.2(b). In this same 1907 paper Einstein also used the equivalence principle with the formula for the Doppler shift of light to deduce that light originating in a strong gravitational field, as at the surface of the sun or a star, would have a lower frequency (or longer wavelength) when observed at a distant point where the gravitational field is weak. We return to this effect below.

In a 1911 paper, 'On the Influence of Gravitation on the Propagation of Light,' Einstein combined his principle of equivalence with Newton's classical theory of gravitation to make a quantitative prediction of the amount by which a beam of light would be bent in passing near the surface of the sun.[4] In 1916 he published the classic 'The Foundation of the General Theory of Relativity.' There he presented a complete reformulation of the theory of gravitation and made three predictions by which his theory could be distinguished from its Newtonian predecessor. Even though here we can discuss Einstein's theory of gravitation only in general terms since its mathematical formulation is quite advanced and complex, the complete theory is a beautiful example of mathematical and logical elegance in its internal structure that carries its own conviction for some. In 1914 Einstein wrote to his friend Besso:

> Now I am fully satisfied, and I no longer doubt the correctness of the whole system, whether the observation of the eclipse succeeds or not.... The sense of the thing is too evident.[5]

At the end of 1915 he wrote to the theoretical physicist Arnold Sommerfeld (1868–1951):

> You will be convinced of the General Theory of Relativity as soon as you have studied it. Therefore I will not utter a word in its defense.[6]

If we refer again to Figure 18.1, we can see another important implication of the equivalence principle. In the situation depicted in (a) of that figure, suppose the cable supporting the elevator were to snap so that the elevator and its contents were to free fall downward. As far as any experiments that an observer could do inside the elevator, this frame would seem in every respect to be an inertial frame. Since the equivalence principle states that locally a uniform acceleration and a uniform gravitational field are equivalent, we conclude that it is always possible (locally) to go to a freely falling frame that is an inertial frame. Because, according to the principle of equivalence, gravitational effects and those due to accelerations cannot be unambiguously separated, we cannot find an absolute inertial frame.

Mach's idea that the inertial frame of the universe is determined with reference to the mass of the fixed stars and Riemann's belief that matter affects the geometry of space that in turn affects the physics taking place in it (see Section 11.4) were influential in Einstein's thinking in arriving at the general theory of relativity. In his original formulation of the general theory of relativity, Einstein took as a cornerstone the requirement that the properties of space should be completely determined by the distribution of matter in the universe. His field equations of general relativity were supposed to incorporate this version of Mach's principle.[7] However, in 1917 the Dutch astronomer, mathematician and cosmologist Willem de Sitter (1872–1934) discovered a 'vacuum' solution to the Einstein field equations in which three-space is flat, but that has an expanding time scale – all with no matter present. Much later, in 1949, the mathematician Kurt Gödel constructed another solution to these field equations in which the (massive) universe as a whole undergoes a detectable (absolute) rotation. Such a solution does not fit well with the original relativity principle that would lead us to expect such a rotation of the frame of the 'fixed stars' (the universe at large) to be undetectable. Both the de Sitter and Gödel solutions show that Mach's principle is not an essential part of general relativity.[8]

General relativity and the expanding universe

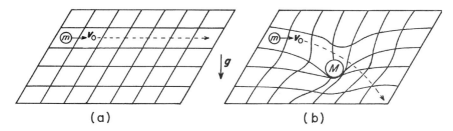

FIGURE 18.3 Geodesic motion in a 'curved' space, giving a 'geometrical' account of gravity

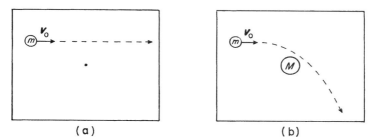

FIGURE 18.4 An equivalent, classical account of gravity in terms of forces in a flat space

Einstein's general theory geometrizes the phenomenon of gravitation because the structure or 'shape' of space affects the motion of bodies through it and that space is in turn affected by the masses located in it. Material particles and light rays then always take the shortest path between two points within the structure of the physical space. These paths are known as *geodesics*. Let us attempt to explain this curvature of space with the following analogy (see Figure 18.3). Suppose that, as in (a) of the figure, we had a flat horizontal rubber sheet with a rectangular cross grid marked on it. If a (small) ball of (negligible) mass m could roll across this sheet without any friction, then, once set in motion with an initial velocity v_0, it would continue to roll in a straight line as indicated by the dotted line. This we would ascribe to inertia, since no net force acts on the ball. Now, as in (b), place a larger sphere M on the rubber sheet. Its weight will stretch the sheet down and distort the grid system.[9] Then the ball m will no longer roll along a straight line, but will dip along the dotted trajectory indicated.

Let us now view these two cases from directly above, assuming we have no depth perception so that the motion appears to take place on a flat surface. The two-dimensional picture would be that of Figure 18.4. We could explain this by saying that in case (a) there were no other masses present to exert a force on m so that the ball continued to move in a straight line. In case (b) the mass M exerted a 'gravitational' attraction on m, just as the sun deflects a planet from a straight-line path onto its trajectory.

Once gravitation was geometrized, Einstein sought to do the same for electrodynamics, but such a unified field theory of both gravity and electrodynamics eluded him until the end of his life, as it has eluded others since him. To a colleague Einstein expressed his reason for this quest.

PART VI   The theory of relativity

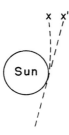

FIGURE 18.5   **The bending of starlight by the sun**

He agreed that the chance of success was very small, but that the attempt must be made. He himself had established his name; his position was assured, and so he could afford to take the risk of failure. A young man with his way to make in the world could not afford to take a risk by which he might lose a great career, and so Einstein felt that in this matter he had a duty.[10]

18.2   EXPERIMENTAL TESTS

We now discuss briefly three empirical tests of the general theory and of the equivalence principle.

1.   *Bending of light rays by a gravitational field*

If we take $E = mc^2$ and think even in classical terms, we might expect light (that carries energy) to have an effective 'mass' $m = E/c^2$ in a gravitational field. Therefore, as indicated in Figure 18.5, when a ray of light, say from a star at X, passes close to a massive body, such as the sun, it should be deflected from its normal straight-line path and would appear to have come from a shifted position, X'. Before he had fully developed his general theory of relativity, Einstein in his 1911 paper used these basically classical ideas to predict the amount of bending expected by the sun and obtained a value of 0.83" (seconds of arc).[11] Later, when he was able to incorporate properly the effects of space curvature, he predicted a value twice as large, 1.7".[12] One might say that the 0.83" deflection represents the 'weight' of light, still in the framework of Newtonian gravitational theory, while the 1.7" deflection has been produced by the effect of the 'curvature of space' that is a feature of general relativity quite foreign to classical physics. This is an extremely small effect to look for astronomically. We now describe how these observations are made.

To simplify matters, consider the earth at two positions six months apart in its orbit around the sun. In position 1 the direction to a fixed star is determined against the background of the night sky, as indicated in Figure 18.6. Later, in position 2, when a ray of light from the same star would just graze the edge of the sun on its way to the earth, the direction to the star is again measured. If, as indicated by the upper solid line in the figure, the light ray is bent, then it will come into the telescope at a steeper angle than it would have had it traveled in a straight line from the star to the earth. The direction to, or position of, that star will appear to have shifted (as

General relativity and the expanding universe

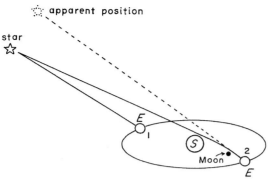

FIGURE 18.6 **The shift of the positions of stars**

indicated by the dotted line in the figure) relative to other stars whose light did not pass close to the sun. The photographic plates exposed at positions 1 and 2 can then be compared for a shift. Of course, under normal conditions the observation of the field of stars from position 2 would be impossible since the brightness of the sun would mask the faint starlight. It is for this reason that the second observation must be made during a total eclipse of the sun by the moon.

Although Einstein's general theory of relativity was published in 1916, no expeditions were organized to those parts of the earth that would experience a total eclipse because the major powers were engaged in World War I. Nevertheless, the Astronomer Royal of England appointed a committee to plan expeditions for making observations during the total eclipse in the spring of 1919, just in case the war should be over by then. The armistice was signed in early November of 1918 and the Royal Society sent one expedition to Sobral in northern Brazil and another to the island of Principe in the Gulf of Guinea, off West Africa. Sending two expeditions to widely separated locations within the zone on the earth where the total eclipse would be visible was a precaution against bad weather that might make observations impossible at one of the sites. It took several months for the expeditions to return to London and for the photographic plates to be compared carefully and possible sources for error to be considered. In early November of 1919, the Royal Society and the Royal Astronomical Society announced the stunning, positive results of both expeditions.[13] Einstein's own reaction was one of joy, but not of surprise. A student who was working for Einstein in 1919 gave the following account.

> Once when I was with Einstein in order to read with him a work that contained many objections against his theory... he suddenly interrupted the discussion of the book, reached for a telegram that was lying on the windowsill, and handed it to me with the words, 'Here, this will perhaps interest you.' It was Eddington's cable with the results of measurement of the eclipse expedition [1919]. When I was giving expression to my joy that the results coincided with his calculations, he said quite unmoved, 'But I knew that the theory was correct', and when I asked, what if there had been no confirmation of his prediction, he countered, 'Then I would have been sorry for the dear Lord – the theory is correct.' ('Da könnt' mir halt der liebe Gott leid tun, die Theorie stimmt doch.')[14]

PART VI   The theory of relativity

FIGURE 18.7   The gravitational redshift far from a star or planet

Similar observations have been repeated several times since and they have given results in essential agreement with the prediction of general relativity.

2. *Gravitational redshift of light*

As we shall see later in more detail, according to quantum theory the energy ε of a beam of light (really, of a single photon) is given as $\varepsilon = h\nu$, where ν is the frequency of the light and h is Planck's constant, the numerical value of which we do not need now. Again using a classical argument and recalling that the Newtonian potential energy in a gravitational field is $V = -GMm/r$, we can compute the change in the ('kinetic') energy of a light beam as it moves from the surface of a body (say the earth or sun) out to infinity (see Figure 18.7). From Planck's $\varepsilon = h\nu$ and the $E = mc^2$ of special relativity, we might (as previously) take the effective mass of the photon to be $m \sim h\nu/c^2$. If we equate the work done by the photon to the change in gravitational potential energy,[15] we obtain

$$\frac{\Delta \nu}{\nu} = \frac{GM}{rc^2} \qquad (18.1)$$

where $\Delta \nu = \nu - \nu'$. Near the surface of the earth, where the gravitational acceleration g is nearly a constant, the frequency shift produced over a height ℓ is (see Figure 18.8)

$$\frac{\Delta \nu}{\nu} = \frac{g\ell}{c^2} \qquad (18.2)$$

This effect has been confirmed by experimental observations.[16]

3. *Advance of the perihelion of Mercury*

According to Newton's law of universal gravitation and his laws of motion, planets orbit the sun in ellipses. However, Einstein's equations predict that these ellipses themselves precess (or rotate). The perihelion, or point of closest approach of the planet to the sun, advances slightly in each revolution as illustrated in Figure 18.9. The value predicted by general relativity is 43″ per century for the planet Mercury, as we saw in Section 11.2. This is over and above all the perturbations produced by

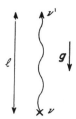

FIGURE 18.8 A terrestrial gravitational redshift

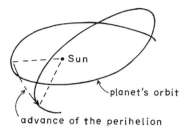

FIGURE 18.9 The precession of Mercury's perihelion

the influence of the other planets and properly accounted for by Newton's law of gravitation.

The actual observations are complicated because they are made from the frame of the earth. As seen from the earth, the perihelion of Mercury's orbit advances about 5,600″ of arc per century.[17] Most of this (5,026″ per century) is accounted for by the precession (or wobble) of the earth's axis of rotation, caused by the gravitational attractions of the sun and of the moon. The period of this precession is about 26,000 years. This leaves 574″ per century of precession of the perihelion of Mercury's orbit to be explained. On the basis of Newtonian gravitational theory, the perturbations on Mercury's orbit due to the other planets in the solar system can be calculated and amounts to 531″ per century. So, 43″ per century remain unaccounted for classically. That is just the amount predicted by the general theory of relativity. A similar advance for the other planets is also predicted (for example, 8.6″ per century for Venus and 3.8″ per century for the earth), but these are extremely small and would be much more difficult to detect.

Even if this internally coherent and beautiful theory, general relativity, did no more than account for these few striking phenomena, it would occupy a place of honor in modern physics. For several decades, until the early 1960s when important new astronomical data became available, many viewed general relativity merely as an arcane curiosity. We have since learned that its range of application goes far beyond these three now classic predictions. General relativity is a central element in our understanding of the universe. To be able to appreciate this, we now turn to a description of the universe as we know it today on the basis of modern astronomical observations and cosmological theories. In this brief summary, we do not always respect the strict time sequence of historical events.

18.3 THE STABILITY OF THE CLASSICAL UNIVERSE

As we saw in Chapter 11, Newton believed that physical space was Euclidean in nature and infinite in extent. Furthermore, in Section 12.2 we referred to some of Newton's Queries from his *Optics* in which he questioned the long-term stability of the universe without the intervention of God (to keep it from collapsing under the mutual gravitational attraction of its parts).[18] In his earlier years Newton held that all of the matter in the universe was distributed throughout only a finite part of the volume of this infinite space and that the material universe was basically static (neither growing nor shrinking in time).[19] One of the problems with only a limited amount of matter in the universe is that any finite quantity of static matter left to itself will necessarily collapse under the mutual gravitational attraction of its mass. Curiously enough, Richard Bentley, the English classicist and clergyman with whom Newton corresponded on the place of God in the universe, found evidence for God's existence, first in the fact that gravity operated between all pairs of bodies in the universe and, second, in that God effectively suspended the law of gravity in the sphere of the fixed stars to prevent the collapse of the universe.[20] Newton himself asked in Query 28 of his *Optics*:

> [W]hat hinders the fixed stars from falling upon one another?[21]

In a General *Scholium* at the end of Book III of the *Principia*, Newton hedged his bets, as it were, by stating that:

> [L]est the systems of the fixed stars should, by their gravity, fall on each other, he [God] hath placed those systems at immense distances from one another.[22]

It was as a result of his correspondence with Bentley in the winter of 1692–3 that Newton revised his position from that of a finite material universe to a picture of the universe with matter uniformly distributed throughout an infinite space. Still, the material universe was to be essentially static in its gross structure. Newton appreciated that there remained a stability problem:

> And much harder it is to suppose all the particles in an infinite space should be so accurately poised one among another as to stand still in a perfect equilibrium. For I reckon this as hard as to make ... an infinite number of [needles] (so many as there are particles in an infinite space) stand accurately poised upon there points.[23]

He attempted to dispense (but not altogether convincingly) with this difficulty by stating in a later edition of the *Principia*:

> [T]he fixt stars, everywhere promiscuously dispersed in the heavens, by their contrary attractions destroy their mutual actions....[24]

Newton saw the actual stability of our universe as evidence of God's action in the world when he stated:

> I would now add that the hypothesis of matters being at first evenly spread through the heavens is, in my opinion, inconsistent with the hypothesis of innate gravity, without a supernatural power to reconcile them; and therefore it infers a Deity.[25]

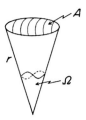

FIGURE 18.10 **The geometry of Olbers' paradox**

Leibniz, too, believed in an infinite universe everywhere populated with matter.

Unfortunately, an infinite universe with stars uniformly distributed through it presented a riddle known as Olbers' paradox, after the eighteenth-century German astronomer Heinrich Olbers (1758–1840). The sky in such a universe would appear everywhere filled with stars, of which our own sun is typical. Therefore, since the surface area of any sphere centered about the earth increases as the square of its radius and since this is just compensated for by an inverse-square decrease in light intensity from these distant stars, our sky, both during the day and at night, should appear as bright as the noonday sun. We can see this in a bit more detail by reference to Figure 18.10. Suppose we look out into an infinite universe through a cone of solid angle Ω, as illustrated in that figure. The number of stars on a surface area A, a distance r away, is proportional to r^2 (since $A = r^2\Omega$, just as the surface area of an entire sphere is $A_{\text{sphere}} = 4\pi r^2$), while the intensity from each star on this surface A decreases as $1/r^2$. Therefore, the net intensity reaching us from an element of area at any distance r is independent of r. Of course, once a given star 'blocks' our line of sight, we no longer receive light from those stars behind it. Hence, we would receive a finite (and constant) amount of light from the (finite) number of stars in our line of sight. But, the night sky is not as bright as the noonday sun. So, an infinite, static, homogeneous universe and a sky that is not always bright presented a contradiction that remained unresolved for over a century. This gave impetus to a search for a model of a finite universe.

We can estimate the gravitational collapse time of such a finite, static universe by considering a uniformly distributed spherical mass M of radius r_0. If we take a mass m initially at rest at a distance r_0 from the center of this sphere (that will itself be collapsing), use conservation of energy and integrate the resulting equation, we find[26]

$$t_{\text{collapse}} = \frac{\pi r_0^{3/2}}{\sqrt{8GM}} \qquad (18.3)$$

Let us take $r_0 = R_\alpha$, the distance to the nearest star, Alpha Centauri, that is 4.3 light years away from the earth. Then, since our sun is the only 'star' within this sphere, we readily compute $t_{\text{collapse}} = 7.89 \times 10^{14}$ s = 25 million years. (A light year is the distance light travels in one year.) Even more simply and directly, we can use Kepler's third law (Eq. (5.4)) and consider a distant star in a highly elliptical orbit.

PART VI The theory of relativity

We can then compute the time it would take such a star to move from its aphelion point, a distance $r_0 = R_a$ away, to its perihelion point (that would be very close to the sun for a highly flattened elliptical orbit). If we take the earth as the other body orbiting the sun in Kepler's third law, we estimate 70 million years as the collapse time. Hence, these two estimates for the collapse time are the same order of magnitude. Newton knew that the distance to the nearest stars was on the order of (in today's terminology) three light years. So, he could have estimated the collapse time as about fifteen million years. Of course, this should have presented no particular problem relative to the accepted ('biblical') age of several thousand years.

18.4 THE EINSTEIN AND FRIEDMANN UNIVERSES

As we saw in Section 18.1, Einstein formulated his general theory of relativity based on a curved space–time. In Chapter 11 we pointed out that prior to the mid-nineteenth century, people believed that the only conceivable three-dimensional space was Euclidean space (flat, with zero curvature). Then mathematicians discovered other possibilities. The Russian Nicholai Lobachevski (1792–1856) and the Hungarian János Bolyai (1802–1860) found a geometry of space with negative curvature, in addition to the positive-curvature one of Riemann that we mentioned in Section 11.5. Two-dimensional cases of each type of curved space are shown in Figure 18.11. A geodesic is the curve of shortest distance between two points on a given surface (or in a given space). In Figure 18.11 three-sided figures ('triangles') have been drawn on each surface. The sides of these triangles are geodesics between the points A, B and C. On the surface of negative curvature the interior angles of such a triangle add up to less than 180°, while on a surface of positive curvature they add up to more than 180°. A plane Euclidean surface has zero curvature and the interior angles of a triangle sum to precisely 180°. As we indicated in Section 11.5, the amount by which the sum of these three angles departs from 180° is a measure of the curvature of the surface (or space).

Einstein found that his original equations of general relativity did not allow a static (or, time-independent) solution with a finite (but nonzero) amount of matter in a spatially finite universe. Therefore, in 1917 he modified the equations by including a term containing what has subsequently been called the cosmological constant.[27] The effect of this additional term was to add to the mutual gravitational attraction between particles a repulsive force sufficient to produce static equilibrium. At the time Einstein did this work, the astronomical data available seemed consistent with a static universe in which there was no overall expansion or contraction of the universe. Of course, local motions, such as those of the planets about the sun, were allowed. This became known as the *Einstein universe*. Immediately there was debate about the physical reasonableness of the cosmological constant. In 1922 the Russian mathematician Alexander Friedmann (1888–1925) demonstrated that the Einstein equations, either in their original form or with the cosmological term, did allow an expanding (or contracting) solution with constant

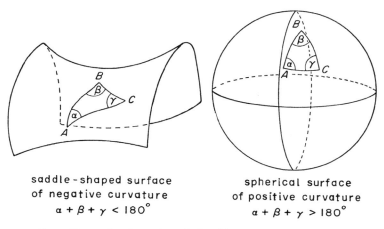

saddle-shaped surface of negative curvature
$\alpha + \beta + \gamma < 180°$

spherical surface of positive curvature
$\alpha + \beta + \gamma > 180°$

FIGURE 18.11 **Geometries of negative and of positive curvature**

curvature (a *Friedmann universe*). At this juncture the universe could have negative (or zero) curvature and be of infinite spatial extent, or have positive curvature and be of finite total volume (increasing or decreasing with time).

In 1927 the Belgian priest and theoretical astronomer Abbé Georges Lemaître (1894–1966), who had been a student at Cambridge University and later obtained his Ph.D. at the Massachusetts Institute of Technology, proved that the finite Einstein universe was unstable. This meant that, if such a universe were disturbed ever so little away from equilibrium, it would expand irreversibly toward an empty universe of constant spatial curvature. Lemaître proposed a primeval fireball (later to become the big bang) theory of the origin of the universe. Here the universe began as a very dense state of electromagnetic radiation and matter and, as it expanded and cooled down in temperature much as a gas would do, galaxies and stars, under the influence of gravity, condensed out of the tenuous matter filling the early universe. In such a Friedmann universe the two possible models – positive curvature, closed, finite volume, pulsating in time; or negative (or zero) curvature, open, infinite volume, expanding – are distinguished by the mean density of matter in the universe, the critical value[28] ρ_c being about 5×10^{-27} kg/m³. This is roughly one hydrogen atom per cubic meter. The observational situation is at present unclear. If one estimates the density of the universe by counting the visible galaxies, one finds[29] $\rho_{visible} \approx 0.10\, \rho_c$. But, as we shall see in the next section, an estimate of ρ by another method indicates that $\rho \cong \rho_c$.

That the difference between a finite universe, that will eventually stop expanding and then contract under gravitational attraction, and an infinite one, that will expand forever at a continually decreasing rate, should depend upon the mean density of matter is reasonable, as can be seen on the basis of the following classical argument (the result of which can be shown to remain valid even in the general theory of relativity). As we stated in Chapter 8, a spherically symmetric mass produces the same gravitational force field outside itself as though all of the mass were concentrated at the center of the sphere. Also, once we are inside a uniform

PART VI   **The theory of relativity**

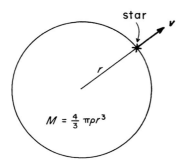

FIGURE 18.12   **Escape velocity**

spherical shell of mass, that mass produces no gravitational force field there. Therefore, if we consider a mass (or star) located at some distance r from any point we choose to take as the center of the universe (see Figure 18.12), then whether or not it has escape velocity (the minimum speed necessary to escape from the gravitational field of the spherical mass) will depend only upon the mass inside this sphere of radius r.[30] If ρ is the mass density of matter in the universe, then we have[31]

$$v_{\text{escape}} = \left(\sqrt{\frac{8\pi G}{3}\rho}\right)r \qquad (18.4)$$

This tells us that, for any fixed distance r, the escape velocity depends only upon the density ρ, as expected.

Of course, it is not immediately clear what significance we are to attach to this distance r from the fictitious 'center' of the universe. It turns out that we can effectively take any point to be the center of our universe. This surprising result is related to the law of expansion of the universe (to which we turn in the next section) and to the so-called *Cosmological Principle*, an hypothesis put forward in seminal form by Einstein in 1931 and so named by Edward Milne (1896–1950) in 1933. This principle, according to which the universe is assumed to be (in its large-scale structure) homogeneous and isotropic, is a general feature of the universe that is presumed to hold in modern cosmology. Here 'homogeneous' means that the universe is the same everywhere (no preferred origin) and 'isotropic' implies that there is no preferred direction. Such a principle is consistent with astronomical observations, since not only is the expansion of space isotropic, but also the distribution of galaxies on the celestial sphere appears to be quite uniform.[32] It is important to appreciate that what is expanding in such universes is space itself.[33] By way of analogy, we may think of a flat rubber sheet (or, perhaps better, the surface of a balloon) as representing space. As the sheet is stretched, even points at rest relative to this surface (or 'space') become a greater distance apart as time goes on.

18.5 HUBBLE'S LAW

From 1912 to 1925 Vesto Slipher (1875–1969) measured the spectral lines of light coming to us from distant galaxies and found those lines to be redshifted. This determination was possible since each element has its own characteristic set of colors (or frequencies of light) that it emits when heated. At the time, it was assumed that this redshifting was a result of the Doppler effect that we discussed in Chapter 17. The shift in these characteristic spectral lines could then be used to deduce the recessional velocity of stars (in these galaxies) emitting this light. The surprising conclusion was that the great majority of distant galaxies that Slipher observed were receding from us. (For a random distribution of motions, one would have expected that about as many galaxies would show blueshifts as redshifts.) Improving on Harlow Shapley's (1885–1972) techniques that superseded direct stellar parallax (see Section 4.4) observations (reliable up to distances of about 300 light years) and brightness measurements (out to distances of about 80 million light years), Edwin Hubble (1889–1953) was able to make independent determinations of the distances to these remote galaxies. In this way a great advance in observational astronomy was made when Hubble conclusively established in 1929 that all of the galaxies in the universe have a recessional velocity that is directly proportional to their distance from us.

$$v_{\text{recession}} \propto \text{distance from earth} \tag{18.5}$$

This is now usually referred to as *Hubble's law*.

However, the Doppler-shift formula applies to the velocity v of a star moving with respect to its background space. Only for the closest stars is the wavelength shift (that can then be either red or blue) due to the local motion of the stars through space. Most of the redshift is due to the 'stretching' of the wave train as space itself expands. The (overall) recessional velocity of distant galaxies comes from the expansion of space (not from the motion of the stars through space). That is, redshifts were observationally found to be proportional to the measured distance as

$$\frac{\lambda - \lambda_0}{\lambda} \propto \text{distance} \tag{18.6}$$

Under the assumption that this is due to the Doppler effect (Eq. (17.1)) for small values of v/c, a law of the form of Eq. (18.5) can be arrived at.[34] But, in our discussion so far, the status of the velocity versus distance law and of the redshift versus distance law are very different: the former being a result of observation and application of the Doppler-shift formula, the latter a direct empirical rule. (In Section 18.A we show how the model of an expanding universe unifies these two laws.) Realize, too, that there is no limit (even c) to recessional velocity due to the expansion of space. In such an expanding universe, where the recessional speed is directly proportional to the distance from the observer, any observer thinks she is at the center of this universe, as indicated in Figure 18.13. In this figure the observer at

PART VI   The theory of relativity

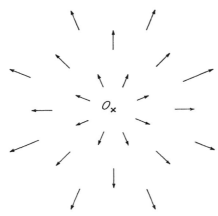

FIGURE 18.13   The expansion of space

O sees all matter receding from her.[35] In Section 18.A at the end of this chapter, we give more details about the basis for Hubble's law as represented in Eq. (18.5).

Returning to our classical analogue of the previous section, notice that Eqs. (18.4) and (18.5) are of the same form. Thus, if ρ is large enough, then the Hubble velocity of Eq. (18.5) can be less than the required escape velocity and the particle will finally stop, after which time it will fall back toward the point at the center of the sphere. Furthermore, on a large scale the universe appears both isotropic and homogeneous at any given instant of time. Only for the most distant galaxies (say about 10^9 light years away) is the recessional velocity somewhat greater than Hubble's law (Eq. (18.5)) would predict. Since light travels at a finite speed, looking at a galaxy far away (that is, receiving now the light that left the galaxy long ago) is equivalent to looking back very far into the past history of the universe. Of course, at any given time, we can (eventually) receive light only from those galaxies such that $v_{\text{recession}} < c$. The distance corresponding to $v_{\text{recession}} = c$ is known as the *Hubble radius*. Since the rate of expansion continually decreases, this Hubble radius continually increases. (For more discussion of this point, see Section 18.A.)

In 1930 Arthur Eddington (1882–1944) realized that the work of Einstein and of Lemaître on an expanding universe provided a natural account of Hubble's law. When Einstein himself was visiting the California Institute of Technology in late 1930 and early 1931, he discussed with scientists there the latest astronomical observations. (Hubble worked at the Mount Wilson Observatory in Pasadena.) In 1931 Einstein abandoned the cosmological constant and adopted the theory of an expanding universe. He later said that the cosmological constant was the biggest blunder of his life.

This fact that the universe is expanding and that, in the distant past, galaxies were receding more rapidly (as evidenced by the observation that very distant galaxies are moving away more rapidly than would be expected from Eq. (18.5)) fits in naturally with the theory that in the past the stars in the universe were nearer to each other and that long ago everything was close together at high density. That is, the universe began with a 'big bang' and has been expanding ever since, although the rate of expansion has been steadily decreasing due to the mutual gravitational attraction of the matter in the universe. On the basis of the big bang theory and of Hubble's law, the age of the universe is estimated to be between eight and twenty billion years. Even if the big bang theory is accepted, there still remains the question of whether the universe is open or closed. Current estimates of the mean density of the visible universe indicate, although not yet conclusively, that our expanding universe is open and infinite so that the expansion will never cease but will continually slow down. However, there is other evidence that conflicts with this presumption of an open universe. The deceleration of the expansion of the universe turns out to depend upon the ratio ρ/ρ_c and the data indicate that ρ is probably somewhat greater than ρ_c.[36] This has led to the conjecture that much of the matter in the universe is 'dark matter' that we cannot see directly, but only detect through its gravitational interaction. Independent evidence for large concentrations of such dark matter is provided by the overall motion of galaxies toward certain locations in the universe.[37] So, this open versus closed status of the universe remains at present undecided.

Is Olbers' paradox (see p. 261) to be resolved by the expanding universe and by the redshift of light? One might at first suppose so, given that the recession of the distant stars certainly does two things. First, since light must travel farther and farther each moment to reach the earth, less light arrives per unit time than would be the case in a static universe. Second, the light is redshifted and hence, as we stated in Section 18.2, has less energy because the energy of a photon is $\varepsilon = h\nu$. The real flaw, though, in the bright-sky argument is that it assumes an infinite universe that has always existed so that light from stars arbitrarily far away can be reaching us at any given moment. But, if the universe had a beginning (or, at least, if the stars were formed only some finite time in the past – typically taken to be several billion years), then there have not been enough stars existing at immense distances (that is, long enough ago) to fill all of the 'dark' regions of the night sky. This effect alone blocks the claim that the night sky in our universe must certainly be bright.[38] So, a dark sky does not necessarily present a contradiction. Depending on the specific distribution, average brightness and average life of the stars in a universe, that universe may be bright or dark at 'night'. In our own actual universe, expansion effects simply make an already dark night sky even darker than it would otherwise be.

18.6 A MODERN MODEL OF OUR UNIVERSE

Let us conclude with a brief description of our universe as we understand it today – modern man's rejoinder to the Newtonian universe, just as the Copernican one replaced Ptolemy's. The earth and the other planets in the solar system orbit about the sun that is a modest star in the heavens. The sun is just one of an immense number of stars in our own galaxy. Our galaxy is a disc-shaped agglomeration of stars. This disc is about 100,000 light years in diameter and has a bulge near its center, as indicated in the side view of Figure 18.14. The entire galaxy rotates about

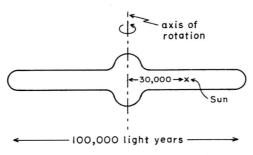

FIGURE 18.14 **The overall geometry of our galaxy**

an axis perpendicular to the disc and through the central bulge. Our sun is located about 30,000 light years from this axis and should take about 200 million years to make one complete revolution about this axis. One of the closest stars to us in our galaxy is Alpha Centauri, about 4.3 light years away. The stars we see in the Milky Way are those stars contained in the plane of the 'pancake' of our galaxy. Our galaxy is one member of a galaxy cluster known as the Local Group containing approximately 20 galaxies. Of these, the only one comparably massive to our own is the Andromeda Nebula, about 2×10^6 light years away from us. There are many other clusters of galaxies in the universe, ranging from Virgo, at 78×10^6 light years, out to Hydra II, at $3,960 \times 10^6$ light years, and others beyond.[39]

In 1965 Arno Penzias (1933–) and Robert Wilson (1936–) discovered what has become known as the cosmic background radiation. It has the wavelength spectrum that a radiating blackbody would have at an absolute temperature $T = 2.7$ K. For this work they received the 1978 Nobel Prize in physics. A hot body in thermal equilibrium with its surroundings emits electromagnetic radiation – so-called *blackbody radiation* – with a characteristic spectrum (see Section 19.1). The big bang theory not only readily explains this observed blackbody radiation, but leads one to expect it. In the early stages of the universe, the density of electromagnetic radiation far exceeded the density of matter. This was the radiation-dominated era of our universe. The blackbody spectrum resulted from this dense, primeval 'fireball' of matter and radiation in thermal equilibrium at a very high temperature. As the universe expanded and became matter dominated, this radiation 'gas' cooled. At the present time this radiation has fallen to a temperature of 2.7 K and is a remnant from an earlier epoch of our universe. An alternative model, the steady-state theory, was proposed in which the universe is infinite, everywhere homogeneous and unchanging in time so that matter must be continuously and spontaneously created to fill the spaces generated by this uniform expansion. In the steady-state theory there is no reason or need for such blackbody background radiation. Therefore, the cosmic background radiation counts at once strongly in favor of the big bang theory and against the steady-state one.

Enormously bright sources that recede from us at great speeds were discovered in 1963–1965. From Hubble's law it was evident that these quasars, as they are called (short for 'quasi-stellar radio source'), were extremely far away (on the order of

10^9–10^{10} light years from us). The existence of so many more very bright sources (hundreds of times brighter than galaxies) far away from us than there are near us indicates that the universe was very different long ago from what it is now. This, again, is taken as evidence consistent with the big bang theory, but not with the steady-state one.

Finally, we mention the existence of rather exotic objects known as black holes. Even in classical Newtonian gravitational theory, if the radius of the earth could be reduced to 9×10^{-3} m (while keeping its mass the same), then the escape speed for an object would be 3×10^8 m/s – the speed of light. Surprisingly enough, an exact general relativistic treatment gives the same numerical value for this radius. Not even light could escape from the gravitational field of the earth, so that there would be no way for the earth to send signals to any other part of the universe. Laplace made a similar observation a century and a half ago:

> A luminous star, of the same density as the earth, and whose diameter should be two hundred and fifty times larger than that of the sun, would not, in consequence of its attraction, allow any of its rays to arrive at us; it is therefore possible that the largest luminous bodies in the universe may, through this cause, be invisible.[40]

Such black holes can be formed according to general relativity when a massive body undergoes gravitational collapse. Any object or light getting too close to a black hole is sucked in and can never escape. Even though such a black hole cannot be 'seen', it will continue to exert an influence on the surrounding universe through its gravitational field. There is a star system in the heavens, Cygnus X-1, that appears to have just one visible member of a star doublet. The other unseen massive member of this doublet is believed to be a black hole. The mass of a black hole does not simply continue to increase as more matter is drawn into it, since quantum-mechanical effects require that a black hole radiates energy and particles. The rate of radiation turns out to increase as the mass decreases, so that an isolated black hole will eventually evaporate.[41]

18.A A DERIVATION OF HUBBLE'S LAW

The relation between the redshift and the expansion of space becomes clear when we realize that coordinate distance and distance as determined by light travel are not the same in an expanding universe. For bodies at rest in space (such as galaxies), the coordinate distance r (as measured, say, by 'markers' or by a grid fixed in or 'painted' onto an expanding space) remains constant. The 'actual' or 'cosmic' distance ℓ (as measured by the time it would take a light ray to travel between two points fixed in (a snapshot of) that space) varies since[42] $\ell = R(t)r$. (Here $R(t)$ is a 'scale factor' that relates these two measures of distance.) If we differentiate this with respect to the time t and then expand about some $t = t_0$, we obtain the familiar form of Hubble's law as an approximation[43]

$$\dot{\ell} \equiv v_{\text{recession}} = H\ell \approx H_0 \ell \tag{18.7}$$

This is just Eq. (18.5). Here $\dot{\ell}$ is *not* the velocity of a body moving through, or relative to, the expanding space, but is the recession velocity as determined by the time of travel of a light signal (sent from the receding star, galaxy, etc. to the observer – us). Furthermore, if we realize that the 'marker' distance between the ends of a wavelength must remain a constant (on the 'rubber sheet'), then we have $\lambda_0/R(t_0) = \lambda/R(t)$. If $R(t)$ is again expanded about t_0, the we find[44]

$$\frac{\lambda - \lambda_0}{\lambda} \approx H_0 \frac{\ell}{c} \tag{18.8}$$

This is the result stated in Eq. (18.6). If the nonrelativistic expression for the Doppler shift is combined with Eq. (18.8), we see that Hubble's law in the familiar form $v_{\text{recession}} = H_0 \ell$ results. Thus, we can understand how the Doppler shift, the effects of the expansion of space and Hubble's law all fit together for galaxies not too far away.

The Hubble radius ℓ_0 is defined by the relation $\dot{\ell} = c \equiv \ell_0 H_0$. The edge of the Hubble sphere is in some ways a horizon to what we can observe. If the rate of expansion were constant (or, if H_0 were truly constant), then the light from any stars or galaxies beyond this Hubble distance would never be able to reach us and we could never see them. However, if the rate of expansion decreases, then some stars that now recede from us with velocities exceeding that of light will eventually move at subluminal velocities relative to us and light from them can eventually reach us. In such a universe, the Hubble radius steadily increases.[45]

FURTHER READING

Michael Crowe's *Modern Theories of the Universe* is a very readable history of cosmological theories in the modern era. Edward Harrison's *Masks of the Universe* gives an insightful analysis, by an eminent astronomer, of cosmological theorizing and of what it actually tells us about our universe, while his 'Newton and the Infinite Universe' recounts Newton's struggle with the stability problem for the classical universe. Owen Gingerich's *Cosmology + 1* consists of a set of *Scientific American* reprints of articles on modern cosmology. Harry Shipman's *Black Holes, Quasars, and The Universe* gives a nonmathematical discussion of modern cosmological theories and of the observational basis for them. Edward Harrison's *Cosmology* is a superb undergraduate textbook on cosmology. In *The First Three Minutes* Steven Weinberg, one of today's leading theoretical physicists, makes accessible to the general reader the widely accepted modern view of the origin of the universe. George Ellis and Ruth Williams' *Flat and Curved Space–Times* presents, at an intermediate level of mathematics, a description and analysis of non-Euclidean geometries of space–time. Samir Bose's *An Introduction to General Relativity* is a compact and clear advanced text on general relativity and cosmological models.

PART VII

The quantum world and the completeness of quantum mechanics

What does not satisfy me in that theory [quantum mechanics], from the standpoint of principle, is its attitude towards that which appears to me to be the programmatic aim of all physics: the complete description of any (individual) real situation (as it supposedly exists irrespective of any act of observation or substantiation).
Albert Einstein, *Reply to Criticisms*

[This] implies the *impossibility of any sharp separation between the behaviour of atomic objects and the interaction with the measuring instruments which serve to define the conditions under which the phenomena appear*. In fact, the individuality of the typical quantum effects finds its proper expression in the circumstance that any attempt of subdividing the phenomena will demand a change in the experimental arrangement introducing new possibilities of interaction between objects and measuring instruments which in principle cannot be controlled. Consequently, evidence obtained under different experimental conditions cannot be comprehended within a single picture, but must be regarded as *complementary* in the sense that only the totality of the phenomena exhausts the possible information about the objects.

...

Indeed the *finite interaction between object and measuring agencies* conditioned by the very existence of the quantum of action entails – because of the impossibility of controlling the reaction of the object on the measuring instruments, if these are to serve their purpose – the necessity of a final renunciation of the classical idea of causality and a radical revision of our attitude towards the problem of physical reality. In fact, as we shall see, a criterion of reality like that proposed by [Einstein] contains – however cautious its formulation may appear – an essential ambiguity....
Niels Bohr, *Discussion with Einstein on Epistemological Problems in Atomic Physics*

Experimental results evidently [imply]... conclusions [that] are philosophically startling: either one must totally abandon the realistic philosophy of most working scientists, or dramatically revise our concept of space–time.
John Clauser and Abner Shimony, *Bell's Theorem: Experimental Tests and Implications*

19

The road to quantum mechanics

The theory of relativity is essentially the culmination of classical physics. Although revisions in our concepts of space and time are necessitated by both the special and general theories of relativity, the notion of causality, in which a cause precedes its effect, remains intact in the relativistic formulations of electrodynamics and mechanics and of gravitation. An upper limit is set on the speed of bodies, and of the propagation of energy, and physical quantities are often defined operationally, but once we adjust ourselves to these new rules, our representations of physical processes proceed pretty much as they did in classical physics. On the other hand, quantum theory seems to require a much more profound philosophical revision of our thought patterns. Even the viability of the usual meaning of a causal connection between one event and another is there called into question.

19.1 SOME HISTORICAL BACKGROUND

We begin with a brief and highly selective review of the experimental and theoretical problems that formed the conceptual backdrop to the emergence of quantum theory. We focus attention first on an important phenomenon – blackbody radiation – that was instrumental for the formulation of quantum mechanics. A blackbody is defined as a surface that absorbs all of the electromagnetic radiation that falls onto it. It is picturesquely so named since any light incident on such a body would be absorbed, making it appear black (at least at typical temperatures in our environment). A blackbody also emits radiation.[1] One can generate blackbody radiation by heating a large cavity to a fixed temperature T, allowing thermal equilibrium to be established and then observing the electromagnetic radiation that escapes from a small hole in the side of this cavity. If the energy density $\rho(\lambda, T)$ is analyzed into its various wavelength (λ) components, experimental curves of the type shown in Figure 19.1 result.[2] Such experimental curves were obtained as early as 1899 by Otto Lummer (1860–1925) and Ernst Pringsheim (1859–1917) and in 1900 by Heinrich Rubens (1865–1922) and Ferdinand Kurlbaum (1857–1927).

Classical physics was not able to account for the shape of the curves in Figure 19.1. Basically, the source of this difficulty classically is the equipartition theorem according to which every degree of freedom of a system (in thermal equilibrium) must share equally in the energy available to the system. To simplify the discussion, let us consider the cavity producing the blackbody radiation to be of dimension ℓ. Standing waves (see Figure 19.2) are possible only for certain wavelengths given by

PART VII   The completeness of quantum mechanics

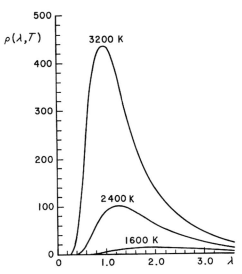

FIGURE 19.1   Blackbody radiation curves for the energy density ρ

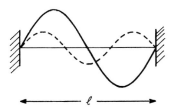

FIGURE 19.2   Standing waves on a string

$\lambda_n = 2\ell/n$ ($n = 1, 2, 3, \ldots$).[3] Each of these is a degree of freedom that should share equally in the energy available and, since there are infinitely many allowed waves of shorter and shorter wavelength, nearly all of the light should be at the short wavelength end of the spectrum. A rigorous classical argument produces the Rayleigh–Jeans formula

$$\rho(\lambda, T) = \frac{8\pi kT}{\lambda^4} \qquad (19.1)$$

where k is Boltzmann's constant. This formula agrees with Figure 19.1 only at long wavelengths. At short wavelengths ($\lambda \to 0$) this becomes infinite (often referred to as the *ultraviolet catastrophe*), as indicated in Figure 19.3. Clearly, Eq. (19.1) does not fit the experimental data. Its roots go back to a paper in 1900 by Lord Rayleigh, but the result of Eq. (19.1) was first properly derived by him and Sir James Jeans (1877–1946) in 1905. The failure of this law was a serious matter since it followed from one of the firmly established laws of classical statistical mechanics, the equipartition of energy. In the quotations for Part VI (*The theory of relativity*) we saw Lord Kelvin's assessment of the situation in a lecture he gave at the Royal

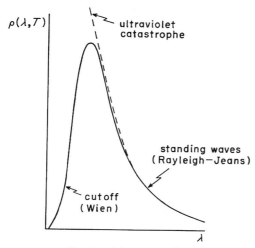
FIGURE 19.3 The ultraviolet catastrophe in the classical theory of blackbody radiation

Institution in 1900. The Maxwell–Boltzmann doctrine of equipartition of energy generated one of the two clouds on the horizon of late nineteenth-century physics.[4]

The other general category of phenomena especially relevant to quantum theory is the pattern of spectral lines emitted by an element. When an electric arc is passed through a sample of gas or vapor, such as hydrogen, the gas glows with a characteristic color. For example, hydrogen is orange and sodium is yellow. Upon closer examination, one finds that only certain frequencies of light are present. Photographs of the spectral lines once they have been separated or resolved by a diffraction grating, or dispersed by a prism, reveal a set of bright lines whose wavelengths are unique to the element or substance emitting the spectrum. Such discrete spectra are characteristic of all of the elements and can be used to 'fingerprint' an unknown substance. These discrete spectra were surprising since, according to classical theory, accelerated charges must emit electromagnetic radiation. If the charges inside matter are in random motion, we would expect electromagnetic radiation of all frequencies to be emitted. This would produce a continuous spectrum. The discrete frequencies of the emission spectra of the elements became associated with certain stable orbits or characteristic vibrations of these charges. In 1885 a Swiss mathematician and secondary school teacher in Basel, Johann Balmer (1825–1898), discovered by trial and error an empirical formula that fit one set of the observed spectral lines of hydrogen. In 1890 Johannes Rydberg (1854–1919) in Lund found a generalization of Balmer's result. We can conveniently summarize this formula for the frequency v of these lines as[5]

$$v = R\left(\frac{1}{m^2} - \frac{1}{n^2}\right) \tag{19.2}$$

Here R is a universal constant (the Rydberg constant), m a fixed integer that characterizes the particular series and n another integer (greater than m) that labels the particular line in a given series.

In unsuccessful attempts to account for these discrete spectra, Kelvin and J. Thomson at the Cavendish Laboratory of the University of Cambridge constructed several classical models of the atom in which the positive charge was assumed to be distributed uniformly throughout the volume of the atom (the so-called Thomson model). However, substantive progress had to await further dramatic experimental results. In 1909 Hans Geiger (1882–1945) and Ernest Marsden (1889–1970) at the University of Manchester, in England, examined the scattering of a beam of charged particles by a thin gold foil. This experiment was suggested by Ernest Rutherford (1871–1937), a professor of physics at Manchester and an experimentalist of outstanding ability and insight who played a central role in the development of atomic and nuclear physics. Rutherford has been referred to as the Faraday of nuclear physics. The basic idea was that the pattern of scattered particles should reflect the structure of and the forces within the atom. Geiger and Marsden used a gold ($_{79}Au^{197}$) foil about 400 atoms thick with alpha (α) particles (He^{++}) as their bombarding projectiles. Most α particles went straight through the foil with no deflection. Since the α particles are very massive (about 7,500 times more so than an electron), they are essentially not deflected at all by electrons. The entire gold nucleus is about 50 times as massive as an α particle. This much was consistent with the results expected on the basis of the Thomson model, since the mass and positive charge were pictured as being uniformly spread out over the volume of the sample. There would be no encounters with single massive objects but only with tenuous bits of the gold atoms. However, occasionally some α particles were observed to be backscattered through angles larger than 90°. This behavior was totally inconsistent with the Thomson model and suggested to Rutherford that the great bulk of the gold atom, along with its positive charge, was concentrated essentially at a point.

Rutherford asked the young theorist Charles Darwin (1887–1962), a grandson of the Darwin who wrote *On the Origin of Species*, to make an exact calculation for the scattering produced by such a nuclear atom. The results confirmed Rutherford's expectations and agreed with experiment. Therefore, physicists found themselves reduced to a nuclear atom model, one not dissimilar to a planet orbiting the sun. Although such a configuration produces a mechanically stable orbit, such a centripetally accelerated electron must radiate electromagnetic energy. This would cause the entire atom to collapse almost immediately as the electron spiraled into the nucleus. Not only that, but the spectrum of electromagnetic radiation emitted would be continuous, not discrete.

How two scientists, Max Planck and Niels Bohr (1885–1962), coped with these riddles and, in the process, laid the foundations upon which modern quantum theory would be built is a familiar story. However, the actual historical route to their discoveries differs significantly from the folklore versions of discovery too often propagated in science textbooks. Let us see how Planck and Bohr proceeded.

19.2 PLANCK'S HYPOTHESIS

In 1900 Planck took a new approach to the blackbody radiation problem. Since Gustav Kirchhoff proved in 1859 that the distribution of radiant energy inside an enclosure at thermal equilibrium is independent of the particular properties of the cavity walls and since the problem of the electromagnetic radiation from a dipole oscillator had been thoroughly studied previously, Planck decided to use a collection of radiating harmonic oscillators in thermal equilibrium as the system giving rise to the blackbody (or cavity) radiation. Furthermore, when his first, direct classical electrodynamic approach proved unsuccessful, he turned to thermodynamics to obtain an expression for the distribution of thermal energy at equilibrium. Planck is usually credited with having obtained a fit to the empirical curve for the spectrum of blackbody radiation by quantizing a set of harmonic oscillators used to represent this radiation field. Even setting aside the controversial question of Planck's real position on the quantization issue,[6] the fact is that he initially produced a two-parameter fit to the empirical curve by means of an ad hoc thermodynamic argument. He modified the classical relations involving the entropy (a measure of the unavailable energy in a system) of the radiation in order to interpolate between the observed behavior (equivalent to the classically expected Rayleigh–Jeans law) at long wavelengths and an ad hoc law, conjectured in 1896 by Wilhelm Wien (1864–1928), that cut off the spectrum to fit the data for short wavelengths (see Figure 19.3). Although the details of his calculation are too technical to present here,[7] the fact is that Planck's original argument consisted essentially of phenomenological curve fitting. He referred to this process of finding a compromise solution as a lucky guess. However, his formula did fit the experimental data exactly for all wavelengths, as Rubens showed the evening following Planck's presentation of his new formula to a meeting of the German Physical Society. At this stage Planck had a formula that fit the data perfectly, but that had no theoretical justification. In his own words:

> But even if the absolutely precise validity of the radiation formula is taken for granted, so long as it had merely the standing of a law disclosed by a lucky intuition, it could not be expected to possess more than a formal significance. For this reason, on the very day when I formulated this law, I began to devote myself to the task of investing it with a true physical meaning. This quest automatically led me to study the interrelation of entropy and probability – in other words, to pursue the line of thought inaugurated by Boltzmann.... After a few weeks of the most strenuous work of my life, the darkness lifted and an unexpected vista began to appear.[8]

Subsequently, Planck returned to the problem of radiating harmonic oscillators in thermal equilibrium and employed statistical–mechanical techniques in which he partitioned the total energy of the system in discrete amounts.[9] He found that he obtained the proper radiation distribution law provided these oscillators could absorb and emit energy only in discrete amounts ε, that are today termed *quanta* (although that name was not used by Planck originally). The energy of these quanta is related to the frequency of the emitted or absorbed radiation as

$$\varepsilon = hv \tag{19.3}$$

Here h is a proportionally constant now known as Planck's constant. Although Planck found it necessary to quantize the emission and absorption of energy, he hesitated extending the concept of energy quantization to the electromagnetic radiation itself. His reason was that Maxwell's electrodynamics, that had been so spectacularly successful, was based upon an electromagnetic field that could carry energy of any continuously varying amount. Today we do quantize the electromagnetic field and its quanta are called *photons*. Under usual (or commonplace) circumstances, a beam of light consists of an enormous number of quanta and, in fact, behaves as a wave phenomenon. Equation (19.3) is completely unexpected classically. Planck fully appreciated the revolutionary nature of his conjecture. On a walk through the Grunewald woods in Berlin, he said to his young son, 'Today I have made a discovery as important as that of Newton.'[10] On December 14, 1900, Planck presented his theory before the German Physical Society. For several years he attempted to fit this new approach into the framework of classical physics, but he finally accepted the reality of quanta. In his opinion, these years were not wasted because this ultimately futile effort taught him the true significance of an elementary quantum of action and the need for a new approach to atomic phenomena.[11]

It is fairly easy to see qualitatively why the assumption of Eq. (19.3) avoids the problems produced by classical physics for cavity radiation. In terms of the wavelength, Eq. (19.3) becomes $\varepsilon = hc/\lambda$. Hence, for any finite total energy contained in the cavity, there is some shortest wavelength that can be excited. This, along with the longest-wavelength constraint implied by Figure 19.2 and the discussion accompanying it in the text, means that few quanta come off at very small or at very large λ. For a given temperature T, each curves peaks about some most probable frequency or wavelength.

19.3 BOHR'S SEMICLASSICAL MODEL

Just as Einstein is identified with the theory of relativity, so Niels Bohr must be seen as the founder and philosophical leader of quantum theory as a branch of physics completely separate from classical physics. While relativity was a theory more nearly exclusively Einstein's than were quantum theory and its major evolution Bohr's, nevertheless Bohr played the central role as the guiding spirit in its development for half a century. After obtaining his Ph.D. in 1911 from the University of Copenhagen, Bohr first went to the Cavendish Laboratory to work with J. Thomson, but found Thomson generally uninterested in the subject of Bohr's Ph.D. dissertation. In 1912 Bohr left Cambridge for Manchester, where Rutherford was developing his own nuclear model of the atom that we discussed earlier. It was during a four-month period in Manchester that Bohr formulated his theory of atomic transitions.

We now outline Bohr's original argument by which he arrived at an explanation of the discrete line spectrum of hydrogen.[12] If (as indicated in Figure 19.4) we consider Rutherford's nuclear atom with an electron of charge $-e$ and mass m in a

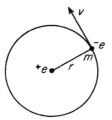

FIGURE 19.4 Bohr's model of the atom

circular orbit of radius r about a massive nucleus of charge $+e$, then the condition for a stable circular orbit is that the electrostatic force of attraction supply the requisite centripetal force. This, plus conservation of energy, allowed Bohr to express the frequency of revolution of the electron in terms of its energy as[13]

$$v = \frac{1}{\tau} = \frac{v}{2\pi r} = \frac{\sqrt{2}(-E)^{3/2}}{\pi e^2 \sqrt{m}} \tag{19.4}$$

Classical electrodynamics requires that such an accelerated electron emits radiation with a frequency equal to the frequency of revolution of the electron given in Eq. (19.4). However, since E could have any (negative) value classically, the atom would radiate at all frequencies thus producing a continuous spectrum – contrary to experiment. In order to obtain a discrete set of stable orbits (that is, just certain values of r), Bohr simply postulated that the electron would be dynamically stable and not radiate any light only when in one of these allowed orbits. There was no classical explanation for this and he offered none. He further assumed that an electron emitted or absorbed radiation only when it made a discontinuous transition from one allowed orbit to another.[14] In accord with Planck's quantization of energy emitted or absorbed by a harmonic oscillator in multiples of hv, Bohr considered an initially free electron at rest (that is, one infinitely far away from the nucleus) falling into an allowed orbit and quantized the energy hv of the photon in terms of the energy of the electron in its final orbit as

$$-E = \frac{n}{2}hv, \; n = 1, 2, 3, \ldots \tag{19.5}$$

The factor of $\frac{1}{2}$ in Eq. (19.5) is crucial and also curious. Bohr effectively argued that the frequency of revolution of the electron was initially zero when it was at rest and later the classical value v of Eq. (19.4) when in its final orbit, so that the average frequency of the electron was just $\frac{1}{2}v$. He took the frequency of the emitted radiation to be some integral multiple of this quantity. Bohr combined Eqs. (19.4) and (19.5) to eliminate v and obtain

$$E_n = \frac{-2\pi^2 me^4}{h^2 n^2}, \; n = 1, 2, 3, \ldots \tag{19.6}$$

as the allowed energies of the electron in the hydrogen atom.

PART VII   The completeness of quantum mechanics

Equation (19.6), along with conservation of energy in the form $(E_i - E_f) = h\nu$ for a transition from an initial level (i) to a final one (f), allows one to obtain the Balmer formula (Eq. (19.2)) for the lines of the hydrogen spectrum.[15] We have presented Bohr's original line of argument, that is different from the way that textbooks usually give Bohr's derivation, in order to show that here, just as in Planck's original ad hoc method of obtaining his blackbody radiation formula, there was a good deal of (not always internally consistent) 'guessing' at the start of the process of groping toward a new theory. This is quite common when groundbreaking advances are made. Very often later treatments are much tidier and omit some of the less pleasing details. In fact, in Bohr's case it seems clear that in 1912–13 he was not certain just what the multiplicative factor should be ($\frac{1}{2}$, 1, 2, ?) in Eq. (19.5) and that it was only after a colleague pointed out Balmer's formula that Bohr realized that numerical agreement with observation could be gotten by using $\frac{1}{2}$. So, the suggestion about the average frequency of the electron being the mean of 0 and ν appears as an afterthought at attempted justification. 'As soon as I saw Balmer's formula, the whole thing was immediately clear to me.'[16] It was only later in the same 1913 paper, communicated to the *Philosophical Magazine* on Bohr's behalf by Rutherford and titled 'On the Constitution of Atoms and Molecules,' that Bohr stated that the quantization condition of Eq. (19.5) was equivalent to quantizing the product mvr, known as the *angular momentum* ℓ.

> While there obviously can be no question of a mechanical foundation of the calculations given in this paper, it is, however, possible to give a very simple interpretation of the result of the calculation of [Eq. (19.5)] by help of symbols taken from the ordinary mechanics. Denoting the angular momentum of the electron round the nucleus by ℓ, ... the result of the calculation can be expressed by the simple condition: that the angular momentum of the electron round the nucleus in a stationary state of the system is equal to an entire multiple of a universal value $[\ell = nh/2\pi]$, independent of the charge on the nucleus.[17]

It is instructive to contrast this with conventional textbook derivations using Bohr's major assumptions. These usually proceed by postulating that, for an electron in a circular orbit about a nucleus: (i) only certain orbits are allowed; (ii) electrons in these orbits do not radiate; (iii) during a transition from one allowed orbit to another, the electrons do emit homogeneous radiation (radiation of one frequency); and (iv) the angular momentum is quantized as

$$\ell = mvr = n\frac{h}{2\pi}, n = 1, 2, 3, \ldots \tag{19.7}$$

The Balmer formula follows at once from this.[18] It is all very efficient, but one has no idea what motivates Eq. (19.7).

The original version of the Bohr model, often referred to as the semiclassical one, works quantitatively only for one-electron atoms such as neutral hydrogen (H), singly ionized helium (He^+), doubly ionized lithium (Li^{++}), etc. The impact of Bohr's work was immediate and extensive, although his radical assumptions remained controversial. The remarkable achievements of obtaining the Balmer formula and the numerical value of the Rydberg constant in terms of fundamental known constants were generally regarded as spectacular and commanded attention. (This and subsequent work on atomic structure and radiation led to his receiving the 1922 Nobel Prize in physics.) Almost immediately after the appearance of Bohr's 1913 paper, attempts at generalizations and refinements were made. The set of rules that evolved for treating atoms is today referred to as the 'old quantum theory' and consisted of essentially three steps to pass from the classical description to the quantum one. Classical mechanics was used to determine the possible motions of the system; quantum conditions were imposed to pick out the actually allowed orbits; finally, energy conservation was employed to fix the frequencies of quanta emitted or absorbed during transitions of an electron from one allowed (or stable) orbit to another. Arnold Sommerfeld, a professor of theoretical physics at the University of Munich, soon became an acknowledged master of this technique. He treated the hydrogen atom relativistically, allowing for the mass variation of the electron in an elliptical orbit passing close to the nucleus. With this he was able to explain some fine experimental details that were discovered when spectral lines that appeared as single lines were found upon closer examination each to consist of several closely spaced lines. However, like Bohr's original model that worked only for single-electron atoms, so Sommerfeld's relativistic calculations produced correct results only for certain spectral lines but not for others. The true origin of this fine structure is something quite different from relativistic mass variation. (Today, the concept of electron spin is an essential ingredient in accounting for such fine structure.)

19.4 ACTUAL DISCOVERIES VERSUS RATIONAL RECONSTRUCTIONS

One may ask why it really matters precisely how these great discoveries (Planck's law and Bohr's model) were made, as long as they were made. If science arrives at the truth, then how can the actual route to this truth (the proverbial 'mountain top') be of any significance as far as the laws or theories themselves are concerned? There is the matter of historical accuracy, but that may be of importance only to historians of science. In Chapter 24 we shall see that the very content of quantum theory may owe something to the actual process of discovery and to the philosophical predilections of some of its founders. However, even on a less controversial note, a knowledge of the process of discovery can often make the discovery itself more understandable.

In the case of Planck's argument for his blackbody radiation law, we see first an ad hoc attempt to concoct an argument – any argument – to provide some justification for his successful curve fitting, and then a scientist forced, almost against his (classically inclined) will, to accept the hypothesis of quantization. This was not some sudden act of discovery ('flash of genius') in which Planck guessed that energy quantization would provide the necessary limitation on the blackbody radiation curve at short wavelengths (that is, an elimination of the ultraviolet catastrophe).

Even more so with the Bohr atom do we see ad hoc moves to obtain a fit to the data (here, the Balmer formula). Bohr's desperate attempt (that is, Eq. (19.5)) to find some energy to quantize becomes understandable when we appreciate that by 1913 Planck's quantization of energy was already well accepted and uncontroversial. Often, good science, like good scientists in their science, is conservative, in that as few new and controversial elements are introduced into a nascent theoretical scheme as are necessary to make some progress with an outstanding problem. Thus, Bohr sought to quantize an energy, and not an angular momentum (that would have had no basis in the then-accepted theory). He was already suspending the laws of electromagnetism (no radiation from the accelerated electron in its orbit) and that was in itself radical enough. If one postulates too many new 'principles', he may be accused of blatantly assuming the result he is attempting to derive or prove. Here Bohr placed his quantization condition (Eq. (19.5)) within a matrix of accepted ideas. (Maxwell had done something similar when he argued for his displacement current within the general framework of an electromagnetic aether, as we saw in Section 14.2.) Therefore, in the cases of Maxwell, of Planck and of Bohr, we can appreciate better the process of the discovery (or, perhaps more appropriately, of the creation) of major scientific theories.

More than just pure logic and empirical evidence were at play in these three exemplary discoveries. Similarly, other factors are often relevant in theory construction and theory choice. Examples of such criteria are fertility, beauty and coherence. While these can be important, they may also be pejoratively biased if they are defined in terms of the successful, victorious or accepted theory and then applied to a competing theory. In Chapter 23 we discuss such criteria when applied to various interpretations of quantum mechanics. However, let us first take up the question of the influence of contingent historical events in the development of quantum theory. For instance, could the philosophical outlooks and backgrounds of the creators of the victorious (Copenhagen) version of quantum mechanics have been important factors?

To set the stage for this, we sketch how the opposing positions – wave- versus matrix mechanics – were arrived at and what some of the arguments were for each view. The one traces its roots back to the nature of electromagnetic phenomena, the other to a study of discrete spectral lines. We shall see that, not only were the general philosophical outlooks of these two groups of key players quite different, but also each group began with very different classes of physical phenomena. The discontinuity versus continuity dichotomy can be seen as contingently rooted in philosophical commitments and in the physical phenomena studied.

19.5 TWO ROUTES TO QUANTUM MECHANICS[19]

We have seen that by the early 1920s a crisis had developed for classical physics in the atomic domain. Its resolution would effect fundamental changes in our view of the physical world. Indeterminism came to play an important role in this. It is worth emphasizing, though, that the possibility of indeterminism being an essential feature of nature was seriously considered even prior to the early part of the twentieth century. By the late nineteenth century, there were already significant, even if not overwhelming, philosophical precedents for the concept of indeterminism (including the possibility of inherent chance) in nature, as opposed to the straightforward determinsim often associated with classical physics. These ideas impressed Poincaré, whose writings on the philosophy of science became part of the conceptual background of the young scientists who would fashion a new quantum mechanics.[20] In a similar vein, Søren Kierkegaard (1813–1855) believed that objective uncertainty can force one to make a leap into the unknown so that decisions cannot always (even in principle) be based on a continuous chain of logic. It is well documented that some of these philosophical views left their mark on Bohr through the teachings of Harald Høffding (1843–1931).[21] For example, one of Høffding's tenets was that in life decisive events proceed through sudden 'jerks' or discontinuities, an idea incorporated into Bohr's view of atomic phenomena.

The point for us is not that such philosophical predilections alone determined the course of quantum theory in the early part of the twentieth century, but rather that these concepts were available to, and in the minds of, the creators of quantum theory, possibly exerting an influence on the choice of the final, 'accepted' form of quantum theory. And, of course, a part of this backdrop was the spirit of what would subsequently be codified as logical positivism, with its emphasis on the central role of the empirical determination or definition of concepts appearing in scientific theories. This late nineteenth- and early twentieth-century heritage of philosophical ideas helped to define the terms and shape the course of the debates that proved to be formative for quantum theory.

We now turn to the two quite distinct historical routes, each of which led to its own formulation of quantum mechanics. Let us begin with matrix mechanics. The small number of central players (Bohr, Heisenberg, Wolfgang Pauli (1900–1958), Pascual Jordan (1902–1980) and Max Born (1882–1970)) involved in the program that led to this version of quantum theory suggests that this was a closed group. This becomes all the more plausible when we realize that Pauli and Heisenberg were both Ph.D. students with Sommerfeld at Munich, each in succession were then Born's assistants at Göttingen and later worked with Bohr at Copenhagen. These two young men were greatly impressed by a lecture given by Bohr at Göttingen in 1922. Jordan was also a student at Göttingen at this time. We have just indicated philosophical factors in Bohr's own background that inclined him toward, or at the very least made him receptive to, a discontinuous structure in nature at the most fundamental level and, eventually, to a doctrine of complementarity between opposites. This element of discontinuous transitions is a central feature in his 1913 semiclassical model for the hydrogen atom (Section 19.3). It was certainly the current language for discussing atomic phenomena in Sommerfeld's school. Discontinuities were the key issue in this formulation of quantum mechanics.[22] Causality, as such, was not the central question initially in the development of this program.

PART VII **The completeness of quantum mechanics**

Largely due to the failure of certain classical approaches, the main players took up various philosophical positions on what was and was not possible in principle. These were not logical or in-principle refutations, but strong, practical beliefs that became dogma. Thus, Bohr's own Ph.D. dissertation argued that the failure of the classical electron theory of metals was attributable to a fundamental insufficiency of the classical principles themselves. Pauli, in his work on general relativity and related field-theory generalizations convinced himself (again, because of a failure) that a continuum field theory, with the particles as singularities, was not possible. In his famous 1921 *Theory of Relativity*, Pauli already was of the opinion that there was no point in discussing quantities that cannot, in principle, be observed experimentally.[23] This certainly has a strong operationalist air about it. By 1923 Pauli required operational definitions[24] of anything used in physics and the replacement of continuous concepts by discrete ones. After all, observations are essentially localized and instantaneous, with measurements being discrete and, he felt, this structure should be carried over into the foundation of any theory that accounts for such observations and measurements. He believed this would require major conceptual revisions. Both Pauli and Heisenberg were involved in Born's program of attempting to apply the old quantum theory, with its orbitals, to molecular systems and the utter failure of his approach convinced them that electron orbitals were meaningless. This desire for a radical conceptual revolution was prevalent in the general cultural milieu of the time and also in Pauli's and Heisenberg's expectations about quantum theory. (Recall Popper's similar observation, from the quotation in Section 3.A, that, after World War I, 'the air was full of revolutionary slogans and ideas, and new and often wild theories.')

The first extensive and consistent mechanics of quantum processes was formulated by Heisenberg in 1925. During the period 1924–7 that he spent at Bohr's Institute in Copenhagen, Heisenberg did some of his most creative work in producing his version of quantum mechanics. He received the 1932 Nobel Prize in physics for this. The mathematical objects (later termed 'operators') that he associated with physical quantities in his new theory had the apparently peculiar property that the answer one got by multiplying two of them together depended upon the order in which the product was taken. That is, the multiplication was not commutative, so that $AB \neq BA$.[25] Because of this mathematical property that most physicists were unfamiliar with at the time, Heisenberg's new mechanics seemed mysterious and was not immediately accepted. But, it was pursued because it held the promise of yielding a complete and consistent new mechanics and of giving answers to questions related to experiments – and, besides, there seemed to be no other viable alternatives just then. It was Max Born who recognized these noncommutative quantities of Heisenberg's 1925 paper as a class of objects (matrices) already well-known to mathematicians. This discrete (as opposed to continuum) mathematics of Heisenberg's theory was well suited to represent the discontinuous (acausal) structure of nature that the Copenhagen school took to be fundamental. Fortunately, we need not pursue this new mathematical subject in order to understand some of the concepts of quantum mechanics since, shortly after Heisenberg's work and quite independently of it, Erwin Schrödinger (1887–1961), starting from a very different point of view, gave an alternative but effectively equivalent formulation of quantum mechanics (this time in terms of a continuous mathematics). Since Schrödinger's theory is more easily discussed in terms of familiar concepts, we now turn to it (and give further details about it in Chapter 20).

Because this second route to quantum mechanics was centered around the work of Einstein, Louis de Broglie (1892–1987) and Schrödinger, let us begin by looking at the roots of Einstein's philosophical commitments. Einstein's general position on foundational questions in physics – relativity, his stance on quantum theory and his long-standing commitment to a unified field theory – can be seen as a search for the rational structure of the external physical world that the physicist tries to capture in a causal space–time theory. That is, Einstein's basic view was that of a rational, causal[26] world that could be comprehended in terms of an objective reality. This concept of causality suggested a continuity of basic physical processes. It is, of course, the intersection of general predilections like these with the puzzles presented by physical phenomena that results in a definite theory or research program.

In 1909 Einstein used the Planck blackbody radiation law and Planck's energy quantization condition to demonstrate mathematically that the radiation governed by these laws exhibits both wave and particle characteristics. At a conference in Salzburg in 1909, he stated:

> It is my opinion, therefore, that the next phase of the development of theoretical physics will bring us a theory of light that can be interpreted as a kind of fusion of the wave and emission theories.[27]

In retrospect, we tend to see this (quite ahistorically) as an early flirtation with the concept of wave–particle duality (to be discussed more later). In a paper on the quantum theory of radiation published in 1917, Einstein showed that, when molecules emit and absorb radiation under the influence of an external radiation field, then momentum and energy must each be conserved, so that such radiation was termed 'needle radiation'. Arthur Compton's (1892–1962) later experimental work on the scattering of radiation by electrons gave support to the hypothesis of free electromagnetic quanta. In this same 1917 paper, Einstein stated that the recoil direction of the molecule that has emitted radiation is '... only determined by "chance", according to the present state of the theory' and that 'the weakness of the theory lies... in the fact... that it leaves the duration and direction of the elementary process to "chance".'[28] Here, as later, Einstein took this to be a shortcoming of the theory, a provisional fault to be overcome in the future.

The next key figure in this 'continuity' school is de Broglie. In 1923 he initiated the theory of wave mechanics in an attempt to understand the dual (wave–particle) nature of Einstein's photon. In his youth, de Broglie was impressed by the well-known formal mathematical analogy between wave optics and classical particle mechanics. Building on this analogy and on some of his own previous work on Einstein's light quanta, de Broglie proposed a model of a particle that follows the trajectory determined by its associated wave. There was a great affinity of views between Einstein and de Broglie. In 1925 Einstein stated his opinion that de

Broglie's ideas 'involve more than merely an analogy.'[29] This notice by Einstein drew attention to de Broglie's work. Indeed, Schrödinger recalled that 'My theory was stimulated by de Broglie's thesis and by short but infinitely far-seeing remarks by Einstein.'[30] In a paper on Einstein's gas theory, Schrödinger concluded that photons can be seen as the energy levels of the aether oscillators, that cavity radiation need not 'correspond to the extreme light-quantum representation' and that:

> This means nothing else but taking seriously the de Broglie–Einstein wave theory of moving particles, according to which the particles are nothing more than a kind of 'wave crest' on a background of waves.[31]

In that same year (1926) Schrödinger exploited Hamilton's analogy between mechanics and optics to give a plausibility argument for his wave equation. Quantization was implemented by boundary conditions imposed on a continuous wave function. Schrödinger shared (with Paul Dirac (1902–1984)) the 1933 Nobel Prize in physics for his formulation of quantum mechanics.

This sketch is intended to indicate that (at least during the initial phase of development just discussed) this small group (Einstein, de Broglie and Schrödinger) shared a commitment to a continuous wave as a basic physical entity subject to a causal description. Visualizability and self-consistency had become accepted hallmarks of classical physical theories. One could well deem it better to have an understandable, if imperfect, theory with a definite, classical type of wave ontology than an abstract theoretical framework with a conceptually opaque ontology based on some concept of wave–particle duality.[32] The position taken by the wave-mechanics school was the more 'natural' (certainly the more conservative) one relative to the then-accepted concepts of classical physics (that is, it represented a less radical departure).

19.6 FORGING THE COPENHAGEN INTERPRETATION

It was the 'collision' between matrix mechanics and wave mechanics that provided the impetus for the formulation of a consistent interpretation of quantum mechanics. There was a sense of crisis since some of the protagonists in this conflict felt that there could be only one correct interpretation of quantum mechanics. Although it is not uncommon for scientists to hold that there is a unique law or theory, Bohr even as a child believed in the uniqueness and necessity of natural laws.[33] Such a belief would justify one in looking for, or attempting to formulate, the one possible correct version of quantum mechanics. Heisenberg's faith in the finality of quantum mechanics was essential for his struggle to fashion the Copehagen interpretation via his uncertainty relations (that we discuss in Section 20.4).[34] Let us now summarize rather briefly how the Copenhagen interpretation came to be formulated under the challenge that Schrödinger's wave mechanics presented to the Göttingen–Copenhagen matrix-mechanics program and how this latter interpretation established its hegemony (or dominance).

One factor relevant to these historical developments was a split in philosophical outlook along generational lines: the 'older' essentially classical, world view of people like Einstein, Schrödinger and de Broglie versus a radically different, eventually indeterministic conception of physical processes engendered by the generally younger generation (Bohr and Born being exceptions here) including Heisenberg, Pauli, Jordan and a new member of the group, Dirac from Cambridge University. The prevalence of an empiricist–operationalist philosophical tendency among Heisenberg, Pauli and Bohr can be traced in part (somewhat ironically, given Einstein's later views) back to Einstein's 1905 relativity papers. This operationalist approach, one aspect of which was an eschewal of unobservable entities in a theory, seems to have made a great impression and to have exerted a profound influence upon young German physicists. Let us now see how such factors help us in understanding the vehemence of the Copenhagen school's reaction against Schrödinger's wave mechanics.

Matrix mechanics was formulated by Heisenberg, and developed by other members of the Copenhagen school, as an essentially abstract mathematical formalism with no physical interpretation. Heisenberg believed that a successful formalism, such as classical mechanics, was of a piece or whole and that it could not be modified in any essential way without destroying the entire structure. (This bears some similarity to the unity of the Aristotelian 'fabric' of the universe that we mentioned in Section 11.1.) He also held the remarkable view that a formalism uniquely determines its proper interpretation. Heisenberg was quite disturbed by the appearance of Schrödinger's apparently very different formalism that was invested with an interpretation in terms of continuous, causal, largely visualizable physical processes. In a sense, the situation became even worse in Heisenberg's view when, also in 1926, Schrödinger demonstrated that the matrix-mechanics and wave-mechanics formalisms were basically mathematically equivalent. It remained essential to find the correct interpretation of matrix mechanics as quickly as possible. This was a major undertaking in Copenhagen with Bohr, Heisenberg and Pauli. Even worse for Heisenberg and his colleagues, the formalism of matrix mechanics did not have many successful applications. In fact, it appeared to be bogged down in a mathematical morass before Schrödinger's wave mechanics allowed theorists to make a stunningly wide variety of well-confirmed calculations.[35] Wave mechanics, not matrix mechanics, was the formalism employed by most theorists. This danger of losing the war on the calculational front threatened further consequences. Heisenberg had personal ambitions for advancement and several chairs in theoretical physics were opening up in Germany. There was a conscious realization by members of the Copenhagen school that control of the future direction of theoretical physics was at stake. This Copenhagen group had the talent, organization and drive to carry the day in establishing the hegemony of the Copenhagen view. Heisenberg's uncertainty relation paper (Chapter 20) was a major step in accomplishing this. They worked in concert, while their opponents (Einstein, Schrödinger, de Broglie) each pulled in his own direction. The influence of the Bohr Institute in Copenhagen was enormous on an entire generation of leading theoretical physicists who passed through it (including most of those who played dominant roles in establishing theoretical physics in the United States).

A crucial encounter occurred at the 1927 Solvay Congress. In that year de Broglie proposed a 'principle of the double solution', in which he suggested a synthesis of the wave and particle nature of matter. At that Fifth Solvay Congress, he presented some of these ideas in a form he termed the pilot-wave theory. Here a physical particle was pictured as being guided by its pilot wave. In discussion at that Congress, Pauli criticized de Broglie's theory on the basis of a specific example and claimed that the pilot-wave theory would not yield the same results as the Copenhagen interpretation. Although he believed he understood the general outlines of a suitable response to Pauli's objection, de Broglie himself felt in later years that he did not respond as convincingly as he might have.[36] Furthermore, neither Einstein nor Schrödinger gave positive support to de Broglie's ideas: Einstein because he did not like the nonlocal (apparently instantaneous action-at-a-distance) nature of the theory and Schrödinger because he wanted a theory based only on waves (not on waves and particles). (By 'nonlocality' we mean here and later instantaneous, long-range action-at-a-distance influences. In Chapters 22 and 23 we return to this apparent feature of the microrealm.) The people who, with the hybrid wave-mechanics/matrix-mechanics formalism, were producing results for problems involving spin (including Heisenberg and Born who also spoke at the 1927 Solvay Congress) strongly favored the indeterministic or noncausal picture. Bohr was for a long time against the concept of the photon (on which some of de Broglie's early work was based), so that de Broglie's ideas had never spread rapidly in the Copenhagen school. The Institute at Copenhagen was a very closed community and those invited there were identified as the 'respectable' theorists. De Broglie was never a member of this group. By 1930, when he wrote a very standard quantum-mechanics book, de Broglie himself changed his mind about the pilot-wave theory. In 1932 the famous mathematician John von Neumann (1903–1957) offered an impossibility proof for *hidden variables* theories (that is, empirically adequate theories containing at-present unknown or 'hidden' parameters whose values would determine the evolution of well-defined physical properties forbidden by standard quantum mechanics). This further confirmed de Broglie's position against his own previous theory. It was not until the early 1960s that John Bell (1928–1990) demonstrated conclusively that von Neumann's theorem was largely irrelevant to the question of hidden variables in quantum mechanics. (We return to these questions in Chapters 22 and 23.)

Apropos of these issues, one can ask what was the warrant for the Copenhagen interpretation being taken as complete and the final word in forbidding even the in-principle possibility of a description of microphenomena that is both causal and pictured in a continuous space–time.[37] One response is that, thus far, experience has shown the consistency of the Copenhagen interpretation and that belief in the ultimate necessity of that interpretation rests on the subjective epistemological criterion of the need for classical concepts to describe the results of measurements and on the indivisibility (and, hence, the discontinuity) of fundamental atomic phenomena. In other words, one simply subscribes to Bohr's act of faith. Bohr's position has been summarized as '[physical] reality is whatever quantum mechanics is capable of describing.'[38] In such a representation, the Copenhagen interpretation defined itself as true and strengthened its hold on physics, rewriting history so that Einstein, de Broglie and Schrödinger largely fade from view, thus leaving Copenhagen as the only intelligible version of quantum mechanics.[39]

So far we have used the term 'Copenhagen' to refer loosely to some common set of principles and commitments shared by the group of physicists associated with Bohr. Differences between them and Einstein, de Broglie and Schrödinger were emphasized. However, there were significant disagreements among members of the Copenhagen school itself on important points of interpretation. In the next chapter we shall see that Born and Heisenberg parted ways on the meaning of the wave function. Both Born's position and the more extreme view taken by Heisenberg are often indifferently identified with the Copenhagen interpretation. Moreover, many physicists seem unaware of any genuine divergence between the positions of Born and Heisenberg. In the next chapter we turn to a study of this not always precisely defined Copenhagen version of quantum mechanics.

FURTHER READING

The best single reference for an historical account of the emergence of quantum theory remains Max Jammer's classic *The Conceptual Development of Quantum Mechanics*. Chapter 1 of Jim Baggott's *The Meaning of Quantum Theory* gives an informal presentation of the road to quantum mechanics. Olivier Darrigol's *From c-Numbers to q-Numbers* is a remarkable work that covers, in great historical and technical detail, the transition from classical to quantum mechanics. Michel Bitbol and Olivier Darrigol's *Erwin Schrödinger* consists of a collection of essays on one route to quantum mechanics. John Heilbron's 'The Earliest Missionaries of the Copenhagen Spirit' is a marvelous essay on the birth and spread of the Copenhagen dogma on quantum theory. Mara Beller's 'Matrix Theory Before Schrödinger' and her 'The Birth of Bohr's Complementarity: The Context and the Dialogues' expose the reader to some of the tangled story of how the Copenhagen program came to dominance. James Cushing's *Quantum Mechanics* considers how we arrived at the 'standard' view on the meaning of quantum mechanics.

20

Copenhagen quantum mechanics

In a physical theory we typically describe a system in terms of its *state*. We specify the relevant physical quantities or variables and then use dynamical laws to find the time evolution of these variables to predict their values in the future. For example, in classical particle mechanics, the state of a system is specified by the positions and velocities (or, equivalently for us, the momenta) of all of the particles in the system (that is, $r(t)$ and $v(t)$ for each particle). Newton's second law, $F = ma$, then determines the time evolution of the variables, or of the state, of the system. For electrodynamics, the state variables are the electric and magnetic fields, E and B, and Maxwell's equations govern the time evolution of these. In classical physics (including relativity here), the state variables that are the central entities of the dynamical equations (typically the positions and momenta of the particles) are also the directly observable physical quantities. That is, the state itself is specified in terms of the observables of the theory. The essential features of the classical world view (as modified by relativity) are that the present state of a system in principle determines its future state (as in the flight of a baseball under the influence of gravity) and that the agency responsible for this event-by-event causal structure must propagate from the cause to the effect at some rate no greater than the speed of light.

In contrast to this situation, the state (or state vector or wave function ψ) of a quantum-mechanical system is a more abstract object and is not itself directly observable. The fundamental dynamical equation of quantum mechanics, the Schrödinger equation (the quantum analogue of Newton's second law), governs the time evolution of the state vector for the system, but this does not itself yield definite values for the positions and momenta of the parts of the system. Rather, the state vector or wave function ψ that is the central entity in the theory allows us to compute only probabilities of various allowed outcomes (the so-called *eigenvalues*) of an experiment or of an observation. While we are able to predict, say, the possible energies associated with the various allowed final states of a system, we cannot (usually) predict which one will actually be observed or produced in an experiment, but just the probabilities of these allowed outcomes. On the standard, or so-called Copenhagen, interpretation of quantum mechanics, we no longer have event-by-event causality and particles do not follow well-defined trajectories in a space–time background. The theory predicts, in general, probabilities, not specific events. In this chapter we use some simple examples to illustrate these generic features of the behavior of quantum systems.

Since we are not going to develop the extensive mathematical tools necessary to treat the Schrödinger equation in all generality, we make use of an analogy in our discussions of some simple quantum systems. It turns out that the time-independent Schrödinger equation that governs systems in stationary states (such as a hydrogen atom with its electron in one of the allowed orbits) is simply the time-independent wave equation familiar from classical physics. Therefore, we exploit our classical intuitions based on wave phenomena to write down the mathematical solutions for some paradigmatic cases. Of course, the ultimate justification for this procedure comes from actually solving the relevant Schrödinger equation (something we do not do here). It is no accident that such formal analogies exist, since Schrödinger was led to his famous equation by working with, and suitably modifying, various classical wave equations.

20.1 SOME SIMPLE QUANTUM-MECHANICAL SYSTEMS

Let us begin by filling in a few technical details relevant to Schrödinger's discovery. As we saw in Section 19.2, work Planck did on the blackbody radiation spectrum eventually led him to postulate the quantization of electromagnetic radiation ('photons' in later terminology). If we combine Planck's relation $E = h\nu$ (Eq. (19.3)) between the energy E and the frequency ν of a photon with the relativistic expression $E = \sqrt{m_0^2 c^4 + p^2 c^2}$ for the energy[1] in terms of the momentum p, then for $m_0 = 0$ (a 'massless particle' such as the photon), we find $E = pc$. From these two expressions for E we deduce (since $\lambda \nu = c$) that

$$\lambda = \frac{h}{p} \tag{20.1}$$

So far we know only that Eq. (20.1) should be expected to hold for light (that is, for photons). However, in the early 1920s, de Broglie postulated an at least formal duality between waves and all matter, according to which a wavelength λ is to be associated (via Eq. (20.1)) with any momentum p. There was a direct experimental confirmation of this hypothesis, but we do not go into that here. De Broglie received the 1929 Nobel Prize in physics for his discovery of the wave aspects of matter.

In 1926 Schrödinger obtained an equation that predicts the wave, or really the wave function $\psi(x, y, z, t)$, associated with any particle, even when that particle is acted on by forces. His equation was to play a role much like Newton's second law $\mathbf{F} = m\mathbf{a}$ in that it allowed quantitative calculations to be made for a large class of problems. One might say that as Galileo, by studying uniform motion, had preceded Newton, so de Broglie, by giving a formula for the wavelength of a free particle, preceded Schrödinger. The precise arguments by which Schrödinger arrived at his wave equation, as well as the equation itself, are too complex mathematically to present here. Just as for light waves, the principle of superposition holds for the wave function ψ so that interference effects between two waves are possible. We illustrate these principles with some simple physical situations.

PART VII The completeness of quantum mechanics

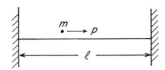

FIGURE 20.1 A mass confined to a line segment

A (nonrelativistic) classical free particle (one subject to no forces) has a total energy (E) that is purely kinetic (K) and that can take on any positive value. More generally, a classical particle whose total energy $E = (K + V)$ is positive (such as an electron that can escape from the attraction of a nucleus and so travel on an open, unbounded trajectory) also has allowed orbits for any positive value of E. When such systems are treated quantum mechanically, the associated wave may encounter barriers or obstacles, so that reflection, transmission and absorption (or diffraction) occur, but there exist solutions to the Schrödinger equation for any positive value of energy. So, classical and quantum-mechanical systems are qualitatively similar in that any positive value of the total energy is allowed.

However, for bound systems, such as the negative-energy states of a hydrogen atom, a distinct quantum-mechanical feature emerges: only certain discrete energies are allowed. Classically, a particle confined to a box can have any energy and momentum. Quantum mechanically, the wavelength must be able to fit between the boundaries and satisfy proper boundary conditions to produce standing waves. (This is similar to the allowed wavelengths for circular Bohr orbits.) For a rigid one-dimensional box (depicted in Figure 20.1) in which the waves cannot penetrate the walls at all, we have for the longest possible wavelength $\lambda_{max}/2 = \ell$ and in general $\lambda_n = 2\ell/n$. (Recall Figure 19.2 and the discussion accompanying it in Section 19.1.) According to de Broglie's hypothesis of Eq. (20.1), the allowed momenta are

$$p_n = \frac{h}{\lambda_n} = \frac{nh}{2\ell} \tag{20.2}$$

We then have for the total energy for the confined 'particle'[2]

$$E_n = \frac{p_n^2}{2m} = \frac{1}{2m}\left(\frac{nh}{2\ell}\right)^2 = \frac{h^2}{8m\ell^2}n^2, \quad n = 1, 2, 3, \ldots \tag{20.3}$$

These energy levels are quantized just as are those of the hydrogen atom. These allowed, or possible, energy values are the eigenvalues for this system. The permitted energy values of Eq. (20.3) have been selected by the boundary conditions imposed on the wave function (here the requirement that the standing waves have nodes at the ends of the 'string' of length ℓ in Figure 20.1). This is what is meant by quantization being effected through boundary conditions that must be satisfied by acceptable solutions to the Schrödinger equation.

Now consider another class of experiments or observations that is very important in atomic physics. Classically, a particle, say an electron, can pass a potential

Copenhagen quantum mechanics

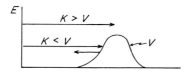

FIGURE 20.2   Classical encounters of a particle with a barrier

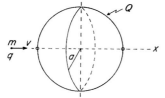

FIGURE 20.3   An electron encounters a charged spherical shell

barrier only if its initial kinetic energy exceeds the height of the barrier, as illustrated in Figure 20.2. As an example of such a potential barrier, take a hollow metal sphere of radius a and positive charge Q with two small holes diametrically opposite each other, as shown in Figure 20.3. If a point particle of positive charge q and mass m is shot along a line passing through these two holes, then the potential $V(x)$ it would 'see' is plotted in Figure 20.4. Classically, if the initial kinetic energy $K_0 = \frac{1}{2}mv^2$ exceeds the maximum potential energy V_{max} of the barrier, then the particle will pass through the sphere and emerge from the other side. If, though, $K_0 < V_{max}$, then the particle will be brought to rest somewhere outside the sphere and then be repelled back. It will never penetrate the potential barrier. However, the solution for the Schrödinger equation for such a situation allows part of the wave function to penetrate the barrier, as indicated in Figure 20.5.[3]

We now discuss some attempts that were made to explain the physical significance of this behavior of the wave function.

20.2   INTERPRETATIONS OF THE WAVE FUNCTION

One of the early interpretations proposed by Schrödinger for the wave function ψ was that the magnitude of its square, $|\psi|^2$, represented the matter density (or possibly the charge density) of the particle. The lower portion of Figure 20.5 shows a difficulty with this interpretation. Since part of the wave function is reflected at the barrier and part transmitted, an electron would have to split up at the barrier, part being reflected and part transmitted. However, experimentally, one always detects a whole electron or none at all, but never a piece of an electron.

Another interpretation, due to Born,[4] that is widely accepted today is that $|\psi(x, y, z, t)|^2$ represents the probability $P(x, y, z, t)$ of finding a particle at a position (x, y, z) at a time t. The probability $P(x)$ can never be negative. This is guaranteed since $P(x)$ is defined as the square of $|\psi(x)|$.[5] In terms of the situation illustrated in Figure 20.5, we would say that the overwhelming probability is that the electron will be reflected from the barrier, but that there is, nevertheless, a finite chance that it will

PART VII The completeness of quantum mechanics

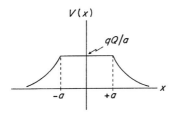

FIGURE 20.4 **An electrostatic potential barrier**

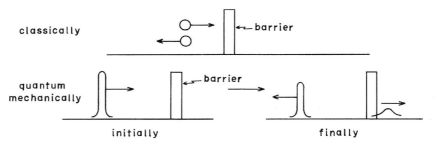

FIGURE 20.5 **Quantum-mechanical barrier penetration**

go through the barrier. Here, again, we have an example of the theory's being able to predict the allowed or possible outcomes of an experiment or observation, but then being able to assign only probabilities to which one result will actually be observed in a given trial or run of an experiment. A somewhat more general way to put this is that, for an ensemble of identically prepared systems (where this ensemble is represented by a given wave function ψ), we are able to predict only the statistical distribution of the measured (or allowed) values that will be obtained for that ensemble. Here a given procedure for measurement is carried out for each member of the ensemble, one-by-one. Quantum mechanics allows us to predict the distribution of outcomes for this long run of ('identical') experiments. For example, if we were to shoot an electron at a barrier (as illustrated in the lower half of Figure 20.5) and repeat this process many times over, then we would find that a certain fraction of the electrons would be reflected and the rest transmitted, but that on any given run an electron would either be reflected or go through the barrier. Born received the 1954 Nobel Prize in physics for this statistical interpretation of quantum mechanics.

Let us apply this probability interpretation to our simple example of Figure 20.1 for the case with $n = 1$. Continuing to exploit our analogy with standing waves on a string of length ℓ, we would expect the standing wave (or wave function, in this case) to be given as (see Figure 20.6)

$$\psi_1(x) = A\sin\left(\frac{2\pi x}{\lambda}\right) = A\sin\left(\frac{\pi x}{\ell}\right) \tag{20.4}$$

and the probability function as

$$P_1(x) = |\psi_1(x)|^2 = A^2\sin^2\left(\frac{\pi x}{\ell}\right) \tag{20.5}$$

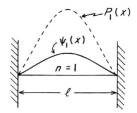

FIGURE 20.6 Ground-state probability distribution ($P_1(x)$) for a confined particle

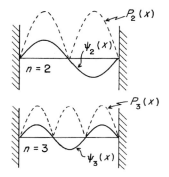

FIGURE 20.7 Higher-level probability distributions ($P_2(x)$ and $P_3(x)$)

Since the total probability of finding the 'particle' somewhere on the line segment of length ℓ must be unity, we require that the total area under the $P(x)$ curve be unity. This can be shown (although we do not do it here) to determine the value of A in Eq. (20.4) so that

$$\psi_1(x) = \sqrt{\frac{2}{\ell}} \sin\left(\frac{\pi x}{\ell}\right) \tag{20.6}$$

Notice that the most likely place to find the particle in the ground state ($n = 1$) is at $x = \ell/2$, although there is a finite probability of finding it anyplace on $0 < x < \ell$, except at the ends ($x = 0$ and $x = \ell$). The corresponding curves for $n = 2$ and $n = 3$ are shown in Figure 20.7. Similarly, an exact quantum-mechanical treatment of the hydrogen atom using the Schrödinger equation yields the probability functions shown in Figure 20.8.[6] Quantum mechanically the most probable location of the electron turns out to be very close to the value $r_{Bohr} = n^2 a_0$ given by the semiclassical Bohr model for the allowed orbits, where a_0 is the Bohr radius. Thus we see that there is some relation between the probabilistic predictions of quantum mechanics and the sharp values expected classically.

The great virtues of the Schrödinger formulation of quantum mechanics are that it allows one to calculate specific solutions for many problems and that the wave function ψ, or more precisely $|\psi|^2$, is an important aid in providing an interpretation of quantum mechanics. Ever since Schrödinger established the formal equivalence between his theory and Heisenberg's in 1926, those two formulations of quantum

PART VII The completeness of quantum mechanics

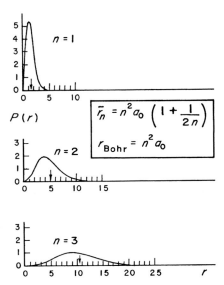

FIGURE 20.8 Hydrogen-atom probability distributions: arrows indicate $\bar{r}_n$ values

mechanics have often been taken to be simply two different mathematical ways of expressing the same physical theory.

20.3 A FUNDAMENTAL DISTINCTION BETWEEN LARGE AND SMALL

We continue our exposition of the standard interpretation of quantum mechanics and we begin to discuss some of the major revisions that this theory is often taken to necessitate in our view of the world. As a vehicle for this, we apply the probabilistic interpretation of quantum mechanics to an experiment in which photon or electron beams interfere with themselves. We center our discussion around excerpts from Dirac's seminal book, *The Principles of Quantum Mechanics*. Not only has this been an influential text for decades, but it is also generally accepted as giving a concise and accurate presentation of the principles of quantum mechanics. Since in later chapters we question some of these interpretative constraints, we want first to present the views of some of the major figures (such as Bohr, Born, Dirac and Heisenberg) on this standard version of quantum mechanics. For that reason we appeal fairly frequently to direct quotations from the writings of these early masters, to document that these scientists did actually hold the, at times extreme, views we claim they did.

In his book, Dirac begins with an analysis of an inherent limitation in the picture of atomic physics as presented by classical physics.

> All material particles have wave properties, which can be exhibited under suitable conditions. We have here a very striking and general example of the breakdown of classical mechanics –

> not merely an inaccuracy in its laws of motion, but *an inadequacy of its concepts to supply us with a description of atomic events.*
>
> The necessity to depart from classical ideas when one wishes to account for the ultimate structure of matter may be seen, not only from experimentally established facts, but also from general philosophical grounds. In a classical explanation of the constitution of matter, one would assume it to be made up of a large number of small constituent parts and one would postulate laws for the behaviour of these parts, from which the laws of the matter in bulk could be deduced. This would not complete the explanation, however, since the question of the structure and stability of the constituent parts is left untouched. To go into this question, it becomes necessary to postulate that each constituent part is itself made up of smaller parts, in terms of which its behaviour is to be explained. There is clearly no end to this procedure, so that one can never arrive at the ultimate structure of matter on these lines. So long as *big* and *small* are merely relative concepts, it is no help to explain the big in terms of the small. It is therefore necessary to modify classical ideas in such a way as to give an absolute meaning to size.[7]

In discussing the breakdown of the classical description of matter, Dirac here emphasizes not so much that empirical facts demand new concepts as the philosophical point that without some fundamental distinction between large and small, any attempt to examine the structure of matter simply multiplies its constituent parts on a smaller and smaller scale without limit. This leads to an infinite regress. The question itself about the smallest particles of matter is never really answered. It becomes unanswerable in terms of our classical, everyday concepts.

As an absolute criterion for the demarcation between the classical and the quantum domains, Dirac suggests a distinction based on the limitation of our ability to observe these objects.

> At this stage it becomes important to remember that science is concerned only with observable things and that we can observe an object only by letting it interact with some outside influence. An act of observation is thus necessarily accompanied by some disturbance of the object observed. We may define an object to be big when the disturbance accompanying our observation of it may be neglected, and small when the disturbance cannot be neglected. This definition is in close agreement with the common meanings of big and small.[8]

Dirac is arguing that an intrinsic distinction between large and small is related to the effects produced on an object when it is observed. We can observe a system only if we interact with it. As a fanciful and simple example of an observation disturbing the system being observed, consider a sealed box with a hamster in it. Let the box be surrounded by a highly lethal gas that kills instantly upon contact. We wish to determine whether the hamster is dead or alive and let it be given that the only means we have of doing this is to open the box to see. Clearly, every time we look into an opened box we will find a dead hamster, no matter what the state of the animal before the box is opened. The act of observing the system (here, the hamster) has forced the system into a given state.

Dirac goes on to stress that, on an atomic scale, this disturbance constitutes an inherent limitation in principle, not just in practice.

> It is usually assumed that, by being careful, we may cut down the disturbance accompanying our observation to any desired extent. The concepts of big and small are then purely relative and refer to the gentleness of our means of observation as well as to the object being described.

PART VII The completeness of quantum mechanics

> In order to give an absolute meaning to size, such as is required for any theory of the ultimate structure of matter, we have to assume that *there is a limit to the fineness of our powers of observation and the smallness of the accompanying disturbance – a limit which is inherent in the nature of things and can never be surpassed by improved technique or increased skill on the part of the observer*. If the object under observation is such that the unavoidable limiting disturbance is negligible, then the object is big in the absolute sense and we may apply classical mechanics to it. If, on the other hand, the limiting disturbance is not negligible, then the object is small in the absolute sense and we require a new theory for dealing with it.[9]

We are being told that there exists an inherent limit to the powers or precision of observation. In the next section we formulate this quantitatively in terms of the Heisenberg uncertainty principle.

We now come to one of the most profound issues in the interpretation of quantum mechanics – that of causality (in the sense of a specific, identifiable cause for each individual effect). Although we do return to this question again in Chapter 23 (but from a very different perspective there), we remind the reader once again that at this point we are simply laying out what is usually taken to be one of the central and necessary lessons of quantum mechanics. As an example of this standard line of argument, Dirac observes:

> A consequence of the preceding discussion is that we must revise our ideas of causality. Causality applies only to a system which is left undisturbed. If a system is small, we cannot observe it without producing a serious disturbance and hence we cannot expect to find any causal connexion between the results of our observations. Causality will still be assumed to apply to undisturbed systems and the equations which will be set up to describe an undisturbed system will be differential equations expressing a causal connexion between conditions at one time and conditions at a later time. These equations will be in close correspondence with the equations of classical mechanics, but they will be connected only indirectly with the results of observations. There is an unavoidable indeterminacy in the calculation of observational results, the theory enabling us to calculate in general only the probability of our obtaining a particular result when we make an observation.[10]

This means that the concept of causality, in the classical deterministic cause and effect sense, must be modified.

Planck, in his *The Concept of Causality in Physics*, makes a similar point.[11] There he tells us that there remains causality, or mathematically deterministic evolution, for the wave function (ψ) itself in the sense that a knowledge of the present value for ψ uniquely determines, via the dynamics of the Schrödinger equation, the value of ψ at any future time. However, this determinism for ψ does not translate into a determinism of, say, the position and momentum of a particle. That is, as we emphasized at the beginning of this chapter, the state of a quantum-mechanical system is specified by ψ, and not by the classical state variables ($r(t)$ and $v(t)$). The rub, of course, is that it is only the classical state variables that are accessible to our (more or less) direct observation. We cannot directly observe ψ. Planck stresses that a definite knowledge of ψ issues (via the probability distribution $P = |\psi|^2$) in only statistical (or probabilistic) predictions for these accessible observables (such as $r(t)$ and $v(t)$). Hence, the argument goes, quantum mechanics does not allow a deterministic description of the evolution of, say, the position and velocity of an individual particle, such as an electron.

20.4 THE UNCERTAINTY RELATION

In this same spirit Heisenberg too felt that, since the mathematical structure of quantum mechanics is so different from that of classical mechanics, it is not possible to interpret quantum mechanics in terms of our commonly understood notions of space and time with classical causality. In 1927 Heisenberg gave a mathematical argument that certain pairs of variables (such as position and momentum for the members of an ensemble of particles whose state is represented by a given wave function ψ) cannot, according to the predictions of quantum mechanics, both (that is, together or 'simultaneously') be ascertained to an arbitrarily high degree of accuracy. This is a statement about the statistics obtained in a series of identical measurements over the ensemble, not a statement about the outcome of a single measurement on a particular member of the ensemble. (Remember, on the Born or probabilistic interpretation, quantum mechanics is only able to make statistical predictions, such as average values for many observations, not outcome-by-outcome predictions.) Subsequently he attempted to use simple thought experiments to make it plausible that the wave–particle duality of light and of matter gives rise to joint uncertainties in our observations on atomic and nuclear systems. Below we return to a discussion of one such thought experiment.

First, though, let us consider a set of measurement results $A_j (j = 1, 2, 3, \ldots, N)$ for an ensemble. We define the average value of these $\{A_j\}$ in the usual way as

$$\langle A \rangle = \frac{1}{N} \sum_{j=1}^{N} A_j \qquad (20.7)$$

and the average value of the squares $\{A_j^2\}$ as

$$\langle A^2 \rangle = \frac{1}{N} \sum_{j=1}^{N} A_j^2 \qquad (20.8)$$

A typical measure of the 'spread' or 'scatter' of these values is given by the root mean square deviation defined as

$$\Delta A = \sqrt{\langle A^2 \rangle - \langle A \rangle^2} \qquad (20.9)$$

If there were no dispersion in the values of the A_j (in other words, if all of the A_j had the same precise value, say A_0), then ΔA would be zero.

As we indicated in Section 19.5, the quantum-mechanical operators corresponding to different observables need not commute (that is, the order in which the observations are made can make a difference in the actual values obtained). For two noncommuting[12] observables A and B, there is in quantum mechanics a general constraint, or lower bound, on how small ΔA and ΔB can be simultaneously for a series of measurements of those variables. This is the Heisenberg uncertainty relation and typically takes the form[13]

$$\Delta A \, \Delta B \geq b\hbar \qquad (20.10)$$

where b is some nonnegative constant (depending on the choice of the operators A

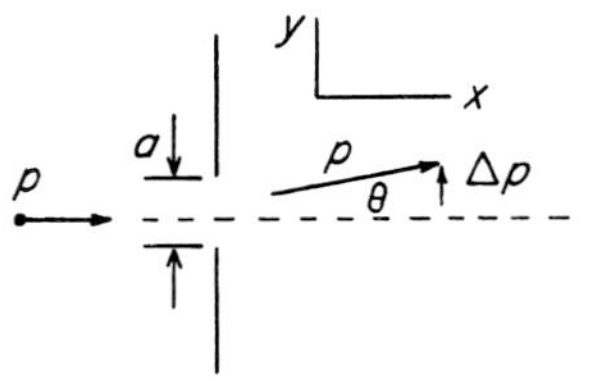

PART VII   The completeness of quantum mechanics

FIGURE 20.9   Joint position–momentum uncertainties upon measurement

and B) and $\hbar = h/2\pi$. For position (x) and momentum (p_x) this becomes

$$\Delta p_x \Delta x \geq \frac{\hbar}{2} \tag{20.11}$$

We emphasize again that the uncertainty relations of Eqs. (20.10) and (20.11) are statistical ones. This is consistent with the Born interpretation of the wave function, according to which ψ represents an ensemble of identical systems, not any individual member of that ensemble.

At this point it is unclear just what restriction Eq. (20.11) places on individual simultaneous measurements of these noncommuting observables. In his famous 1927 'uncertainty' paper, Heisenberg argued that these statistical spreads (the Δx and Δp_x in Eq. (20.11)) are produced by ineliminable disturbances at the level of individual measurements. This effectively assumes that ψ is the wave function for each individual member of the ensemble separately. Such an interpretation is different from and goes considerably beyond Born's statistical interpretation.[14] Commiting oneself to ψ as the state of an individual system can lead to serious problems, as we show when we consider the Schrödinger 'cat' paradox in Section 21.4.

Arguments similar to Heisenberg's are typically given in physics textbooks.[15] Let us examine one representative of these thought experiments. Suppose we consider passing a small particle (say an electron) through a narrow slit, as shown in Figure 20.9.[16] The uncertainty in its y-coordinate is then $\Delta y = a$. However, since the particle has a wave character associated with it, this wave is diffracted, the first minimum occurring at $\sin\theta = \lambda/a$ (just as in the optical diffraction of a light wave by a slit). This means that the probability distribution for the particle will be confined to the wedge-shaped region of half-angle θ, since the probability of finding the electron at a given position is proportional to the square of this diffracted wave function. The de Broglie relation of Eq. (20.1) gives the momentum as $p = h/\lambda$. From Figure 20.9 we see that the uncertainty in the y-component of the particle's momentum can be expressed, with the aid of Eq. (20.1), as $\Delta p_y \approx p \sin\theta = p \lambda/a$ so that $\Delta p_y \Delta y \cong h$, consistent with Eq. (20.11). In this relation Δp_y and Δy stand for the magnitudes of the uncertainties that are produced by the measurement process. This, of course, is a heuristic argument only, not a general proof.

When $\lambda/\ell \ll 1$, where λ is the de Broglie wavelength of an object and ℓ is a typical dimension of its environment, we are in the classical domain (since diffraction effects will be negligible) and Newtonian mechanics applies. In our world $h = 6.67 \times 10^{-34}$ J s so that everyday phenomena do not appear quantum-mechanical in nature. However, if our world were such that $h = 1$ J s, then life would be very different. For example, if we confined an object of mass 10^{-3} kg to a line segment of 10^{-2} m, its minimum allowable energy would be given by Eq. (20.3) with $n = 1$ and this would correspond to a velocity of $v \approx 50$ km/s. Such fluctuations would be a pronounced phenomenon and would directly affect our perception of the world.

Another version of the Heisenberg uncertainty relation

$$\Delta E \, \Delta t \geq \frac{\hbar}{2} \tag{20.12}$$

relates the energy resolution to the time over which the observation takes place. For instance, if a particle lives only a very short time, then its energy (or mass) cannot be known with great precision.

The Heisenberg *uncertainty principle* refers collectively to all of these (and other) versions of the uncertainty relations. These are formal statements of the inherent and irreducible imprecision with which we can determine certain pairs of observables in a series of measurements.

20.5 PHOTON INTERFERENCE – THE DOUBLE SLIT

We now return to Dirac's text to examine in detail the effect that a measurement has upon microscopic systems. The vehicle for this discussion is basically the double-slit interference experiment.

> We shall discuss the description which quantum mechanics provides of the interference of photons. Let us take a definite experiment demonstrating interference. Suppose we have a beam of light which is passed through some kind of interferometer, so that it gets split up into two components and the two components are subsequently made to interfere. We may... take an incident beam consisting of only a single photon and inquire what will happen to it as it goes through the apparatus. This will present to us the difficulty of the conflict between the wave and corpuscular theories of light in an acute form.
>
> ...
>
> [W]e must now describe the photon as going partly into each of the two components into which the incident beam is split. The photon is then, as we may say, in a translational state given by the superposition of the two translational states associated with the two components.... For a photon to be in a definite translational state it need not be associated with one single beam of light, but may be associated with two or more beams of light which are the components into which one original beam has been split. In the accurate mathematical theory each translational state is associated with one of the wave functions of ordinary wave optics, which wave function may describe either a single beam or two or more beams into which one original beam has been split. Translational states are thus superposable in a similar way to wave functions.[17]

The type of situation Dirac has in mind here is one in which interference effects are

PART VII   The completeness of quantum mechanics

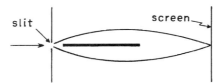

FIGURE 20.10   An interference experiment for a split beam of photons

produced by a beam of photons that is split and then recombined, as indicated in Figure 20.10.[18] However, it is crucial to appreciate that, as experiments have shown to be the case, just one photon at a time goes through a slit. For example, a very weak light source can be used to illuminate a double slit and then a photomultiplier tube can be used to detect these single photons as they arrive at the screen. The distribution of observed photons still follows the double-slit diffraction pattern expected from optics, even though just one photon at a time is detected at a given location on the screen, as we discuss in more detail below.

Dirac continues:

> Let us consider now what happens when we determine the energy in one of the components. The result of such a determination must be either the whole photon or nothing at all. Thus the photon must change suddenly from being partly in one beam and partly in the other to being entirely in one of the beams. This sudden change is due to the disturbance in the translational state of the photon which the observation necessarily makes. It is impossible to predict in which of the two beams the photon will be found. Only the probability of either result can be calculated from the previous distribution of the photon over the two beams.[19]

Here we are told that an energy observation on one of these split beams forces the photon into one or the other of the beams (or paths). The theory can only assign a probability to which beam the photon will appear in (that is, we have a statistical interpretation). This sudden and discontinuous change of the state (ψ) of a quantum-mechanical system upon observation or measurement is an example of one of the central and long-standing conceptual difficulties of the standard interpretation. It is termed the 'measurement problem' and we discuss it further in Section 21.3.

As a concrete example of the type of situation that Dirac has in mind in the passages quoted above, suppose we consider the following sequence of experiments that we could perform, as illustrated in Figure 20.11.[20] Let us send a beam of particles through the aperture shown on the left in the figure. After they go through this opening, they then pass through either slit 1 or slit 2 in the second plate. Finally, we detect or count the number of particles (say, electrons) arriving at various positions on the 'screen' at the right. As we would expect on the basis of classical particle physics, nearly all of the particles that manage to arrive at the screen would impinge at those points directly opposite either slit 1 or slit 2 in the middle plate. In fact, if we cover up slit 2, then we obtain the dashed distribution curve labeled N_1. Similarly, if we close slit 1, we obtain N_2, while if both 1 and 2 are open we find N'_{12}. Furthermore, it turns out that N'_{12} is just the sum of curves N_1 and N_2 ($N'_{12} = N_1 + N_2$), as we would expect classically for particles. There are no interference effects.

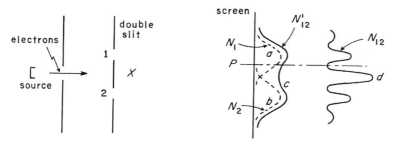

FIGURE 20.11 Double-slit interference patterns

Next, let us repeat our experiment, but this time using waves rather than particles. When both slits are open, then – just as we would expect from classical optics – we observe the double-slit interference pattern I_{12} (that we have labeled as N_{12} in the figure for reasons that will become evident shortly). When slit 2 is closed, we obtain a single-slit diffraction pattern I_1 (of the same shape as the curve labeled N_1 in the figure) and, when slit 1 is closed, the pattern I_2. Notice that, unlike the situation for classical particles, these intensities do not simply add. Rather, $I_{12} \neq I_1 + I_2$ so that there are interference effects, as is typically the case for wave phenomena. Although the intensities I_1 and I_2 do not add together in any simple fashion to produce I_{12}, the amplitudes of the wave disturbance generated by each slit do obey the superposition principle $A_{12} = A_1 + A_2$, where the intensity of the light or wave phenomenon is given as $I_{12} \propto |A_{12}|^2$. In comparing these two experiments we see that I_1 and N_1 are similar, as are I_2 and N_2, but that I_{12} (or N_{12}) and N'_{12} are very different.

Now let us discuss a situation that will bring out the quantum-mechanical features of a system. We again consider the arrangement of Figure 20.11, but this time we use a beam of electrons incident upon the slit arrangement. At the position of the 'screen' we have a detector that we can move up and down to see whether or not an electron arrives at a given location on the screen. We send one electron after another through the double-slit apparatus and count the relative number of electrons that arrive at various positions on the screen. What we find experimentally is that the distribution curve N_{12} follows the interference pattern for a double slit. At each position on the screen we always detect either a whole electron or none and each electron sent through the apparatus appears at one and only one point on the screen. In this sense, electrons are acting as particles. But, the curve that indicates the likelihood or probability of detecting an electron at a given location on the screen follows the interference pattern for a wave. Hence, the electron also has a wave character here. Next, if we close slit 2, then the distribution of detected electrons becomes N_1 (or N_2 if we close slit 1). Again, notice that $N_{12} \neq N_1 + N_2$, just as was the case for wave interference. It also seems strange that, if we place our electron detector at point P in Figure 20.11, then when both slits are open we practically never find an electron there, whereas when we cover slit 2, we detect many electrons there.

Is the electron 'really' a wave or a particle? Let us attempt to discover through

PART VII The completeness of quantum mechanics

which slit a given electron goes as it passes through the double slit. Before that instrument is turned on, we detect curve d of Figure 20.11 when both slits are open. Once the light is on so that we see every electron as it passes through the double slit, the pattern switches to c of that figure where $N'_{12} = N_1 + N_2$. That is, once we 'force' the electrons to act as particles and definitely pass through one slit or the other, the probability curve becomes that for particles. If we now decrease the intensity of the light at X so that we begin to miss some of the electrons going through the double slit, the detection pattern begins to change. As we continue to decrease the intensity of the light to zero, the curve varies continuously from c back to d.[21]

We have no way to predict where a given electron will land on the screen. Quantum mechanically, we can only give the probability of detecting an electron at a specified position on the screen. This is often taken to imply that there is no law of motion that will determine the trajectory of a single electron, given only its initial conditions as it leaves the source. (We come back to this issue in Chapter 23.) In any event, whether or not we observe the electron going through one of the slits does affect the interference pattern that appears on the screen. This is one instance of the general result of the standard interpretation of quantum mechanics that in atomic physics there are no laws by which we can predict from initial conditions the exact future locations of events at some point in space. There is a lack of absolute predictive power.

In the next few chapters we take up the question of whether or not it might be possible to have a more complete and more understandable view of the physical microworld than that provided by this standard, or Copenhagen, interpretation of quantum mechanics.

FURTHER READING

Chapter 2 of Jim Baggott's *The Meaning of Quantum Theory* gives an intermediate-level, very readable account of the physical implications of some of the mathematics used in elementary quantum mechanics. Sandro Petruccioli's *Atoms, Metaphors and Paradoxes* reconstructs Bohr's conceptual development of the new physics. Chapter 6 of Richard Feynman's *The Character of Physical Law* contains a now classic discussion of the quantum-mechanical double-slit experiment.

21

Is quantum mechanics complete?

When we are presented with apparently bizarre properties of microsystems governed by quantum mechanics (as in our discussion of the double-slit experiment in the previous chapter), we naturally ask whether it might be possible to find another theory, or at least another story to go with the equations, that would make more sense to us. We might seek a theory and a view of fundamental physical processes that would accord more with our classically based, everyday common sense. This typically takes the form of the question of the possibility of the existence of a more detailed description of atomic processes than that afforded by the wave function (that is, by the Copenhagen interpretation of the previous chapter). In this chapter we discuss some of the central interpretive problems of the Copenhagen version of quantum mechanics. Then, in the next two chapters, we look at the severe, general restrictions that are placed on any such completion of quantum mechanics and at one particular extension of quantum mechanics. First, though, we begin here with the struggle that ensued over the status of causality in quantum theory.

21.1 THE COMPLETENESS OF QUANTUM MECHANICS

Although the Heisenberg uncertainty principle (Section 20.4) and the lack of absolute predictive power are an inherent feature of quantum mechanics, there have been attempts to preserve an in principle completely deterministic structure for physics. Even classically, in a macroscopic sample of gas there are so many molecules that it would be hopeless, as a practical matter, to predict the future locations of all these molecules from Newton's laws of motion. Nevertheless, the underlying structure is a completely deterministic one. (In Chapter 12 we saw a similar situation for classical deterministic chaos.) According to the standard interpretation of quantum mechanics, things are very different. In principle, there is no deterministic scheme to predict the exact future trajectory of an electron. Still, the question naturally arises whether one can construct some completely deterministic theory that will give the same predictions or results as quantum mechanics. Can there be as yet undiscovered properties of particles (often called hidden variables) whose values do determine the future behavior of these particles? Can the indeterminacy we observe be simply a result of our ignorance of the values of such hidden variables? (We return to one such theory in Chapter 23.)

The two different schools of thought on this question are usually identified in their origins with Bohr and with Einstein. Early in the history of quantum mechan-

PART VII   The completeness of quantum mechanics

ics, Bohr stated his belief that there could be no complete separation of the phenomena and the observer. In 1927 the International Congress of Physics was held in Como, Italy, just one hundred years after the death of that city's most famous scientist, Alessandro Volta (1745–1827). There Bohr gave the first thorough presentation of his ideas on complementarity and articulated what has become known as the Copenhagen interpretation of quantum mechanics. (Although we give further illustrations of the meaning of the term later in the chapter, by 'complementarity' here we mean Bohr's doctrine that mutually exclusive (in the sense of not simultaneously physically realizable) characteristics (such as wave–particle or position–velocity) are needed for a complete description of a physical system.) In this interpretation, the space–time description of a physical process is exclusive of and complementary to a strict cause–effect description.

> On one hand, the definition of the state of a physical system as ordinarily understood, claims the elimination of all external disturbances. But in that case, according to the quantum postulate, any observation will be impossible, and, above all, the concepts of space and time lose their immediate sense. On the other hand, if in order to make observation possible we permit certain interactions with suitable agencies of measurement, not belonging to the system, an unambiguous definition of the state of the system is naturally no longer possible, and there can be no question of causality in the ordinary sense of the word. The very nature of the quantum theory thus forces us to regard the space–time coordination and the claim of causality, the union of which characterizes the classical theories, as complementary but exclusive features of the description, symbolizing the idealization of observation and definition respectively.[1]

By the 'quantum postulate' Bohr means the irreducible, ineliminable and uncontrollable minimum (but necessarily nonzero) disturbance that is caused when one system interacts with another. In the second quotation for Part VII (*The quantum world and the completeness of quantum mechanics*) we saw how Bohr framed this. The central importance of the relation between the observer and the phenomena was further elaborated by Bohr. Today we refer to this dependence of the outcome of a measurement upon the means used to effect it as *contextuality*.

Difficulties and paradoxes seem to arise when we ask questions like, 'Really, what are the simultaneous values of x and p_x for an electron?' or 'What slit did the electron really go through in passing from the source to the detector in an arrangement of the type shown in Figure 20.11?' As we indicated in Chapter 20, we are always frustrated by the experimental results when we attempt to answer this last question. The Copenhagen school of quantum mechanics tells us that we can never answer such questions. Still, one's first reaction is that the answer must exist even if we are at the present time (or even forever) incapable of finding that answer. We do not feel satisfied with the statement that, in principle, we can know *either* the momentum *or* the position of an electron; or that the apparently obvious proposition that the electron had to go through either the first slit or the second slit is operationally meaningless. We can observe the electron as it leaves the source in Figure 20.11 and then as it arrives at the screen on the right, but we can assign no reality to its location or behavior between these two observations unless we modify the experimental arrangement to observe it there, in which case we change the outcome of the experiment. That is what the Copenhagen interpretation tells us. In expounding this view Werner Heisenberg observed:

> In classical physics science started from the belief... that we could describe the world... without any reference to ourselves.... Its success has led to the general ideal of an objective description of the world.... Does the Copenhagen interpretation of quantum theory still comply with this ideal? One may perhaps say that quantum theory corresponds to this ideal as far as possible.... [I]t starts from the division of the world into the 'object' and the rest of the world, and from the fact that at least for the rest of the world we use the classical concepts in our description. This division is arbitrary and historically a direct consequence of our scientific method; the use of the classical concepts is finally a consequence of the general human way of thinking.[2]

Albert Einstein was unwilling to accept quantum mechanics as anything more than a highly successful provisional theory because he believed that an objective physical reality exists whether or not anyone observes it or interacts with it. In a 1926 letter to his lifelong friend Born, who held views very different from his own on quantum mechanics, Einstein expressed his feelings against this essentially random element in quantum theory as:

> Quantum mechanics is certainly imposing. But an inner voice tells me that it is not yet the real thing. The theory says a lot, but does not really bring us any closer to the secret of the 'old one'. I, at any rate, am convinced that *He* is not playing at dice.[3]

21.2 THE BOHR–EINSTEIN CONFRONTATIONS

Bohr and Einstein confronted each other on this issue at both the Fifth (1927) and Sixth (1930) Solvay Congresses. During these conferences Einstein attempted to show that quantum mechanics contained internal inconsistencies. Near the beginning of the general discussion session of the 1927 meeting, he considered an electron passing through a small hole and being detected on a screen, as shown in Figure 21.1. Before any observation is made, there is, according to quantum mechanics, at nearly every point on the screen a nonzero probability of detecting an electron. However, once the electron has been detected at, say, point A, there will then be absolutely zero probability of finding the electron at any other point B. Einstein argued that, if one were to say that the electron had been virtually present everywhere over an appreciable portion of the screen before the observation but that the probability at B had been instantaneously affected by an observation at A, then this would require an action at a distance that relativity is usually taken to rule out (since an effect ought not be propagated instantaneously between two spatially separated points). On the other hand, he contended, if there did exist some actual trajectory along which the electron proceeded through the slit to point A on the screen and if quantum mechanics were incapable of yielding that information, then quantum mechanics would be an incomplete theory. This theme of the alleged incompleteness of quantum mechanics was one to which Einstein would return often during his life. In the view of Bohr, Born, Dirac, Heisenberg and nearly all of

PART VII The completeness of quantum mechanics

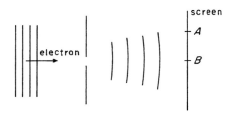

FIGURE 21.1 Einstein's thought experiment for wave-function diffraction

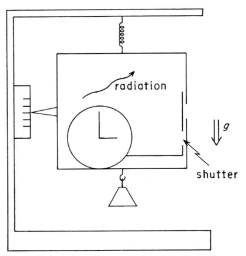

FIGURE 21.2 Einstein's attempt to circumvent the uncertainty principle

the other quantum theorists, Einstein's argument missed the point since the wave function did not, in their opinion, represent anything like an ordinary wave propagating in a space–time background. In the course of the discussion among Bohr, Einstein and the other participants, a double-slit arrangement like the one we studied in Chapter 20 was brought up. This entire debate on the completeness issue was pretty much a standoff, since each side remained entrenched in its position and neither could be dislodged with a telling counter argument.

At this same 1927 meeting Einstein also attempted to construct a thought experiment that would prove that the Heisenberg uncertainty relations could be circumvented. However, Bohr was able to find a flaw in the argument and maintain the validity of the uncertainty principle. The same type of debate resumed at the next Solvay Congress in 1930. This time Einstein had concocted a clever thought experiment that would violate the energy–time version (Eq. (20.12)) of the uncertainty relation. Basically what he proposed was to have a box filled with low-density electromagnetic radiation and equipped with a shutter driven by a clock inside the box, as illustrated in Figure 21.2. The clock would be set to open and close the shutter very quickly in a time Δt so that only a single photon would escape from the box. This time interval Δt could be set as accurately as desired. Since $E = mc^2$, simple accurate weighings of the radiation-filled box before and after the photon had been emitted would determine the energy difference ΔE that would be the energy of the ejected photon. Because Δt and ΔE could each be independently determined to any degree of accuracy, the simultaneous values of ΔE and Δt could be made so small that $\Delta E \Delta t < \hbar/2$, in contradiction to Heisenberg's uncertainty relation. Bohr spent a sleepless night looking for an effective response to this challenge and he finally found one. The resolution of the apparent paradox lay in Einstein's own general theory of relativity according to which the change in the gravitational field produced by the loss of mass $\Delta m = \Delta E/c^2$, when the photon escapes from the box, affects the rate at which the clock keeps time.[4] Bohr was able to show quantitatively that this effect exactly restores the uncertainty relation. Henceforth, Einstein no longer claimed that quantum mechanics was inconsistent, but concentrated on the incompleteness question.

21.3 THE MEASUREMENT PROBLEM

In order to be able to state precisely what the collapse of the wave function, or the measurement problem, of standard quantum mechanics is, we begin with the definition of a few terms used in the formal structure of quantum theory. There are operators A that represent physical observables. The *eigenstates* ψ_j of these operators are the states (or wave functions) of the physical system under consideration when the observable is one of the allowed eigenvalues λ_j. These operators A,[5] the eigenstates ψ_j and the eigenvalues λ_j are related as[6]

$$A\psi_j = \lambda_j \psi_j \tag{21.1}$$

We now use this formal machinery to describe a measurement quantum mechanically.

As indicated in Figure 21.3, the particular measurement process we consider in some detail consists of a magnetic field $\boldsymbol{B}$ (in, say, a Stern–Gerlach apparatus) to measure the spin of an electron.[7] The basic idea in this, as in a typical measurement, is to couple a microsystem (here, the electron) to a macrosystem (here, the apparatus) and to arrange things so that one can learn something about the value of a variable of the microsystem (say, a spin component) by examining the state of the macrosystem after the two systems have interacted. Often we speak of the 'readout' or the 'printout' of the apparatus. If the measurement is to be reliable, then we want the final state of the macrosystem to be strongly correlated (ideally, 100%) with the final state of the microsystem. It would also be nice if the measurement on the microsystem were faithful in that it did not disturb the state of the microsystem. Of course, we want the state of the macrosystem to be affected by the interaction. As in Figure 21.3, we use ψ to denote the wave function of the microsystem (here, the electron) and ϕ that of the macrosystem. For two noninteracting subsystems, the total wave function for the entire system is represented by the product[8]

$$\Psi(\text{total wave function}) = \psi\phi \tag{21.2}$$

PART VII The completeness of quantum mechanics

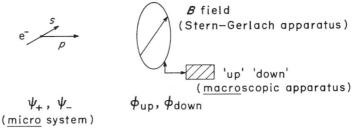

ψ_+, ψ_- ϕ_{up}, ϕ_{down}
(micro system)

FIGURE 21.3 An illustration of the measurement process

In the present case, there are two independent wave functions, ψ_+ and ψ_-, corresponding to spin up and spin down, respectively, along an axis (that we here take to be the direction of the magnetic field B).[9] Similarly, ϕ_{up} and ϕ_{down} represent the wave functions for the apparatus corresponding, respectively, to its being in the state indicating spin up (say, a readout of $+1$) or spin down (-1). By ϕ_0 we mean the initial state of the apparatus (that is, no reading). To begin, let us assume that, before the interaction, we know the electron to have spin up (along the vertical axis of the apparatus, as defined by the B field). Then, prior to any interaction between the micro- and the macrosystems (at $t = 0$), the wave function for the combined system is simply

$$\Psi(t = 0) = \psi_+ \phi_0 \tag{21.3}$$

The Schrödinger equation governs the time evolution of the state vector $\Psi(t)$.[10] In particular, as the electron passes through the magnetic field of the apparatus, there is an interaction, or coupling, between the microsystem and the macrosystem. If this is a decent measuring device, $\Psi(t)$ evolves continuously into the final state (after the electron has left the magnetic field of the apparatus)

$$\Psi(t) \xrightarrow[t \to +\infty]{} \Psi_{out}(t) = \psi_+ \phi_{up} \tag{21.4}$$

Similarly, if we begin in the state $\psi_- \phi_0$, then the asymptotic final state will be

$$\psi_- \phi_0 \xrightarrow[t \to +\infty]{} \psi_- \phi_{down} \tag{21.5}$$

If this were the end of the story, there would be no measurement problem because we would have given a quantum-mechanical description of the measurement process in which we successfully used a macrosystem (the apparatus) to interrogate the microsystem (the electron) without changing the state of the microsystem.

However, we usually do not know the state of the microsystem to be either 'up' or 'down' prior to the measurement (or else there would be no need to make a measurement). Rather, the initial state of the electron is represented by the superposition

$$\psi_0 = \alpha \psi_+ + \beta \psi_- \tag{21.6}$$

where $|\alpha|^2 + |\beta|^2 = 1$. Since the initial state of the apparatus is still ϕ_0, the initial state of the combined system is

Is quantum mechanics complete?

$$\Psi(t=0) = (\alpha\psi_+ + \beta\psi_-)\phi_0 \tag{21.7}$$

Since the Schrödinger equation is *linear*,[11] it follows from Eqs. (21.4)–(21.6) that the states for the micro- and macrosystems become *entangled* as

$$\Psi_0 \equiv (\alpha\psi_+ + \beta\psi_-)\phi_0 \xrightarrow{t \to +\infty} \Psi_{out} = \alpha\psi_+\phi_{up} + \beta\psi_-\phi_{down} \tag{21.8}$$

Therefore, we are left with no definite state for the macroscopic apparatus. In other words, the state of Eq. (21.8) represents a situation in which the measuring apparatus is in a superposition of 'up' and 'down' ($+1$ and -1) states. However, we never observe everyday physical objects in such superpositions. So, it would seem that the theory of quantum mechanics makes predictions that are grossly at odds with observations and, thus, should be rejected out of hand. Not so. The standard, or Copenhagen, interpretation of quantum mechanics invokes the 'reduction' or 'collapse' of the wave packet upon observation as

$$\alpha\psi_+\phi_{up} + \beta\psi_-\phi_{down} \xrightarrow{discontinuously} \begin{cases} \psi_+\phi_{up} \text{ with probability } |\alpha|^2 \\ \quad\quad -\text{OR}- \\ \psi_-\phi_{down} \text{ with probability } |\beta|^2 \end{cases} \tag{21.9}$$

That is, the theory appends to the Schrödinger-equation time development of Eq. (21.8) the ad hoc rule that, upon observation, the state vector uncontrollably reduces or 'jumps' from the superposition of Eq. (21.8) to one or other of the actually observed states on the right side of Eq. (21.9).

This is the measurement problem in quantum mechanics. It is not peculiar to the specific example we used here, but is characteristic of all quantum-mechanical measurement processes. As we shall see in the next section, this is related to the completeness of quantum mechanics. If quantum mechanics gives a complete description of a physical system, then one must take this reduction seriously as an actual physical process and account for it in terms of physical interactions. There has been no general, successful resolution along such lines.

21.4 SCHRÖDINGER'S CAT PARADOX

In 1935 Schrödinger formulated his famous 'cat paradox' to illustrate the incompleteness of quantum mechanics.[12] This paper came about as a result of correspondence Schrödinger had with Einstein on the topic of the Einstein–Podolsky–Rosen paper that we discuss in detail in the next chapter. Schrödinger asks us to consider an initially live cat placed into a steel box that is then sealed. With him is a sample of a radioactive element whose probability of random decay in one hour is exactly $\frac{1}{2}$. Things are so arranged that if an atom of the element does decay, then a device will shatter a vial of hydrocyanic acid and kill the cat. Otherwise, the cat continues to live. When the problem begins, the wave function for this combined system (cat + atom) necessarily corresponds to a live cat. However, as time evolves, there is

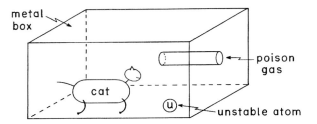

FIGURE 21.4 Schrödinger's 'cat' paradox

some finite probability that the atom has decayed, so that the cat is dead. Suppose that, at the end of one hour, the wave function for the system is an equal mixture or superposition of the state for a live cat and that for a dead cat. In this case a macrosystem, the cat, would be thrown into one state or another by our mere act of observation provided the wave function gives a complete, objective description of reality. Schrödinger did not believe this to be reasonable. In replying to him on this, Einstein wrote in 1939:

> I am as convinced as ever that the wave representation of matter is an incomplete representation of the state of affairs, no matter how practically useful it has proved itself to be. The prettiest way to show this is by your example with the cat (radioactive decay with an explosion coupled to it). At a fixed time parts of the ψ-function correspond to the cat being alive and other parts to the cat being pulverized.
>
> If one attempts to interpret the ψ-function as a complete description of a state, independent of whether or not it is observed, then this means that at the time in question the cat is neither alive nor pulverized. But one or the other situation would be realized by making an observation.
>
> If one rejects this interpretation then one must assume the ψ-function does not express the real situation but rather that it expresses the contents of our knowledge of the situation. This is Born's interpretation, which most theorists today probably share. But then the laws of nature that one can formulate do not apply to the change with time of something that exists, but rather to the time variation of the content of our legitimate expectations.
>
> Both points of view are logically unobjectionable; but I cannot believe that either of these viewpoints will finally be established.[13]

Figure 21.4 is a schematic of Schrödinger's thought experiment.[14] The point of the exercise is to have us consider what the standard quantum-mechanical description of this situation would be. If we let $\Psi(t)$ be the wave function for the composite system (cat + atom), ϕ that for the cat and ψ that for the atom, then the initial wave function for the combined system is given as

$$\Psi_0 = \Psi(t=0) = \phi_{live}\psi_{atom} \qquad (21.10)$$

Under the action of the Schrödinger equation, this evolves in time into the entangled state (recall Eq. (21.8))

$$\Psi(t) = \alpha(t)\phi_{live}\psi_{atom} + \beta(t)\phi_{dead}\psi_{decay} \qquad (21.11)$$

According to the usual quantum-mechanical rules for computing probabilities, the probability or chance of our finding a live cat in the box once we open it is[15]

$$P_{\text{live}}(t) = |\langle\phi_{\text{live}}\psi_{\text{atom}}|\Psi(t)\rangle|^2 = |\alpha(t)|^2 \sim e^{-t/\tau_0} \qquad (21.12)$$

while that for finding a dead cat is

$$P_{\text{dead}}(t) = |\langle\phi_{\text{dead}}\psi_{\text{decay}}|\Psi(t)\rangle|^2 = |\beta(t)|^2 \sim 1 - e^{-t/\tau_0} \qquad (21.13)$$

Here $\Psi(t)$ represents the state of the system before we look – a superposition of a live cat and of a dead cat. After we look, the state vector has been (magically?) reduced to either a live cat or a dead cat.

An obvious rhetorical question now presents itself. What does the wave function represent – our state of knowledge of the system (in which case quantum mechanics is incomplete) or the actual physical state of the system (in which case there must be a sudden change of the system upon our observation of it)? Appreciate that the incompleteness option is compatible with Born's statistical interpretation (Section 20.2), so that the collapse of the wave function corresponds merely to our revising our knowledge about the state of the cat. Nothing physical happens to the cat by our becoming aware that it is alive or dead. It is then an additional and logically independent restriction to claim that a more complete description than this statistical one is not possible. On the other hand, if the wave function represents the state of an individual cat (the completeness option – recall Heisenberg's position on ψ in Section 20.4), then a dramatic physical change in the cat accompanies the collapse of the wave function. While a sudden change of the physical state just by observation may not appear too bothersome at the microlevel, it is disturbing and appears highly improbable at the level of macroscopic systems, such as a cat. Schrödinger seems to bank on our visceral (negative) reaction to the suggestion that our mere act of observing the cat actually itself produces a live cat or a dead cat (that is, either kills the cat or not). Thus, Schrödinger has produced an argument (but not, of course, a logical impossibility proof) against the completeness of quantum mechanics (that is, against the Copenhagen interpretation of quantum mechanics).

21.5 DIRAC ON THE EFFECT OF MEASUREMENT

Let us now return to Dirac's own example of the split-beam experiment of Figure 20.10 to see how he accounts for the behavior observed for the photon (or electron) beam.

> One could carry out the energy measurement without destroying the component beam by, for example, reflecting the beam from a movable mirror and observing the recoil. Our description of the photon allows us to infer that, *after* such an energy measurement, it would not be possible to bring about any interference effects between the two components. So long as the photon is partly in one beam and partly in the other, interference can occur when the two beams are superposed, but this possibility disappears when the photon is forced entirely into one of the beams by an observation. The other beam then no longer enters into the description of the photon, so that it counts as being entirely in the one beam in the ordinary way for any experiment that may subsequently be performed on it.[16]

That is, after such an energy measurement or observation along one of the two

PART VII The completeness of quantum mechanics

paths, no interference is any longer possible. Although the system may appear in either of two states (or 'components') before the measurement, nature has (in the image suggested by Dirac) been forced to 'make a choice' when observed. Since the system is thereafter in a definite component, no subsequent interference with the other component is possible. The 'collapse' of the wave function has taken place. (This is the same type of loss of interference that occurs in the double-slit arrangement of Figure 20.11, once we determine through which slit an electron or photon passed – say by turning on the light at X to observe the electron or photon.)

> On these lines quantum mechanics is able to effect a reconciliation of the wave and corpuscular properties of light. The essential point is the association of each of the translational states of a photon with one of the wave functions of ordinary wave optics. The nature of this association cannot be pictured on a basis of classical mechanics, but is something entirely new. It would be quite wrong to picture the photon and its associated wave as interacting in the way in which particles and waves can interact in classical mechanics. The association can be interpreted only statistically, the wave function giving us information about the probability of our finding the photon in any particular place when we make an observation of where it is.
>
> Some time before the discovery of quantum mechanics people realized that the connexion between light waves and photons must be of a statistical character. What they did not clearly realize, however, was that the wave function gives information about the probability of *one* photon being in a particular place and not the probable number of photons in that place. The importance of the distinction can be made clear in the following way. Suppose we have a beam of light consisting of a large number of photons split up into two components of equal intensity. On the assumption that the intensity of a beam is connected with the probable number of photons in it, we should have half the total number of photons going into each component. If the two components are now made to interfere, we should require a photon in one component to be able to interfere with one in the other. Sometimes these two photons would have to annihilate one another and other times they would have to produce four photons. This would contradict the conservation of energy. The new theory, which connects the wave function with probabilities for one photon, gets over the difficulty by making each photon go partly into each of the two components. Each photon then interferes only with itself. Interference between two different photons never occurs.[17]

The essential point being made is that the probabilities under discussion here are probabilities for single photons (or electrons), not for aggregates of them. This interpretation of probability is necessary, Dirac tells us, in order to avoid contradictions with energy conservation. Even though this wave–particle 'duality' is in principle universal and extends to all types of systems, we are not commonly aware of it in everyday experience because the proportionality constant between mass and frequency is so small (since $m = h\nu/c^2$):

> The association of particles with waves discussed above is not restricted to the case of light, but is, according to modern theory, of universal applicability. All kinds of particles are associated with waves in this way and conversely all wave motion is associated with particles. Thus all particles can be made to exhibit interference effects and all wave motion has its energy in the form of quanta. The reason why these general phenomena are not more obvious is on account of a law of proportionality between the mass or energy of the particles and the frequency of the waves, the coefficient being such that for waves of familiar frequencies the associated quanta are extremely small, while for particles even as light as electrons the associated wave frequency is so high that it is not easy to demonstrate interference.[18]

Finally, Dirac emphasizes that these superpositions in the wave function ψ give rise, not to superpositions resulting in intermediate properties (say, gray from black and white) of the systems observed, but rather to intermediate values for the probabilities of occurrence of one or another of the basic possibilities (black or white in our example here).

> Let us take any atomic system, composed of particles or bodies with specified properties (mass, moment of inertia, etc.) interacting according to specified laws of force. There will be various possible motions of the particles or bodies consistent with the laws of force. Each such motion is called a *state* of the system.
>
> . . .
>
> When a state is formed by the superposition of two other states, it will have properties that are in some vague way intermediate between those of the two original states and that approach more or less closely to those of either of them according to the greater or less 'weight' attached to this state in the superposition process. The new state is completely defined by the two original states when their relative weights in the superposition process are known....
>
> . . .
>
> The non-classical nature of the superposition is brought out clearly if we consider the superposition of two states, *A* and *B*, such that there exists an observation which, when made on the system in state *A*, is certain to lead to one particular result, *a* say, and when made on the system *B* is certain to lead to some different result, *b* say. What will be the result of the observation when made on the system in the superposed state? The answer is that the result will be sometimes *a* and sometimes *b*, according to a probability law depending on the relative weights of *A* and *B* in the superposition process. It will never be different from both *a* and *b*. *The intermediate character of the state formed by superposition thus expresses itself through the probability of a particular result for an observation being intermediate between the corresponding probabilities for the original states, not through the result itself being intermediate between the corresponding results for the original states.*[19]

In the last paragraph here Dirac gives an example of a system that is represented by a superposition of two states *A* and *B*. In Sections 21.3 and 21.4 we gave concrete illustrations of such superpositions as represented in the formalism of quantum mechanics.

FURTHER READING

Bernard d'Espagnat's 'The Quantum Theory and Reality' is a nontechnical presentation of several of the counterintuitive features of the quantum world. Chapter 3 of Jim Baggott's *The Meaning of Quantum Theory* discusses informally the Copenhagen interpretation and the Bohr–Einstein debate. Chapters 8 through 10 of Edward MacKinnon's *Scientific Explanation and Atomic Physics* treat these issues in more depth. Andrew Whitaker's *Einstein, Bohr and the Quantum Dilemma* is a nonmathematical but very thorough and insightful discussion of the conceptual issues that separated Bohr and Einstein when it came to quantum mechanics. Max Jammer's *The Philosophy of Quantum Mechanics* explores in depth the philosophical implications of quantum theory and sets them in historical context.

PART VIII

Some philosophical lessons from quantum mechanics

[I] think I can safely say that nobody understands quantum mechanics.
Richard Feynman, *The Character of Physical Law*

Quantum mechanics [is] that mysterious, confusing discipline, which none of us really understands but which we know how to use. It works perfectly, as far as we can tell, in describing physical reality, but it is a 'counter-intuitive discipline'....
Murray Gell-Mann, *Questions for the Future*

[The quantum] postulate implies a renunciation as regards the causal space-time co-ordination of atomic processes.
Niels Bohr, *Atomic Theory and the Description of Nature*

It should be emphasized, however, that the probability function does not in itself represent a course of events in the course of time. It represents a tendency for events and our knowledge of events. The probability function can be connected with reality only if one essential condition is fulfilled: if a new measurement is made to determine a certain property of the system.

. . .

[T]he idea of an objective real world whose smallest parts exist objectively in the same sense as stones or trees exist, independently of whether or not we observe them ... is impossible....
Werner Heisenberg, *Physics and Philosophy*

[In] quantum theory it is the *principle of causality*, or more accurately that of *determinism*, which must be dropped and replaced by something else.... We now have a *new form* of the law of causality.... It is as follows: if in a certain process the initial conditions are determined as accurately as the uncertainty relations permit, then the probabilities of all possible subsequent states are governed by exact laws.
Max Born, *The Restless Universe*

Bohr's ... approach to atomic problems ... is really remarkable. He is completely convinced that any understanding in the usual sense of the word is impossible.
Erwin Schrödinger, *letter to Wilhelm Wien on October* 10, 1926

It is therefore not, as is often assumed, a question of a re-interpretation of quantum mechanics, – the present system of quantum mechanics would have to be objectively false, in order that another description of the elementary processes than the statistical one be possible.
John von Neumann, *Mathematical Foundations of Quantum Mechanics*

I am, in fact, firmly convinced that the essentially statistical character of contemporary quantum theory is solely to be ascribed to the fact that this (theory) operates with an incomplete description of physical systems.
Albert Einstein, *Reply to Criticisms*

It is clear that [the results of the double-slit experiment] can in no way be reconciled with the idea

PART VIII **Some philosophical lessons from quantum mechanics**

that electrons move in paths.... In quantum mechanics there is no such concept as the path of a particle.

Lev Landau and Evgenii Lifshitz, *Quantum Mechanics: Non-Relativistic Theory*

But in 1952 I saw the impossible done. It was in papers by David Bohm. Bohm showed explicitly how parameters could indeed be introduced, into nonrelativistic wave mechanics, with the help of which the indeterministic description could be transformed into a deterministic one. More importantly, in my opinion, the subjectivity of the orthodox version, the necessary reference to the 'observer,' could be eliminated.

Moreover, the essential idea was one that had been advanced already by de Broglie in 1927, in his 'pilot wave' picture.

But why then had Born not told me of this 'pilot wave'? If only to point out what was wrong with it? Why did von Neumann not consider it? More extraordinarily, why did people go on producing 'impossibility' proofs, after 1952, and as recently as 1978? When even Pauli, Rosenfeld, and Heisenberg, could produce no more devastating criticism of Bohm's version than to brand it as 'metaphysical' and 'ideological'? Why is the pilot wave picture ignored in the text books? Should it not be taught, not as the only way, but as an antidote to the prevailing complacency? To show that vagueness, subjectivity, and indeterminism, are not forced on us by experimental facts, but by deliberate theoretical choice?

John Bell, *Speakable and Unspeakable in Quantum Mechanics*

22

The EPR paper and Bell's theorem

When it was first put forward, special relativity struck many people as counterintuitive and possibly inconsistent, as we saw when we considered the twin paradox in Chapter 17. Quantum mechanics had a similar effect, only to a much greater and longer lasting degree, from the time of its inception in the mid 1920s until the present. This reaction to quantum theory was not confined to an initial confusion that often accompanies a new subject, but remained a life-long puzzle for Einstein, to mention only the most prominent opponent of what he took to be the reigning orthodoxy. In this chapter we study a famous attempt by Einstein and two of his colleagues to show that quantum mechanics was, if not logically inconsistent, then at least an incomplete theory. This charge of incompleteness is just that raised by Schrödinger's cat paradox (Section 21.4).

22.1 THE EPR PARADOX

In 1935 Einstein, Boris Podolsky (1896–1966) and Nathan Rosen (1909–1995) (hereafter referred to as EPR) published a paper titled 'Can Quantum-Mechanical Description of Physical Reality Be Considered Complete?'. They take as their criterion for the completeness of a theory that it contains terms corresponding to every relevant entity found in reality. For example, in the classical description of a system, such as a planet going around the sun, there are symbols representing the position r, the momentum p and so forth for the various parts of the system. EPR stipulate that if a complete theory allows the value of one of these variables or quantities to be predicted with no uncertainty and without disturbing the system under consideration, then that quantity has a real existence in nature and must actually have the predicted value. On the basis of these criteria, they outline the logic of their argument.

> In a complete theory there is an element corresponding to each element of reality. A sufficient condition for the reality of a physical quantity is the possibility of predicting it with certainty, without disturbing the system. In quantum mechanics in the case of two physical quantities described by non-commuting operators, the knowledge of one precludes the knowledge of the other. Then either (1) the description of reality given by the wave function in quantum mechanics is not complete or (2) these two quantities cannot have simultaneous reality. Consideration of the problem of making predictions concerning a system on the basis of measurements made on another system that had previously interacted with it leads to the result that if (1) is false then (2) is also false. One is thus led to conclude that the description of reality as given by a wave function is not complete.[1]

The logical structure of the proof is simply that (1) and (2) cannot both be false (since, as we shall see, this implies a contradiction for quantum mechanics). EPR then construct a specific quantum-mechanical example in which a denial of (1) (that is, assuming that quantum mechanics is complete) necessarily entails a denial of (2) (that is, implies that certain pairs of quantities can have simultaneous physical reality). Since, as we have just stated, (1) and (2) cannot both be false, the denial of (1) must not be allowable. Hence, quantum mechanics is incomplete. It is important to point out that EPR assume that action at a distance does not exist. Such an assumption, that one event in a certain region of space cannot immediately affect another event separated from it, is often referred to as the *locality* assumption: no instantaneous action at a distance. Later, we consider a simplified version of the EPR experiment.

In that same year and in the same journal Bohr responded with an article[2] bearing the same title as that of the EPR paper. Perhaps even more explicitly than EPR had done, he certainly denied the possibility of any actual physical action at a distance (nonlocality). The thrust of his rejoinder was that their criterion of physical reality was ambiguous since reality can be defined only in terms of measurement that requires a specification of the apparatus to be used. For Bohr, actual access to the knowledge of the value of an 'element of reality' is taken to be a criterion for the very existence of that element of reality. He claimed that with this refinement the EPR argument could not be pushed through. However, the issue seems never to have been resolved conclusively since one's criterion for the physical reality of a quantity remains debatable.

Einstein never changed his opinion that quantum mechanics was an incomplete theory. Many years later Heisenberg summarized Einstein's view of quantum mechanics as follows.

> Einstein agreed with Born that the mathematical formulation of quantum mechanics, developed in Göttingen and consolidated further in Cambridge and Copenhagen, correctly described the phenomena within the atom. He may also have been willing to admit, for the time being at least, that the statistical interpretation of Schroedinger's wave function, as formulated by Born, would have to be accepted as a working hypothesis. But Einstein did not want to acknowledge that quantum mechanics represented a final, and even less a complete, description of these phenomena. The conviction that the world could be completely divided into an objective and a subjective sphere, and the hypothesis that one should be able to make precise statements about the objective side of it, formed a part of his basic philosophical attitude. But quantum mechanics could not satisfy these claims, and it does not seem likely that science will ever find its way back to Einstein's postulates.[3]

Heisenberg went on to say that Einstein appeared unwilling, or perhaps unable, to modify radically his own conceptual framework in order to accommodate the standard view of quantum phenomena (that is, the Copenhagen interpretation).

Einstein himself returned to this theme often in his correspondence with Born. While he admitted that the statistical interpretation that Born was the first to introduce proved very useful, Einstein nevertheless felt that the inability of quantum mechanics to represent real, continuous events in a space–time background counted heavily against it. In 1947 he wrote:

> [I] am quite convinced that someone will eventually come up with a theory whose objects, connected by laws, are not probabilities but considered facts, as used to be taken for granted until quite recently. I cannot, however, base this conviction on logical reasons, but can only produce my little finger as witness, that is, I offer no authority which would be able to command any kind of respect outside of my own hand.[4]

Here, as in similar correspondence a year later to Born,[5] Einstein maintained his belief that an acceptable physical theory would ultimately have to conform to certain general ideals. Among these, in his opinion, was the requirement that entities exist locally in space and time, being independent of what is real in a far-separated region of space. Hence, what actually exists at some point should not depend upon what measurements might be performed at some distant location. So, even more than a dozen years after the appearance of the EPR paper, Einstein still found unacceptable any type of instantaneous change produced by distant measurements: 'My instinct for physics bristles at this.'[6]

In a sense, that Einstein should be such an adamant opponent of quantum mechanics is somewhat paradoxical since it was he who introduced the notion of probability as an essential feature of the quantum theory of atomic processes in his own 1917 paper.[7] Einstein was not the only founder of quantum mechanics who rejected the Copenhagen or probabilistic interpretation of that theory. De Broglie at first attempted to construct a deterministic theory to explain his wave–particle duality, but after the 1927 Solvay Congress he became an adherent of the Copenhagen school. Later still, in 1952, for reasons connected with the hidden variables theory that we discuss in Chapter 23, he reverted to a belief in an ultimately causal description of nature. Schrödinger, the founder of wave mechanics in its modern form, once said to Bohr, 'If one has to stick to this damned quantum jumping, then I regret having ever been involved in this thing.'[8]

In light of the EPR argument, deterministic hidden-variables theories might seem to be a serious possible alternative to quantum mechanics. Basically what one could do is to assume that there exists a set of variables, or as yet undiscovered properties, of a system and that the exact space–time behavior of the system is causally determined by the values of these 'hidden' variables. Since we do not yet know the precise values of these variables, there will exist an uncertainty or spread in the values of, say, the position and momentum of a particle. By postulating a sufficiently large number of such hidden variables, it would appear as though we could always account for the results predicted by quantum mechanics. Since quantum mechanics does seem to be a highly successful theory because it agrees with experiment, the idea is to produce a completely deterministic and causal hidden-variables theory that will be able to mimic quantum mechanics in all empirical predictions. The introduction of such a large number of hidden variables may seem to be a high price to pay to maintain locality and realism, but there does not at first sight appear to be any logical inconsistency in doing so. We return to this question after we examine the EPR argument itself in some detail.

PART VIII Some philosophical lessons from quantum mechanics

22.2 AN ANALYSIS OF THE EPR PAPER

Although the EPR paper has proven to be one of the most important papers in the foundations of quantum mechanics, the mathematics it employs is fairly elementary and the line of argument is relatively easy to follow.[9] EPR begin by defining a complete theory as one that contains an element corresponding to each element of reality (in other words, there is a mapping or correspondence). A necessary condition X for such completeness would mean that, if a theory is complete, then X must obtain (or, completeness $\Rightarrow X$). Conversely, a sufficient condition X would mean that, if X holds, then the theory is complete ($X \Rightarrow$ completeness). Their necessary condition for the completeness of a physical theory is that every element of physical reality has a counterpart in the physical theory. Therefore, the question arises about what the elements of physical reality are. EPR take as their sufficient condition for an element of physical reality:

> If, without in any way disturbing a system, we can predict with certainty (i.e., with probability unity) the value of a physical quantity, then there exists an element of physical reality corresponding to this physical quantity.[10]

This 'prediction' is, of course, within the context of a theoretical framework.

If the standard, or Copenhagen, interpretation of quantum mechanics is correct, then ψ is a complete characterization of the state of a system. For a system in an eigenstate ψ of an operator A corresponding to the (definite) eigenvalue a of the observable (that is, $A\psi = a\psi$, recall Eq. (21.1)), one can say that a system in such a state ψ corresponds to an element (a) of physical reality (since then $\Delta A = 0$). In fact, a system can be in an eigenstate of A ($A\psi = a\psi$) if and only if the value of the observable has a sharp value ($\Delta A = 0$). This is often referred to as the eigenvalue–eigenstate link in quantum mechanics: a definite value for an observable A if and only if the state ψ of the system is an eigenstate of A. Now if an element of reality exists, then (as EPR stipulate) it has a definite value. According to their criterion of completeness, this value must be represented in the description provided by the theory. For quantum mechanics, this description requires a specification of the wave function for the system. But a definite value a for an observable implies that the wave function must be an eigenstate of the operator A corresponding to that value. Therefore, completeness and reality entail that quantum mechanics would predict a sharp value for this observable ($\Delta A = 0$). So far, there is no problem for just one observable that has a sharp value.

However, the same type of argument we have just given for a single element of reality would apply as well if two quantities corresponding to noncommuting operators were to have simultaneous reality. This combined system would have to be in a state that is simultaneously an eigenstate of two noncommuting operators. This is equivalent to requiring that these two noncommuting observables both have sharp values simultaneously. Now consider the observables x and p_x. We have seen that the calculational rules of quantum mechanics imply that the values of x and p_x cannot both be sharp in any state at all (because $[x, p_x] = i$ (see note[11])). (This is just

322

the content of the Heisenberg uncertainty relation of Section 20.4.) Hence, if quantum mechanics is complete, then such observables cannot simultaneously have sharp values. All of this amounts to the result that the joint assumptions both of the completeness of quantum mechanics and of the simultaneous reality of observables corresponding to certain noncommuting operators lead to a logical contradiction in at least one case. So, by the rules of logic (recall Section 1.2), at least one of these assumptions must be false.

In summary then, the key to the EPR argument is the claim that at least one of the following must be true: (i) quantum mechanics is not complete, or (ii) observables corresponding to certain noncommuting operators cannot have simultaneous reality. We have just argued that not both of these can be false simultaneously. Here we have an inclusive (but not an exclusive) logical disjunction (that is, both could be true). The completeness of quantum mechanics plus the criterion of reality implies a contradiction in the case EPR consider.

From there, the actual EPR proof proceeds as follows. In quantum mechanics, the states for an interacting system become entangled, as we saw in our discussion of the measurement problem (see Eqs. (21.8) or (21.11)). Let us choose a set of states $\{u_n(x_1)\}$ such that

$$Au_n(x_1) = a_n u_n(x_1) \tag{22.1}$$

and consider the state Ψ for a composite system, whose two subsystems 1 and 2 are noninteracting after the initial preparation of the combined system. To make this argument particularly tight, we can take the subsystems 1 and 2 to be far apart when the measurements are made. The collapse of Ψ upon the measurement of the value of A is (recall Eq. (21.9))

$$\Psi(x_1, x_2) = \sum_{n=1}^{\infty} \psi_n(x_2) u_n(x_1) \xrightarrow{\text{measurement}} \psi_k(x_2) u_k(x_1) \tag{22.2}$$

Let B be an operator that does not commute with A (that is, $[A, B] \neq 0$) and the set $\{v_s(x_1)\}$ be its eigenstates

$$Bv_s(x_1) = b_s v_s(x_1) \tag{22.3}$$

If we instead choose to measure the value of the observable B, rather than that of A, then we have (according to Eq. (21.9))

$$\Psi(x_1, x_2) = \sum_{s=1}^{\infty} \phi_s(x_2) v_s(x_1) \xrightarrow{\text{measurement}} \phi_r(x_2) v_r(x_1) \tag{22.4}$$

The crucial locality assumption that EPR make is that, once systems have separated, they can no longer (instantaneously) influence each other – especially when they are far apart. That is, EPR assume determinate (or definite) values of their elements of reality, independent of the act of observation (something Bohr would not accept), and the absence of instantaneous, long-range influences. In EPR's specific example of the momenta and positions of a two-particle system,[12] ψ_k and ϕ_r

are eigenstates of the two noncommuting observables p_2 and x_2, respectively, for system 2. So, by choosing to measure $A = p_1$ (on system 1), they may predict with certainty, and without in any way disturbing system 2, the value of p_2; or they may choose to measure $B = x_1$ and then infer the value of x_2.[13] Therefore, (by the EPR criterion) p_2 and x_2 both correspond to elements of physical reality and, hence, both wave functions ψ_k and ϕ_r, that are different, belong to the same reality. But, that leads to the same contradiction we discussed earlier.[14]

The structure of this part of their argument can be summarized as: quantum mechanics is complete (that is, $\sim$(i)) implies that certain noncommuting observables have simultaneous reality ($\sim$(ii); or, $\sim$(i) $\Rightarrow$ $\sim$(ii)). But, as we have already seen, $\sim$(i) and $\sim$(ii) together is not possible. Therefore, (i) must be true. Appreciate that the very formalism of quantum mechanics itself is used to establish the incompleteness of quantum mechanics (assuming, of course, no action at a distance).

22.3 BELL'S THEOREM

There is a long and involved history of mathematical arguments that attempted to prove the impossibility of the existence of any hidden-variables theory that could reproduce all of the predictions of quantum mechanics.[15] In 1965 the theoretical physicist John Bell of the European Center for Nuclear Research (CERN) in Geneva, Switzerland, proved a remarkable theorem. We state the content of this theorem by means of a simple and idealized thought experiment. A derivation of one version of this theorem is given in the next section. Here we first discuss the meaning of this result.

Suppose that two measurements are made at the spatially separated positions A and B as indicated in Figure 22.1. Let the measurements be simultaneous in our frame. For example, one might be detecting the momenta or spins of two separate particles as they reach A and B after they originated in a reaction at P. What Bell showed is that if there is an objective reality (that is, a truth of the matter whether or not we observe it: results are *determinate*)[16] and if the result (for an individual outcome) obtained at A is independent of the choice of what one decides to measure at B (made, say, by picking a particular orientation of the detector at B), then no hidden-variables theory could give the same results (that is, predictions) as quantum mechanics in all possible situations. This independence of two simultaneous events occurring at spatially separated points is what we earlier referred to as the locality assumption, since it means, in effect, that something done at B could not instantaneously affect a measurement at A when A and B are separated by a distance d. Simply put, no determinate, local hidden-variables theory can agree with all of the predictions of quantum mechanics.

Furthermore, Bell established that certain experiments could distinguish between these two theories. When experiments of the type suggested by him and by later workers were actually performed in the laboratory, the results agreed with the

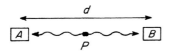

FIGURE 22.1   An Einstein–Podolsky–Rosen (EPR) correlation experiment

quantum-mechanical predictions, not with those of local hidden-variables theories.[17] Therefore, a completely deterministic basis for physics, in the sense that the outcome of a given observation can always be predicted with certainty, does not seem possible if we maintain a causality that embodies the locality assumption. A recent paper reviewing this hidden-variables question summarizes the situation as follows.

> Realism is a philosophical view, according to which external reality is assumed to exist and have definite properties, whether or not they are observed by someone. So entrenched is this viewpoint in modern thinking that many scientists and philosophers have sought to devise conceptual foundations for quantum mechanics that are clearly consistent with it. One possibility, it has been hoped, is to reinterpret quantum mechanics in terms of a statistical account of an underlying hidden-variables theory in order to bring it within the general framework of classical physics. However, Bell's theorem has recently shown that this cannot be done. The theorem proves that all realistic theories, satisfying a very simple and natural condition called locality, may be tested with a single experiment against quantum mechanics. These two alternatives necessarily lead to significantly different predictions. The theorem has thus inspired various experiments, most of which have yielded results in excellent agreement with quantum mechanics, but in disagreement with the family of local realistic theories. Consequently, it can now be asserted with reasonable confidence that either the thesis of realism or that of locality must be abandoned. Either choice will drastically change our concepts of reality and of space–time.[18]

There is a long-standing debate about whether the dilemma of having to choose between such a reality and locality can be resolved definitely in favor of nonlocality.[19] In the next section we consider one type of determinate reality – a deterministic one. Our reason for this restriction is that the commitments made are then clearly defined and the resulting proof is quite simple. Even though there are more general proofs (that is, ones making assumptions less drastic than determinism), the argument we give is sufficient for the topics we discuss in Chapters 23 and 24. Furthermore, the move to inherently probabilistic (or stochastic) theories arguably makes the situation even worse with regard to avoiding nonlocality.[20] It does not seem unreasonable, then, that a fundamental lesson to be learned from these considerations is that any theory accounting for violations of Bell's inequality must be nonlocal.[21] To the extent that a type of nonlocality turns out to be a feature of the world, one would have an effective rejoinder to the arguments that Maxwell gave (toward the end of the nineteenth century) in favor of locality (see Section 13.A).

22.4 A DERIVATION OF BELL'S THEOREM

The version of the EPR type of experiment we discuss is due to David Bohm (1917–1992)[22] – hence the new abbreviation EPRB. Figure 22.2 is a schematic of the

PART VIII Some philosophical lessons from quantum mechanics

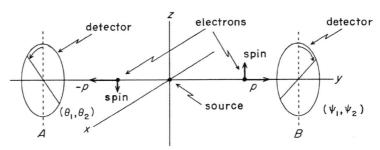

FIGURE 22.2 Bohm's version of the EPR experiment

experiment.[23] A source, such as an unstable atom, is located at the origin of the coordinate system. When the atom decays, it emits two electrons[24] that travel off along the horizontal or y-axis with equal, but oppositely directed momenta. These electrons have spins. For simplicity, we take these spins to be aligned at right angles to the momenta. (That is, the spins lie in a vertical plane (x–z) parallel to the plane of the detectors in Figure 22.2.) Detectors at positions A and B can measure the spins of the electrons arriving there. Let us accept as an empirical fact that when the spin of an electron is measured along any fixed axis perpendicular to the y-axis, its spin will always be up or down along that fixed axis. We take $+1$ to correspond to an up value and -1 to a down value.

The observer at A can set her detector in one of two orientations (θ_1 or θ_2), while the observer at B has the choices ψ_1 or ψ_2. Each reading for each observer is either $+1$ or -1 on each run of the experiment. We denote these values for the results as r. More particularly:

$r_{A_k}(\theta, \psi) = \pm 1$ – result at detector A for the kth pair of electrons when setting θ (that is, θ_1 or θ_2) has been chosen at A and ψ (that is, ψ_1 or ψ_2) at B

A similar notation holds for $r_{B_k}(\theta, \psi)$. Here $k = 1, 2, \ldots, N$ labels the particular repetition of the experiment. There is a total of N repetitions. An average value, or expectation value, is defined as (recall Eq. (20.7))

$$\langle r_A(\theta, \psi) \rangle \equiv \frac{1}{N} \sum_{k=1}^{N} r_{A_k}(\theta, \psi) \xrightarrow[N \to \infty]{} 0 \qquad (22.5)$$

The far right-hand side of this equation simply indicates that the outcomes at station A consist of a random string of $+1$s and -1s for this experiment. (A similar statement holds for station B.) We are interested in the average value of the quantities $r_{A_k}(\theta, \psi) r_{B_k}(\theta, \psi)$. This is the correlation

$$\langle r_A(\theta, \psi) r_B(\theta, \psi) \rangle \equiv \frac{1}{N} \sum_{k=1}^{N} r_{A_k}(\theta, \psi) r_{B_k}(\theta, \psi) \qquad (22.6)$$

The concept of a correlation is actually quite familiar in everyday contexts. For example, if we examine a sample of the population to see who is or is not a smoker

(r_A) and who does or does not have lung cancer (r_B), we find that there is a (positive, but not perfect) correlation between those who smoke and those who have lung cancer. To press the analogy, let us assign to r_A the value $+1$ for smoking, -1 for not smoking, and to r_B the value $+1$ for having lung cancer, -1 for not having it. Then, for a sample of N people, we could define a correlation $\langle r_A r_B \rangle$, much as in Eq. (22.6). If everyone who smoked had lung cancer and no one who did not smoke had lung cancer, then there would be a perfect correlation, $\langle r_A r_B \rangle = +1$. If no one who smoked had lung cancer and everyone who did not smoke had lung cancer, then there would be perfect anticorrelation, $\langle r_A r_B \rangle = -1$. In general, a correlation must lie between -1 and $+1$. Quite often we interpret strong correlations in terms of local causal relations between two variables.

Many common correlations between distant events can be understood in terms of local causes only. For example, suppose that two people, say A and B, are flat broke and a third party puts a quarter into one person's slacks pocket and nothing into the other person's pocket and that neither person knows who has been given the quarter. If A travels to New York and B to San Francisco, each can only say that there is a 50% chance of her having the coin in her pocket. Let A now look in her pocket and find the quarter there. She knows immediately that B, thousands of miles away, has no quarter in her pocket. However, there has been no mysterious action at a distance in this case, since the correlation of only one person having a quarter was built into the system at the beginning when A and B were together. The point about the EPRB correlations we are now discussing is that, as we shall see shortly, they cannot be explained in terms of such local interactions.

To return now to our EPRB example represented by Eq. (22.6), the actually observed outcome of such an experiment (or, equivalently, the quantum-mechanical prediction for the correlation) is (for a spin-zero state)

$$\langle r_A(\theta, \psi) r_B(\theta, \psi) \rangle_{QM} = -\cos(\theta - \psi) \tag{22.7}$$

We simply state this result and do not derive it here. For our present purposes, Eq. (22.7) could be taken as a phenomenological representation of the data, with no commitment to any particular theory. However, in Section 22.A we do indicate how this correlation can be computed quantum mechanically. Let us now impose locality and determinism. If the outcome $r_{A_k}(\theta, \psi)$ at location A is predetermined by the state of the system and if the choice (ψ) that the observer at B makes cannot affect the result $r_{A_k}(\theta, \psi)$ obtained by the observer at A, then, in fact, $r_{A_k}(\theta, \psi)$ can only depend upon θ so that we should write $r_{A_k}(\theta)$. Similarly, we use $r_{B_k}(\psi)$ for the result obtained at station B.

For any actual outcome $r_{A_k}(\theta_1) r_{B_k}(\psi_1)$ on, say, the kth run, we could contemplate another possible experiment in which the choice ψ_2 (rather than ψ_1) had been made by the experimenter at B (while the experimenter at A still made choice θ_1). The outcome (on that run) would then have been $r_{A_k}(\theta_1) r_{B_k}(\psi_2)$, where we would expect (because of locality and determinism) that $r_{A_k}(\theta_1)$ would still have the same value in this (possible) alternative experiment that it has in the actual one. We do not know, however, the value of $r_{B_k}(\psi_2)$ (that is, whether it is a $+1$ or -1). Similarly, we could

PART VIII Some philosophical lessons from quantum mechanics

consider other possible experiments with corresponding results (on a run-by-run basis) $r_{A_k}(\theta_2)r_{B_k}(\psi_1)$ and $r_{A_k}(\theta_2)r_{B_k}(\psi_2)$. We can then ask whether or not there is any conceivable set (or collection) of outcomes $\{r_{A_k}(\theta_1)r_{B_k}(\psi_1)\}$, $\{r_{A_k}(\theta_1)r_{B_k}(\psi_2)\}$, $\{r_{A_k}(\theta_2)r_{B_k}(\psi_1)\}$ and $\{r_{A_k}(\theta_2)r_{B_k}(\psi_2)\}$ such that the experimental correlation of Eq. (22.7) would be satisfied. Realize that each $r_{A_k}(\theta)$ and $r_{B_k}(\psi)$ is $+1$ or -1. Now, because of this, observe that with the definition

$$P_k \equiv r_{A_k}(\theta_1)[r_{B_k}(\psi_1) + r_{B_k}(\psi_2)] + r_{A_k}(\theta_2)[r_{B_k}(\psi_1) - r_{B_k}(\psi_2)] \qquad (22.8)$$

it follows that $|P_k| \equiv 2$. This is an algebraic identity holding for all values of k. Define (for these four possible experiments) a correlation function R as

$$R \equiv |\langle r_A(\theta_1)r_B(\psi_1)\rangle + \langle r_A(\theta_1)r_B(\psi_2)\rangle \\ + \langle r_A(\theta_2)r_B(\psi_1)\rangle - \langle r_A(\theta_2)r_B(\psi_2)\rangle| \qquad (22.9)$$

From the definition of Eq. (22.6) we see that this can be written as

$$R = \left|\frac{1}{N}\sum_{k=1}^{N} P_k\right| \leq \frac{1}{N}\sum_{k=1}^{N} |P_k| = 2 \qquad (22.10)$$

Equation (22.10) states that for any local, deterministic theory, R must be less than or equal to 2.

Next, let us choose $\theta_1 = \psi_1 = 0$, $\theta_2 = -\psi_2 = \pi/3$ (or $60°$) and use the empirical (or the quantum-mechanical) expression of Eq. (22.7) to evaluate R_{QM} from Eq. (22.9) as

$$R_{QM} = \left| -1 - \cos(60°) - \cos(60°) + \cos(120°) \right| \\ = \left| -1 - \frac{1}{2} - \frac{1}{2} - \frac{1}{2} \right| = |-2.5| = 2.5 \qquad (22.11)$$

But Eqs. (22.10) and (22.11) together imply that $2.5 \leq 2$. This is a clear contradiction, showing that the assumptions of locality and of determinism cannot both be compatible with any possible sequence of results or outcomes of these experiments (the likes of which have been performed). Which do you want to give up: locality or determinism? This particular proof is due to the physicist Asher Peres (1934–).[25] It is significant, given the nature of this dilemma, that Peres' article is titled 'Unperformed Experiments Have No Results.'[26]

In Eq. (22.11) we have a clear contradiction between any deterministic local theory and (at least) one prediction of quantum mechanics. However, the reader ought not get the impression that local deterministic theories cannot reproduce any of the predictions of quantum mechanics. To illustrate this point, let us set $\theta_1 = \psi_1 = 0$, $\theta_2 = -\psi_2 = \phi$, so that R becomes a function of ϕ.[27] Figure 22.3 is a plot of $R(\phi)$ versus ϕ. The Bell limit is $R = 2$. Notice that violations of the Bell limit occur only in the range $0 < \phi < 90°$ (with the maximum violation for $\phi = 60°$, as in Eq. (22.11)), but not in the range $90° < \phi < 180°$. It could be possible to construct

328

The EPR paper and Bell's theorem

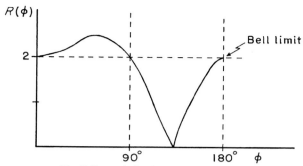

FIGURE 22.3 The Bell correlation for the EPRB experiment

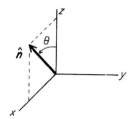

FIGURE 22.4 Detector alignment in a spin measurement

local deterministic models that do agree with quantum mechanics in the latter region.

22.A A CALCULATION OF THE EPRB CORRELATIONS

In this appendix we provide the technical details needed to derive the quantum-mechanical correlation given in Eq. (22.7). The state vector for the spin-zero state of two electrons is

$$\Psi = \frac{1}{\sqrt{2}}\left[\begin{pmatrix}1\\0\end{pmatrix}\otimes\begin{pmatrix}0\\1\end{pmatrix} - \begin{pmatrix}0\\1\end{pmatrix}\otimes\begin{pmatrix}1\\0\end{pmatrix}\right] \qquad (22.12)$$

The state vectors for spin up and down, respectively, along the unit vector $\hat{n}$ of Figure 22.4 (refer to Figure 22.2 as well) are given as

$$\phi_+(\theta) = \begin{pmatrix}\cos(\theta/2)\\ \sin(\theta/2)\end{pmatrix} \qquad \phi_-(\theta) = \begin{pmatrix}-\sin(\theta/2)\\ \cos(\theta/2)\end{pmatrix} \qquad (22.13)$$

$$\hat{n} = (\sin\theta, 0, \cos\theta), \qquad \hat{n}\cdot\boldsymbol{\sigma} = \begin{pmatrix}\cos\theta & \sin\theta\\ \sin\theta & -\cos\theta\end{pmatrix} \qquad (22.14)$$

$$\hat{n}\cdot\boldsymbol{\sigma}\,\phi_+(\theta) = \phi_+(\theta) \qquad \hat{n}\cdot\boldsymbol{\sigma}\,\phi_-(\theta) = -\phi_-(\theta) \qquad (22.15)$$

Here $\hat{n}\cdot\boldsymbol{\sigma}$ is the projection of the spin operator along the unit vector $\hat{n}$. That these two states are indeed the eigenstates of this projection operator can be verified by direct matrix multiplication. The final states of the electrons at stations A and B of Figure 22.2 are given as $\phi^A_\pm(\theta)\otimes\phi^B_\pm(\psi)$ and the joint probabilities as $p^{AB}(\pm\pm|\theta,\psi)$ $=|\langle\Psi|\phi^A_\pm(\theta)\otimes\phi^B_\pm(\psi)\rangle|^2$. Direct calculation then yields

$$p^{AB}(++|\theta,\psi) = p^{AB}(--|\theta,\psi) = \frac{1}{2}\sin^2\left(\frac{\theta-\psi}{2}\right) \qquad (22.16a)$$

$$p^{AB}(+-|\theta,\psi) = p^{AB}(-+|\theta,\psi) = \frac{1}{2}\cos^2\left(\frac{\theta-\psi}{2}\right) \qquad (22.16b)$$

Finally, the correlation function is simply

$$\begin{aligned}\langle r_A(\theta,\psi)\, r_B(\theta,\psi)\rangle_{\mathrm{QM}} &= p^{AB}(++|\theta,\psi)(+1)(+1) + p^{AB}(--|\theta,\psi)(-1)(-1) \\ &\quad + p^{AB}(+-|\theta,\psi)(+1)(-1) + p^{AB}(-+|\theta,\psi)(-1)(+1) \\ &= \sin^2\left(\frac{\theta-\psi}{2}\right) - \cos^2\left(\frac{\theta-\psi}{2}\right) \\ &= -\cos(\theta-\psi) \end{aligned} \qquad (22.17)$$

This is the result stated in Eq. (22.7) of the text.

FURTHER READING

Chapter 4 of Jim Baggott's *The Meaning of Quantum Theory* gives a nicely written and accurate overview of the EPRB experiment, of Bell's theorem and of its experimental tests. James Cushing and Ernan McMullin's *Philosophical Consequences of Quantum Theory* contains a set of essays on various aspects of the implications of Bell's theorem. Arthur Fine's *The Shaky Game* is a superb analysis of Einstein's views on quantum theory, including the historical background of and the philosophical issues related to the EPR paper. Michael Redhead's *Incompleteness, Nonlocality, and Realism* has become the standard technical reference on the philosophical aspects of quantum mechanics, such as nonlocality, Bell's theorem and hidden variables. Finally, John Bell's incomparable *Speakable and Unspeakable in Quantum Mechanics* gathers together in one place all of Bell's major essays on the foundations of quantum mechanics – truly a 'must read' for anyone wanting an insightful guide through the subtleties of quantum theory by an unquestioned master.

23

An alternative version of quantum mechanics

In previous chapters we presented and discussed the standard, generally accepted Copenhagen interpretation of quantum mechanics according to which there exists in nature at the most fundamental level an irreducible and ineliminable indeterminism. Although we saw that what historically constituted the Copenhagen interpretation of quantum mechanics is difficult to specify with precision, three central commitments do characterize it. (1) In general, no particle trajectories can exist in a space–time background. (2) No deterministic description of fundamental physical phenomena is possible. (3) There exists in the laws of the fundamental physical phenomena of nature an essential and ineliminable indeterminism or probability (unlike the probability of classical physics, that there reflects our ignorance of the finer details of complex physical phenomena). A flavor of these features of the Copenhagen interpretation of quantum mechanics can be gotten from the quotations at the beginning of Part VIII (*Some philosophical lessons from quantum mechanics*). On this view, it is in principle impossible to predict, say, the exact future behavior of an electron (that is, to give its position and velocity as functions of the time). According to the Copenhagen school, there can be no causal description of microphenomena in terms of a continuous space–time background (as there is in classical physics). It is generally believed that a causal interpretation of quantum mechanics is impossible, although no proof of this exists.

There is, however, a logically consistent causal interpretation[1] of quantum mechanics, due to Bohm, in which a microentity, such as an electron in a double-slit experiment, does follow a specific and well-defined trajectory (through just one of the slits) from the source to the screen (recall Figure 20.11). In this picture an electron consists of a wave and a particle, in the sense of being a particle accompanied by a wave. The pilot, or guiding, wave (a term originally due to de Broglie in a different context in the 1920s) exerts an influence on the particle. The wave goes through both slits, but the particle through only one. The Heisenberg uncertainty relations then arise because of the disturbance produced on the microsystem during the process of interaction with macroscopic objects (such as the slits or a measuring device). It is important to appreciate that this interpretation of quantum mechanics is based on precisely the same formalism (or set of mathematical equations) as is the usual Copenhagen interpretation so that there would seem to be no possibility of an experimental decision between these two interpretations or pictures of the world. How the Copenhagen interpretation came to be accepted to the virtual exclusion of the causal interpretation is a fascinating story, some of which we tell in Chapter 24.[2]

The point of these comments is to make the reader aware of the possibility of

PART VIII   Some philosophical lessons from quantum mechanics

interpretations of quantum mechanics alternative to the commonly accepted Copenhagen one. We ought not underestimate, though, the stringent constraints placed on any interpretation of quantum mechanics by Bell's theorem (discussed in the previous chapter). For instance, Bohm's interpretation is nonlocal (as, in some sense, is the Copenhagen interpretation) and that is a radical departure from the previous classical world view. This and the next chapter are concerned with the question of whether or not all of the general philosophical commitments made by the Copenhagen school were really warranted, let alone demanded, by the facts.

23.1 AN OVERVIEW

We characterized the standard, or Copenhagen, view of quantum mechanics – the one almost universally accepted by practicing physicists and often by philosophers of science concerned with such issues – as requiring complementarity (say, wave–particle duality), inherent indeterminism at the most fundamental level of quantum phenomena and the impossibility of an event-by-event causal representation in a continuous space–time background. However, quantum mechanics as a theory has two components (as does any modern theory in physics): a formalism and an interpretation. Very loosely, the *formalism* refers to the equations and calculational rules that prove empirically adequate (essentially, 'getting the numbers right') and the *interpretation* refers to the accompanying representation the theory gives us about the physical universe (the story that goes with the equations or what our theory 'really' tells us about the world). Since a (successful) formalism does not uniquely determine its interpretation, there may be two radically different interpretations (and ontologies) corresponding equally well to one empirically adequate formalism.

Let us expand a bit upon this distinction between these two components of a theory. We do not mean to imply that this division of a theory into the two components of formalism plus interpretation is necessarily unique, complete or exhaustive. For our purposes here we need only the recognition that a formalism and an interpretation are two distinct, even if related, parts of a theory. An interpretation is based on a (necessarily) incomplete examination of a formalism, since it is not possible to apply a given formalism to all conceivable situations and experiments (either actual or of the 'thought' variety) in arriving at an interpretation of that formalism. Our 'intuition' about the world is based very largely on those relatively few (but hopefully 'typical') cases or problems we can solve (often exactly). Thus, in the case of classical mechanics applied to the motion of a planet (mass m) about the sun (mass M), we might in thumbnail sketch represent the appropriate formalism in terms of Newton's second law ($F = ma$) and of his inverse-square law of universal gravitation. On the basis of many applications (basically the two-body problem and the perturbations thereof), we then develop an interpretation of the nature of the world governed by this system of laws. The picture, or folklore belief, that emerges historically is that of a completely deterministic, (in principle) predictable physical universe. This intuition is based on the class of problems (today termed 'integrable') that yielded to the analytical tools available. However, as we saw in Chapter 12, the lesson of modern chaos theory that has emerged in the last couple of decades or so is that such integrable dynamical systems are quite atypical of classical mechanical systems. We now appreciate that a 'typical' mechanical system can exhibit chaotic behavior so that we have really no predictive power about its long-term future behavior. Our intuition about the nature of classical mechanical systems was seriously wrong for about 300 years. The formalism (or equations) of classical mechanics has not changed, but, for many people, the interpretation most definitely has (although one could still accept an ontological determinism). With this as an elementary illustration of the difference between a formalism and its interpretation, let us now turn to the case of quantum mechanics.

23.2 THE COPENHAGEN INTERPRETATION

Entire books have been written on the formalism of (nonrelativistic) quantum mechanics and we intend here only to sketch in the briefest form, in terms of a few simple rules, the types of postulates that are usually employed in making quantum-mechanical calculations. No claim is made that these postulates are complete, independent or the most general ones possible. They are intended only as an illustration of a formal structure when a state vector ψ can be used to represent a specific physical situation.

 i a state or wave function (ψ) for the system.[3]
 ii a dynamical equation (the Schrödinger equation[4]), giving the time evolution of the state ψ under the influence of the Hamiltonian H for the physical system.
 iii a correspondence between (Hermitian) operators A and physical observables a.[5]
 iv a rule for computing average values for a series of observations of a.[6]
 v a projection postulate, or collapse of the wave function, (either explicitly or effectively assumed) upon measurement.[7]

From this formalism follows the Heisenberg uncertainty, or indeterminacy, principle. This is related to Bohr's complementarity principle. For our purposes here, a special case of complementarity that will serve as an illustration is the well-known wave–particle duality according to which a physical system (say, an electron or a photon) behaves either as a wave or as a particle, depending upon the context or environment. Applications of the formalism of quantum mechanics to (idealized) position–momentum measurements, double-slit arrangements and the like lead to a picture, or interpretation, in which definite space–time trajectories cannot be maintained, specific possessed values of observables (such as all components of the spin) are not possible at all times, and event-by-event causality must be abandoned (to be replaced, perhaps, by 'statistical causality', whatever that may be). Furthermore, the process of measurement assumes a central and highly problematic role in nature (that is, the projection postulate or collapse of the wave function), as does the passage to a classical limit (in terms of an underlying physical ontology). An examination of the formalism in specific EPRB correlation-type experiments shows the *nonseparable* nature of the theory (that is, the types of entangled quantum states like those in Eq. (21.8) or in the sums of Eqs. (22.2) and (22.4)) and this gives rise to correlations that imply the existence of nonlocal influences between spatially separated regions. So, on the Copenhagen interpretation of quantum mechanics, physical processes are, at the most fundamental level, both inherently indeterministic and nonlocal. The ontology of classical physics is dead.

23.3 A LOGICALLY POSSIBLE, EMPIRICALLY VIABLE ALTERNATIVE

Since the formalism of quantum mechanics is not identical with, or need not include, the Copenhagen interpretation of that formalism, let us outline an alternative, equally as empirically adequate, interpretation of that same formalism. Once the existence of such an interpretation is established, we then turn to the question of its historical origin and of its fate. Perhaps the most direct way to introduce this is to discuss Bohm's causal interpretation.[8] Mathematical details aside (these can be found in Section 23.A at the end of this chapter), what Bohm did was to take the Schrödinger equation and, by means of a mathematical transformation, rewrite it in an equivalent form similar to Newton's second law of motion $\boldsymbol{F} = m\boldsymbol{a}$, where $\boldsymbol{F}$ is now determined by both the usual classical potential V and a new quantum potential U (see Eqs. (23.7) and (23.4) in Section 23.A).

What is relevant about these results for our purposes is that the dynamics of quantum mechanics can be put into the 'Newtonian' form $\boldsymbol{F} = m\boldsymbol{a}$ (Eq. (23.7)) and given a causal interpretation in which microscopic particles, such as electrons, follow well defined trajectories in space–time. However, because of the influence of the quantum potential, these trajectories are very sensitive to the initial conditions $(\boldsymbol{r}_0, \boldsymbol{v}_0)$ of the particles. (Although Bohm's original papers were written in 1952, well before the advent and popularity of modern chaos theory, his general approach and several of his insights are forerunners of, and certainly consonant with, this current field of activity.) Let us parse Bohm's causal interpretation in the following way. The wave function ψ represents the effect of the environment on the microsystem under consideration (here, a particle of mass m). This ψ is a solution to the Schrödinger equation and yields the quantum potential U (via Eq. (23.4)). That is the fundamental role of ψ in this theory. The causal interpretation and the standard Copenhagen one are based on the same formalism and are indistinguishable in their predictions if the following three assumptions are made:

  i the field ψ satisfies the Schrödinger equation (Eq. (23.1)).
  ii the particle velocity is restricted by the 'guidance condition' $\mathbf{v} = (1/m)\nabla S$ (Eq. (23.6)).[9]

iii the precise location of a particle is not predicted or controlled, but has a statistical (ensemble) distribution according to the probability density $P(x, t) = |\psi(x, t)|^2$.

These are logically independent assumptions. In particular, notice that ψ plays very different roles conceptually in (i) and (iii). In our gloss of the formalism presented above, (i) and (ii) would be taken as representing the quantum dynamics of a microsystem (influenced by a quantum potential through the wave function ψ). In response to why it should in addition happen that the probability density P has the value $|\psi|^2$, Bohm[10] gave an argument to show that any initial P such that $P \neq |\psi|^2$ could be 'driven' to $P = |\psi|^2$ by random interactions and by the quantum dynamics ((i) and (ii) above), much as an arbitrary initial distribution in (classical) statistical mechanics is driven to an equilibrium (Maxwell–Boltzmann) one through random interactions. Hence, (iii) need not be viewed as an ad hoc assumption made just to produce agreement with the Copenhagen version of quantum mechanics. Of course, from a purely logical point of view, one could simply demand (iii) by fiat, as is essentially done for the Copenhagen interpretation. Specifically, Bohm himself says of (iii):

> [W]e do not predict or control the precise location of the particle, but have, in practice, a statistical ensemble with probability density $P(x) = |\psi(x)|^2$. The use of statistics is, however, not inherent in the conceptual structure, but merely a consequence of our ignorance of the precise initial conditions of the particle.[11]

It is not our purpose here to discuss in great detail the question of the empirical indistinguishability of this and of the standard interpretation of quantum mechanics. Such details can be found elsewhere.[12] We have indicated how radically different interpretations can be based on the standard formalism of quantum mechanics. It is also worth pointing out that there is no measurement problem in this causal interpretation and no collapse of the wave function, although all of the standard results, such as the Heisenberg uncertainty relations, still obtain.[13] That is, there is no ontological rift between the classical and quantum worlds or domains.[14] Furthermore, we can characterize the classical regime by the condition that the quantum potential (Eq. (23.4)) be negligible. Since this involves the wave function ψ, it becomes the property that the wave function must satisfy for a classical domain (a much more coherent criterion than the often stated, but unrealizable and conceptually ill-defined, 'limit' such as $\hbar \to 0$).

Another aspect of the causal quantum mechanics program has been various attempts to provide a physical underpinning for that interpretation. This project has proven to be quite difficult and it is important to distinguish these efforts from the logical exercise of Bohm's interpretation (outlined above). We do not have space here to discuss these other progams. At present, none is complete and without its problems – any more than is standard quantum mechanics.

PART VIII   **Some philosophical lessons from quantum mechanics**

23.4 THE VALUE OF AN ALTERNATIVE INTERPRETATION

Now that we have in hand an actual alternative version of quantum mechanics, we can ask what the value of the exercise is. One might feel that there is little point to it since 'Copenhagen' works and is consistent. But, even if this latter claim is accepted, the question of understandability remains. In his classic 1952 paper, Bohm concisely and elegantly states his own assessment of this situation.

> The usual interpretation of the quantum theory is self-consistent, but it involves an assumption that cannot be tested experimentally, *viz.*, that the most complete possible specification of an individual system is in terms of a wave function that determines only probable results of actual measurement processes. The only way of investigating the truth of this assumption is by trying to find some other interpretation of the quantum theory in terms of at present 'hidden' variables, which in principle determine the precise behavior of an individual system, but which are in practice averaged over in measurements of the types that can now be carried out. In this paper and in a subsequent paper, an interpretation of the quantum theory in terms of just such 'hidden' variables is suggested. It is shown that as long as the mathematical theory retains its present general form, this suggested interpretation leads to precisely the same results for all physical processes as does the usual interpretation. Nevertheless, the suggested interpretation provides a broader conceptual framework than the usual interpretation, because it makes possible a precise and continuous description of all processes, even at the quantum level.[15]
>
> . . .
>
> As a matter of fact, whenever we have previously had recourse to statistical theories, we have always ultimately found that the laws governing the individual members of a statistical ensemble could be expressed in terms of just such hidden variables.[16]
>
> . . .
>
> The usual physical interpretation [that is, in its finality and completeness] . . . presents us with a considerable danger of falling into a trap, consisting of a self-closing chain of circular hypotheses, which are in principle unverifiable if true.[17]

Does the Copenhagen interpretation give us a description of the world that we can understand in any meaningful sense of that term? That is certainly open to debate. The quest for a more nearly understandable world view can be a motivating factor in seeking another interpretation of the quantum formalism. These issues are taken up at greater length in the next chapter. Alternative versions of quantum mechanics ought at least to be made known, rather than simply dogmatically dismissed out of hand.

In this spirit, let us return to the simple distinction we made in Section 23.1 between two components of a scientific theory: its formalism and its interpretation. Both standard quantum mechanics and Bohm's version use exactly the same set of rules for predicting the values of quantities that can actually be observed. The formalism of quantum mechanics has been so stunningly successful – beyond argument the most quantitatively successful theory in the history of physics – that we simply accept it in our argument as being completely empirically adequate for any set of physical phenomena to which it has ever been applied in the many decades since its inception (around, say, 1930). The physical interpretation refers to what the theory tells us about the underlying structure of these phenomena (the corresponding picture story about the furniture of the world – an ontology). The point at issue is not the formalism, but rather the interpretation. With regard to the latter, there are quantities (such as the trajectory for a particle) that conceptually have in-principle well-defined meanings for Bohm, but not for Copenhagen. These need not be accessible to observation.

The standard, Copenhagen interpretation of quantum mechanics requires a discontinuous and inherently uncontrollable change, jump or collapse of the state vector ψ upon observation or measurement of the physical system under consideration. An implication of this is that there is an essential and ineliminable indeterminacy at the most fundamental level of physical processes. If taken as telling us something about the nature of our physical world, the Copenhagen interpretation of quantum mechanics requires an ontology that is irreducibly indeterministic so that in principle we cannot, for a microsystem, speak of the continuous evolution of possessed values or about a trajectory in a space–time background.

On the other hand, we know that in Bohm's interpretation of precisely the same formalism, microsystems evolve completely deterministically. The particles follow definite, if at times highly irregular, trajectories in a space–time continuum. There is no collapse of the wave function upon observation, but we simply discover where the particle is.[18] The Heisenberg uncertainty relations become limitations on the accuracy of our measurements due to the effects of the quantum potential. We recover all of the statistical predictions of standard quantum mechanics. On this interpretation, a microsystem behaves much as a classical chaotic system.

Since Bohm's theory is deterministic and makes all of the same predictions as standard quantum mechanics, it follows from Bell's theorem (Section 22.3) that Bohm's theory must be nonlocal. This it clearly is because of the influence of the quantum potential: a change in the environment necessitates an immediate change of the wave function everywhere in space, in turn modifying the quantum potential (U of Eq. (23.4)) everywhere and instantaneously. At first sight this nonlocality in Bohm's theory might suggest a conflict with the first signal principle of special relativity. But this is not so, since it can be proven that these long-range correlations (see Eq. (22.7) for an example) cannot be used for signaling.[19]

So, the Bohm interpretation can be seen as demoting special relativity from the status of an absolutely universal foundational theory and instead demands relativistic invariance only of the 'observational' content of a physical theory. And, why is this so bad? After all, Einstein's postulates for special relativity (that is, form invariance of the laws of physics – no possible detection of an absolute velocity – and the constancy of the speed of light for all inertial observers) were themselves based on observational consequences. We mean this in the sense that these postulates were made to account for the results of experiments, but that they were not themselves demanded by experiment. In this regard recall our comments in Section 16.3 about the conventionality of Einstein's criterion for simultaneity. There we saw that other conventions, such as one allowing absolute simultaneity among all inertial frames, are possible, but lead to identical empirical results. It is worth keeping this option in mind, because there may be no acceptable way to combine full relativistic invariance (at the level of the equations of the theory) with quantum mechanics.[20]

So, Bohm does allow us to be realistic about actual space–time trajectories, but at a price. All interpretations of quantum mechanics have price tags attached. In terminology that originated in a different context, we might say that these interpretations or world views get entheorized into a general network of concepts and commitments and one must arrange for oneself the most acceptable trade-off.[21] Let us remember, though, that one purpose of Bohm's 1952 papers was to show that trajectories are possible and that, therefore, part of the Copenhagen dogma (completeness and, hence, the alleged impossibility of such trajectories) is false. That it certainly did.

23.5 EXPLANATION VERSUS UNDERSTANDING

We now take up the question of the type of access that quantum mechanics gives us to the working of the physical world at the most fundamental level. This will be done in part by contrasting the classical and quantum views of reality. Consider three levels on which scientific theories function: (i) empirical adequacy, (ii) (formal) explanation and (iii) understanding. These are three distinct goals of a scientific theory, even though it may not always be possible to draw a sharp dividing line between one of these hierarchical levels and the next. The distinctions and relations we discuss may apply only to the physical sciences, the area of science from which examples will be chosen.

Since we are concerned with the difference between an explanation and an understanding of physical phenomena, let us first distinguish between the terms 'explain' and 'understand' as we use them. The nature of explanation in science is a topic much discussed by philosophers of science.[22] Consider explanation and scientific understanding with the aim of deciding what an explanation is and what about it gives us understanding. Our 'formal explanation' (or just 'explanation' for short) is basically an explanation in terms of logical entailment and is essentially equivalent to the concept of explanation in the deductive-nomological (D-N) or covering-law model. (This is a standard term in the philosophy of science for an explanation that proceeds by the logical deduction of a result from some hypothesis, or 'law', that has been accepted.) The ultimate facts doing the explaining can be quite strange and unfamiliar, as in quantum mechanics. Such an explanation, while quite precisely definable and objective, does not in itself give us understanding of the phenomena subsumed under the law in question. Much of scientific explanation is in fact just such formal explanation and does not necessarily provide us with any sense of understanding.

There are certain pragmatic aspects of an explanation that seem required for understanding. Unification is an important element. A global explanation aids understanding since it reduces the number of independent phenomena that we must accept as given. While unification and reduction (to a few simple, irreducible 'givens' or hypotheses, as in atomic and elementary-particle theory) are important characteristics of an explanation that can produce understanding, these are not enough in themselves. It is precisely with quantum effects, such as the EPRB correlations, that this becomes especially clear, as we argue in the next section. What are some of the features that have been common in producing a sense of understanding in past theories? Understanding in our sense would be a pragmatic bonus, going beyond the purely epistemic, that some theories possess and that may be relevant for our acceptance of them. It is admittedly contingent because it depends upon certain conditions by which we, as humans, are able to have a sense of understanding. Not surprisingly, causality and causal mechanisms are often essential features of an explanation that issues in understanding. Of course, we have no guarantee that we can always find an explanation of physical phenomena that will produce understanding.

Our argument really begins from the intuition, based on experience and on the history of physics, that understanding of physical phenomena involves picturable, physical mechanisms and processes. The historian and philosopher of science William Whewell (1794–1866), in his *History of the Inductive Sciences* (first published in 1837), adumbrated such a position when he distinguished between a law and a cause or between the formal and the physical stages in the development of a theory. Similarly, we saw at the end of Section 16.3 that Einstein contrasted principle theories (such as special relativity, based as it is on the general principles of the form invariance of the laws of physics and on the invariance of the speed of light) with constructive theories (such as the Lorentz aether-based theory of the electron). He saw the virtues of the former to be epistemic security (of the deduced consequences once the empirically based principles are accepted) and generality of applicability, while the latter provided clarity of comprehension (our understanding), often in terms of specific physical models. This is very much the difference between formal explanation and understanding.

As a simple illustration, suppose that (once we have learned, observed or been told that the orbit of a planet is a plane curve) we are given (not necessarily on the basis of a derivation from any theory) Kepler's second law of planetary motion in mathematical form.[23] It gives the distance r (measured from the focus at which the sun is located) of the planet from the sun in terms of the angular location (θ) of the planet. This formula is an empirically adequate representation of the observational data, but (at this level) we have not the slightest idea why this particular form of relation should obtain. It is a straightforward (but not wholly trivial) exercise to prove that Newton's second law of motion plus his law of universal gravitation together can be solved to yield this result. (Recall Chapter 9.) Such a deductive argument certainly provides a formal explanation, but it gives us no understanding of what physical process causes the planet to follow an elliptical orbit. One attempt at a causal explanation would be to invoke the notion of (instantaneous) gravitational action at a distance. It seems implausible that anyone understood such action at a distance, for over two hundred years. In a sense, Einstein's general theory of relativity provided an understandable (picturable) causal explanation in terms of a curved space–time background (whose specific structure is determined by the distribution of masses, as impressionistically depicted in Figures 18.3 and 18.4) through which gravitons (or gravitational waves) propagate to transmit physical influences of one mass upon another. In some loose sense, this space–time plays the role of an 'aether' in acting as a background through which disturbances can propagate at a finite velocity.

Let us now consider specifically the situation for quantum phenomena as illustrated by the well-known EPRB thought experiment (recall Figure 22.2). There, two observers, in spatially separated regions, each can make one of two choices (for their instrument settings) about what to observe and for each setting there are two possible outcomes (spin up or spin down). As we have seen, Bell's work showed that no determinate, local theory can account for the experimental outcome that is predicted by quantum mechanics.[24] Much subsequent analysis has sharpened our understanding of precisely what must be assumed to obtain Bell-type inequalities. It turns out that, for the EPRB situation, the Bell inequalities are the necessary and sufficient conditions on the joint distributions for a local common-cause explanation (of the type we considered in the two people–one quarter example in Section 22.4) of the actually observed correlations.[25] Since these inequalities are violated, then, at least for this one experimental arrangement (and there are others), no local common-cause explanation is possible. A reasonable list of exhaustive explanatory resources for the observed results is (i) mere chance or coincidence; (ii) a direct causal link between the two spatially separated stations; (iii) a local common cause explanation located in the common past events of the system and apparatus. The last, (iii), is blocked by the violation of the Bell inequalities, (ii) (it is often claimed) by the first signal principle of special relativity and (i) we put aside for the moment as no explanation at all.

In this present case, the distinctions we have in mind among empirical adequacy, explanation and understanding can be seen by starting with the correlation $<r_A r_B(\theta, \psi)>$ of Eq. (22.7) for the EPRB experiment of Figure 22.2. Recall that there A and B refer to the two spatially separated stations and θ, ψ to the directions (the first at A, the second at B) along which the spins are observed. Given the data for a series of such experiments, a modern-day Ptolemy (without use of quantum mechanics or of any other theory) might find, within experimental errors, the purely phenomenological representation of Eq. (22.7) for the data (much as we saw in Section 19.2 how Planck discovered his radiation law). This representation is surely empirically adequate. In fact, such curve-fitting is little more than equivalent to the data itself. At this stage, though, we have neither an explanation of these results nor any understanding of how they came about. Next, if someone gives us the formalism of quantum mechanics (perhaps as represented at the beginning of Section 23.2), then we can formally explain the results of Eqs. (22.16a) and (22.16b) by performing a calculation of the type given in Section 22.A. We thus have a formal explanation of the joint distributions in Eq. (22.7), but we have no understanding (in terms of underlying physical processes) of how these events are produced in nature. In other words, an intelligible interpretation of the formalism is missing.

23.6 ATTEMPTS AT UNDERSTANDING QUANTUM MECHANICS

What is it about the formalism of quantum mechanics that makes it so difficult to tell a story that we feel we understand about fundamental physical phenomena? The heart of the problem is the entanglement (or nonseparability) of quantum states (see Eqs. (21.8), (22.2) and (22.4)) that gives rise to the measurement problem (Section 21.3) and to nonlocality (for example, the long-range quantum correlations, such as those of Eq. (22.7)). This entanglement makes it impossible to assign independent properties to an arbitrary isolated physical system once it has interacted with another system in the past – even though these two sytems are no longer interacting. This was illustrated with a specific example in Section 21.3 for a measurement process, but such entanglement is a generic feature of quantum systems (except under very special circumstances that do not usually obtain). Within the framework of standard quantum mechanics, there seems to be no way to account for the classical behavior and appearance of everyday objects that do, in fact, exhibit such possessed properties.[26]

We now indicate some of the suggestions that have been made to cope with these technical and conceptual difficulties, all broadly within the standard, or Copenhagen, interpretation of quantum mechanics, where indeterminism is taken to be essential and ineliminable. We saw some moves – specifically, direct causal links and local common-cause explanations – that are usually assumed blocked in any attempt to produce an understanding of quantum phenomena.[27] To provide some understanding of the (formal) explanatory scheme of quantum mechanics is a very desirable goal, one pursued mainly by the *scientific realist*, who believes that our successful scientific theories give us a reliable view even of the entities in the microrealm. It seems to be a part of our human nature (at the very least in the more recent Western tradition) to want something additional that goes beyond mere formal explanation. There have been many, often realist, attempts at interpreting quantum mechanics. Let us see whether some of the current ones succeed in providing us with any understanding of the phenomena.

One could advocate a quantum realism in which one takes seriously an 'ontic blurring' of many of the variables that can be observed in a quantum system (the lack of simultaneously sharply observable values of, say, the position and the momentum of a particle). The term 'quantum particle' can be introduced for an object that exhibits wave–particle duality to distinguish it from the traditional 'classical particle'. The nonseparability characteristic of quantum systems can be seen as an indication of the 'holistic character' of such systems. In a sense, what has been done is to take some of the unique aspects of the quantum formalism and then assign names and ontic status to them. This in itself does not produce any sense of understanding of the physical phenomena. If we say that quantum realism is that realism required by quantum mechanics, we have not thereby helped anyone to comprehend just what quantum realism is as a picture of the world.

Some claim that we need a new concept of causality, but it is not clear just what that is. Heisenberg long ago suggested that we introduce a new class of physical entity, *potentia*, into our theory (and into our ontology). This is a move that promises to help our understanding of how definite results appear or actualize in one localized region of space–time. That is an attempt to cope with an aspect of what is commonly referred to as the reduction of the wave packet in quantum mechanics (see Eq. (21.9)) that is the most basic problem for the formalism of quantum mechanics, as standardly interpreted. Others have proposed a modification of the Schrödinger equation that gives a unified (basically wave-ontology) treatment of macro- and microphenomena, including the EPRB correlations. However, this formal resolution of the measurement problem leaves us, at the level of the actual physical phenomena, with the same (physics) puzzle about how correlations of the EPRB type are produced in spatially separated systems. In light of this, one can opt for a type of relational holism in which the objects have inherent relations among themselves, with these relations persisting across arbitrarily great distances. Even aside from the serious problem that would remain of the passage from the microrealm, where relational holism supposedly holds, to the macrorealm, where it is not obviously prevalent, does the assertion of the phenomenon of relational holism provide us with a concept we can understand for physical systems?

A very different type of move takes the point of view that, since direct, local causal links and local common causes are impossible for quantum correlations, these quantum correlations themselves ought to be taken as the irreducible brute facts of the primitive givens upon which we build any further explanations. This would sidestep the issue of understanding. One characterization of quantum mechanics is:

> 'It [quantum mechanics] is ... the blackest of black-box theories; a marvelous predictor but an incompetent explainer.'[28]

Such an attempt to place the brute facts first, as a basis for an explanation, is not inconsistent with a claim that we cannot understand quantum phenomena. So, if we must buy wholeheartedly into an indeterministic world view, it may still be that we cannot understand this world view.

However, by taking position to be the only possessed property (or, the only observable) and by accepting the inseparable nonlocality of quantum phenomena, Bohm's theory avoids the measurement problem and has an ontology that produces no conceptual rift with many of our classically conditioned representations of a real world with its objectively possessed properties. This yields a specific, picturable, more nearly traditional (in some features almost 'classical') interpretation of quantum mechanics. It is a completely deterministic theory in which particles have actual, objectively existing trajectories at all times and they interact via a quantum potential. The price is nonlocality (direct, instantaneous, long-range interactions), but of a fairly benign variety that produces no empirical conflict with relativity. This model, one might claim, produces more of a sense of understanding than other attempts that we have just mentioned.

23.A SOME MATHEMATICAL DETAILS OF BOHM'S THEORY[29]

What Bohm did can be represented basically as the following. Beginning with the (nonrelativistic) Schrödinger equation (that is accepted, not derived, there)

$$i\hbar \frac{\partial \psi}{\partial t} = -\frac{\hbar^2}{2m}\nabla^2 \psi + V\psi \tag{23.1}$$

one defines two real functions $R(x, t)$ and $S(x, t)$ as $\psi(x, t) = R \exp(iS/\hbar)$. Substitution of this decomposition into Eq. (23.1) and separation of the real and imaginary parts of the resulting expression yields

$$\frac{\partial R}{\partial t} = -\frac{1}{2m}\left[R\nabla^2 S + 2\nabla R \cdot \nabla S\right] \tag{23.2}$$

$$\frac{\partial S}{\partial t} = -\left[\frac{(\nabla S)^2}{2m} + V - \frac{\hbar^2}{2m}\frac{\nabla^2 R}{R}\right] \tag{23.3}$$

The *quantum potential* U is defined as

$$U \equiv -\frac{\hbar^2}{2m}\frac{\nabla^2 R}{R} \tag{23.4}$$

With the definition $P = R^2 = |\psi|^2$, Eq. (23.2) can be rewritten as

$$\frac{\partial P}{\partial t} + \nabla \cdot \left(P\frac{\nabla S}{m}\right) = 0 \tag{23.5}$$

If we identify a velocity field **v** as

$$\mathbf{v} = \frac{1}{m}\nabla S \tag{23.6}$$

then Eq. (23.5) is just the continuity equation. Equation (23.6) is often referred to as the 'guidance' condition in Bohm's theory.[30] This defines the possible trajectories for the particles. We can interpret P as the probability density for the distribution of particles. With the identification of the momentum as $\mathbf{p} = m\mathbf{v}$ and Eq. (23.6) for **v**, we easily show that Eq. (23.3) can be rewritten as[31]

$$\frac{d\mathbf{p}}{dt} = -\nabla(V + U) \tag{23.7}$$

so that $d\mathbf{p}/dt = \mathbf{F}$, where $\mathbf{F}$ is the gradient of the potential energy $(V + U)$. This potential energy now includes the familiar 'classical' potential energy V as well as the 'quantum' potential energy U. The quantum potential of Eq. (23.4) introduces highly nonclassical, nonlocal effects.

PART VIII Some philosophical lessons from quantum mechanics

FURTHER READING

Chapter 5 of Jim Baggott's *The Meaning of Quantum Theory* is a nice discussion of several of the currently available alternatives to the Copenhagen version of quantum mechanics. James Cushing's *Quantum Mechanics*, especially Chapters 2 through 5, contains extensive discussion of several topics covered in the present chapter. David Bohm's 'A Suggested Interpretation of the Quantum Theory in Terms of "Hidden" Variables, I and II' is the original locus for what has become known as Bohmian mechanics and contains many penetrating observations about the foundations of quantum mechanics. David Bohm and Basil Hiley's *The Undivided Universe* is a general presentation, in some technical detail, of Bohm's version of quantum theory. Peter Holland's *The Quantum Theory of Motion* is a very competently done advanced text on Bohm's theory. Antony Valentini's *On the Pilot-Wave Theory of Classical, Quantum and Subquantum Physics* treats many aspects of Bohm's theory and its extensions at an advanced level, yet in a very readable and thought-provoking style. The volume *Bohmian Mechanics and Quantum Theory*, edited by James Cushing, Arthur Fine and Sheldon Goldstein, has essays on many facets of Bohm's program.

24

An essential role for historical contingency?

In previous chapters we argued that the empirically successful formalism of quantum mechanics supports equally well two mutually incompatible general ontologies: the inherently indeterministic and generally accepted Copenhagen interpretation and Bohm's completely deterministic one. This underdetermination of the interpretation by the formalism is not simply an apparent one of two equivalent theories since there is no way to translate the terms of one of these ontologies into those of the other. This case could be seen as presenting a challenge to the scientific realist who seeks from successful scientific theories a true representation of the world. Furthermore, we suggest that the actual historical competition between these theories and the selection of Copenhagen over Bohm illustrate that such an historically contingent process is not meaningfully distinct from the rational reconstruction and logical judgment of the victorious theory.[1] What is deemed successful and put on offer by the scientific community for the philosopher of science to justify after the fact is itself a contingent and nonunique product.

24.1 UNDERDETERMINATION

The origin of the underdetermination of scientific theories, what we refer to here as the so-called Duhem–Quine thesis,[2] is typically located in Pierre Duhem's *The Aim and Structure of Physical Theory* (first published in 1906). There Duhem was quite explicit about what he took to be the basis for judging whether or not a given physical theory is acceptable.

> The sole purpose of physical theory is to provide a representation and classification of experimental laws; the only test permitting us to judge a physical theory and pronounce it good or bad is the comparison between the consequences of this theory and the experimental laws it has to represent and classify.[3]

However, when one attempts to move beyond mere successful prediction of and accounting for empirical regularities to the level of warranting hypotheses by means, say, of a crucial experiment or a decisive disjunction between or among a putatively exhaustive set of alternatives, Duhem argued that we are necessarily doomed to failure since 'the physicist is never sure he has exhausted all the imaginable assumptions.'[4] When a theory is contradicted by an experimental result, Duhem tells us, it is a conjunction of hypotheses that is refuted. That is, other assumptions or auxiliary hypotheses are needed, in addition to the theoretical assumption of main interest, to make a prediction or to make a calculation practicable. For example, in doing elementary projectile-motion calculations (to test Newton's second law of motion, say), one typically neglects the effects of air resistance. At times, this can lead to predictions that disagree with observation (as illustrated at the end of Section 6.5). Recall, too, the complex situation with the Kaufmann experiments (see Section 15.5). Different scientists may choose to modify different hypotheses when such disagreements occur. For Duhem, all are equally justified logically, as long as they all save the phenomena:

> The methods [physicists] follow are justifiable only by experiment, and if they both succeed in satisfying the requirements of experiments each is logically permitted to declare himself content with the work that he has accomplished.[5]

Of course, not even Duhem claimed that we are left hopelessly adrift in uncertainty. In fact, he observed, we use good sense to decide which of two or more approaches to theory modification is preferable or acceptable. In more current jargon, we might say that nonevidential and nonlogical criteria are used to eliminate underdetermination in specific cases. While we can, and do, make such moves, there is a problem here.

> But these reasons of good sense do not impose themselves with the same implacable rigor that the prescriptions of logic do. There is something vague and uncertain about them; they do not reveal themselves at the same time with the same degree of clarity to all minds.[6]

We return later to the status of such nonevidential criteria. This is a key issue in the underdetermination debate.

While Duhem took the appropriate unit of appraisal to be at least a scientific theory, Willard Quine held that 'the unit of empirical significance is the whole of science.'[7] Any scientific theory must meet the hard boundary conditions of physical reality, but 'there is much latitude of choice as to what statements to reevaluate in the light of any single contrary experience.'[8] As for the role of common sense, or germaneness in his terminology, Quine saw 'nothing more than a loose association reflecting the relative likelihood, in practice, of our choosing one statement rather than another for revision in the event of recalcitrant experience.'[9] This is consonant with his picture 'of the conceptual scheme of science as a tool, ultimately, for predicting future experience in the light of past experience.'[10] For him, not only microentities, but physical objects generally, are introduced into a theory as a conceptual or pragmatic convenience:

> But in point of epistemological footing the physical objects and the gods [of Homer] differ only in degree and not in kind. Both sorts of entities enter our conception only as cultural posits. The myth of physical objects is epistemologically superior to most in that it has proved more efficacious than other myths as a device for working a manageable structure into the flux of experience.[11]
>
> . . .
>
> Each man is given a scientific heritage plus a continuing barrage of sensory stimulation; and the considerations which guide him in warping his scientific heritage to fit his continuing sensory promptings are, where rational, pragmatic.[12]

If one could make a case for radical and universal underdetermination, then there would arguably exist a serious problem for the ability of science to select a theory that would give us a reliable representation of the physical world in terms of hypothesized microentities. However, even if underdetermination could be shown not to hold as a universal claim (for every conceivable theory), it could turn out to hold contingently for our best scientific theory. Such possible cases of individual underdetermination might have wide-ranging implications for science as an enterprise. For example, if one took a reductionist view of scientific theories (explaining successful scientific theories in all areas of the physical sciences in terms of a single underlying foundational theory – such as quantum mechanics), then an essential underdetermination in that foundational theory could effect those theories built on it.

Let us summarize the problem that the Duhem–Quine underdetermination thesis poses for the scientific realist.[13] Succinctly put, if there are two equally empirically adequate successful scientific theories, agreeing on all possible empirical tests and so being observationally indistinguishable, that support radically different, incompatible ontologies, then such a situation may frustrate the scientific realist in her search for the correct scientific theory that gives, even within limits set by reasonable caveats, a true picture of the world. As long as the two theories under consideration differ in relatively minor respects regarding their ontologies, one can simply decide to bracket these as inessential differences. The case we discussed in the last few chapters involves one theory that represents the fundamental physical processes in the world as being inherently and irreducibly indeterministic and another theory that is based on an absolutely deterministic behavior of the physical universe. This would not appear to be a minor or irrelevant difference.

It is perhaps worthwhile here to enter a disclaimer or two. We are not centrally concerned with the question of the refutation of scientific theories. In response to the question of whether or not theories can be refuted, we would be willing to give away a 'yes' answer in the following sense. Even if one grants that there are many theories that can reasonably be rejected on evidential grounds (those are the easy cases), there still remain other cases, we claim, in which viable, fertile theories have been rejected. Our position is not a radical one that claims that practical underdetermination always exists in all cases,[14] but rather that there are some situations – at least one important instance – in which genuine underdetermination does exist. The case discussed in the previous chapter and in this one does not concern just the mere compatibility of two essentially different theories with the presently available data, but involves a much more deep-seated indistinguishability. While a choice can be, and has been, made on the basis of nonevidential criteria, the question must then be faced of the basis for such criteria and of the role historically contingent factors played in fashioning them. One must resist an urge to seek resolution of the underdetermination problem in terms of future developments that may take place in science. That alone would be more a declaration of belief than an argument.

The basic issue here is a belief in the at least effective uniqueness of a correct scientific theory, with the selection process being objective and not involving in any ineliminable fashion subjective criteria such as coherence, beauty, simplicity or minimum mutilation (of the previously accepted theoretical structure).[15] By coherence we most specifically do not mean just lack of logical contradiction, since both theories we are discussing are logically consistent and neither is pejoratively ad hoc in nature. Scientists typically take for granted the practical uniqueness of successful scientific theories. For instance, Einstein allowed the theoretical (or logical) possibility of more than one empirically adequate theory, but then went on to declare that at any given time the 'world of phenomena' (that sounds pretty objective) uniquely determines one theory as superior to all others.[16] Similarly, Heisenberg, in his retrospective reconstruction of what was going on in Copenhagen in 1926, states:

> I wanted to start from the fact that quantum mechanics as we then knew it already imposed a unique physical interpretation of some magnitude occurring in it... so that it looked very much as if we no longer had any freedom with respect to that interpretation. Instead, we would have to try to derive the correct general interpretation by strict logic from the ready-to-hand, more special interpretation.[17]

It is just this even practical or effective, as opposed to merely conventional, uniqueness that we are questioning.

24.2 A DILEMMA FOR THE REALIST

Now this situation represents, in a sense, a double threat to the scientific realist. To begin with, the almost universally accepted Copenhagen interpretation has traditionally been a serious challenge to a realistic construal of quantum mechanics (as we indicated in Section 23.6). The core of the difficulty is the measurement problem, one entailment of which is that a physical system cannot even in principle possess definite, but merely unknown to us, values of all physically observable attributes (such as position, velocity, all components of the spin, etc.). That is, a simple 'ignorance' interpretation is not consistent with the assumed completeness of quantum mechanics. Taking standard quantum theory seriously as an actual representation of the physical world at the level of individual microentities requires that we accept a rather bizarre ontology (if, indeed, there is any genuine ontology at all). The measurement problem has been around now for many years (since 1930 or so) and has resolutely defied any successful, generally accepted solution. This provides effective ammunition for the antirealist who begins at the level of microphenomena, accepts Copenhagen quantum mechanics as the fundamental and exact theory of all physical processes, and then throws down the challenge to the realist to construct a coherent ontology consistent with the demands of quantum theory.[18] That is, such an antirealist begins his argument in the microrealm, extrapolates to the macrorealm of everyday experience and leaves the ensuing conundrum at the doorstep of the realist. In fairness, though, we must point out that the realist has relatively easy going in the domain of macrophenomena (everyday objects, bacteria, dinosaurs, etc.),[19] but then encounters difficulties in carrying these explanatory resources down to the domain of microphenomena.

At first sight, the Bohm interpretation of quantum mechanics would seem to offer consolation and a potentially powerful means of rebuttal of antirealism to the realist. This interpretation, that represents a microentity as a wave and a particle, rather than as a wave or a particle as Copenhagen does (via complementarity), lends itself readily to a realist construal of even fundamental physical processes that develop completely deterministically in a continuous space–time background. While there are some highly nonlocal, nonclassical effects present, this is also true for the Copenhagen interpretation. Although this Bohm interpretation is consonant with, and even conducive to, a realist position, it is empirically indistinguishable from the ontologically incompatible Copenhagen interpretation. So, the realist may have little ground for requiring a realistic interpretation other than by predilection or fiat. Realism is in double jeopardy here: Copenhagen is not conducive to realism, while Bohm, that provides a consistent realistic interpretation, presents an underdetermination dilemma and thus blocks the realist from achieving the desired goal. In other words, if one can erect mutually incompatible ontologies on a given formalism, then that can pose a genuine problem for the realist. A proponent of realism would have to prove or argue strongly for a claim that, once genuinely different ontologies are proposed, it will be possible to extend the formalism along different lines, because distinct ontologies involve distinct physical magnitudes. This may happen, but need it?

Of course, there is one obvious move still open to the realist at this point. She can claim that any two theories that are empirically indistinguishable can, essentially by definition, differ only in inessentials. Apropos of this, Schrödinger expressed the belief that it is impossible to decide on the basis of observation whether the world is basically deterministic or indeterministic. (This is reminiscent of our discussion toward the end of Section 12.3.) Which description one uses is a purely pragmatic matter, dictated by convenience. In his 1929 inaugural address before the Prussian Academy of Sciences, Schrödinger claimed: 'The most that can be decided [on this issue of determinism versus indeterminism] is whether the one or the other concept leads to the simpler and clearer survey of all the observed facts'.[20] So, the scientific realist could write off the difference between indeterminism and determinism in the ontology of the world as an insignificance,[21] but that would indeed be strange for one concerned with a reliable and meaningfully complete picture of the world. Furthermore, this radical conceptual difference between inherent indeterminism and absolute determinism does not allow one to conceive of a 'dictionary' that would map the language of one of these theories onto the other (that is, to map some central concept onto its negation). That effectively blocks one obvious route of escape from this dilemma.

Since empirical adequacy and logical consistency together do not alone provide sufficient criteria to choose between the two theories presented here, one may wish to enlarge these criteria to include factors such as fertility, beauty, coherence, naturalness and the like. Certainly, the scientific community quite early on did make a decisive selection in favor of the Copenhagen interpretation over the rival interpretations present at that time (around 1927): Schrödinger's wave picture and de Broglie's pilot-wave model (that was a precursor to Bohm's conceptually similar interpretation some twenty-five years later). One theory was in fact chosen. Could the actual criteria used have been objective, or at least unchanging in the sense that they are not in any essential way unstable products of historical contingency or accident? Either of these theories passes well a test for fertility, in the sense of possessing the internal resources to cope with anomaly and new empirical developments that actually occurred, as well as for suggesting new avenues for research and generalization.[22] In Chapter 19 we indicated that the historical record shows that key motivating factors, for certain crucial assumptions about the features that an acceptable theory must have, were based upon the philosophical predilections of the creators of the Copenhagen version of quantum mechanics and upon highly contingent historical circumstances that could easily have been otherwise.

For example, toward the end of Section 19.6 we suggested that, in the collision between the two forms of quantum mechanics, the Copenhagen group essentially declared itself victorious and defined its interpretation as the only possible empirically adequate one. In this process, dissident views were discouraged and history was retold so as to make it seem that the Copenhagen interpretation had been inevitable and that workers outside the fold had played no key role in the development of the theory.[23] Part of the reason that this could happen was that most practicing scientists care more about the calculational success and empirical adequacy of a theory than they do about matters of what is often seen as so much excess baggage going along with the formalism. The scientific community was by and large content to leave the 'philosophical' issues to Bohr and his colleagues. Besides, it is often unclear, especially in the early stages of development of a major new theoretical framework, just what is interpretation and what is central to the calculational rules (what we have previously termed the formalism). One can get an idea of some external factors that were present when the Copenhagen interpretation was being formulated:

> [T]he Pauli letters testify to the tremendous psychological *Zwang* ('constraint', another term of the Copenhagen art) under which the quantum physicists around Born and Bohr labored. Born himself oscillated between modest self-confidence and deep self-doubt; Bohr was frequently ill; Pauli hovered on the verge of breakdown; while Heisenberg, maintaining a healthy flippancy, succeeded in rescuing them all. Their correspondence is filled with strong words expressing despair, misery, and resignation, or joy, hope, and elation.... [T]hese romantic expressions... suggest the frame of mind that soon generated a religion of complimentarity....[24]

In addition, there were other personal and sociological factors, such as professional advancement and the determination of the future of physics, that lent urgency to the Copenhagen school's establishing the hegemony of its interpretation.[25]

What is the relevance for the scientific realist of this claimed underdetermination rooted in historical contingency? Well, if the historical record indicates that a rearrangement of highly contingent factors could plausibly have led to a radically different scientific theory and world view being accepted as correctly and uniquely representing the physical world, then one can reasonably question the value of the philosopher's rational reconstruction to pass judgement on a scientific theory. As a means of discovering and understanding the rationality of science, philosophers typically analyze the successful theory already accepted by the scientific community. They doubtless have the ability to reconstruct rationally, and hence to legitimate, any theory that has already survived the scrutiny of the scientific community. But for historical contingency, though, they might find themselves doing just as well reconstructing and justifying an essentially different, equally successful and widely accepted theory. So, what is the value of the exercise, except as a check for noncontradiction? Each of these reconstructions could be equally rational, but there would not necessarily be any rational means to choose between them.

24.3 AN ALTERNATIVE HISTORICAL SCENARIO?[26]

Now that we have two conceptually very different versions of quantum mechanics that are both internally logically consistent and that have proven to be empirically indistinguishable, we might naturally ask how it happened that one of these theories was accepted to the exclusion of the other. If neither logical considerations nor empirical ones were sufficient to select one over the other, there must have been other factors that came into play. To see what these were and how they influenced the final form of the accepted theory, we must go back to the historical development of quantum theory and to the cultural background out of which it emerged, a story that we outlined in Sections 19.5 and 19.6. We saw that the nascent causal program (due to de Broglie) was left stillborn after the 1927 Solvay Congress and the Copenhagen interpretation established its dominance. Over the years, beliefs hardened into a dogma that a consistent and empirically adequate causal version of quantum mechanics was impossible, even though no conclusive proof existed for this.

There matters essentially stood until 1952, when Bohm published two papers on a causal interpretation of quantum mechanics. (We discussed this in Chapter 23.) Initially, de Broglie was against Bohm's ideas (that were similar to his own pilot-wave theory of 1927) and he raised objections like those that had been brought against his own. Interestingly enough, when Bohm sent Pauli a copy of the paper in which Bohm showed Pauli's objections to the causal interpretation to be specious, Pauli never responded. Furthermore, Pauli's views on the nature of science and its relation to his conception of God (basically, a cosmic bookkeeper who enforced statistical causality) made it inconceivable to him that anything like a return to a 'classical' world with causality and picturable, continuous processes in space–time was either possible or anything less than a loss of nerve and a return to darkness. Bohm did, as we have seen, produce a causal version of quantum mechanics – one capable of a realistic interpretation with a largely classical (micro-) ontology. Bohm's work of the early 1950s reconverted de Broglie to his former ideas. For de Broglie the issue at stake was not classical determinism, but rather the possibility of

PART VIII Some philosophical lessons from quantum mechanics

a precise space–time representation for a clear picture of microprocesses. In this, his expectations were similar to Einstein's. De Broglie felt that classical Hamilton–Jacobi theory[27] provided an embryonic theory of the union of waves and particles, all in a manner consistent with a realist conception of matter. With the concept of the quantum potential, one could envisage a model for fundamental processes. This strong commitment by de Broglie to a realistic interpretation of the quantum formalism is consistent with his own high estimate of the work of the philosopher Émile Meyerson, as stated by de Broglie in a preface to Meyerson's *Essais* (1936). Meyerson attempted to dispel the positivist bias and held that the goal of science is an ontological one.

Two additional facts are relevant for the story we are suggesting. First, Bohm's program has been generalized to the relativistic case, so that it cannot be claimed to be limited to nonrelativistic phenomena alone and, hence, incapable of the broad empirical adequacy of the standard program.[28] Second, subsequent to Bohm's papers, it was demonstrated that a single particle subject to Brownian motion, with a diffusion coefficient ($\hbar/2m$) and no friction, and responding to imposed forces in accord with Newton's second law, $F = ma$, obeys exactly the Schrödinger equation.[29] Although there is randomness in such a theory, a radical departure from classical physics is unnecessary so that the resulting theory is probabilistic in a classical way.

In light of these technical and conceptual resources inherent in the causal program, we can fashion a highly reconstructed but entirely plausible bit of partially 'counterfactual' history as follows (all around 1925–1927). Heisenberg's matrix mechanics and Schrödinger's wave mechanics are formulated and shown to be mathematically equivalent. Study of a classical particle subject to Brownian motion, about which Einstein surely knew something (given his famous work on that topic[30]), leads to a classical understanding of the already discovered Schrödinger equation. A stochastic theory underpins this interpretation with a visualizable model of microphenomena and, so, a realistic ontology remains viable. Since stochastic mechanics is difficult to handle mathematically, study naturally turns to the mathematically equivalent linear Schrödinger equation. Hence, the Dirac transformation theory[31] and an operator formalism are available as a convenience for further development of the mathematics to provide algorithms for calculation.

At this point it is worthwhile to review in summary fashion Einstein's central philosophical commitments. In Section 19.5 we saw that his view of the physical world was a rational, causal one. He had a strong belief in a realistic world view in which microentities have a continuous, objective, observer-independent existence. Two other core commitments for him were event-by-event causality (essentially determinism) and locality. These were three *desiderata* that he felt any candidate for a fundamental physical theory should satisfy. Now the entanglement or nonlocality of this stochastic mechanics formalism would soon have become apparent. Einstein would have rejected any model with this property.[32] Now let us return to our counterfactual scenario.

A Bell-type theorem is proven and taken as convincing evidence that nonlocality

352

is present in quantum phenomena. A no-signaling theorem[33] for quantum-mechanical correlations is established and this puts to rest Einstein's objections to the nonseparability of quantum mechanics. The important point here is the following. If one considers a system S consisting of two subsystems S_1 and S_2 that are spatially separated at some time, then Einstein felt that '... the real factual situation of the system S_2 is independent of what is done with the system S_1, which is spatially separated from the former.'[34] Einstein worried what it would even mean to do science if this were not the case. But, a no-signaling theorem would have shown that relativity could be respected at the practical or observational level and that the nonlocality present in nature was of a benign variety. This could reasonably have been enough to overcome his objections to the nonlocal nature of a de Broglie–Bohm interpretation of the formalism of quantum mechanics. Because the stochastic theory is both nonlocal and indeterministic, whereas the de Broglie–Bohm model is nonlocal only and still susceptible to a realistic interpretation, Einstein might have made the transition to the latter type of theory.

That is, these developments, that could, conceptually and logically, have taken place around 1927, could have overcome the resistance of Einstein and of Schrödinger to supporting a de Broglie–Bohm program. Already in 1926 Erwin Madelung (1881–1972) had the same equations Bohm would employ in 1952, but his interpretation was very different from Bohm's and did not carry conviction. Bohm's interpretation would certainly have been possible in 1927. These models and theories could be generalized to include relativity and spin. The program is off and running. Finally, this causal interpretation can be extended to quantum fields.[35]

So, if, say in 1927, the fate of the causal interpretation had taken a very different turn and been accepted over the Copenhagen one, it would have had the resources to cope with the generalizations essential for a broad-based empirical adequacy. We could today have arrived at a very different world view of microphenomena. If someone were then to present the merely empirically equally as adequate Copenhagen version, with all of its own additional counterintuitive and mind-boggling aspects, who would listen? It is essential to appreciate that this story is neither ad hoc (in the sense of these causal models having as their sole justification an origin in successful results of a rival program) nor mere fancy, since all of the technical developments discussed here actually exist in the physics literature. However, Copenhagen got to the top of the hill first and, to most practicing scientists, there seems to be no point in dislodging it.

24.4 INTERNAL VERSUS EXTERNAL EXPLANATIONS

It is not our intention here to argue in favor of a Bohm type of interpretation of quantum mechanics over the standard Copenhagen one. Rather, our interest has been to see whether a very different choice from the actual historical one might reasonably have been made. In the present case, we saw that neither internal factors (such as logical consistency or empirical adequacy) alone nor external ones (say, sociological or psychological) alone are sufficient to account for the wide acceptance of the Copenhagen world view in place of a causal one. While just facts do not uniquely constrain theory construction and selection, neither can one's predilections by themselves enforce just any theory.[36] Nature provides often tight constraints, but there still remains latitude in theory choice – here, an interpretation or world view. The actual course of theory construction and selection is a rich and involved one with many overlapping factors. Science, even in its products or laws, remains historical or contingent in an essential manner. How things might have gone a very different way at certain crucial junctures and why they did not may be as important as the reasons for the 'right' choices that science has made. We are not particularly uncomfortable with a lack of inevitability in other areas of history. That point is nicely made in the review of a book that examines whether it was inevitable that the Confederacy in the South should lose the United States Civil War to the North in the late nineteenth century.

> Inevitability is an attribute that historical events take on after the passage of sufficient time. Once the event has happened and enough time has passed for anxieties and doubts about how it was all going to turn out to have faded from memory, the event is seen to have been inevitable. Different outcomes become less and less plausible, and before long what did happen appears to be pretty much what had to happen. To argue about what might have happened or whether and why the presumably inevitable turned out to be thought so strikes many people as a waste of time.[37]

Herbert Butterfield (1900–1979), a Cambridge professor of modern history, wrote on this same theme: 'the whig interpretation of history.'[38] By this he means the tendency we have to study events of the past with reference to, or from the pejoratively biased perspective of, the present[39] – as, for example, a retrospective account of the sixteenth-century struggle with Catholics one might find written by descendants of the victorious Protestants, but from an 'enlightened' nineteenth-century perspective of the Whigs. It is no simple task to free oneself from deeply ensconced views that are taken to be true.[40]

> [T]his whig tendency is so deep-rooted that even when piece-meal research has corrected the story in detail, we are slow in re-valuing the whole and reorganising the broad outlines of the theme in the light of these discoveries....[41]

The corrective for such skewed history is not another equally skewed account from an opposing perspective, but rather a balanced presentation and interpretation of the events and of their causes. Because all interesting and truly informative history is unavoidably written from some perspective and after a selection of facts to be presented, it is all the more necessary that we attempt to indicate explicitly what the sympathies of the author are and to beware of how they might influence the representation of past events. While such objectivity should be the goal, it is never fully realizable in practice.

This is similar in spirit to Francis Bacon's admonition that we not impose on the facts of nature more order than actually exists there (that is, that we not force what we observe to conform to our expectations or predilections).

> The human understanding, from its peculiar nature, easily supposes a greater degree of order and

unity in things than it really finds; and although many things in nature be *sui generis* and most irregular, will yet invent parallels and conjugates and relatives, where no such thing is.

...

The human understanding, when any proposition has been once laid down (either from general admission and belief, or from the pleasure it affords), forces everything else to add fresh support and confirmation; and although most cogent and abundant instances may exist to the contrary, yet either does not observe or despises them, or gets rid of and rejects them by some distinction, with violent and injurious prejudice, rather than sacrifice the authority of the first conclusions. It is well answered by him who was shown in a temple the votive tablets suspended by such as had escaped the peril of shipwreck, and was pressed as to whether he would then recognize the power of the gods, by an inquiry. But where are the portraits of these who have perished in spite of their vows?[42]

In this chapter we indicated the prevalence of a certain whiggish tendency in commonly accepted, often folklore, versions of the history of quantum mechanics that are too frequently presented to students, either informally in lectures or as asides in textbooks. There the Copenhagen interpretation is made to seem inevitable and the only one logically consistent with the empirical facts. At a minimum, the reader should now be aware that the actual historical record is more complex than that.

FURTHER READING

On some of the contingencies in the early course of quantum theory, see Franco Selleri's *Quantum Paradoxes and Physical Reality*, especially Chapters 1, 2 and 7. Chapters 10 and 11 of James Cushing's *Quantum Mechanics* develop in more detail the counterfactual historical scenario sketched in this chapter. Herbert Butterfield's *The Whig Interpretation of History* is the classic study of the influence that our present knowledge of affairs has on our attempts to reconstruct past events, especially when the events under study are relevant to how we arrived at today's 'knowledge'.

PART IX

A retrospective

To the question, "What is Maxwell's theory?" I know of no shorter or more definite answer than the following: – Maxwell's theory is Maxwell's system of equations.
Heinrich Hertz, *Electric Waves*

[T]here is, in my opinion, a right way, and ... we are capable of finding it. Our experience hitherto justifies us in believing that nature is the realisation of the simplest conceivable mathematical ideas. I am convinced that we can discover by means of purely mathematical constructions the concepts and the laws connecting them with each other, which furnish the key to the understanding of natural phenomena. Experience may suggest the appropriate mathematical concepts, but they most certainly cannot be deduced from it. Experience remains, of course, the sole criterion of the physical utility of a mathematical construction. But the creative principle resides in mathematics. In a certain sense, therefore, I hold it true that pure thought can grasp reality, as the ancients dreamed.

...

The belief in an external world independent of the perceiving subject is the basis of all natural science. Since, however, sense perception only gives information of this external world or of 'physical reality' indirectly, we can only grasp the latter by speculative means. It follows from this that our notions of physical reality can never be final. We must always be ready to change these notions – that is to say, the axiomatic sub-structure of physics – in order to do justice to perceived facts in the most logically perfect way. Actually a glance at the development of physics shows that it has undergone far-reaching changes in the course of time.
Albert Einstein, *Ideas and Opinions*

The scientist, however, cannot afford to carry his striving for epistemological systematic [too] far. He accepts gratefully the epistemological conceptual analysis; but the external conditions, which are set for him by the facts of experience, do not permit him to let himself be too much restricted in the construction of his conceptual world by the adherence to an epistemological system. He therefore must appear to the systematic epistemologist as a type of unscrupulous opportunist: he appears as *realist* insofar as he seeks to describe a world independent of the acts of perception; as *idealist* insofar as he looks upon the concepts and theories as the free inventions of the human spirit (not logically derivable from what is empirically given); as *positivist* insofar as he considers his concepts and theories justifed *only* to the extent to which they furnish a logical representation of relations among sensory experiences. He may even appear as *Platonist* or *Pythagorean* insofar as he considers the viewpoint of logical simplicity as an indispensable and effective tool of his research.
Albert Einstein, *Reply to Criticisms*

25

The goals of science and the status of its knowledge

So far in this book we have studied several major advances in the history of physics. We can now look back and ask what the relation of physics to the other natural sciences is and attempt to draw some lessons about the nature of science and how it operates. For example, what are the hallmarks of science and what, if anything, makes scientific knowledge different from other types of knowledge? The traditional view of science, supported by nearly all scientists and by many philosophers of science, has been to treat the scientific knowledge finally arrived at as being valid largely independently of the details of how it was obtained, as long as it passes the scrutiny of logical analysis and agreement with further observations. Such a position has been called into question by a considerable body of recent work in the history and philosophy of science. We did touch on certain sociological or public aspects of science in our historical presentation of the formulation of quantum theory in Chapter 19. In this chapter we focus on attempts to identify a strictly rational side of science and to reconcile this with the larger social context within which science functions. We use physics as our prototypical science and also discuss at some length the views of Albert Einstein on reality, on a theory of knowledge and on the method and goals of science. Our first task is to delimit certain features of physics for consideration.

25.1 EINSTEIN ON SCIENCE AND ITS GOALS

Our reasons for considering Einstein's opinions on these philosophical matters were well summarized by Philipp Frank, a theoretical physicist turned philosopher of science and a one-time associate and friend of Einstein in Berlin.

> As a matter of fact, Einstein's physical theories have played a great part in the history of contemporary philosophy, a very great part indeed. We can venture on good grounds the statement that no 'professional philosopher' of the 20th century was quoted so frequently by other philosophers as Einstein was. He was, measured by professional standards, an 'amateur' in philosophy, but a 'philosopher' in the literal sense of the word that means a 'lover of wisdom.'[1]

On the occasion of Max Planck's sixtieth birthday in 1918, Einstein delivered an address in his honor before the Physical Society in Berlin. There Einstein gave a fine statement of the goal of physics and of the reasons for which people do physics.[2] In his view, the scientist, like the poet and philosopher, strives to create a simplified, yet representationally adequate, and intelligible picture of the world. What characterizes the physicist's work is the rigorous precision of mathematics, at the cost of great reduction in the scope of the problems considered. The hope, in Einstein's opinion, is that the laws thus fashioned will be universally valid for all physical phenomena and that, from them, one might be able in principle to account for every natural process (life included). That this approach severely restricts its subject matter to those relatively simple situations that it can study thoroughly and in complete mathematical detail is a point made also by a modern theoretical physicist with an interest in the philosophy of science.

> The most astonishing achievements of science, intellectually and practically, have been in *physics*, which many people take to be the ideal type of scientific knowledge. In fact, physics is a very special type of science, in which the subject matter is deliberately chosen so as to be amenable to quantitative analysis.[3]
>
> . . .
>
> Physics defines itself as the *science devoted to discovering, developing and refining those aspects of reality that are amenable to mathematical analysis.*[4]

The theoretical physicist builds idealized models that can be formulated and treated precisely mathematically. Such models can never give a complete description of reality, for then they would be far too complicated to treat in full mathematical detail. The real art in building or conceiving such models is to incorporate into them a few of the truly essential properties of the real system so that the model will still mirror reality well enough to be able to make predictions, upon suitable mathematical analysis, that will be of physical interest. One major difference between the 'games' played by theoretical physicists and those played by pure mathematicians is that, aside from meeting the demands of internal consistency and mathematical rigor, a physical model must ultimately also meet the inflexible boundary condition of agreeing with physical reality. A simple example of a model is that of an ideal gas consisting of point molecules in the kinetic theory of gases. We are able to treat that model fairly completely with just elementary mathematics and thereby to explain the gross properties of real gases over a wide range of pressures and temperatures. Still, the model is not a perfect reflection of reality and ceases to be valid, for instance, at high densities. Einstein emphasized that, as physical models and theories manage to organize more and more of the phenomena of reality in terms of fewer and fewer premises, they become ever more abstract.[5]

Like Planck (see Section 3.6), Einstein believed in the reality of an objective world that exists independently of our perception of it.[6] This was greatly different from Bohr's conception of reality (see Section 21.1). Also like Planck, Einstein combined elements of Mach's positivism and Poincaré's conventionalism, as indicated by a famous passage from his Herbert Spencer Lecture, 'On the Method of Theoretical Physics,' delivered at Oxford University in 1933 (see the quotations for Part I (*The scientific enterprise*)).[7] There Einstein sketched the system of theoretical physics as consisting of concepts, fundamental laws relating those concepts and the predictions made on the basis of these laws. While the structure of the system is the product of reason (his 'free inventions of the human mind'), the observed phenomena of nature must find their place among the concepts and relations of this theoretical structure. The success of such a representation is the ultimate justification for the utility of this overall mathematical framework. Thus, in Einstein's opinion (see the first set of Einstein quotations for Part IX (*A retrospective*)), this combination of creative principles, mathematics and empirical constraints allow pure thought to lay hold of reality.[8] This conception of physics as a search for a succession of theories that might approach ever more closely to ultimate but incomplete truth is similar to Planck's view of a metaphysical reality that we constantly strive for but never reach (recall Section 3.6). The object of science is to coordinate our sense experiences and to fit them into a logical system to produce a unity in our world picture and to do this on the basis of the fewest premises possible (that is, simplicity becomes a criterion for a satisfactory physical theory, as we saw in other contexts in Section 12.1). For Einstein, this process also brings us closer to a true understanding of reality.[9]

These ideas of Einstein's characterize physics in terms of a logically simple and coherent set of axioms from which we may deduce predictions that agree with observations. Even if, for the sake of discussion, we accept this as a reasonable description of science, then how do we judge whether or not a succession of scientific theories represents progress toward truth?[10] A simple and obvious minimal requirement would be that each succeeding theory should explain or account for more than (or at least as much as) its predecessor on the basis of ever simpler axioms.[11] However, a quantitative criterion for the logical simplicity of the axioms is not easy to formulate. As we saw previously, there is a considerable subjective element in judgments of simplicity or of elegance. Much of the modern philosophy of science is concerned with discerning and arguing for objective criteria for evaluating competing theories or progress in science. If there were a valid inductive logic, this problem of confirmation would be obviated, since then each induction to the next level of generalization would necessarily lead us to a broader, more fundamental theory. While it was recognized, at least from the time of Hume, that we do not possess such an absolutely certain inductive logic, there were attempts in the twentieth century to establish criteria, even quantitative ones, by which to assign a degree of reliability to a given generalization or law based on a specific set of data. The philosophers Hans Reichenbach (1891–1953) and Rudolf Carnap, among others, tried to formulate mathematical theories of the probability of the truth of a theory. However, these formalisms do not provide a workable basis on which to judge complex scientific theories during their development, nor do they correspond to the way in which scientists actually judge theories.

25.2 A REDUCTIONIST PROGRAM

Toward the end of his address on behalf of Planck, Einstein argued that the idealized knowledge that physics gains of certain sectors of reality can be used as the basis, in principle, for a description of any natural phenomenon.[12] Let us now see how far such a program can plausibly be implemented in regard to the other natural sciences, such as geology, chemistry and biology.

Geology would appear to be an example of a natural science that is explicable in terms of the laws of classical physics. Given certain initial conditions of the earth's crust, the system developed according to basic mechanical laws under the influence of the stresses generated by heating and cooling. However, the system is so large and complex that a mathematical calculation based on Newton's laws alone would be hopeless. Modern plate tectonic theory, according to which the earth's surface is divided into about six major rock segments that extend downward through the earth's crust into its upper mantle, was first put forward in 1912 by Alfred Wegener (1880–1930). This hypothesis of the continental drift was proposed largely on the basis of the fit between the eastern coastline of South America and the western coastline of Africa. Subsequent geological evidence was sufficiently great that by the 1960s the idea was accepted of the former existence of a hypothetical southern hemisphere supercontinent, known as Gondwanaland, that included South America, Africa, India, Australia and Antarctica. Such a conjecture is an example of a theory that can be understood in terms of basic physical laws – here great forces acting on gigantic land masses over eons of geological time to produce a slow separation of the continents – although physics could not have predicted such a phenomenon *ab initio* because of the tremendous complexity of the system. We meet a similar situation in chemistry and in biology. A certain amount of empirical data is necessary to show which macrosystems (here, the continents), consisting of innumerable microsystems (the molecules making up the continents), are the proper entities to study in a phenomenological model. The detailed path from the microscopic level is too involved for the theory to be able to reconstruct at the present time, if ever.

In Chapter 20 we saw that quantum mechanics, in the form of the Schrödinger equation, is able to give a detailed, quantitative treatment of the simplest of physical systems, a single hydrogen atom. Involved numerical calculations based on the Schrödinger equation and the Pauli exclusion principle account for the shell structure of the electrons in the elements. When we turn to the simplest diatomic gas molecule, that of hydrogen, H_2, the mathematical problem is extremely complex, although reliable numerical predictions can still be obtained with the use of modern electronic computers. As we extend our considerations to even simple chemical compounds, such as water (H_2O), the situation soon devolves into hopeless complexity from a computational point of view. The exact quantum-mechanical equations can be written down, but they cannot be solved in any reliable quantitative fashion. (Of course, as time goes on one may expect ever more complex systems to yield to quantitative analysis, but there can still be a limit to how far this can go.) At best, only qualitative predictions and trends are possible. There are often useful phenomenological models that are based on large subsystems, much as in the continental drift model in geology. While there may be no reasonable doubt that the basic laws of physics – mainly quantum mechanics – can in principle account for all of chemistry, this claim cannot be subjected to rigorous and thorough test because of purely technical and computational barriers. It is, therefore, arguable the extent to which chemistry can be said to reduce to applied physics. Most of the laws discovered in chemistry have a directly empirical basis, as opposed to having been derived from basic, established laws of physics.

When we come to modern biophysics, the picture is even more clouded. We discuss briefly just one aspect of this field. In 1943 Schrödinger, then at the Dublin Institute for Advanced Studies, delivered a series of lectures at Trinity College, Dublin, on the physical aspect of the living cell. A year later these were published as the book *What Is Life?* and gave impetus to the investigation of whether or not the basic mechanisms of biology could be explained in terms of the laws of chemistry and of physics alone. At that time the problem of the mechanism of biological replication seemed beyond understanding so that Schrödinger concentrated on the way in which genetic information is stored and on the riddle of how this information could remain so stable from generation to generation. One well-known means of avoiding sizable statistical fluctuations away from an equilibrium state is by having the system made up of a huge number of molecules. For a collection of, say, n gas molecules, the probability or chance for a significant fluctuation away from equilibrium is of the order of $1/\sqrt{n}$. Thus, for a system of 100 molecules, sizable fluctuations would be on the order of $1/\sqrt{100}$ or 10%, while for $n = 10^6$, it would be 10^{-3} or 0.1%. This feature is quite general so that if we want extreme stability or reliability, we need a very large system. However, Schrödinger knew that the genes, that are found in the nucleus of a cell and that carry the genetic information, are so small that they could consist at most of a few million atoms. The fluctuations expected classically were too large to account for the extreme stability of the structure of the gene. Using a model of the gene proposed by the physicist Max Delbrück (1906–1981), Schrödinger argued that responsible both for the stability of the gene and for its random mutations was the same mechanism – namely, quantum mechanics. Because of the quantum nature of the atoms constituting the gene, the possible energy levels of this system are discrete. The gene gains its stability from the fact that, at normal temperatures, thermal energy is insufficient to change the gene structure by putting it into another allowed configuration of its molecules. The inherent quantum-mechanical randomness by which a gene accepts or rejects the energy offered to it by X rays or γ rays is a source of chance mutations that occur.

In the same pioneering work Schrödinger gave his answer to *What Is Life?*

> What is the characteristic feature of life? When is a piece of matter said to be alive? When it goes on 'doing something', moving, exchanging material with its environment, and so forth, and that for a much longer period than we would expect an inanimate piece of matter to 'keep going' under similar circumstances.[13]

He defined death as a state of maximum entropy in which a system is in thermodynamic equilibrium with its environment so that no further observable changes take place. A living organism avoids this decay to a state of maximum entropy (or disorder) by consuming 'negative entropy' in the form of highly ordered substances (food) that also yield energy. That is, since the natural processes in an isolated living organism tend to increase entropy (or disorder) and thus degrade the system, Schrödinger pictured a living system as maintaining its overall order (keeping its net entropy low) by taking in negative entropy (order) from its environment. We return to the connection between life and entropy below.

While Delbrück's quantum-mechanical molecular model of the gene accounted for the stability of genetic information, it did not in itself offer an explanation for the replication of this information. However, his 1945 lectures on biophysics at Cold Spring Harbor in New York did initiate a program that eventually led to an understanding of the genetic code. Delbrück shared the 1969 Nobel Prize in physiology as a result of his contribution to this work. The riddle of the genetic code was eventually solved by James Watson (1928–) and Francis Crick (1916–) at the Cavendish Laboratory in 1953, for which they shared the 1962 Nobel Prize in physiology. In his *Chance and Necessity*, Jacques Monod (1910–1976), a 1965 Nobel Laureate in physiology for his work on the theory of genetic replication, explicated the physical theory of heredity based solely upon organic chemistry and quantum mechanics. He is careful to make the point that, while the behavior of living organisms cannot be deduced from these first principles, they can be explained by them. Watson is even more explicit in his view of the physical basis of living organisms.

> The growth and division of cells are based upon the same laws of chemistry that control the behavior of molecules outside of cells. Cells contain no atoms unique to the living state; they can synthesize no molecules which the chemist, with inspired, hard work, cannot some day make. Thus there is no special chemistry of living cells. A biochemist is not someone who studies unique types of chemical laws, but a chemist interested in learning about the behavior of molecules found within cells (biological molecules).[14]

Monod argues (see Section 25.7) that of the three hallmarks of living organisms – teleonomy (apparent goal-directed action), autonomous morphogenesis (the internal determination of form and growth), and reproductive invariance (the exact replication of its own kind) – reproductive invariance is the most fundamental in terms of which the other two are explained. This reproductive invariance – always subject to the random fluctuations or mutations mentioned earlier – is accounted for by the genetic coding carried in the double helix of DNA (deoxyribonucleic acid) that is self-replicating (reproductive invariance) and that via RNA (ribonucleic acid) transfers the coded genetic information to the cell's protein-forming structures to cause development (autonomous morphogenesis). These processes that are described in terms of the three-dimensional structure of the molecules involved and in which no new physical or chemical laws are operative are universal in the biosphere of the earth:

> Today we know that from the bacterium to man the chemical machinery is essentially the same, in both its structure and its functioning.[15]

In the picture Monod paints for us, the quantum-mechanically rooted perturbations or mutations are random (chance), but once caused they are locked into the genetic matrix and reproduced (necessity). It is these mutations or variations of the stable background matrix that are subjected to the natural selection of the environment, whence comes the apparent development and goal-directed activity (teleonomy) of an organism or species. All of these processes take place in strict accord with the laws of thermodynamics. In particular, the total entropy of any closed system never decreases so that, if a living organism increases its order, then its environment suffers the requisite increase in disorder. These thermodynamic statements admit of precise, quantitative and empirical verification, so that Schrödinger's view of a living organism as a consumer of negative entropy has been vindicated.

We have outlined a chain of argument connecting three of the basic sciences as

$$\text{physics} \rightarrow \text{chemistry} \rightarrow \text{biology (living organisms)}$$

While the degree of plausibility granted to this line of reasoning may vary according to your own philosophical predilections, even the most optimistic supporter must admit that the argument is not as tight and gap-free as in the examples typical of the physics we have discussed in earlier chapters. Perhaps the more timid or conservative would hesitate to take the next, obvious step with Salvador Luria (1912–1991), one of the recipients of the 1969 Nobel Prize in physiology.

> Life evolved, reached its present state, and will further develop by the creative interplay of the chemical workings of the genetic material and historical workings of the natural forces that favor now one species, now another, promoting any biochemical invention that provides increased fitness. Stupendous devices such as the brain and mind of man are biochemical inventions as challenging and as mysterious as those that produced the equally stupendous social organization of insects. To the scientist, the uniqueness of man is purely a biological uniqueness rather than a superposition of something nonbiological – soul or spiritual essence – upon the workings of biological evolution. The nature of the mechanisms responsible for these highly complex phenomena still escapes the biologist, but he is confident that this will not always be so.[16]

Luria here mentions the 'historical workings of the natural forces that favor now one species, now another.' The essential role of contingency in the evolutionary chain of events that brought us to our present state is also emphasized by Stephen Gould (1941–), the paleontologist and evolutionary biologist, in his recent writing[17] on our understanding of the early record of the evolution of life on this planet.

25.3 STYLES OF SCIENTIFIC INFERENCE

Whether or not we have any concerns about the viability of a complete reductionist program, we can still ask about the reliability of the knowledge that science gives us in some specific discipline, such as physics. How is it that we infer or warrant our successful scientific theories? In Section 3.5 we outlined four general models or methodologies of inference to arrive at 'true' statements.[18]

In the *intuitive–deductive* (or *axiomatic*) method one begins with a set of axioms that are claimed to be self-evident, or intuitively obvious, and to these applies strict logical deduction to arrive at valid conclusions about nature and reality. This is the method often associated with Aristotle (Chapter 2) and in a later age with Descartes

(Chapters 1 and 3). As we saw several times in earlier chapters, the difficulty with this approach is that pure intuition and self-evidentness have in fact proven to be unreliable guides to theories that correspond to the realities of nature with all its complexities. Similar shortcomings became evident for the *inductive* method in which one begins with specific observations and data and then attempts to use only those facts to reach a general law. In Chapter 3 we presented some of Hume's objections to such a straightforward program of induction.

Newton (Chapter 7) was quite explicit and positive about his own use of induction. However, in his science he in effect also implicitly combined this with a *retroductive*, or *hypothetico-deductive*, strategy that has proven to be one of the major tools employed in modern science. Here one accepts a set of hypotheses and deduces logical consequences from them. Correct predictions are taken as a warrant for the accuracy of the hypotheses, although, of course, this line of argument cannot prove the truth of the premises. One can reasonably claim that Gilbert and Harvey, like Newton, at least implicitly employed the hypothetico-deductive method as a major tool of scientific advance. Boyle certainly applied this line of argument to support an atomistic and mechanical natural philosophy.[19] We have seen (Chapter 11) that his influence, along with the philosophy of More and of Barrow, became evident in the natural philosophy of Newton that dominated science for nearly two hundred years. A key element of reality that emerged from this Newtonian tradition was a mechanistic view with atoms possessing simple quantitative properties (mass, location, etc.) and interacting causally in accord with definite mathematical laws and equations of motion. This mechanistic philosophy (Chapter 12) replaced the organismic philosophy of the Middle Ages (Chapter 2). Newtonian physics was so successful in the practical sphere that it engendered a philosophical outlook that claimed that the laws of classical physics were necessarily or a priori true, as in some sense held by the philosopher Immanuel Kant (1724–1804).[20] (Of course, the eventual overthrow of Newtonian mechanics and gravitational theory dramatically exposed the wishful nature of that project.)

The *falsificationist* methodology (Chapter 3) holds that the only meaningful test of a theory is the degree to which it is capable of being subjected to specific comparisons with reality in such a fashion that its predictions may prove to be incorrect. Once a theory has been falsified, it is rejected and is replaced by another that is then subjected to testing. In this way science progresses from one theory to a more encompassing one. Even though such a criterion appears clear-cut in the abstract, science does not, in fact, operate on the basis of strict falsificationism, since scientists do not scrap a theory immediately once it has come into conflict with a single set of experimental data. (Recall Chapter 15 and the Kaufmann experiments.) Nevertheless, the falsification criterion does play an important, though not an exclusive, role in the practice of science.

None of these general methodologies of inference, though, has ever provided an instrument sufficient to warrant with certainty or to single out uniquely laws and theories. Additional guides are necessary for us to select one scientific theory over another. Quite often these are aesthetic criteria. We more than once drew attention to Galileo's statement that the language of science is mathematics. Similarly, Kepler independently recognized that mathematical harmonies existed in the motions of the heavenly bodies, as exemplified by his third law relating the periods and mean radii of the planets in their orbits about the sun (see Chapter 5). For him these harmonies were indicative of a rational structure and design of the universe and were the cause of the phenomena of nature. He also articulated the criterion that:

> [O]f a number of variant hypotheses about the same facts, that one is true which shows why facts, which in the other hypotheses remain unrelated, are as they are, *i. e.*, which demonstrates their orderly and rational mathematical connexion.[21]

According to Kepler, mathematical order was found in the real world. This marked a significant transition from the qualitative descriptions and explanations of Aristotelianism to quantitative properties characterizing nature and expressible in precise mathematical relations. Faith in the truth and reliability of the exact laws of nature was also insisted upon by Galileo. In this scheme, mathematics became another important aid in – but certainly not a tool sufficient on its own for – the discovery of new laws and theories.

By the end of the nineteenth century, classical physics was encountering difficulties of its own (for example, the aether of Chapter 13), so that some began to question the absolute necessity of the attendant mechanistic philosophy. Kirchhoff, who subscribed to a phenomenological methodology, believed that the job of physics was to provide a complete, accurate and simple description of the phenomena of nature, rather than an understanding, either in terms of models or in terms of principles a priori necessary or self-evident to the human mind. Hertz shared a similar view. (On this, see the first quotation for Part IX (*A retrospective*).) Mach explained this idea that physical laws are simply convenient summaries of phenomena and developed a positivistic criterion according to which knowledge of facts must be based solely upon the 'positive' data of sense experience without any extraneous theoretical concepts. In such a positivistic philosophy, all metaphysical concepts that transcend any possible observation or sense experience must be eliminated from our description of nature. It was on this basis that Mach criticized Newton's absolute space (Chapter 11).[22]

About this same time, in the United States, the philosopher Charles Peirce (1839–1914), the founder of pragmatism, independently formulated what can be characterized as a philosophy with positivistic overtones – in the sense that the practical or operational meaning of general theoretical concepts were to be based on observed facts rather than defined in terms of even more general metaphysical principles. In a similar spirit, the important school of logical positivism, or logical empiricism, known as the Vienna Circle, was established in Vienna in the 1920s. This movement was founded by Moritz Schlick and included Rudolph Carnap, Philipp Frank, Kurt Gödel, Carl Hempel (1905–) and Hans Reichenbach.[23] The special and general theories of relativity often served as exemplars of that school's ideology, although Einstein himself rejected much of the positivist doctrine. The resulting theory of scientific methodology held that the logical structure of a theory could be analyzed independently of its specific subject matter, just as the logical structure of a syllogism does not depend on the particulars of its terms (Chapter 1), and that the only meaningful propositions are those based on sense experience, so that metaphysical concepts become untenable.

The French mathematician and theoretical physicist Poincaré took a very different position, however. Although impressed with the criticisms of the philosophical status of Newtonian physics made by Kirchhoff, Hertz and Mach, he felt that the general laws of physics could not be simply direct summaries of observations and data, because the distance from immediate sense perceptions to these abstract laws is so great. Poincaré's writings on this subject were very influential and led to what is termed *conventionalism*. This sees the principles of science as free creations of the human mind that are neither true nor false a priori, but some of which may be useful in organizing the experiences of reality. As an example, the axioms either of Euclidean or of any particular non-Euclidean geometry need not be true in the real world (see Chapters 11 and 18). However, once we establish the convention that a physical light ray travels along a straight line (or a geodesic), then one of these systems becomes relevant to physical phenomena. Similarly, in mechanics Poincaré viewed the law of inertia as a convention or definition stipulating that uniform, straight-line motion was the criterion for force-free motion. (Recall Poincaré's analysis of the logical structure of classical mechanics in Section 7.5.)

> [W]e shall see that the principles of [mechanics], although more directly based on experience, still share the conventional character of the geometrical postulates.[24]

With regard to these hypotheses that are really conventions, Poincaré wrote:

> [S]uch conventions are the result of the unrestricted activity of the mind, which in this domain recognizes no obstacle.... [B]ut let us clearly understand that while these laws are imposed on *our* science, ... they are not imposed on Nature. Are they then arbitrary? No, for if they were, they would not be fertile. Experience leaves us our freedom of choice, but it guides us by helping us to discern the most convenient path to follow.[25]

In Chapter 24 we saw that the French theoretical physicist–philosopher Duhem also believed that the aim of physical theories is to offer a condensed representation of phenomena, rather than to give an explanation of them in terms of 'true' theoretical constructs.

> We have proposed that the aim of physical theory is to become a natural classification, to establish among diverse experimental laws a logical coordination serving as a sort of image and reflection of the true order according to which the realities escaping us are organized.[26]

What we are left with after discussing all of these methodologies, criteria and constraints is a toolbox of instruments, each more or less useful in different circumstances for advancing the scientific enterprise, but none alone or in combination constituting 'the' method of science. Once again, the corrigibility of scientific knowledge becomes evident. It might seem reasonable, then, to ask how we at least have access to scientific knowledge that we can reasonably say is well confirmed.

25.4 A PARADOX OF CONFIRMATION

However, establishing even a reasonable qualitative criterion for the corroboration of a scientific theory is no simple matter. Suppose that one way or another – by intuition, induction, good luck or any combination of these and more – we have formulated a physical theory or hypothesis and that we now wish to find out whether it is supported by physical reality. We deduce logical consequences from the hypothesis and compare these with the data of reality. If these predictions disagree with experience, then we know the hypothesis is incorrect. What do we say, though, if the predictions agree with experience? Our inclination is to claim that, while we have not proven the hypothesis in the strict logical sense of the term, the data do confirm (warrant, support, etc.) the hypothesis. A reasonable criterion would seem to be the following. If the data agree with the predictions, they confirm the hypothesis; if they disagree with the predictions, they disconfirm (actually in that case, disprove) the hypothesis; and, if the hypothesis makes no prediction about a certain class of data, then those data are neutral (or irrelevant) with regard to the hypothesis. On the basis of such a criterion, the more confirming evidence the better, even if, as pointed out previously, we are unable to assign a useful and realistic quantitative measure of this degree of confirmation.

As reasonable as all of this may appear, the contemporary philosopher of science Carl Hempel showed that such a criterion for the confirmation of an hypothesis leads to puzzling results.[27] His example is often referred to as the paradox of the ravens and runs as follows. Suppose we have the proposition that 'All ravens are black'. This has the form $p \Rightarrow q$, where p stands for ravens and q for black. By our confirmation criterion above, if we find a black raven, we have a confirming instance of this; if we find a non-black raven, the hypothesis p cannot be correct; but if we find a non-black, non-raven (say, a green frog), then we have a piece of neutral data. However, $p \Rightarrow q$ is logically equivalent to $\sim q \Rightarrow \sim p$: 'All non-black (things) are non-ravens.' Each of these statements logically implies the other. Now the existence of a green frog is, according to our confirmation criterion, a confirming instance of $\sim q \Rightarrow \sim p$ and, hence, of $p \Rightarrow q$. While there is no strict logical paradox or contradiction in all of this, it certainly does appear highly counterintuitive, since we would not normally agree that the existence of a green frog confirms the hypothesis that all ravens are black. By Hempel's argument we see that a large body of apparently neutral evidence has become confirmatory evidence.

In order to make it clear that all of this is not just a bit of philosophical whimsy, let us recast these considerations in terms of a statement about a physical theory. In Chapter 1 we took the example 'If Newton's law of gravitation governs a planet orbiting the sun, then the orbit of the planet is an ellipse.' Since it is true that there are perturbations due to the effects of the other planets, we restrict our hypothesis here to a single planet orbiting the sun. Then, if we see a planet orbiting the sun and know it should be governed by Newton's law of gravity and are further able to determine that the orbit is indeed an ellipse, then we have a confirming instance of our hypothetical proposition ($p \Rightarrow q$). Similarly, if we know that a certain planet orbiting the sun should be governed by Newton's law of gravitation, but find that its orbit is not elliptical, then we have a piece of disconfirming evidence (p and $\sim q$) for the proposition. So far, there is no problem. However, just as in the paradox of the ravens, $p \Rightarrow q$ is logically equivalent to $\sim q \Rightarrow \sim p$. This proposition says that anything that need not follow an elliptical orbit about the sun (say, for example, a green frog just sitting somewhere in an otherwise empty universe) must also not be a planet orbiting the sun governed by Newton's law of gravitation (the green frog will do well enough again). Of course, many things besides green frogs will do to confirm that $\sim q \Rightarrow \sim p$ and hence, by our confirmation criterion, that $p \Rightarrow q$. But this does seem nonsensical, since science does not operate in such a fashion in attempting to find supporting evidence for a scientific theory.

The lesson to be drawn from all of this is that the confirmation criteria actually used in science are not as simple as the naive one that seemed so apparently reasonable. Hempel's is not the only paradox of confirmation discussed in the philosophy of science[28] and much has been written in an attempt to resolve these questions. The problem of a realistic and generally accepted criterion of confirmation remains unsolved. It is precisely because of the complex nature of scientific practice and advance that some modern philosophers of science have claimed that it is essential to consider not only the objective evidence and theories of science, but also the interaction of the scientist with this evidence and these theories and with other scientists. This school of the sociology of knowledge holds that there is an ineliminable sociological component to the very content or core of scientific knowledge.[29] (Chapter 24 examined this question in the context of quantum mechanics.)

25.5 THE PARADIGM MODEL OF SCIENCE

A large part of the effort of current philosophy of science is devoted to constructing a fairly faithful description or model of what science is and how it has in fact functioned historically. Thereafter, a more optimistic goal might be a normative set of criteria (such as those candidates considered in the previous two sections) whereby a scientific theory being studied could be evaluated against a standard. The former project (of description) could turn out to be doable, even if the more overarching (normative) one were impossible. After all, not all philosophers of science would agree that science does indeed progress toward an ever truer picture of the world. For some, successive theories simply encompass or organize more and more data within a convenient scheme that need not give us an explanation of those data in terms of more fundamental entities in reality. So, in this and the next section we discuss two current models of science that are at least putatively descriptive.

In his very influential schema, Thomas Kuhn (1922–1996) stresses the public and sociological aspects of science in his division of the history of science into periods of normal science separated by relatively rare periods of revolution. During normal periods the overwhelming majority of scientists accept a common paradigm (or picture, theory, etc.), solve problems with it, seek more confirming instances of it and indoctrinate their students with it.

> [T]hree classes of problems – determination of significant fact, matching of facts with theory, and articulation of theory – exhaust, I think, the literature of normal science, both empirical and theoretical.[30]

When a well-established and venerable paradigm repeatedly fails to meet a serious challenge – 'anomalies' in Kuhn's terms – and a sufficiently large fraction of the practicing scientists becomes convinced that the current paradigm is unworkable, science enters a period of turmoil or revolution during which a new paradigm is sought. (We have certainly seen examples of that previously in this book.) Once another successful paradigm has been formulated, science returns to its normal mode of operation. Scientists are forced, against their collective will as it were, to make the transition from one period of normal science to the next.

> As in manufacture so in science – retooling is an extravagance to be reserved for the occasion that demands it. The significance of crises is the indication they provide that an occasion for retooling has arrived.[31]

In Kuhn's model of the working of science, the logic, or context, of discovery and that of justification are not easily, if at all, separable.[32] The process of making scientific discoveries is intimately connected with that of justifying the laws and generalizations thus arrived at, so that one ought not attempt to evaluate the theory abstractly without due consideration for the influence of the scientist and of the scientific community upon the products of science.

> [T]his issue of paradigm choice can never be unequivocally settled by logic and experiment alone....[33]

Kuhn is not alone in stressing the importance of the sociological and psychological aspects of science, in addition to its philosophical one. As just one example, a theoretical physicist with an interest in the structure and working of the modern scientific enterprise has characterized science as follows.

> [T]he goal of science is a consensus of rational opinion over the widest possible field.[34]
>
> ...
>
> [S]cientific knowledge cannot be justified or validated by logic alone.[35]

From a Kuhnian perspective, science is not claimed to progress toward any absolute truth. The language used for communication in producing a consensus is mathematics that is itself exact and unambiguous, but this alone does not guarantee the truth of the message being communicated. Certain present-day philosophers of science feel that science cannot be studied in any meaningful fashion without taking into account as some of its essential components the basic thought patterns available to the human mind[36] and sensory-motor apparatus by which we perceive the external world.[37] Just as in some of the more extreme readings of Bohr's Copenhagen interpretation of quantum mechanics the observer or knower plays an essential role in determining the outcome of an experiment (or the knowledge obtainable), so the workings of the human mind and body may be in principle essential determinants of the possible knowledge that can exist about the external world. It may be effectively impossible to uncouple knowledge from the knower and treat it independently, so that the ideal of abstract, objective knowledge may be illusory.

Of course, there are dissenting views. While it is possible that the knower–knowledge system may be this intimately bound together, other schools in current philosophy of science take science to be an essentially objective, rational process and hold that these irrational (or, perhaps, only nonrational), extra-logical elements are external to the scientific enterprise and can be minimized in a discussion of science and its progress. A major proponent of this view was Karl Popper with his methodological falsification, or falsifiability criterion, that we discussed previously.[38]

25.6 AN ECLECTIC DESCRIPTION OF SCIENCE

So far in this chapter we have given a brief overview of some of the major hallmarks of science, its goals and practice, and illustrated a few of the models by which philosophers attempt to understand science. No definite methodology or set of criteria for theory evaluation emerged from this, because an exclusive and exhaustive set does not seem to exist. Nevertheless, we are able to give a reasonable summary characterization of the scientific enterprise.

Science attempts to reduce the myriad of physical phenomena to a few basic, simple questions or problems and then to construct theories that can explain these basic facts. In physics, an explanation consists of accounting for a set of phenomena in terms of a theory or law. A good theory must be internally consistent (free of logical contradictions), produce results or predictions that are correct (that agree with physical reality) and possess simplicity in its structure and premises. The hypothetico-deductive method (via falsifiability) is centrally important in judging theories. Models and analogies (via induction) play a large role to aid one in guessing useful hypotheses. Scientific progress and the growth of knowledge consist of having ever more general (more abstract and simpler) theories that can account for ever more phenomena in terms of fewer and fewer assumptions. Nevertheless, theories and scientific knowledge remain provisional in character, are influenced by historical and sociological factors, and are always subject to revision. Our theories give a representation of reality in terms of convenient concepts and conventions constructed by the human mind.

We can incorporate key features of several of previous attempts at theories of science into a composite description of how the scientific enterprise seems to function in practice. As a general conceptual matrix, we suggest a three-level, or hierarchical, scheme of the practice, methods and goals of science.[39] Here the lowest level, that of practice, is guided or informed by the accepted methods of science in order to direct the entire process toward the goals of science. Under practice, with its facts, laws and theories, we include the overwhelming bulk of scientific activity that encompasses experimental work and the construction of laws and theories. For many this may seem to be just about the whole of science. Shifts at this level (such as the succession from classical mechanics to relativity or to quantum mechanics) are familiar enough and few, if any, would argue for an absolute stability there. The next level, or layer, is that of methodology or rules by which one judges theories and constructs. For example, the hypothetico-deductive method, induction, generally accepted modes of explanation and rules of inference are found there. Formal logical structure and the canons of rationality – such as what counts as 'good reasons' – are located here as well. This level is relatively more stable than that of practice, but it is not unchangeable. At the highest level, or at the innermost layer, are the goals or aims of science – such as, once upon a time, truth. Throughout this book we encountered illustrations of changes at all three of these levels. (Examples of change or development in the levels of methods and goals are provided – among others discussed in this book – by the 'time line' of philosophers from Plato and Aristotle up through Popper and Quine as illustrated in Section 3.6.)

What is central to this model or representation of science is that none of these levels or components is immune to change, either in principle or in fact.[40] This triad is mutually coupled and evolving. The outer layers of practice are subject to more rapid variation, but these changes also affect the inner core. This coupled network of practice–methods–goals evolves over time under the pressure of constraints from the real world (such as experiment), as well as for external (sociological) factors. In this evolution of the coupled components of our triad, practice is typically the overriding and ultimate determiner of this evolution. That is, new innovations in methodology and in goals are often ultimately rooted in successful scientific practice. This means that no methodological principles (peculiar to science) can be prescribed ahead of time. They must be discovered or recognized in science as it is practiced. Furthermore, the historical record shows that this exercise of abstracting from past (or current) scientific practice does not lead to an invariant (or forever-after fixed) complete characterization of science. Rather, it would seem that the practice, methods and goals of science are contingent, as is any historical entity.[41]

There may not (and there certainly need not) be a single rationality that is the hallmark or essence of all science. However, just because we are unable to give a neat, compact representation of so inherently complex an activity as science does not mean that this evolving network of practice, methods and goals does not have enough coherence to generate useful knowledge. Even a superficial glance over those aspects of the history of physics that we have presented in earlier chapters provides convincing testimony to the enormous power and productivity of this imperfect instrument called science.

Perhaps the better part of wisdom would be to end the book on this note. Instead, in the final section that follows, we give an example of one dominant view of the world that is based on modern science. The problematic aspect of doing this is that we raise several important issues that are left unresolved here. On the other hand, a virtue of such a move may be that it leaves the reader to contemplate a quandary in which modern science arguably leaves us and for which that science may be unable to provide a resolution.

PART IX   **A retrospective**

25.7   ONE MODERN WORLD VIEW BASED ON SCIENCE

In order to be able to present a picture of the world that many people associate with modern science, let us begin by considering just what is meant by the term *world view*. Sigmund Freud in his *New Introductory Lectures on Psychoanalysis* discusses the question of a *Weltanschauung* (that is, a world view) and the relation of science to it.

> 'Weltanschauung' is, I am afraid, a specifically German concept, the translation of which into foreign languages might well raise difficulties. If I try to give you a definition of it, it is bound to seem clumsy to you. In my opinion, then, a *Weltanschauung* is an intellectual construction which solves all the problems of our existence uniformly on the basis of one overriding hypothesis, which, accordingly, leaves no question unanswered and in which everything that interests us finds its fixed place. It will easily be understood that the possession of a *Weltanschauung* of this kind is among the ideal wishes of human beings. Believing in it one can feel secure in life, one can know what to strive for, and how one can deal most expediently with one's emotions and interests.
>
> . . .
>
> But the *Weltanschauung* of science already departs noticeably from our definition. It is true that it too assumes the *uniformity* of the explanation of the universe; but it does so only as a programme, the fulfillment of which is relegated to the future. Apart from this it is marked by negative characteristics, by its limitation to what is at the moment knowable and by its sharp rejection of certain elements that are alien to it. It asserts that there are no sources of knowledge of the universe other than the intellectual working-over of carefully scrutinized observations – in other words, what we call research – and alongside of it no knowledge derived from revelation, intuition or divination. It seems as though this view came very near to being generally recognized in the course of the last few centuries that have passed.[42]

Notice Freud points out that science does not give us an all-embracing and complete world view in the classical sense of that term. We do not get back to the totally integrated fabric of, say, the Aristotelian universe. The main reason for this is that fundamental to all scientific enterprises is a belief in what Jacques Monod termed the postulate of objectivity. By this he means that nature is *objective* (obeying definite laws that are not dependent on any final causes or 'purpose'), not *projective* (evolving in accord with a plan or toward a set goal, as preconceived by us or by some other intelligence), and that the only valid avenue to knowledge is an objective confrontation of nature.[43] A strict adherence to this postulate (itself an act of faith) logically precludes an unquestioning acceptance of many other beliefs that presume a particular explanation for certain questions. Monod argues that we cannot gain true knowledge about nature by attempting to interpret phenomena in terms of final causes or of a teleonomic principle (as, for example, did Aristotle). The crux of the matter is that there is a tension between an a priori commitment to an overarching religious or philosophical conceptual matrix and a simultaneous commitment to open intellectual inquiry – as illustrated in Chapter 10 with the trial of Galileo. One can hope that the two are compatible, but that in itself is an act of faith, not an argument.[44]

We turn now to a brief description of one scientific world view found in Monod's modern work *Chance and Necessity*. We begin by characterizing three terms that he uses.

> *postulate of objectivity* –the systematic denial that 'true' knowledge can be obtained by interpreting phenomena in terms of final causes – or, of 'purpose'. That is, objective knowledge is provisional and can be obtained only through science, not through metaphysics or religion.
>
> *vitalist theories* –those that imply a radical distinction between living beings and the inanimate world.
>
> *animist theories* –those that posit a universal teleonomic principle (that is, that all beings are endowed with a purpose or project and hence act for an end) responsible for the course of affairs throughout the cosmos as well as within the sphere of living things.

In Monod's representation, vitalist theories (or philosophical systems) are those that postulate a different force or set of laws governing living beings – laws over and above the ones for inanimate things. In light of the advances of modern science, he discards such systems and turns his attention to the much more prevalent animist systems that, while they postulate no special laws peculiar to the sphere of life, do assume that there exists a purposeful force or principle that guides all activity and development toward a goal or end. It is this goal-directedness, or teleonomic principle, that Monod takes as providing the basis for an explanation in terms of purpose and for moral values in the traditional religious and philosophical systems. As we saw earlier in this chapter, his explanation of the apparent goal-directedness in living organisms is that chance variations get locked into the invariant replication of biological systems and those that survive environmental selection continue to be reproduced, giving rise to a sense of purposeful development or necessity (growth). He also feels that science, having as its only guide objective knowledge, must deny this animist tradition since it assumes an answer or system of values, rather than constructing these on the basis of 'true' (for him, scientific) knowledge. A basic conflict for us today is the following. Monod believes that we have a biological or genetic need (as the result of evolution) for the all-encompassing explanations provided by animist theories. Science cannot yield these since its knowledge is always provisional, rather than being based on or guaranteed by absolute first principles. (Recall Freud's discussion of *Weltanschauung* above.) The dilemma for us is that we accept scientific practice (in our minds) because of its success in producing new goods and technologies, but we do not accept (in our hearts) the more profound philosophical implications of a rejection of the animist tradition. However, once this practice is accepted, the basic assumption of science (the postulate of objectivity) in fact produces an evolution of thought in the direction of there being just one source of truth. Monod sees traditional philosophical and religious systems as necessarily hostile to this core of science. This has created a love–hate relationship toward science for many people today. We feel isolated in our ultimate responsibility for making our own choices and creating a purpose for existence.⁴⁵ Monod's sense of *angst* is quite similar to that expressed by Bertrand Russell in the quotation for Part IV (*A perspective*).

Here we have a tension: we see as a material – perhaps even as intellectual – good (the fruits of science and technology) something that is in conflict with our psychological or spiritual needs. A society can feel trapped between the two, unable effectively to choose one over the other. As examples, we mention two episodes in fairly recent history in which there was an attempt to turn away from science and technology. Mahatma Gandhi (1869–1948) insisted that Indians hand weave their own cloth, rather than use the output of British mills, because he appreciated that one culture cannot, in practice, take over just one dominant characteristic or product (here, technology) from a rival and radically different culture (the West) and still avoid being invaded by other, undesirable aspects (such as materialism). Science and technology have their own inescapable wider influences on any culture that accepts them. Similarly, in Germany during the Weimar Republic after the crushing defeat suffered in World War I, there was a reaction against science (the failure of which was seen by many as the reason for Germany's defeat) and a rise of irrationalism. The teaching of mathematics and science was reduced and replaced by a return to the traditional emphasis on 'cultural' education (as in the USA today?). This resulted from a strongly felt sense of disappointment with a society supposedly largely based on science.

And yet, in Weimar Germany, in Gandhi's India and in the West today, science and technology generate their own apparently irresistible momentum for 'progress'. It is by no means evident that such 'progress' is necessarily beneficial for mankind. (For example, are we better off for the acceleration, caused by the widespread use of computers, of the already hectic pace of modern life?) Nevertheless, an individual, company or nation simply cannot afford not to join the rush, lest it be swept aside into a backwater. This is not meant to be a statement supporting any type of technological determinism, but rather a reflection of a felt 'psychological' or 'social' compulsion – given our present technologically based societies in the West and ever more in much of the rest of the world. There is certainly no reason in principle that we could not make other choices. However, it remains to be seen whether or not we can collectively resist the lure of the availability of more and more material goods. One difficulty is that the long-range effects – for net gain or loss – of technological advances are often not foreseeable before those consequences are virtually irreversibly upon us.

We offer this world view based on physical science as yet another illustration of the long chain of interaction between philosophy and science – a chain reaching back to antiquity. There is little reason to expect that Monod's scheme represents the end of the chain.⁴⁶

FURTHER READING

Edwin Burtt's well known *The Metaphysical Foundations of Modern Physical Science* is a nice, if somewhat dated, background piece on the broad philosophical underpinnings of modern physics. The commentary and selected readings in Joseph Kockelmans' *Philosophy of Science* provide historical background from the late sixteenth century (with Kant) to the early twentieth century (with Whitehead and Bridgman). Albert Einstein's essays 'Principles of Research' and 'Physics and Reality' are beautifully, at times almost poetically, written, philosophically sensitive statements about the nature and goals of science.[47] Arthur Fine's 'Einstein's Realism', that is Chapter 6 of *The Shaky Game*, gives a careful and insightful discussion of Einstein's views on the relation of our physical theories to the world they represent. Janet Kourany's *Scientific Knowledge* is a very useful collection of essays on basic issues in the current philosophy of science. Thomas Kuhn's *The Structure of Scientific Revolutions* has remained the most influential single book in the philosophy of science for decades and is essential reading for some understanding of the sociological turn that much modern scholarship in the field has taken. A direct descendant of Karl Popper's falsification methodology and one that attempts in that tradition to emphasize the rationality of science is Imre Lakatos' methodology of scientific research programs (MSRP) as ably set forth in useful essays and case studies in Imre Lakatos' 'Falsification and the Methodology of Scientific Research Programmes,' 'History of Science and Its Rational Reconstructions' and *The Methodology of Scientific Research Programmes*, in Imre Lakatos and Alan Musgrave's *Criticism and the Growth of Knowledge*, in Colin Howson's *Method and Appraisal in the Physical Sciences*, and in Gerard Radnitzky and Gunnar Andersson's *Progress and Rationality in Science*. Larry Laudan's *Science and Values* criticizes both the Kuhnian and the Popper–Lakatos schools in the philosophy of science and offers his own 'reticulated' model of scientific practice (that we outlined in Section 25.6). Erwin Schrödinger's *What Is Life?* is an early and influential attempt to account for life on the basis of the laws of physics alone. Jacques Monod's *Chance and Necessity*, written with Gallic verve, makes a rational, and at the same time an impassioned, case for a purely physical explanation of life. For a modern defense of a Darwinism that claims a universe with no basic design, see Richard Dawkins' *The Blind Watchmaker*. Christian De Duve's *Vital Dust* forcefully argues that life is a necessary and inescapable result of the basic physical laws of nature, yet curiously eschews Monod's cold and atheistic view of the universe. Finally, Gerald Holton's *Einstein, History and Other Passions* is a collection of papers on science and on its relation to the late twentieth-century world.

Notes

The citation abbreviations DSB, GB and M are explained in the *General References* section of the book.

PART I

Jevons 1958, 228.
Einstein 1954b, 271–2.
Popper 1965, 278, 280.
Quine 1990, 1, 5, 12, 13.

CHAPTER 1

1. Jefferson 1952, 1. (GB **43**, 1)
2. Aristotle 1942b, Book I, Chapter 1, 184^a (10–24). (GB **8**, 259)
3. Aristotle 1942a, Book I, Chapter 1, 24^b (18–21). (GB **8**, 39)
4. Of course, the philosophically meticulous will (quite correctly) point out that the type of necessity involved here is moral (or ethical), not logical, as in, say, 'If a figure is a circle of radius R, then its area is πR^2.' In this latter example, the conclusion follows by logical deduction from the very properties of a circle. Similarly, two paragraphs further down in the text, our illustration employing planets and gravity involves physical necessity, but arguably not logical necessity. Having now entered these caveats, the point remains that once one buys into some framework of constraints (be they logical, physical or moral) as represented by a hypothetical syllogism, certain implications follow. That is the point of all of these examples.
5. Descartes 1977a, 9. (GB **31**, 5)
6. Herschel 1830, 80.
7. Planck 1949, 84–5.
8. Holton 1978, Chapter 2.
9. White 1934, vii–viii.
10. For a dramatic illustration of the essential role played by historical contingency in the evolution of early life forms, see Gould (1989).
11. In this book we attempt to respect informally a distinction between the terms 'law' and 'theory'. The former usually refers to some statement of (often hopefully universal) regularity governing a class of phenomena – such as Boyle's law for (ideal) gases, $PV = const$ – whereas the latter typically denotes a larger framework within which laws are situated and explained – such as statistical mechanics.
12. Descartes 1977a, Rules II and III, 3–8. (GB **31**, 2–4)

CHAPTER 2

1. Rose 1886, 449.
2. Plato 1892, [30] 450. (GB **7**, 448)
3. Aristotle 1942c, Book III, Chapter 2, 300^a (20–31), 301^b (17–23). (GB **8**, 391, 393)
4. Here and later we use 'a priori', usually in a pejorative sense, to refer to arguments or reasoning based on unquestioned, allegedly self-evident or purportedly logically necessary premises.
5. Aristotle 1942b, Book IV, Chapter 8, 216^a (14–16). (GB **8**, 295)
6. From a modern perspective, consider the motion of a body that starts from rest and is acted upon by a constant force F as it covers a distance x in a time t. Then, if v_f is the final velocity, $\bar{v} = (v_f/2)t = (F/2m)t$ and $x = \bar{v}t$. Here the average velocity is proportional to the force.
7. Dugas 1988, 21.
8. Cooper 1935, 14, 20, 36.
9. Lucretius 1952, Book II, par. 225, 18. (GB **12**, 18)
10. Aristotle 1942b, Book IV, Chapter 8, 215^a (19–20). (GB **8**, 294)
11. Dijksterhuis 1986, 39.

12. Bacon 1952, Book I, Aphorism 95, 126. (GB **30**, 126) In addition, see Aphorisms 1, 14, 19, 22, 26, 45, 46, 73 and 104 for Bacon's view of what the method of science ought to be.
13. Aristotle 1942c, Book III, Chapter 2, 301^a (24–34). (GB **8**, 392)
14. Aristotle 1942c, Book IV, Chapter 1, 308^a (29–33), and Chapter 2, 309^b (11–15). (GB **8**, 399, 401)
15. Aristotle 1942c, Book I, Chapter 6, 273^b (31)–274^a (2). (GB **8**, 365)
16. Aristotle 1942c, Book I, Chapter 8, 277^a (28–30). (GB **8**, 368)
17. Aristotle 1942b, Book IV, Chapter 8, 215^a (29–30), 215^b (1–12), 216^a (11–20). (GB **8**, 295)
18. Recall (Section 2.3) that for Aristotle the weight of a body derived from proximity to its natural place.

CHAPTER 3

1. Harvey 1952, 319. (GB **28**, 319)
2. Newton 1934, Book III, Rule IV, 400. (GB **34**, 271)
3. Planck 1949, 33–4; see also 105–8.
4. Watkins 1978, 23–43, especially 24–5.
5. Hume 1902, 73–7. (GB **35**, 476–7)
6. Mill 1855, Book III, Chapter II, 171–2; Chapter III, 183–5.
7. Popper 1968, 33–7.

PART II

Galilei 1946, Third Day, 154, 155, 160. (GB **28**, 200, 202)
Galilei 1967, 33–4, 193–4.

CHAPTER 4

1. Our account in this and the next chapter is a sketch of a few aspects of the classic study by Kuhn (1957) that should be consulted for more details on the topics covered here.
2. Eratosthenes knew the arc length AB of Figure 4.2 to be 5,000 stadia (by measurement) (see Heath (1981b, Vol. II, p. 107)). This would make the circumference of the earth $50 \times 5,000 = 250,000$ stades (or *stadis*). Unfortunately, it remains somewhat uncertain how many feet are in a stadium. For example, Pliny (23–79 A.D.) gives an estimate equivalent to 516.73 ft [157.50 m]. On this basis, Eratosthenes' value for the circumference of the earth turns out to be 24,466 miles [3.93×10^4 km].
3. Archimedes 1897, 221–2. (GB **11**, 520)
4. See Kuhn (1957, 52–3) for Vitruvius' (first century B.C.) argument by analogy for this belief.
5. Kuhn 1957, 80 *ff.*
6. Ptolemy 1952, Book I.2, 7. (GB **16**, 7)
7. Ptolemy 1952, Book III.3, 86–7. (GB **16**, 86–7)
8. Ptolemy 1952, Book IX.5, IX.6, 291, 293–4. (GB **16**, 291, 293–4)
9. Lakatos and Zahar 1978.
10. Plato 1892, [33–4] 452–3. (GB **7**, 448–9)
11. Quoted in Duhem (1969, 5).
12. Ptolemy 1952, Book IX.2, 270. (GB **16**, 270)
13. Gardner 1983.
14. Copernicus 1952, Introduction, 505–6. (GB **16**, 505–6)
15. See 'Aristarchus on the Sizes and Distances of the Sun and Moon' in Heath (1981a, 351–414, and the commentary on 328–36). We follow the discussion given there and in Kuhn (1957, 274–8).
16. Of course, at the time of Aristarchus, trigonometry had not been invented (see Heath 1981a, 328). A later astronomer Hipparchus was an important figure in the development of this field of mathematics and is the first person whose use of trigonometry is supported by documentary evidence (see Heath 1981b, Vol. II, 257; Cohen and Drabkin 1948, 82–5). Aristarchus' actual argument consisted of placing limits on the ratio R_e/R_m as follows (see Heath 1981a, 330–1, 353). In Figure 4.8, let us label as β the acute angle that is the complement of α. Since Aristarchus took α to be 'less than a quadrant [that is, 90°] by one-thirtieth of a quadrant' (Heath 1981a, 353), or $\alpha = 90° - 3° = 87°$, it follows that $\beta = 3°$. If s denotes the length of the circular arc (on a circle of radius R_e) from the earth to the moon, then $s = R_e\beta = R_e(0.0524)$, where β is given in radians. If we take $s \approx R_m$ (that is, if we approximate the chord length R_m by the arc length s), then we find $R_e/R_m = 19.1$. The value obtained by using trigonometric functions is $R_m = 2R_e\sin(\beta/2) = 0.0524\,R_e$, that yields the same numerical ratio for R_e/R_m. In fact, Aristarchus was able to limit this ratio as lying between 18 and 20.
17. These equalities are approximate because the

Notes

vertical sides of the intermediate and larger triangles of Figure 4.10 are not quite r_e and r_s, respectively (since these radii are perpendicular to the *tangents* to the circles representing the earth and sun). See also Heath (1981a, 330, especially Figure 14), Copernicus (1952, figure on p. 711) and Armitage (1938, 128, Figure 23).

CHAPTER 5

1. Copernicus 1952, Introduction, 507. (GB **16**, 507)
2. Copernicus 1959, 57.
3. Copernicus 1952, Book I.8, 519. (GB **16**, 519)
4. Copernicus 1952, Book I.4, 513–4. (GB **16** 513–14)
5. Copernicus 1952, Book I.9, 521. (GB **16**, 521)
6. A diagram much like our Figure 5.3 appears in Copernicus (1952, 778). There he gives the value $\alpha \approx 45°$ for Venus.
7. Our Figure 5.4 is adapted from Figure 36(b) of Kuhn (1957, 176). For Mars, the argument corresponding to our Figure 5.4 appears in Copernicus (1952, 775–7). On this complex method of Copernicus', see Armitage (1938, 135–46).
8. Copernicus 1952, Book I.10, 526. (GB **16**, 526) Our Figure 5.5 is adapted from the corresponding figure in Copernicus (1952, 526).
9. Neugebauer 1968, especially 92–6.
10. Kepler 1937, Pt. II, Sec. 19, 178; 1992, 286; also quoted in Koestler (1959, 322).
11. Kepler 1937, Pt. IV, Sec. 58, 366; 1992, 575.
12. Kepler 1952a, Book V.3, 975. (GB **16**, 975)
13. Kepler 1952a, Book V.4, 983. (GB **16**, 983) Our Figure 5.9 is adapted from the corresponding figure in Kepler (1952a, 983).
14. Kepler 1952b, Book V.3, 1020. (GB **16**, 1020)
15. Heath 1981b, Vol. I, 244.

CHAPTER 6

1. This is to paraphrase Lakatos (1970, 133).
2. Cooper 1935, 47; Cohen and Drabkin 1948, 220.
3. Cohen and Drabkin 1948, 222–223.
4. Galilei 1946, Third Day, 147–8. (GB **28**, 197)
5. Galilei 1946, Third Day, 162. (GB **28**, 203)
6. It would seem at first sight that $\Delta v \propto \Delta x$ is equally as simple as $\Delta v \propto \Delta t$. However, Galileo argues that, for free fall starting from rest, $v_f \propto x$ leads to a contradiction. In modern notation we can easily see why, since that assumption would imply that $dv/dt \propto dx/dt = v$ and this, in turn, would mean that no continuous motion could ensue from an initial state in which $v = 0$. See Galilei (1946, 161). (GB **16**, 203)
7. Galilei 1946, Third Day, 154–5. (GB **28**, 200)
8. There has been a good deal of controversy among historians of science about which of these inclined-plane experiments, if any at all, Galileo actually performed. The manuscript evidence now tends to support the contention that he did perform this often-quoted one described in Galilei (1946, Third Day, 171–2. (GB **28**, 208)). See Segre (1980, especially 242–4).
9. McMullin 1967b, 27–31.
10. Descartes 1905, 62–4; 1977b, 267 (titles of propositions only).
11. Blackwell 1977, 574.
12. Franklin 1976.
13. Galilei 1946, First Day, 62. (GB **28**, 158)
14. Cooper 1935, 51–2.
15. Feinberg 1965. Here we summarize the calculations contained in this reference.
16. Adler and Coulter 1978; Casper 1977.
17. Feinberg 1965.
18. Casper 1977.
19. Adler and Coulter 1978.
20. Galilei 1946, First Day, 60–1. (GB **28**, 157–8)
21. Galilei 1946, Third Day, 234–5. (GB **28**, 238)
22. Galilei 1946, Third Day, 236–7. (GB **28**, 239) Our Figure 6.6 is adapted from Figure 106 of Galilei (1946, 236).

PART III

Galilei 1967, 234.
Newton 1934, Book I, Law I, Law II, 13; Book III, General *Scholium*, 546–7. (GB **34**, 14, 371)

CHAPTER 7

1. A photograph of this portrait can be found in Westfall (1980b, 482).
2. Westfall 1962, 172.
3. Westfall 1980a, 109.
4. The summary chronology given here of Newton's early mathematical work is based on Whiteside (1967–1976, Introductions to Vols. I–III) and on Cohen (1970, 45–53).
5. This is the so-called *brachistochrone* problem

of finding the shape of a smooth curve between any two points (at different heights and both in the same vertical plane) along which a body acted on only by gravity will descend in the shortest time. (For the technical details of a solution in modern notation, see, for example, Cushing (1975, 240–1).)

6. More 1934, 475.
7. Manuel 1968, 388–9.
8. Bacon 1952, Book I, Aphorisms 104 and 106, 128. (GB **30**, 128)
9. Newton 1934, Preface to the first edition, xvii–xviii. (GB **34**, 1–2)
10. Newton 1934, Book III, 398–400. (GB **34**, 270–1)
11. Newton 1934, Book III, General *Scholium*, 547. (GB **34**, 371) (M, 93)
12. Newton 1952, Book III, Pt. 1, Query 28, 528. (GB **34**, 528)
13. Newton 1934, Book I, Prop. 69, *Scholium*, 192. (GB **34**, 131)
14. It is easy for a modern reader to equate Newton's method of fluxions with differential calculus as we learn it today. We are used to taking limits, such as $\Delta x \to 0$. Leibniz worked with such infinitely small quantities, while for Newton there existed just one infinitesimal increment o ($\sim \Delta x$) that was not taken to the limit zero. I thank Dr. Whiteside for pointing this out to me.
15. Quoted in Cohen (1971, 295). Although the form of the *Principia* may appear to be that of classical Greek geometry, the geometry actually used often employs limit-increment arguments that were foreign to 'classical' geometry. Thus, the dichotomy sometimes claimed between Newton's analysis of discovery and synthesis of presentation is not that between calculus and classical geometry nor is it a sharp one since many of the techniques of analysis and discovery were synthesized into his presentation.
16. Cohen 1971, 295.
17. Cohen 1971, 296. However, in spite of these statements by Newton, no manuscript evidence exists to support the claim that inverse fluxions were used in first proving these two laws of Kepler's (Cohen 1971, 79–81). Newton's purpose seems to have been to establish a claim prior to Leibniz's on the discovery of the calculus.
18. In 1691 Newton wrote a work with a similar title, *De Quadratura Curvarum*. This was a more general treatise than the *Tractatus* appended to the *Optics*.
19. Newton 1934, Book I, Def. I, 1. (GB **34**, 5) (M, 31)
20. Newton 1934, Book I, Def. II, 1. (GB **34**, 5) (M, 32)
21. Newton 1934, Book I, Law I, 13. (GB **34**, 14) (M, 36)
22. Newton 1934, Book I, Law II, 13. (GB **34**, 14) (M, 37)
23. Truesdell 1960–2, 23.
24. Newton 1934, Book I, Corollaries I and II, following Law III in Book I, 15. (GB **34**, 15) (M, 38) Our Figure 7.1 is adapted from the corresponding figure of Newton (1934, 15).
25. Newton 1934, Book I, Law III, 13. (GB **34**, 14) (M, 37)
26. Newton 1934, Book I, discussion following Law III, 13–14. (GB **34**, 14) (M, 37)
27. Newton 1934, Book I, 26. (GB **34**, 22–3)
28. Mach 1960, 303.
29. Poincaré 1952, 110.

CHAPTER 8

1. We mean this to apply mainly to the early propositions of Book III where Newton used a chain of inductive and deductive reasoning.
2. Newton 1934, Book III, Phenomena I–VI, 401–5. (GB **34**, 272–5). Note that the 'circumjovial planets' and the 'circumsaturnial planets' are what we would refer to as 'moons'.
3. Newton 1934, Book III, Props. 1–3, 406. (GB **34**, 276)
4. Westfall 1971, 358.
5. Huygens 1934, 366. (M, 28)
6. Jammer 1961, 62.
7. Jammer 1957, 109–10.
8. Jammer 1961, 62.
9. More 1934, 290, 294.
10. Newton 1934, Book I, Prop. 4, 45. (GB **34**, 35)
11. Westfall 1980b, 148–51; Herivel 1965, 7–12; Westfall 1971, 343, 353–5.
12. Westfall, 1980b, 148–51.
13. Newton 1934, Book I, Prop. 4, *Scholium*, 47. (GB **34**, 37)
14. Newton 1934, Book III, Prop. 4, 407. (GB **34**, 277)
15. Newton 1934, Book III, discussion following Prop. 4, 408. (GB **34**, 277–8)

381

Notes

16. Newton 1934, Book III, Prop. 7, 414–15. (GB **34**, 281–2)
17. Newton 1934, Book III, General *Scholium*, 546–7. (GB **34**, 371)
18. Letter to Bentley in Thayer (1953, 54).
19. Cohen 1970, 62–4.
20. Whiteside 1964.
21. Newton 1934, Book III, discussion following Prop. 8, 415–16. (GB **34**, 282)
22. Cohen 1970, 61; Westfall 1971, 461.
23. Cohen 1970, 62–4; Westfall 1971, 357–9.
24. Cohen 1970, 62; Westfall 1971, 424; Cohen 1981.
25. Newton 1934, Book I, Prop. 70, 193. (GB **34**, 131)
26. A geometrical proof of this result is given in Cushing (1982, 625).
27. Cushing 1982, 625–6.
28. Newton 1934, Book I, Prop. 71, 193. (GB **34**, 131)
29. Newton 1934, Book III, Prop. 8, 415. (GB **34**, 282)
30. Newton 1934, Book I, Prop. 73, 196. (GB **34**, 133)
31. Newton 1934, Book I, discussion following Def. 1, 1. (GB **34**, 5)
32. Newton 1934, Book II, Prop. 24, 303. (GB **34**, 203)
33. In Eq. (8.6) we have supplied the proportionality constant 2π. This follows directly from solving the dynamical equation $\ddot{\theta} = -(w/m_i\ell)\theta$ for small angular displacements θ of the pendulum about the vertical.
34. Newton 1934, Book III, discussion following Prop. 6, 411. (GB **34**, 279)

CHAPTER 9

1. Newton 1934, Book III, Prop. 13, 420. (GB **34**, 286)
2. Newton 1934, Book I, Prop. 1, 40. (GB **34**, 32)
3. Newton 1934, Book I, Prop. 2, 42. (GB **34**, 34)
4. Details of these geometric arguments in modern notation can be found in Cushing (1982). Our Figures 9.4 and 9.5 are adapted, respectively, from Figures 2 and 4 of Cushing (1982, 621, 622).
5. The concept of curvature was not a new one even in Newton's day, since it can be found in the *Conics* of Apollonius of Perga (262–200 B.C.) (see Heath 1896, Book V, Props. 51 and 52, 168–78).
6. Huygens 1920, 387–405. Although Huygens formulated his criterion for curvature in 1659, he did not publish it until 1673 in his *Horologium Oscillatorium* (see Huygens 1934, Part III, Prop. II, 225–42.).
7. Westfall 1980b, 111–12; Whiteside 1967–76, Vol. I, 146, 289–92, Vol. III, 152. In modern notation this radius of curvature ρ can be expressed as $\rho = (1 + z^2)^{3/2}/|\Delta z/\Delta x|$, where $z = \Delta y/\Delta x$ is the tangent to the curve.
8. The expression for ρ in the previous Note and Eq. (9.2) lead to Eq. (9.3). See Cushing (1982, 621–2) for a derivation of this result.
9. Newton 1934, Book I, Prop. 11, 56. (GB **34**, 42)
10. Newton 1934, Book I, Prop. 17, 65. (GB **34**, 48)
11. Cushing 1982, 623.
12. For a very different perspective on the significance of this corollary, see Weinstock (1982).
13. Newton 1934, Book I, Corollary to Prop. 13, 61. (GB **34**, 46)
14. The overall logical structure of this argument of ours from Proposition 11 on can be summarized as follows: (i) $F = ma$ governs all dynamics; (ii) $r = r(t)$ is a conic $\Rightarrow f(r) \propto 1/r^2$; (iii) for a fixed (and given) force center, $(r_0, \theta_0, \alpha_0, \rho_0) \Rightarrow$ a unique conic; (iv) $(r_0, v_0, \lambda$ $[f(r) = -\lambda/r^2]) \Rightarrow$ a unique conic that satisfies $F = ma$ for all times; (v) for any values of r_0, v_0 and λ, a conic section solution has been exhibited; (vi) therefore, a $1/r^2$ force always has a conic section for an orbit. In fact, once a solution has been exhibited, one could even make the plausible assumption that the physics of the situation will do the rest (recall Figure 9.3 and the accompanying discussion at the beginning of this section); (vii) r_0 and v_0 uniquely specify a solution to $F = ma$. This technique of first exhibiting a solution and then arguing that it is unique is quite familiar in physics and in mathematics.
15. Newton 1934, Book I, Prop. 42, 133. (GB **34**, 91)
16. Newton 1934, Book III, Prop. 2, 406. (GB **34**, 276)
17. Newton 1934, Book I, Prop. 15, 62. (GB **34**, 46)
18. See Cushing (1982, 624) for the necessary geometrical details.
19. For a particularly economical derivation of Kepler's three laws from Newton's, see Vogt (1996).

20. Moulton 1914, 431–2.
21. Pedersen 1993, 391.
22. Pedersen 1993, 381.
23. Melchior 1966, 2.
24. Pedersen 1993, 316.
25. For Galileo's views on the ocean tides, see Finocchiaro (1989, 119–33).
26. Galilei 1967, 416–17.
27. Kepler 1992, 55–6.
28. Galilei 1967, 419–20.
29. Galilei 1967, 424.
30. Galilei 1967, 426.
31. Galilei 1967, 432.
32. See also Finocchiaro (1989, 127–9).
33. Newton 1934, Book III, Props. 24 and 37, 435–40 and 479–81, respectively. (GB **34**, 296–99, 324–6)
34. Newton 1934, Book I, Prop. 66, 173–89. (GB **34**, 118–28)
35. This follows since $x = [m/(M + m)]/R_m$, $M = 5.98 \times 10^{24}$ kg, $m = 7.34 \times 10^{22}$ kg, $R_m = 60\, r_e$, where we have used data from Table 5.2.
36. That is, $\tan\phi = r_e/x = (m_0 g - N_4)/(Gmm_0/R'^2)$. With use of Eq. (9.9), we obtain Eq. (9.12).
37. For simplicity we have neglected the tidal effects of the sun on the earth's tides, even though these can be appreciable.
38. For the identification of Leuconia as the Philippines, see Quirino (1963, 69). I thank Ms Sarah Tyache of the Map Room in the British Museum for this reference.
39. Newton 1934, Book III, Prop. 24, 439–40. (GB **34**, 298–9)
40. Newton 1934, *The System of the World in Mathematical Principles*, 587.
41. Peacock 1855, 202–3, 170. (Whittaker 1973 Vol. I, 101–2) On the famous double-slit (or double-pinhole) experiment, see Young (1845, 364–5). (M, 310–11).

PART IV

Russell 1917, 47–8.

CHAPTER 10

1. For a more complete chronology of these events, see Finocchiaro (1989, 297–308).
2. Quoted in Fahie (1903, 150).
3. Quoted in Drake (1957, 162–4).
4. Quoted in Drake (1957, 166–7).
5. Ptolemy 1952, Book XIII.2, 429. (GB **16**, 429)
6. Drake 1957, 272.
7. *The Jerusalem Bible* 1966, 286–7.
8. Drake 1957, 211–12.
9. Drake 1957, 212–13.
10. Quoted in de Santillana (1955, 126).
11. De Santillana 1955, 162–5; Drake 1967, 62; Galilei 1967, 491.
12. Galilei 1967, 464.
13. Westfall 1989, 43–52, 74.
14. Westfall 1989.
15. Westfall 1989, 59.
16. Whitehead 1967, 12–13.
17. Drake 1957, 175, 177, 179–80, 181, 183–4, 186, 188–9, 193–4, 206–7.

CHAPTER 11

1. Rohrlich and Hardin 1983, 604–5.
2. Smith 1974, 276.
3. Donne 1959, Devotion 17, 108.
4. Lovejoy 1936, 38, 46, 49–50.
5. Drake 1957, 237–8.
6. Quoted in Thayer (1953, 6–7).
7. Quoted in Thayer (1953, 7–8).
8. The actual observations made by Roemer involved the eclipses of the satellites of Jupiter and were not as simple as Figure 11.1 would imply, but the basic principle was the same. See Huygens (1952, Chapter I, 556–7) for details. (GB **34**, 556–7) (M, 335–7)
9. Lucretius 1952, Book I, par. 958, 12–13. (GB **12**, 12–13)
10. Aristotle 1942b, Book IV, Chapter 1, 208[b] (9–12). (GB **8**, 287)
11. Descartes 1977b, Pt. II, Principle IV, 255.
12. Descartes 1977b, Principle XVI, 262.
13. *The Jerusalem Bible* 1966, Psalm 139, 922.
14. Quoted in Koyré (1957, 148).
15. Newton 1934, Preface to the first edition, xvii. (GB **34**, 1)
16. Newton 1934, Book I, *Scholium*, 6–7, 10. (GB **34**, 8–11) (M, 33–5)
17. Newton 1934, Book I, *Scholium*, 10–11. (GB **34**, 11–12) (M, 35–6)
18. One could also 'extend' Newton's argument as follows. Once the water is rotating, suddenly stop the bucket and let the water continue to rotate. In the beginning of the experiment, when the bucket is moving relative to the water with its surface still flat, and now at the end, when the water with its concave surface is

383

Notes

moving relative to the bucket, the physical situations are different (since the shape of the water's surface has changed), even though the relative rotational velocities are the same (or, actually, merely reversed in direction). This would indicate that the circular motion was absolute, and not simply relative, motion. See Jammer (1969, 107–8).

19. Mach 1960, 284; see also 279.
20. Newton 1934, Book III, Hypothesis I, 419. (GB **34**, 285)
21. Newton 1952, Book III, Pt. 1, Query 28, 529. (GB **34**, 529)
22. Newton 1952, Book III, Pt. 1, Query 31, 542. (GB **34**, 542)
23. Thayer 1953, 46.
24. On the history of this concept of space as a principle of individuation, see Howard (1997).
25. See Einstein's discussion of these two views of space in Jammer (1969, xiii–xiv).
26. Galilei 1967, *Foreword* (A. Einstein), xiii.
27. Bloch 1976, 27.

CHAPTER 12

1. It is not our intention here to imply that there is any simple line of historical developments that leads compellingly from the early philosophers of Miletus to the atomists (Leucippus and Democritus). Rather, the point is that in retrospect one can see a certain consonance in some of these related concepts.
2. Newton 1934, Book III, Rule I, 398. (GB **34**, 270)
3. Meyerson 1930.
4. Newton 1952, Book III, Pt. 1, Query 28, 529. (GB **34**, 529) (My italics.)
5. Newton 1952, Book III, Pt. 1, Query 31, 542. (GB **34**, 542) (My italics.)
6. Laplace 1902, 3–4.
7. Quoted in Newton (1934, Appendix, 677); Laplace 1886, vi–vii.
8. Quoted in Newton (1934, Appendix, 677).
9. This material on the central-force problem and simple perturbations therefrom is standard fare in intermediate textbooks on classical mechanics. See, for example, Becker (1954, 237–9).
10. Several examples of equations governing such systems can be found in Schuster (1988, 7, 17), in Rasband (1990, 4) and in Guckenheimer and Holmes (1983, 67, 82, 92).
11. The technical books on chaos cited at the end of this chapter contain detailed material on the examples mentioned here. See Briggs (1987) for a discussion of laboratory demonstrations of chaotic behavior for easily constructed systems.
12. Technically, this is a separation of trajectories in *phase space* (as opposed to ordinary space), a term to be defined in Section 12.5.
13. As an illustration of how this connection between properties of solutions to differential equations and discrete maps can come about, consider a set of differential equations that determine a trajectory $[x(t), y(t), z(t)]$ that moves about in a bounded region in three-dimensional space. Now pass a plane through this region and ask for the intersection of the system trajectory with this plane. If the motion of the system point is such that the trajectory continues to pierce this plane repeatedly as time goes on, then we can treat these successive points of intersection (or of recurrence) as a map. If (u_n, v_n) are the coordinates of this point of intersection of the system trajectory on our plane on the nth recurrence, then we can write the coordinates of the next recurrence as the map $(u_{n+1}, v_{n+1}) = f(u_n, v_n)$. The long-term behavior of this map gives us information about long-term behavior of the system trajectory.
14. Figures 12.2, 12.3 and 12.4 are adapted from Figures 2, 4 and 5, respectively, of Jensen (1987, 170, 174, 175), although this material on the logistic map is fairly common and considerably older, as indicated in Schuster (1988, 4). See also Rasband (1990, Chapter 2).
15. That is, the maximum value of y in Eq. (12.5) occurs at $x = \frac{1}{2}$ for any value of a, as is evident from Figure 12.2. This maximum value is just $a/4$.
16. Jensen 1987; Rasband 1990, 13.
17. By 'stable' here we mean stable against small perturbations away from the fixed point.
18. In a random walk (on a line or in a plane), successive steps of equal magnitude are taken, but the direction of each step is chosen at random (say, by flipping a coin). One can then ask for the probability of finding the walker at some given position after N steps have been taken.
19. Ford 1983, 43.
20. For a somewhat reserved evaluation of the

fundamental significance of chaos theory, see Dresden (1992). We have avoided any mention of fractals in our discussion of chaos because that topic adds little to the understanding of the conceptual issues involved in deterministic chaos. That is, fractals can provide showy graphics, but it is unclear that they actually help one in grasping the concept of chaos at this level of discourse.

PART V

Russell 1917, 33–4.

CHAPTER 13

[1] This is a type of 'screening' effect produced by certain filters, such as those employed in many sunglasses today. The concept of polarization will become clearer shortly, once we have discussed the nature of transverse waves.
[2] McMullin 1978a, Sections 4.5 and 4.6.
[3] A longitudinal wave is one in which the medium of transmission undergoes small (usually oscillatory) displacements in the direction of propagation of the wave itself.
[4] A transverse wave is one in which the medium of transmission undergoes small (usually oscillatory) displacements in a plane perpendicular to the direction of propagation of the wave itself.
[5] Faraday 1952, 'On the Physical Lines of Magnetic Force,' 816. (GB **45**, 816) (M, 506)
[6] Faraday 1952, 'On the Physical Lines of Magnetic Force,' 818. (GB **45**, 818) (M, 510)
[7] Faraday 1952, par. 1729, 530. (GB **45**, 530)
[8] Faraday 1952, par. 3075, 759. (GB **45**, 759)
[9] Faraday 1952, par. 1164 and 1165, 441. (GB **45**, 441)
[10] Faraday 1952, par. 3302, 831. (GB **45**, 831)
[11] Quoted in Whittaker (1973, Vol. I, 241–2).
[12] Maxwell 1954, viii–ix.
[13] Cardwell 1972, 176–82. Our Figure 13.7 is adapted from Figure 33 of Cardwell (1972, 177).
[14] Quoted in Whittaker (1973, Vol. I, 250).
[15] Maxwell 1890, Vol. II, 775.
[16] Maxwell 1890, Vol. I, 489–90.
[17] Maxwell 1890, Vol. II, 311, 312, 313, 320–1.

CHAPTER 14

[1] Maxwell 1954, Vol. II, Article 786, 435–6. Here ε_0 and μ_0 are related, respectively, to the k_1 and k_2 in our preceding discussion. The precise relations among these various constants can be found in Jackson (1975, 813–17). See also Section 14.A. Historically, there were several different sets of units that were used in electricity and magnetism and this accounts for the variation in notation.
[2] My reconstruction here is based on Bork (1963) and on Bromberg (1967). Buchwald (1985, 20, 23, 29) takes a somewhat different point of view, claiming that in Maxwell's own conception charge and current are produced by the electric field, rather than vice versa as we see things today.
[3] Maxwell 1954, Vol. I, Article 111, 166.
[4] There is a long and intriguing history of other possibilities, such as an aether partially or wholly dragged along by the motion of a body through it. However, all of these led to predictions at odds with experiment. By the time of the Michelson–Morley experiment, a static aether seemed the only remaining possibility.
[5] Zahar 1976, especially 221–30.
[6] Lorentz n.d., 5–6.
[7] Quoted in Whittaker (1973, Vol. II, 30).
[8] Lorentz 1952, 229–30.
[9] In brackets {} we also list the laws of electricity and magnetism in a form displaying explicitly the separate 'electrostatic' constant $1/4\pi\varepsilon_0 = 9 \times 10^9$ and the 'magnetostatic' constant $\mu_0/4\pi = 10^{-7}$, where $1/\mu_0\varepsilon_0 = 9 \times 10^{16}$ m^2/s^2 = $(3 \times 10^8$ m/s$)^2$. (Here we use MKSA units; see Jackson (1975, 817).) The notations ε_0 and μ_0 are not quite those found in Maxwell (1954), but otherwise the identifications made here are correct.
[10] It is, actually, ahistorical to list Lorentz's force law with Maxwell's equations, since Lorentz lived later in the nineteenth century than did Maxwell. However, the content of this equation was known to Maxwell.
[11] Equation (14.13c), the vector identity
$\nabla \times (\nabla \times A) = \nabla(\nabla \cdot A) - \nabla^2 A$,
and Eq. (14.13b) yield Eq. (14.14).
[12] This is readily seen by defining $r = x - ct$, $s = x + ct$ and then transforming Eq. (14.16) to $\partial^2 f/\partial r \partial s = 0$. Direct (partial) integration shows that $f(r, s) = g(r) + h(s)$. A similar discussion holds for the original three-dimensional problem of Eq. (14.15).

Notes

13. The point is that the divergence of the curl of any vector field A is identically zero:

$$\nabla \cdot (\nabla \times A) \equiv \frac{\partial}{\partial x_i}((\nabla \times A)_i) = \frac{\partial}{\partial x_i}\left(\varepsilon_{ijk}\frac{\partial A_k}{\partial x_j}\right)$$

$$= \varepsilon_{ijk}\frac{\partial^2 A_k}{\partial x_i \partial x_j} = \varepsilon_{jik}\frac{\partial^2 A_k}{\partial x_j \partial x_i} = -\varepsilon_{ijk}\frac{\partial^2 A_k}{\partial x_i \partial x_j} = 0$$

CHAPTER 15

1. Figure 15.1 is adapted from Figure 34-2 of Boorse and Motz (1966, 511).
2. Miller 1981, 48.
3. Miller 1981, 107.
4. From Eq. (14.10) we have $\boldsymbol{B} = (q/(4\pi c)) \times (\boldsymbol{v} \times \boldsymbol{r}/r^3)$ so that $B = ev\sin\theta/(4\pi c r^2)$, $r > a$ (in terms of the geometry of Figure 15.2). In order to facilitate comparison with the original papers, we write all electrodynamics expressions in Heaviside–Lorentz units in this chapter. For a comparison of this with the MKSA system see Jackson (1975, 818).
5. This figure, and subsequent ones in this chapter, are adapted from Cushing (1981).
6. Details of the calculations and of other material treated in this chapter, as well as a complete set of references, can be found in Cushing (1981).
7. Goldberg 1970–1, 7–25, especially 15; Miller 1981, Chapter 1.
8. Again, Miller (1981, 226–35, 335–52) and Cushing (1981) should be consulted for specific details of these experiments.
9. The Lorentz force (Eq. (14.13f)) provides the centripetal force for motion on a circle of radius ρ as $mv^2/\rho = e(v/c)B$ so that $\rho = (mcv)/(eB)$.
10. This follows from the last equation in the previous note and from the fact that $\bar{z} \propto 1/\rho$ (see Cushing (1981, Eqs. (3.11) and (3.27)). The inverse proportionality between $\bar{z}$ and ρ obtains because ρ is large compared to the dimensions of the apparatus (and this, in turn, follows from the large value of the speed v). In terms of the origin (0, 0, 0) in Figure 15.5, the equation of the circular trajectory for the electron then becomes $(x - (x_1/2))^2 + (z - \rho)^2 = \rho^2$. For the point $(x_2, \bar{z})$ this reduces (to the same order of approximation) to $(x_2 - (x_1/2))^2 = 2\bar{z}\rho$. Realize, here and later, that the mass m of the electron depends only on its speed v so that $m(v)$ remains a constant along the trajectory as long as v remains constant – whatever the specific dependence of m on v happens to be.
11. In terms of the strength E of the electric field, we have as the equation of motion here $d^2y/dt^2 = (e/m)E$.
12. Use of the relation $dy/dt = \dot{x}dy/dx = v\,dy/dx$ allows one to integrate the dynamical equation from the previous note to obtain $y = (eE)/(2mv^2)x(x - x_1)$ in the region between the condenser plates and, finally, $\bar{y} = (eE)/(2mv^2)x_1(x_2 - x_1)$ at the position of the photographic plate (see Figure 15.6). In deriving this result, it is important to keep in mind that y versus x is a straight line for $x_1 \leq x \leq x_2$.
13. Realize that $\beta = v/c$ here (not to be confused with the β in the expression 'β ray').
14. Again, see Cushing (1981) for details.
15. This and other tables in this chapter are abbreviated versions of Kaufmann's published tables. See Cushing (1981) for more complete tables. We omit Bucherer here.
16. Kaufmann 1902, 56. (My translation.)
17. It is important to note that Kaufmann weighted the first entry in Table 15.2 with a factor of $\frac{1}{2}$ and the last two with a factor of $\frac{1}{4}$ each.
18. Kaufmann 1905, 956. (My translation.)
19. The logical and philosophical implications of Planck's reappraisal has been carefully and elegantly discussed by Zahar (1978).
20. Planck 1906a. The suitably modified form of Newton's second law is there given as

$$\frac{d}{dt}\left(\frac{m_0 \boldsymbol{v}}{\sqrt{1-\beta^2}}\right) = q\left(\boldsymbol{E} + \frac{\boldsymbol{v}}{c} \times \boldsymbol{B}\right).$$

21. Planck 1906a, 136. (My translation.)
22. Planck 1906b.
23. See Cushing (1981) for these details.
24. See Cushing (1981, Eq. (4.36) and the discussion leading up to it).
25. This is not difficult to understand, since we saw in Note 10 above and in the accompanying text that $\bar{z}$ alone determines the radius of curvature ρ and this, in turn, yields $p = m(v)v$ (as shown in Note 9 above).
26. This was $e/m_0 = 1.878 \times 10^7$ emu/g in electromagnetic units.
27. While this is not meant to be obvious, it is

plausible from Eq. (15.4). See Cushing (1981, 1143–5, especially Eqs. (4.34) and (4.35)) for details of Planck's analysis.

28. Even Eqs. (15.3) and (15.4) make this fairly evident since they imply that $\bar{z}/\bar{y} \propto \beta$.
29. Planck 1907. Here α is in V/cm.
30. It was $e/m_0 = 1.72 \times 10^7$ emu/g.
31. Einstein 1907, 439. (My translation.)
32. Bestelmeyer 1907.
33. Bucherer 1909.
34. Neumann 1914.
35. Figure 15.8 is based on Figure 18 of Neumann (1914).
36. Guye and Lavanchy 1915.
37. Similar things have happened in more recent times. For example, the Feynman theory of β decay is a dramatic case in which the theory, when first proposed, disagreed with at least three experiments, all of which subsequently proved to be incorrect. See Feynman and Gell-Mann (1958).
38. This dependence of the meaning or significance of data on a theoretical background (used in the analysis of the data) is sometimes referred to as the theory-ladenness of observation.
39. Actually, Eq. (15.5) stands for the three equations $p_j = \partial L/\partial v_j, j = 1, 2, 3$ (or, x, y, z).
40. We omit here the entire discussion of the confusion that arose because of the use of both longitudinal and transverse masses. See Cushing (1981) for details.
41. Here we discuss only Abraham's and Lorentz's models, but not Bucherer's, since the latter was before too long eliminated from serious contention.

PART VI

Kelvin 1901, 1–2.
Einstein 1949a, 31, 33.

CHAPTER 16

1. Quoted in Clark (1972, 22).
2. Frank 1947, 49–50.
3. Einstein 1949a, 9, 11.
4. Einstein 1982.
5. Frank 1947, 75.
6. Frank 1947, 232.
7. Einstein 1949a, 21.
8. Einstein 1949a, 53.

9. That is, if one takes an arbitrary $f(x)$ that is a function of x only, but not of t as well, and substitutes it into Eq. (14.16), there is a contradiction (unless $f(x)$ does not depend upon x either).
10. Einstein 1905a, 891–2. The quotation here is taken from Einstein (n.d.a, 37–38).
11. Einstein 1905a, 893–4. The quotation here is taken from Einstein (n.d.a, 39–40).
12. Selleri (1994) has recently raised this question anew.
13. See Selleri (1994). In particular, one possible set is nearly identical to the Lorentz transformations of Eq. (16.5) below, but with the important difference that the last equation is replaced by $t' = \gamma^{-1} t$ so that there is now an absolute simultaneity among all inertial frames.
14. See, for example, Tangherlini (1961) or Mansouri and Sexl (1977a, 1977b, 1977c).
15. Einstein 1905a, 895. The quotation here is taken from Einstein (n.d.a, 41).
16. Zahar 1989, 28.
17. Einstein 1905a, 921. The quotation here is taken from Einstein (n.d.a, 65).
18. Einstein 1954d, 228.
19. From $(c\Delta t'/2)^2 = \ell^2 + (v\Delta t'/2)^2$ we obtain $\Delta t' = (2\ell/c)/\sqrt{1 - (v/c)^2} = \Delta t/\sqrt{1 - (v/c)^2}$ (see Figure 16.4).
20. Incidentally, the visual appearance of a rapidly moving object turns out to be rotated, but not contracted (Weisskopf 1960).
21. Frisch and Smith 1963.
22. Fock 1959, 12–16.
23. That is, $u'_x = dx'/dt' = (dx'/dt)/(dt'/dt)$ and $u'_y = dy'/dt' = (dy'/dt)/(dt'/dt)$. Direct differentiation of Eqs. (16.5) then yields Eqs. (16.6).
24. Here $B(r)$ is the static magnetic field produced by the bar magnet (in the rest frame of the magnet).

CHAPTER 17

1. Since $\lambda v = c$, it follows that $1/t = (1 - (v/c))v$.
2. From Eq. (16.1) we have $t'/t = \sqrt{1 - (v/c)^2}$. With this and the result of the previous Note, Eq. (17.1) follows at once.
3. The only change in the previous argument is that the time t (as determined by A) that it takes for two successive wave fronts to pass

the moving observer B now becomes
$t = (\lambda + v(\cos\alpha)t)/c$.

4. Of course, Eq. (17.4) is derived classically for the situation in which the source A is at rest with respect to the medium (the aether). If both the source A and the receiver B move (each measured with respect to the aether), then the classical expression for the Doppler shift, to first order in (v/c) where v is the relative velocity of A with respect to B, agrees with the first-order relativistic expression. However, this is no longer true once higher-order terms, such as $(v/c)^2$, are retained.

5. In Section 16.4 we arrived at the time-dilation effect via a thought-experiment argument, but we could also have obtained Eq. (16.1) directly from the last of Eqs. (16.5).

6. We simply state this result here with no attempt at a derivation. (See Einstein (n.d.a, 51–4) or Jackson (1975, Eqs. (11.149) and (7.11)) for details.) The central point once again, though, is that the results of Eqs. (17.5) follow from the demand of the form invariance of Maxwell's equations – that is, from one of the two relativity postulates.

7. See Jackson (1975, 236).

8. For convenience we use units (such as Gaussian) for which $|E| = |B|$ for the plane-wave configuration we are considering here.

9. Einstein 1905a, 914. The quotation here is taken from Einstein (n.d.a, 58).

10. The more familiar form of Newton's second law, $F = ma$, is valid only when the mass m does not vary with the velocity v. Since the momentum p is defined as $p = mv$ and the acceleration a as $a = dv/dt$, $F = ma$ and $F = dp/dt$ are then equivalent. However, when $m = m(v)$, as in the relativistic case, then it is the form $F = dp/dt$ that is correct. In Eq. (17.9) $v \times B$ is the cross, or vector, product of two vectors, where $(v \times B)_x = v_y B_z - B_y v_z$, with cyclic permutations for the y- and z-components of this vector.

11. Einstein 1905b, 640–1. Reference is to Einstein (n.d.b, 70–1).

12. Einstein 1905b, 641. The quotation here is taken from Einstein (n.d.b, 71).

13. Chang 1993. See Debs and Redhead (1996) for a particularly clear exposition of this problem.

14. Recall that special relativity allows one to synchronize clocks in a common inertial frame. What it does not allow one to do (according to Einstein's convention on this) is to establish an absolute simultaneity among different inertial frames.

15. For a very different (minority) view on the twin paradox, see Sachs (1971) and the preponderance of strong (negative) response to his position that followed.

16. Miller 1981, 261–2.

17. Fock 1959, 212–14. Actually, we have rewritten Fock's expression (his Eq. (62.16)) to obtain our Eq. (17.16) for easy comparison with our Eq. (17.15). Fock's argument is given to order $(v/c)^2$ and, hence, only Eq. (17.17) is strictly valid.

18. Hafele (1972); Hafele and Keating (1972).

19. As we pointed out in Section 16.3, it is perfectly consistent to arrange to have an absolute simultaneity among inertial frames.

20. For differing views on this see, for example, Maxwell (1985) and Stein (1991).

CHAPTER 18

1. Of course, strictly speaking, the M on the right side of this equation should also have a suffix 'g', but we are focusing discussion here on m to make the relevant point.

2. Einstein 1907. Pais (1982, Chapter 9) recounts Einstein's discovery of the equivalence principle and discusses the contents of this 1907 paper.

3. Einstein and Infeld 1938, 214–22.

4. This (Einstein 1911) and some of the other of Einstein's important papers on general relativity that we mention in this chapter can be found in Lorentz et al. (n.d.).

5. Quoted in Clark (1972, 222).

6. Quoted in Clark (1972, 252).

7. Hon 1996.

8. Harrison 1981, 183. This de Sitter (1917) solution is for a nonvanishing cosmological constant (see Section 18.4 below).

9. Of course, a shortcoming of this analogy is that it employs the weight of the larger ball (acting down, perpendicularly to the plane of the undistorted sheet) in order to 'explain' the effect of motion on the sheet in geometrical terms. Nevertheless, this analogy may be visually helpful.

10. Quoted in Clark (1972, 493–4); see also Whitrow (1973, xii).

11. Einstein 1911, 908. This reference corresponds to Einstein (n.d.c, 108).
12. Einstein 1916, 822. This reference corresponds to Einstein (n.d.d, 163). See Fock (1959, 200–2) for a modern derivation of this result.
13. Dramatic accounts of these measurements and of the meeting at which the results were announced can be found in Eddington (1920, 114–16) and in Whitehead (1967, 10–11).
14. Quoted in Holton (1968, 653) and in Bernstein (1976, 144).
15. That is, $h\Delta v = -\Delta \varepsilon = \Delta V = (GM/r)/(\varepsilon/c^2)$. This is only approximate since we have assumed that the fractional change in frequency is small.
16. Pound and Rebka 1960.
17. Clemence 1947.
18. Newton 1952, Book III, Queries 28 and 31, 529 and 542. (GB 34, 529, 542)
19. Harrison 1986.
20. Koyré 1957, 187–8.
21. Newton 1952, Book III, Part I, Query 28, 529. (GB 34, 529)
22. Newton 1934, Book III, General *Scholium*, 544. (GB 34, 370)
23. Thayer 1953, 50–1.
24. Newton 1934, Book III, Prop. XIV, Cor. II, 422. (GB 34, 287)
25. Thayer 1953, 57.
26. From conservation of energy $[\frac{1}{2}m\dot{r}^2 - (GMm/r) = -(GMm/r_0)]$ we can solve for $\dot{r} = dr/dt$ and integrate the result from r_0 to 0 to arrive at Eq. (18.3).
27. Einstein 1917b.
28. Shipman 1976, 268–78; Gott et al. 1977, 86; Bose 1980, 91–2; Harrison 1981, 298.
29. Shipman 1976, 268–78; Gott et al. 1977, 90; Bose 1980, 92.
30. Realize that all of the matter in our model universe is expanding, including that beyond the mass m (say, a star).
31. Since the mass within this sphere of radius r is just the density ρ times the volume of that sphere and since the escape velocity is given as $\sqrt{2GM/r}$, we obtain Eq. (18.4) immediately. The escape velocity is defined by the condition that the particle's total energy (the sum of its kinetic energy and of its gravitational potential energy) is zero, so that it would just come to rest at infinity (where the gravitational potential energy is also zero).
32. Gott et al. 1977, 83; Harrison 1981, 89–92.
33. Harrison 1981, Chapter 10.
34. From Eq. (17.1) for small (v/c) we have $(\lambda_0/\lambda) \approx 1 - (v/c)$. In Eq. (18.6) λ_0 is the wavelength of the light when it was emitted from a distant galaxy (long ago) and λ its wavelength upon reception here on earth. (This convention differs from that often used in cosmology texts.)
35. We can again use a classical analogue to see why this is reasonable. Suppose that, relative to some observer O, all bodies in the heavens recede from O with a velocity **v** proportional to the (vector) distance **r** from O to that body ($\mathbf{v} = \alpha \mathbf{r}$). Now consider two such receding bodies A and B so that $\mathbf{v}_A = \alpha \mathbf{r}_A$ and $\mathbf{v}_B = \alpha \mathbf{r}_B$. Then the velocity of A relative to B is just $\mathbf{v}_{AB} \equiv \mathbf{v}_A - \mathbf{v}_B = \alpha(\mathbf{r}_A - \mathbf{r}_B) = \alpha \mathbf{r}_{AB}$. But, this argument can be made about any origin O we choose.
36. Bose 1980, 92.
37. Waldrop 1991.
38. See Harrison (1981, 255–9) for a discussion of these points and for other factors that vitiate the usual bright-sky paradox.
39. Shipman 1976, 132. These distances, especially to the more distant objects, are only approximate and one often finds differences by factors of two or three for the same galaxy in different references.
40. Quoted in Eddington (1926, 6).
41. Bose 1980, 77; Harrison 1981, 361.
42. This follows from the so-called Robertson–Walker metric for a homogeneous, isotropic universe in which $ds^2 = g_{\mu\nu}dx^\mu dx^\nu \to c^2dt^2 - R^2(t)dr^2 \equiv c^2dt^2 - d\ell^2$. (Actually, this is the Robertson–Walker metric only for a special case, but one that will do for illustrative purposes here. See Bose (1980, 79–87) for more details.) In this equation t is a cosmic time for such a universe in a particular frame that is comoving with the expansion of the universe (Bose 1980, 81; Harrison 1981, 216). The geodesics for light rays are defined by $ds = 0$.
43. That is, $\dot{\ell} = \dot{R}r = (\dot{R}/R)\ell \equiv H\ell$ and then $R(t) \approx R(t_0)[1 + H_0(t - t_0)]$, where $H_0 \equiv \dot{R}(t_0)/R(t_0)$.
44. Since $ds = 0$ for the geodesic of light, we have $(t - t_0) \approx \ell/c$. This, the expansion given in the previous Note, and the relation $\lambda_0/\lambda = R(t_0)/R(t)$ then lead to Eq. (18.8).
45. For a careful discussion of various types of

389

Notes

horizons in an expanding universe, see Harrison (1981, Chapter 19).

PART VII

Einstein 1949b, 667.
Bohr 1949, 210, 232–3.
Clauser and Shimony 1978, 1881.

CHAPTER 19

1. Of course, the frequency of this radiated energy need not be the same as that of the absorbed radiation. So, if the temperature were such that the peak in the curve of Figure 19.1 occurred well below the visible region and light in the visible region were shone on a blackbody, then that blackbody would indeed appear black.
2. Our Figure 19.1 is adapted from Figure 2-1 of Leighton (1959, 61). Since this figure (like 19.3) is here intended to illustrate the qualitative behavior of the blackbody radiation curves, units are not explicitly displayed for ρ and λ.
3. That is, an integral number (n) of half-wavelengths ($\lambda/2$) must fit onto the length ℓ.
4. Jammer 1989, 11.
5. The frequency v and the wavelength λ are related as $\lambda v = c$, where c is the speed of light.
6. Kuhn 1978.
7. For specifics, see Jammer (1989, 7–18) and Klein (1962). Some of the most relevant of Planck's original papers can be found in English translation in ter Haar (1967). For a very different, and somewhat controversial, interpretation of Planck's work, see Kuhn (1978). Actually, Planck seems to have been unaware of the Rayleigh–Jeans formula, although he did know about the experimental data for long-wavelength cavity radiation.
8. Quoted in Klein (1962, 468).
9. Klein (1962) and Kuhn (1978) differ considerably on the arbitrariness (and even on the internal consistency) of Planck's counting procedure, given that he was allegedly following Boltzmann's method in statistical mechanics.
10. Quoted in Clark (1972, 95).
11. Planck 1949, 45.
12. A detailed historical case study of this can be found in Heilbron and Kuhn (1969).
13. Stability of a circular orbit requires that $mv^2/r = e^2/r^2$ so that $K = \frac{1}{2}mv^2 = \frac{1}{2}e^2/r$ and, hence, $E = K + V = -\frac{1}{2}e^2/r$. This then leads directly to the result stated in Eq. (19.4). Notice that Eq. (19.4) remains valid for elliptical orbits (the case actually considered by Bohr) as well, as follows from Kepler's third law, $\tau = 2\pi a^{3/2} m^{1/2}/e$ and the relation $a = -e^2/(2E)$, both of which are standard results from classical mechanics (see Goldstein (1950, 79–80)).
14. Bohr 1913, 7.
15. Bohr's actual route from Eq. (19.6) to the Balmer formula of Eq. (19.2) was more indirect than this because of the inconsistency he had generated by his quantization condition of Eq. (19.5) that would represent n quanta each of frequency $v/2$, rather than one quantum of frequency $n(v/2)$ (as needed to account for the observed pattern of spectral lines). See Heilbron and Kuhn 1969, 269–77.
16. Quoted in Jammer (1989, 77).
17. Bohr 1913, 15. Later in this same paper (pp. 24–25) Bohr finally gave his general rule for 'quantizing' the orbital angular momentum.
18. See the relations given at the beginning of Note 13 above.
19. See Jammer (1989, Chapter 5) and Cushing (1994, Chapter 6) for a more complete story of the discovery of these two versions of quantum mechanics.
20. Jammer 1989, Section 4.2.
21. Faye 1991.
22. Beller 1983a, 155*ff*.
23. Pauli 1981, especially 4, 206.
24. Here, as previously in the text, by 'operational definitions' we mean definitions in terms of the actual operations or procedures by which a quantity could be measured or determined.
25. A familiar example of such noncommutativity is provided by the vector (or 'cross') product of two vectors **A** and **B**, since $A \times B = -B \times A \neq B \times A$. As a simple illustration (from the realm of everyday phenomena) of a situation in which the outcome does depend upon the order in which two operations are performed, consider first pulling the trigger on a loaded pistol and then pointing the gun to your head, versus doing things in the reverse order.
26. Of course, for Einstein 'causal' meant local causal. That is, Einstein was certainly

committed to locality as a basic principle of physics, so that he would not countenance instantaneous action at a distance.
27. Klein 1964, 5.
28. Einstein 1917a, 128. The quotation here is taken from van der Waerden (1967, 76).
29. Quoted in Jammer (1989, 258). An analogy that de Broglie believed important is the following. The wave equation governing classical optics can be rewritten in a form that yields geometrical optics in a suitable limit. This limiting form of the optical equations is mathematically similar to the (Hamilton–Jacobi form of the) equations for particle mechanics (that is, Newton's second law of motion). Because geometrical (or ray) optics follows as a limiting case of wave optics, de Broglie and, later, Schrödinger hoped that classical particle mechanics would be a limiting case of a wave mechanics they were seeking. See Goldstein (1950, Section 9-8) and Cushing (1994, 106 and references cited there) on the analogy between optics and particle mechanics and on the role it played in fashioning wave mechanics.
30. Klein 1964, 4.
31. Klein 1964, 43.
32. Hendry 1984, 7.
33. Bohr 1985, xix.
34. Beller 1985, especially 340.
35. Beller 1983b; Cassidy 1992, Chapter 11.
36. Cushing 1994, 118–23. For an assessment of de Broglie's early pilot-wave theory that differs substantially from the usual one implied here and in Cushing (1994), see Valentini (1997).
37. Heilbron 1988, 203–4.
38. Heilbron 1988, 211. The quoted statement was made by Nathan Rosen.
39. Heilbron 1988, 219.

CHAPTER 20

1. This follows from $E = m_0 c^2 + K$, plus Eqs. (17.11) and (17.12).
2. Until now in our examples we have adopted the usual convention that the potential energy must vanish at spatial infinity. This is always possible in physically realistic situations, since we may define the potential energy to be zero at any reference point we choose. However, for the potential well with infinite barriers that we are now discussing, it is much more convenient to take $V = 0$ inside the 'box' and $V = \infty$ at the walls. With this convention, the allowed, bound-state energies are positive.
3. Our Figure 20.5 is adapted from Figure 5 of Goldberg et al. (1967, 184).
4. Born 1926, 863.
5. Even though P and ψ generally depend on the time t, as well as on the spatial variables, we often suppress the time dependence to simplify the notation. For convenience we also often write just x to represent the full set of spatial variables (x, y, z).
6. Our Figure 20.8 is adapted from Figure 10-2 of Eisberg (1961, 306).
7. Dirac 1958, 3.
8. Dirac 1958, 3.
9. Dirac 1958, 3–4.
10. Dirac 1958, 4.
11. Planck 1949, 135–8.
12. Two operators A and B are said to *commute* if $AB = BA$. We can characterize this property in terms of the *commutator* $[A, B]$ of these operators, where $[A, B] \equiv AB - BA$. If A and B commute, then $[A, B] = 0$; otherwise $[A, B] \neq 0$. For example, in quantum mechanics p_x is represented by the operator $-i d/dx$ so that $[x, p_x]f = if$, where $f(x)$ is any differentiable function of x. It is a straightforward mathematical result (that we simply state here, but do not prove) that there is no possible state (or ensemble) for which Δp_x and Δx both vanish. This is the mathematical basis for Eq. (20.11).
13. If $[A, B] = C$ (where C is in general also an operator), then the precise mathematical statement of Eq. (20.10) is $\Delta A \, \Delta B \geq |<C>|/2$. Whether or not the b of Eq. (20.10) in the text can ever be zero (for some quantum state ψ) depends on the form of C. When C is simply a number (really, a constant multiple of the identity operator), as in $[x, p_x] = i$, then b can never be zero (see Eq. (20.11)).
14. For a discussion of differing interpretations of quantum mechanics by various members of the Copenhagen school, see Cushing (1994, 27–32).
15. See, for example, Leighton (1959, 86–8) or Eisberg (1961, 156–9).
16. In his 'uncertainty' paper, Heisenberg (1927) actually used a now famous 'supermicroscope' thought experiment, but the general nature of

Notes

the argument is the same as the one we use here. (See Wheeler and Zurek (1983, 62–84, especially 64).)

17. Dirac 1958, 7–8.
18. Figure 20.10 is highly schematic and is meant to represent any arrangement in which a beam is split, sent along two different paths, and then recombined at some point (on a 'screen' or at a detector). A double-slit configuration, such as that sketched in Figure 20.11, is a common implementation of this idea.
19. Dirac 1958, 8.
20. This type of thought experiment is also discussed in Feynman (1965, 127–48). Our Figure 20.11 is adapted from Figure 30 of Feynman (1965, 136).
21. A modern and refined version of this 'double-slit' experiment has recently been performed and it again exhibits single-photon interference (see Godzinski 1991).

CHAPTER 21

1. Quoted in Jammer (1989, 366).
2. Heisenberg 1958, 55–6.
3. Born 1971, 91.
4. Hughes 1990.
5. The usual restriction on these operators representing physical observables is that they be Hermitian ($A = A\dagger$) so that the eigenvalues will be real quantities.
6. As an example of Eq. (21.1), consider the differential operator $A = -i d/dx$, the eigenstate $\psi = e^{ipx}$, and the eigenvalue $\lambda = p$.
7. Our Figure 21.3 is adapted from Figure 3.1 of Cushing (1994, 35).
8. This product form follows from the fact that the Hamiltonian H in the Schrödinger equation ($i\partial\Psi/\partial t = H\Psi$) for two noninteracting systems (denoted by 1 and 2) becomes the sum of two commuting terms ($H = H_1 + H_2$). Here, as often in illustrations, we have chosen units such that Planck's constant has the value $\hbar = 1$.
9. The magnitude of the spin of the electron is measured to be $\hbar/2$. If we choose suitable units, we can rescale this to the value 1. It is a fact of nature (incorporated into the formalism of quantum mechanics) that the measured value of this spin in any direction is always either $+1$ (up) or -1 (down). These are the only allowed (eigenvalues) for this spin,

so that there are just two eigenstates: ψ_+ corresponding to $\lambda_+ = +1$ and ψ_- corresponding to $\lambda_- = -1$.

10. The formal solution to $i\partial\Psi/\partial t = H\Psi$ is given as $\Psi(t) = e^{-iHt}\Psi(0)$. The proof of this formal solution is straightforward (by direct differentiation of this solution and substitution into the Schrödinger equation) and can be found in nearly any standard book on quantum mechanics. However, all that we need for our purposes here is that such a solution exists. We do not require an explicit representation of this solution in terms of functions, matrices and column vectors.
11. That is, $H(\Psi + \Phi) = H\Psi + H\Phi = i\partial\Psi/\partial t + i\partial\Phi/\partial t = i\partial(\Psi + \Phi)/\partial t$. This means that the sum of two solutions is itself a solution. In general, an operator A is said to be linear if $A(\phi + \psi) = A\phi + A\psi$ for all ϕ and ψ.
12. Schrödinger 1935. See p. 812 (or Wheeler and Zurek 1983, 157) for the original statement of the cat paradox.
13. Przibram 1967, 35–6.
14. Our Figure 21.4 is adapted from Figure 3.2 of Cushing (1994, 38).
15. The notation $<u|v>$ in Eq. (21.12) stands for the projection of the state vector u onto the state vector v. This gives the component of u contained in v. It is similar to the scalar (dot or inner) product $\hat{i} \cdot r = x$ where $\hat{i}$ is a unit vector in the x-direction, r is the position vector and x is the component of r in the x-direction. Here we are simply assuming, or taking for granted, that the nuclear decay process follows the usual exponential decay law $N(t) = N_0 \exp(-t/\tau_0)$ for such spontaneous disintegrations, where τ_0 is essentially the half-life of the sample. (Technically, the half-life is defined as $0.693\tau_0$.) See the discussion of μ-meson decay in Section 16.4.
16. Dirac 1958, 8–9.
17. Dirac 1958, 9.
18. Dirac 1958, 9–10.
19. Dirac 1958, 11, 13.

PART VIII

Feynman 1965, 129.
Gell-Mann 1981, 169–70.
Bohr 1934, 53.
Heisenberg 1958, 46, 129.

Born 1951, 155, 163–4.
Moore 1989, 228.
von Neumann 1955, 325.
Einstein 1949b, 666.
Landau and Lifshitz 1977, 2.
Bell 1987, 160.

CHAPTER 22

[1] Einstein *et al*. 1935, 777.
[2] Bohr 1935.
[3] Born 1971, ix–x.
[4] Born 1971, 158.
[5] Born 1971, 164–5.
[6] Born 1971, 164.
[7] Einstein 1917a.
[8] Quoted in Jammer (1989, 344).
[9] The reader may find it helpful to read the paper itself (Einstein *et al*. 1935) before proceeding with the analysis of this section.
[10] Einstein *et al*. 1935, 777.
[11] This is easily seen if we recall that in quantum mechanics p_x is represented by the operator $-\mathrm{i}\mathrm{d}/\mathrm{d}x$ so that $[x, p_x]f = \mathrm{i}f$, where $f(x)$ is any differentiable function of x. Remember that there can be no state for which Δx and Δp_x both vanish, so that x and p_x cannot simultaneously have sharp values. See also Notes 12 and 13 of Chapter 20.
[12] See Eqs. (9)–(18) in Einstein *et al*. (1935).
[13] In a compact mathematical form, we can see this as follows. Use the fact that $[x_j, p_i] = \mathrm{i}\delta_{ij}$, define two operators as $P = p_1 + p_2$, $X = x_1 - x_2$ and then show by direct calculation that $[X, P] = 0$. Therefore, both P and X can be measured or predicted simultaneously. So, we may measure p_1 and x_2 (that do commute with each other) to predict p_2 and x_1, thereby obtaining knowledge of the values of noncommuting observables (the pair p_1 and x_1 or the pair p_2 and x_2).
[14] Notice that, if one accepts quantum mechanics as complete and therefore takes seriously the option that the state of the system was actually changed (in an arbitrarily short period of time) by the distant choice made of what to measure, then the EPR argument shows that the completeness of quantum mechanics implies the nonlocality of that theory.
[15] For a discussion of this, see Cushing (1994, Chapter 8).
[16] A deterministic world would furnish an example of such a situation.
[17] The actual experimental situation is more complex than this simple statement might lead one to conclude. For a technical discussion of the issues involved, see Clauser and Shimony (1978).
[18] Clauser and Shimony 1978, 1883.
[19] Stapp (1989; 1990) has claimed that nonlocality is necessitated by the violation of Bell's inequality. Clifton *et al*. (1990) and Dickson (1993) argue that Stapp's proof is invalid. Maudlin (1994, 121; 1996, especially 285–6) argues that any physically adequate theory that accounts for the observed violations of Bell's inequality must be nonlocal. This represents only a sampling of papers on this question – for more, see the references contained in the works cited here.
[20] Maudlin 1994, 135–40. For example, taking the probabilities for various occurrences as fundamental and as determined by a set of local hidden variables leads as well to contradictions with the predictions of quantum mechanics.
[21] Maudlin (1994, 121; 1996, especially 285–6).
[22] Bohm 1951, 614–23.
[23] Our Figure 22.2 is adapted from Figure 3 of Cushing and McMullin (1989, 4).
[24] There is nothing special about electrons for this discussion. In fact, in many of the actual experiments performed, photons were employed.
[25] Peres 1978.
[26] That is, this dilemma has been generated by assuming (at least tacitly) that there is a definite (and fixed) outcome at one station whatever setting might be chosen at the other station.
[27] $R_{\mathrm{QM}}(\phi) = |1 + 2\cos\phi - \cos(2\phi)| = 2|1 + \cos\phi - \cos^2\phi|$. To extend the plot of Figure 22.3 to the entire range $0 \leq \phi \leq 2\pi$, we could make use of the fact that $R(\phi)$ is symmetric about $0°$ and $180°$:
$R(\phi) = R(-\phi) = R(2\pi - \phi)$.

CHAPTER 23

[1] In keeping with our parsing of a theory into a formalism and an interpretation later in this section, we should refer here to the Bohm theory, rather than just to the Bohm

interpretation. That is, even though the formalisms may be the same for Copenhagen and for Bohm, the interpretations are different and, hence, so are the theories. However, it has become so common in the literature to refer to the Bohm interpretation and the Copenhagen interpretation that we simply follow this standard practice.

2. Detailed references for, and further discussions of, the issues raised in this and in the next chapter are given in Cushing (1994, especially Chapters 5–7 and 11). I thank the University of Chicago Press for permission to modify previously published material for use in this chapter.

3. More precisely, ψ is a vector (in a Hilbert space $\mathcal{H}$) representing the state of the physical system.

4. As we stated previously, this is $i\hbar\partial\psi/\partial t = H\psi$, where the Hamiltonian is $H = K + V$, with K being the kinetic energy and V the potential energy of the system.

5. These physical observables a can take on only the eigenvalues a_j where $A\psi_j = a_j\psi_j$.

6. This is given as the expectation value of A, namely $\langle\psi|A|\psi\rangle$.

7. Just as in Eq. (21.9), this is $\psi = \sum_k a_k\psi_k \to \psi_j$.

8. Bohm 1952.

9. Here ∇ is the gradient operator whose components are $\nabla_j = \partial/\partial x_j$, so that in (ii), for instance, $v_x = (1/m)\partial S/\partial x$ where $S = S(x, y, z, t)$.

10. Bohm 1953. For a discussion of some recent work on this question, see Cushing (1994, Chapter 9).

11. Bohm 1952, 171.

12. Bohm 1952; Bohm et al. 1987. See also, Bohm and Hiley (1993), Holland (1993) and Cushing (1994).

13. See Cushing (1994, Chapter 4) for Bohm's account of measurement.

14. It is sometimes claimed that Ehrenfest's theorem allows one to recover Newton's second law of motion ($ma = \mathbf{F}(\mathbf{r})$) from quantum mechanics and the Schrödinger equation. However, that theorem only states that $m(d^2 <\mathbf{r}(t)> /dt^2) = <\mathbf{F}(\mathbf{r})>$, where $<\mathbf{r}> = <\psi|\mathbf{r}|\psi>$. It is generally not true that $<\mathbf{F}(\mathbf{r})> = \mathbf{F}(<\mathbf{r}>)$. So, even for the time evolution of the expectation (or 'average') value of $<\mathbf{r}(t)>$, we do not recover the classical equation of motion.

15. Bohm 1952, 166.

16. Bohm 1952, 168.

17. Bohm 1952, 169.

18. In the Bohm theory, all measurements are ultimately measurements of position. For example, instrument readings finally come down to something like the position of a needle on a meter face.

19. Shimony 1984. See also Cushing (1994, especially Section 10.4.2) on the question of relativity in Bohm's theory.

20. Maudlin 1994, 220.

21. Fine 1986a, 87.

22. Detailed references for this material can be found in Cushing (1991).

23. This is admittedly a highly 'reconstructed' form of the empirical data summarized by Kepler's second law. However, we believe this is a fair enough simplification to make our point.

24. Experiments have been performed to test the Bell inequality and, although the detailed analysis of these actual experiments is quite complicated, the overwhelming consensus is that the experimental results violate the Bell inequality. For details on one of the most important of these experiments, see Aspect et al. (1982).

25. Fine 1982a.

26. There are several formally highly developed attempts at alternative interpretations (aside from the Copenhagen and Bohm versions), such as the many-worlds, consistent histories, decoherence, modal, and spontaneous collapse ones that we do not discuss here. Our purpose in this section is to illustrate some of the key conceptual problems that any interpretation must address. All of these other schemes just mentioned arguably fail to resolve satisfactorily the difficulties we raise in this section. See Albert (1992) and Baggott (1992, Chapter 5) for discussions of several of these attempts.

27. We say 'usually assumed' because one of the background assumptions typically accepted in such discussions is that the special theory of relativity necessarily excludes instantaneous connections among spatially separated events. For an enlightening and careful analysis of this question of quantum nonlocality see Maudlin (1994).

28. Fine 1982b, 740. This use of 'predict' is akin to our 'explain' and this 'explain' to our 'understanding'.
29. Good overviews of the technicalities of Bohm's theory are given in Dürr et al. (1996) and in Valentini (1996).
30. In terms of the usual quantum-mechanical probability density $P = |\psi|^2$ and the current density $j = (i\hbar/2m)(\psi\nabla\psi^* - \psi^*\nabla\psi)$, this velocity is just $v = j/P$.
31. The formal manipulations required to make this definition are

$$\frac{dp}{dt} = v_j \frac{\partial}{\partial x_j}(\nabla S) + \frac{\partial}{\partial t}(\nabla S) = \frac{1}{m}\nabla S \cdot \nabla(\nabla S) + \frac{\partial}{\partial t}(\nabla S)$$
$$= \nabla\left[\frac{1}{2m}(\nabla S)^2 + \frac{\partial S}{\partial t}\right]$$

plus use of Eq. (23.3).

CHAPTER 24

1. The reader should be aware that some of the opinions expressed in this chapter are more speculative or controversial than most of those found in the book thus far. Much in this chapter is a summary of an extensive case study to be found in Cushing (1994, especially Chapters 8–11). I thank the University of Chicago Press for permission to modify previously published material for use in this chapter.
2. There are actually distinctions to be made among Duhem's thesis, Quine's thesis and the thesis of underdetermination (see, for example, Harding 1976). However, we simply use the expression 'Duhem–Quine thesis' and the term 'underdetermination' indifferently here. I thank Yuri Balashov for pressing me on this point.
3. Duhem 1974, 180.
4. Duhem 1974, 190.
5. Duhem 1974, 217.
6. Duhem 1974, 217.
7. Quine 1951, 39.
8. Quine 1951, 39–40.
9. Quine 1951, 40.
10. Quine 1951, 41.
11. Quine 1951, 41.
12. Quine 1951, 43.
13. Ben-Menahem 1990.
14. Laudan and Leplin (1991) argue against such radical and universal underdetermination.
15. Ben-Menahem 1990, 267.
16. Einstein 1954e, 226.
17. Heisenberg 1971, 76.
18. van Fraassen 1980.
19. McMullin 1984.
20. Hanle 1979, 268.
21. Harmke Kamminga has kindly informed me that Russell (1917, Chapter IX, especially 199–208) also argued for the epistemological (and empirical) indistinguishability between determinism and indeterminism and used this indistinguishability to argue that Laplacian determinism is vacuous. I thank her for this interesting and relevant reference.
22. For a detailed historical study of this question, see Cushing (1994, especially Chapter 11).
23. Heilbron 1988, 219.
24. Heilbron 1985, 391.
25. Beller 1983b; Cassidy 1992, Chapters 12 and 13.
26. Selleri 1990; Cushing 1994.
27. Hamilton–Jacobi theory is simply a different, more general form of Newton's second law of motion. When the dynamical equations of classical mechanics are expressed in Hamilton–Jacobi form, they bear a marked resemblance to the equations that Bohm obtained for quantum mechanics. De Broglie was also aware of this similarity. See Note 29 in Chapter 19 and also Cushing (1994, Chapter 5, Appendix 2) for details.
28. While these theories (for relativistic particles and for quantum fields) are noncovariant (that is, they do not have the same form in all inertial frames), their predictions for observations are identical to those of standard relativisitc quantum mechanics and quantum field theory (see Cushing 1994, Section 10.4).
29. Nelson 1966; 1967. See Cushing (1994, 159–62) on the relation between Nelson's stochastic mechanics and standard quantum mechanics.
30. Einstein 1905c; 1908; 1926.
31. This is a formal mathematical technique that allows one to express the equations of quantum mechanics in different bases or representations (one of which is Heisenberg's matrix mechanics and another of which is Schrödinger's wave mechanics).
32. See Cushing (1994, Section 8.3 and references therein) on Einstein's own 1927 hidden-variables theory that he rejected because of certain unphysical nonlocality features.

Notes

33. This refers to the proof that the entanglement of quantum states, while responsible for the long-range correlations like those of Eqs. (22.16a) and (22.16b), cannot be used to transmit information via these quantum-mechanical correlations. See Cushing (1994, Chapter 10, Appendix 2).
34. Einstein 1949a, 85.
35. Bohm 1952; Bohm et al. 1987.
36. Cushing 1990b, Chapter 10.
37. vann Woodward 1986, 3.
38. Butterfield 1965. I thank Yuri Balashov for impressing upon me the relevance of Butterfield's position for the discussion of this section.
39. Butterfield 1965, 11.
40. The reader has doubtless found numerous examples of my having fallen into just such a trap in this book.
41. Butterfield 1965, 5.
42. Bacon 1952, Aphorisms 45 and 46. (GB **30**, 110)

PART IX

Hertz 1900, 21.
Einstein 1934, 36–7, 60 (1954b, 274; 1954c, 266).
Einstein 1949b, 684.

CHAPTER 25

1. Frank 1949, 350.
2. Einstein 1954e, especially 225–6.
3. Ziman 1978, 9.
4. Ziman 1978, 28.
5. Einstein 1954f, 282.
6. Born 1971, 170.
7. Einstein 1954b, especially 272, 274.
8. Einstein 1954b, 274.
9. Einstein 1954g, especially 322; 1954e, especially 226.
10. Here we are addressing the descriptive versus the normative aspects of the philosophy of science.
11. Even this is not always the case in the history of physics.
12. Einstein 1954e, especially 225–6.
13. Schrödinger 1944, 74.
14. Watson 1970, 68.
15. Monod 1971, 102.
16. Luria 1973, 7.
17. Gould 1989.
18. McMullin 1978b, especially 232–3.
19. Laudan 1981, 9, 25, 34–44.
20. Friedman 1992, 143, 171.
21. Burtt 1927, 54.
22. Frank 1947, 36–44.
23. Strictly speaking, Hempel and Reichenbach were not members of the original group in Vienna in the 1920s, but of a cognate group in Berlin during that period. However, they were certainly part of this 'movement' by the late 1920s.
24. Poincaré 1952, xxvi.
25. Poincaré 1952, xxiii.
26. Duhem 1974, 31.
27. Hempel 1945.
28. Goodman 1946; 1965, 72–83.
29. Bloor 1976; Pickering 1984.
30. Kuhn 1970, 34.
31. Kuhn 1970, 76.
32. Kuhn 1970, 8–9.
33. Kuhn 1970, 94.
34. Ziman 1978, 3.
35. Ziman 1978, 99.
36. Holton 1973.
37. Weimer 1975.
38. One of Popper's followers, Imre Lakatos, particularly emphasized the importance of the scientific research program (that is essentially a succession of related theories fashioned via falsification) as the proper unit of appraisal in science, rather than single scientific theories as in a simple falsificationist model. See Howson (1976) and Lakatos (1978) for extensive discussions of Lakatos' work on the methodology of scientific research programs (MSRP).
39. Much of the sketch we give here is a gloss on Larry Laudan's reticulated, or triadic, model of science (Laudan 1984, especially Chapter 3; Cushing 1990b, Chapter 10, particularly 287–8).
40. We can, if we wish, exclude from such change those constraints on scientific practice and reasoning already present in the normal, common-sense demands placed on everyday argument (such as the usual rules of deductive reasoning). Surely such characteristics, even if taken to be invariant, are not peculiar to the method of science and, hence, cannot be used as some singular hallmark of science (Cushing 1990a).
41. Fine 1986b.
42. Freud 1965, 158–9.

[43] Monod 1971, 21. By 'objective' here Monod means with no prior commitment, on the part of the scientist, to what the implications 'must' be relative to some overall purpose or goal of the functioning of the universe.
[44] This necessity of examining nature objectively has also been underscored by Richard Feynman (1965, 147–8), a Nobel Laureate in physics.
[45] These views are compellingly expressed in Monod (1971, 169–72).
[46] For a view that assigns very different weights from Monod's to the 'chance' versus 'necessity' aspects of the emergence of life, see De Duve (1995).
[47] But for copyright difficulties, these important essays by Einstein would have been reproduced in this book.

General references

Here are a few readily available references for general background information and for primary historical source material.

The New Encyclopaedia Britannica, 15th Edition (Encyclopaedia Britannica, Chicago, 1977). The *Micropaedia* is a convenient source for biographical material on philosophers and scientists.

Robert M. Hutchins (ed.), *Great Books of the Western World* (Encyclopaedia Britannica, Chicago, 1952). In these fifty-four volumes can be found (in English) the major works of many of the greatest thinkers in our Western tradition. Whenever possible, we have used this as the source of the direct quotations given in this book. In the *Notes* this source is abbreviated as GB, followed by the relevant volume and page numbers.

Charles C. Gillispie (ed.), *Dictionary of Scientific Biography*, 14 Vols. (Charles Scribner's Sons, New York, 1970–6). This provides exhaustive and authoritative biographical material on virtually every figure, major and minor, in the history of science and summarizes the scientific contributions of each. In the *Notes* this source is abbreviated as DSB, followed by the relevant volume and page numbers.

Morris R. Cohen and I. E. Drabkin (eds.), *A Source Book in Greek Science* (Harvard University Press, Cambridge, MA, 1948). The sections on mathematics, astronomy and physics provide a useful reference to the work of the ancients on these subjects.

William F. Magie, *A Source Book in Physics* (Harvard University Press, Cambridge, MA, 1963). This contains very brief biographical sketches of many scientific figures from approximately 1600 up to the year 1900, as well as excerpts from their works. To increase access to original historical material for this important period, we have, in addition to citing the primary sources, also given the reference to this work whenever possible. These are listed in parentheses as M, followed by the relevant page numbers. At times the text as quoted in our chapters will differ slightly from the references to Magie due to variations in the translations from the original.

Paul Edwards (ed.), *The Encyclopedia of Philosophy*, 8 Vols. (Macmillan Publishing Co., New York, 1967). Here one finds, in essays written by today's experts, biographical information on major philosophers from ancient to modern times and summary discussions of their positions.

Anton Pannekoek, *A History of Astronomy* (Dover Publications, New York, 1989). This provides a good summary of developments in astronomy from ancient to modern times.

Shmuel Sambursky, *Physical Thought From the Presocratics to the Quantum Physicists* (Pica Press, New York, 1975). This contains excerpts of philosophical writings with introductory essays on the various historical periods.

Sir Edmund Whittaker, *A History of the Theories of Aether and Electricity*, 2 Vols. (Humanities Press, New York, 1973). This is a thorough and masterful presentation of the history of electricity and magnetism, atomic physics and nuclear physics, gravitation and quantum mechanics, from antiquity up to 1925. Whittaker was an outstanding theoretical physicist and the various historical developments are treated in great technical detail.

Henry A. Boorse and Lloyd Motz (eds.), *The World of the Atom*, 2 Vols. (Basic Books, New York, 1966). The central interest of these two marvelous reference volumes is the history and content of atomic, nuclear, and quantum theory. There are biographical essays on many scientists from approximately 1600 to the present, descriptions and analyses of their work and excerpts (all in English) of their writings.

Bibliography

Adams, C. and Tannery, P. (eds.) (1905), *Oeuvres de Descartes* (Leopold Cerf, Paris).

Adler, C. G. and Coulter, B. L. (1978), Galileo and the Tower of Pisa Experiment, *American Journal of Physics* **46**, 199–201.

Albert, D. Z. (1992), *Quantum Mechanics and Experience* (Harvard University Press, Cambridge, MA).

Archimedes (1987), *The Sand-Reckoner* in Heath (1897), pp. 221–32.

Aristotle (1942a), *Prior Analytics* in Ross, Vol. I.

Aristotle (1942b), *Physics* in Ross, Vol. II.

Aristotle (1942c), *On the Heavens* in Ross, Vol. II.

Aristotle (1942d), *Nicomachean Ethics* in Ross, Vol. V.

Armitage, A. (1938), *Copernicus: The Founder of Modern Astronomy* (George Allen & Unwin, London).

Aspect, A., Dalibard, J. and Roger, G. (1982), Experimental Tests of Bell's Inequalities Using Time-Varying Analyzers, *Physical Review Letters* **49**, 1804–7.

Bacon, F. (1952), *Novum Organum*, Book I in Hutchins, Vol. 30.

Baggott, J. E. (1992), *The Meaning of Quantum Theory* (Oxford University Press, Oxford).

Ball, R. S. (1921), *Time and Tide, A Romance of the Moon* (Macmillan Publishing Co., New York).

Barone, M. and Selleri, F. (eds.) (1994), *Frontiers of Fundamental Physics* (Plenum Publishers, New York).

Becker, R. A. (1954), *Introduction to Theoretical Mechanics* (McGraw-Hill Book Co., New York).

Beer, A. (ed.) (1968), *Vistas in Astronomy*, Vol. 10 (Pergamon Press, New York).

Bell, J.S. (1987), *Speakable and Unspeakable in Quantum Mechanics* (Cambridge University Press, Cambridge).

Beller, M. (1983a), *The Genesis of Interpretations of Quantum Physics, 1925–1927*. Unpublished Ph.D. Dissertation, University of Maryland.

Beller, M. (1983b), Matrix Theory Before Schrödinger, *Isis* **74**, 469–91.

Beller, M. (1985), Pascual Jordan's Influence on the Discovery of Heisenberg's Indeterminacy Principle, *Archive for History of Exact Sciences* **33**, 337–49.

Beller, M. (1992), The Birth of Bohr's Complementarity: The Context and the Dialogues, *Studies in History and Philosophy of Science* **23**, 147–80.

Ben-Menahem, Y. (1990), Equivalent Descriptions, *British Journal for the Philosophy of Science* **41**, 261–79.

Bernstein, J. (1976), *Einstein* (Penguin Books, New York).

Bestelmeyer, A. (1907), Spezifische Ladung und Geschwindigkeit der durch Röntgenstrahlen erzeugten Kathodenstrahlen, *Annalen der Physik* **22**, 429–47.

Biagioli, M. (1993), *Galileo Courtier: The Practice of Science in a Culture of Absolutism* (University of Chicago Press, Chicago).

Bitbol, M. and Darrigol, O. (eds.) (1992), *Erwin Schrödinger: Philosophy and the Birth of Quantum Mechanics* (Editions Frontiers, Gif-sur-Yvette Cedex, France).

Blackwell, R. J. (1977), Christiaan Huygens' *The Motion of Colliding Bodies*, *Isis* **68**, 574–97.

Blake, R. M., Ducasse, C. J. and Madden, E. H. (eds.) (1960), *Theories of Scientific Method: The Renaissance Through the Nineteenth Century* (University of Washington Press, Seattle, WA).

Bloch, F. (1976), Heisenberg and the Early Days of Quantum Mechanics, *Physics Today* **29** (12), 23–27.

Bloor, D. (1976), *Knowledge and Social Imagery* (Routledge & Kegan Paul, London).

Bohm, D. (1951), *Quantum Theory* (Prentice-Hall, Englewood Cliffs, NJ).

Bohm, D. (1952), A Suggested Interpretation of

the Quantum Theory in Terms of 'Hidden' Variables, I and II, *Physical Review* **85**, 166–93.

Bohm, D. (1953), Proof That Probability Density Approaches $|\psi|^2$ in Causal Interpretation of the Quantum Theory, *Physical Review* **89**, 458–66.

Bohm, D. and Hiley, B. J. (1993), *The Undivided Universe: An Ontological Interpretation of Quantum Theory* (Routledge, London).

Bohm, D., Hiley, B. J. and Kaloyerou, P. N. (1987), An Ontological Basis for the Quantum Theory, *Physics Reports* **144**, 321–75.

Bohr, N. (1913), On the Constitution of Atoms and Molecules, *Philosophical Magazine* **26**, 1–25.

Bohr, N. (1934), *Atomic Theory and the Description of Nature* (Cambridge University Press, Cambridge).

Bohr, N. (1935), Can Quantum-Mechanical Description of Physical Reality Be Considered Complete?, *Physical Review* **48**, 696–702. (Reprinted in Wheeler and Zurek (1983, 145–51).)

Bohr, N. (1949), 'Discussion with Einstein on Epistemological Problems in Atomic Physics' in Schilpp, pp. 199–241.

Bohr, N. (1985), *Collected Works*, Vol. 6 (North-Holland Publishing Co., Amsterdam).

Boorse, H. A. and Motz, L. (eds.) (1966), *The World of the Atom*, Vols. I and II (Basic Books, New York).

Bork, A. M. (1963), Maxwell, Displacement Current, and Symmetry, *American Journal of Physics* **31**, 854–9.

Born, M. (1926), Zur Quantenmechanik der Stossvorgänge, *Zeitschrift für Physik* **37**, 863–7. (Appears in English translation as 'On the Quantum Mechanics of Collisions' in Wheeler and Zurek (1983, 52–5).)

Born, M. (1951), *The Restless Universe* (Dover Publications, New York).

Born, M. (1971), *The Born–Einstein Letters* (Walker and Company, New York).

Bose, S. K. (1980), *An Introduction to General Relativity* (Wiley Eastern Limited, New Dehli.

Brackenridge, J. B. (1995), *The Key to Newton's Dynamics* (University of California Press, Berkeley, CA).

Briggs, K. (1987), Simple Experiments in Chaotic Dynamics, *American Journal of Physics* **55**, 1083–9.

Brody, B. A. (ed.) (1970), *Readings in the Philosophy of Science* (Prentice-Hall, Englewood Cliffs, NJ).

Bromberg, J. (1967), Maxwell's Displacement Current and his Theory of Light, *Archive for History of the Exact Sciences* **4**, 218–34.

Bucherer, A. H. (1909), Die experimentelle Bestätigung des Relativitätsprinzips, *Annalen der Physik* **28**, 513–36.

Buchwald, J. Z. (1985), *From Maxwell to Microphysics* (University of Chicago Press, Chicago).

Buchwald, J. Z. (1989), *The Rise of the Wave Theory of Light* (University of Chicago Press, Chicago).

Burtt, E. A. (1927), *The Metaphysical Foundations of Modern Physical Science* (Harcourt Brace & Co., New York).

Butterfield, H. (1965), *The Whig Interpretation of History* (W. W. Norton & Co., New York).

Cardwell, D. S. L. (1972), *Turning Points in Western Technology* (Science History Publications, New York).

Caspar, M. (ed.) (1937), *Johannes Keplers Gesammelte Werke* (C. H. Beck'sche Verlagsbuchhandlung, Munich).

Casper, B. M. (1977), Galileo and the Fall of Aristotle: A Case of Historical Injustice?, *American Journal of Physics* **45**, 325–30.

Cassidy, D. C. (1992), *Uncertainty: The Life and Science of Werner Heisenberg* (W. H. Freeman and Company, New York).

Chang, H. (1993), A Misunderstood Rebellion: The Twin-Paradox Controversy and Herbert Dingle's Vision of Science, *Studies in History and Philosophy of Science* **24**, 741–90.

Clagett, M. (1959), *The Science of Mechanics in the Middle Ages* (University of Wisconsin Press, Madison, WI).

Clark, R. W. (1972), *Einstein: The Life and Times* (Avon Books, New York).

Clauser, J. F. and Shimony, A. (1978), Bell's Theorem: Experimental Tests and Implications, *Reports on Progress in Physics* **41**, 1881–927.

Clemence, G. M. (1947), The Relativity Effect in Planetary Motions, *Reviews of Modern Physics* **19**, 361–4.

Clifton, R. K., Butterfield, J. N. and Redhead, M. L. G. (1990), Nonlocal Influences and Possible Worlds – A Stapp in the Wrong

Bibliography

Direction, *British Journal for the Philosophy of Science* **41**, 5–58.

Cohen, I. B. (1970), 'Isaac Newton' in Gillispie, Vol. X, pp. 42–101.

Cohen, I. B. (1971), *Introduction to Newton's 'Principia'* (Harvard University Press, Cambridge, MA).

Cohen, I. B. (1981), Newton's Discovery of Gravity, *Scientific American* **244** (3), 166–79.

Cohen, M. R. and Drabkin, I. E. (1948), *A Source Book in Greek Science* (McGraw-Hill Book Co., New York).

Cooper, L. (1935), *Aristotle, Galileo, and the Tower of Pisa* (Cornell University Press, Ithaca, NY).

Copernicus, N. (1952), *On the Revolutions of the Heavenly Spheres* in Hutchins, Vol. 16.

Copernicus, N. (1959), *Commentariolus* in Rosen.

Crowe, M. J. (1990), *Theories of the World from Antiquity to the Copernican Revolution* (Dover Publications, New York).

Crowe, M. J. (1994), *Modern Theories of the Universe: From Herschel to Hubble* (Dover Publications, New York).

Crutchfield, J. P., Farmer, J. D., Packard, N. H. and Shaw, R. S. (1983), Chaos, *Scientific American* **255** (6), 46–57.

Cushing, J. T. (1975), *Applied Analytical Mathematics for Physical Scientists* (John Wiley & Sons, New York).

Cushing, J. T. (1981), Electromagnetic Mass, Relativity, and the Kaufmann Experiments, *American Journal of Physics* **49**, 1133–49.

Cushing, J. T. (1982), Kepler's Laws and Universal Gravitation in Newton's *Principia*, *American Journal of Physics* **50**, 617–28.

Cushing, J. T. (1990a), Is Scientific Methodology Interestingly Atemporal?, *British Journal for the Philosophy of Science* **41**, 177–94.

Cushing, J. T. (1990b), *Theory Construction and Selection in Modern Physics: The S Matrix* (Cambridge University Press, Cambridge).

Cushing, J. T. (1991), Quantum Theory and Explanatory Discourse: Endgame for Understanding?, *Philosophy of Science* **58**, 337–58.

Cushing, J. T. (1994), *Quantum Mechanics: Historical Contingency and the Copenhagen Hegemony* (University of Chicago Press, Chicago).

Cushing, J. T., Fine, A. and Goldstein S. (eds.) (1996), *Bohmian Mechanics and Quantum Theory: An Appraisal* (Kluwer Academic Publishers, Dordrecht).

Cushing, J. T. and McMullin, E. (eds.) (1989), *Philosophical Consequences of Quantum Theory: Reflections on Bell's Theorem* (University of Notre Dame Press, Notre Dame, IN).

Darrigol, O. (1992), *From c-Numbers to q-Numbers* (University of California Press, Berkeley, CA).

Darwin, G. H. (1898), *The Tides and Kindred Phenomena in the Solar System* (Houghton, Mifflin and Co., Boston).

Daston, L. (1988), *Classical Probability in the Enlightenment* (Princeton University Press, Princeton, NJ).

Dawkins, R. (1986), *The Blind Watchmaker* (W. W. Norton & Co., New York).

Debs, T. A. and Redhead, M. L. G. (1996), The Twin 'Paradox' and the Conventionality of Simultaneity, *American Journal of Physics* **64**, 384–92.

De Duve, C. (1995), *Vital Dust* (Basic Books, New York).

Defant, A. (1958), *Ebb and Flow: The Tides of Earth, Air, and Water* (University of Michigan Press, Ann Arbor, MI).

Densmore, D. (1995), *Newton's Principia: The Central Argument* (Green Lion Press, Santa Fe, NM).

Descartes, R. (1905), *Principia Philosophiae* in Adams and Tannery, Vol. VIII.

Descartes, R. (1977a), *Rules for the Direction of the Mind* in Haldane and Ross.

Descartes, R. (1977b), *The Principles of Philosophy* in Haldane and Ross.

Diacu, F. and Holmes, P. (1996), *Celestial Encounters: The Origins of Chaos and Stability* (Princeton University Press, Princeton, NJ).

Dickson, M. (1993), Stapp's Theorem Without Counterfactual Commitments: Why It Fails Nonetheless, *Studies in History and Philosophy of Science* **24**, 791–814.

Dijksterhuis, E. J. (1986), *A Mechanization of the World Picture* (Princeton University Press, Princeton, NJ).

Dirac, P. A. M. (1958), *The Principles of Quantum Mechanics*, 4th edn (Oxford University Press, Oxford).

Dobbs, B. J. T. (1975), *The Foundations of Newton's Alchemy* (Cambridge University Press, Cambridge).

Donne, J. (1959), *Devotions Upon Emergent Occasions* (University of Michigan Press, Ann Arbor, MI).

Drake S. (ed.) (1957), *Discoveries and Opinions of Galileo* (Anchor Books, Garden City, NY).

Drake, S. (1967), 'Galileo: A Biographical Sketch' in McMullin (1967a), pp. 52–66.

Dresden, M. (1992), Chaos: A New Scientific Paradigm – or Science by Public Relations?, Parts I and II, *The Physics Teacher* **30** (1), 10–14, (2), 74–80.

Dugas, R. (1988), *A History of Mechanics* (Dover Publications, New York).

Duhem, P. (1969), *To Save the Phenomena* (University of Chicago Press, Chicago).

Duhem, P. (1974), *The Aim and Structure of Physical Theory* (Atheneum, New York).

Dürr, D., Goldstein, S. and Zanghi, N. (1996), 'Bohmian Mechanics as the Foundation of Quantum Mechanics' in Cushing *et al.*, pp. 21–44.

Earman, J. (ed.) (1983), *Testing Scientific Theories* (University of Minnesota Press, Minneapolis, MN).

Earman, J. (1986), *A Primer on Determinism* (D. Reidel Publishing Co., Dordrecht).

Earman, J. and Norton, J. (eds.) (1997), *The Cosmos of Science: Essays of Exploration* (University of Pittsburgh Press, Pittsburgh) (to be published).

Eddington, A. S. (1920), *Space, Time, and Gravitation* (Cambridge University Press, Cambridge).

Eddington, A. S. (1926), *The Internal Constitution of the Stars* (Cambridge University Press, Cambridge).

Edwards, P. (ed.), *The Encyclopedia of Philosophy*, 8 Vols. (Macmillan Publishing Co., New York).

Einstein, A. (n.d.a), 'On the Electrodynamics of Moving Bodies' in Lorentz *et al.*, pp. 37–65.

Einstein, A. (n.d.b), 'Does the Inertia of Body Depend upon its Energy-Content?' in Lorentz *et al.*, pp. 67–71.

Einstein, A. (n.d.c), 'On the Influence of Gravitation on the Propagation of Light' in Lorentz *et al.*, pp. 97–108.

Einstein, A. (n.d.d), 'The Foundation of the General Theory of Relativity' in Lorentz *et al.*, pp. 109–64.

Einstein, A. (n.d.e), 'Cosmological Considerations on the General Theory of Relativity' in Lorentz *et al.*, pp. 174–88.

Einstein, A. (1905a), Zur Elektrodynamik bewegter Körper, *Annalen der Physik* **17**, 891–921. (Appears in English translation in Einstein n.d.a.) See also Miller (1981, 392–415).

Einstein, A. (1905b), Ist die Trägheit eines Körpers von seinem Energiegehalt abhängig?, *Annalen der Physik* **17**, 639–41. (Appears in English translation in Einstein n.d.b.)

Einstein, A. (1905c), Die von molekularkinetischen Theorie der Wärme geforderte Bewegung von in ruhenden Flüssigkeiten suspendierten Teilchen, *Annalen der Physik* **17**, 549–60. (Appears in English translation in Einstein (1926).)

Einstein, A. (1907), Über das Relativitätsprinzip und die aus demselben gezogenen Folgerungen, *Jahrbuch der Radioaktivität und Elektronik* **4**, 411–62.

Einstein, A. (1908), Elementare Theorie der Brownschen Bewegung, *Zeitschrift für Elektrochemie* **14**, 235–9. (Appears in English translation as 'The Elementary Theory of Brownian Motion' in Boorse and Motz (1966, 587–96).)

Einstein, A. (1911), Über den Einfluss der Schwerkraft auf die Ausbreitung des Lichtes, *Annalen der Physik* **35**, 898–908. (Appears in English translation in Einstein n.d.c.)

Einstein, A. (1916), Die Grundlage der allgemeinen Relativitätstheorie, *Annalen der Physik* **49**, 769–822. (Appears in English translation in Einstein n.d.d.)

Einstein, A. (1917a), Zur Quantentheorie der Strahlung, *Physikalische Zeitschrift* **18**, 121–8. (Appears in English translation as 'On the Quantum Theory of Radiation' in van der Waerden (1967, 63–77) and in ter Haar (1967, 167–83).)

Einstein, A. (1917b), Kosmologische Betrachtungen zur allgemeinen Relativitätstheorie, *Sitzungsberichte der Preussischen Akademie der Wissenschaften* 1917, Pt. I, 142–52. (Appears in English translation in Einstein n.d.e.)

Einstein, A. (1926), *Investigations on the Theory of the Brownian Movement* (Methuen & Co., London).

Einstein, A. (1934), *The World As I See It* (Covici Friede Publishers, New York).

Einstein, A. (1949a), 'Autobiographical Notes' in Schilpp, pp. 1–95.

Bibliography

Einstein, A. (1949b), 'Remarks to the Essays Appearing in this Co-Operative Volume' in Schilpp, pp. 663–88.

Einstein, A. (1954a), *Ideas and Opinions* (Crown Publishers, New York).

Einstein, A. (1954b), 'On the Method of Theoretical Physics' in Einstein (1954a), pp. 270–6.

Einstein, A. (1954c), 'Maxwell's Influence on the Evolution of the Idea of Physical Reality' in Einstein (1973a), pp. 266–70.

Einstein, A. (1954d), 'What Is the Theory of Relativity?' in Einstein (1954a), pp. 227–32.

Einstein, A. (1954e), 'Principles of Research' in Einstein (1954a), pp. 224–7.

Einstein, A. (1954f), 'The Problem of Space, Ether, and the Field in Physics' in Einstein (1954a), pp. 276–85.

Einstein, A. (1954g), 'Physics and Reality' in Einstein (1954a), pp. 290–323.

Einstein, A. (1982), How I Created the Theory of Relativity, *Physics Today* **35** (8), 45–7.

Einstein, A. and Infeld, L. (1938), *The Evolution of Physics* (Simon and Schuster, New York).

Einstein, A., Podolsky, B. and Rosen, N. (1935), Can Quantum-Mechanical Description of Physical Reality Be Considered Complete?, *Physical Review* **47**, 777–80. (Reprinted in Wheeler and Zurek (1983, 138–41).)

Eisberg, R. M. (1961), *Fundamentals of Modern Physics* (John Wiley & Sons, New York).

Ellis, G. F. R. and Williams, R. M. (1988), *Flat and Curved Space-Times* (Clarendon Press, Oxford).

d'Espagnat, B. (1979), The Quantum Theory and Reality, *Scientific American* **241** (5), 158–81.

Fahie, J. J. (1903), *Galileo, His Life and Work* (John Murray Publishers, London).

Faraday, M. (1952), *Experimental Researches in Electricity* in Hutchins, Vol. 45.

Faye, J. (1991), *Niels Bohr: His Heritage and Legacy* (Kluwer Academic Publishers, Dordrecht).

Feinberg, G. (1965), Fall of Bodies Near the Earth, *American Journal of Physics* **33**, 501–2.

Feynman, R. (1965), *The Character of Physical Law* (The MIT Press, Cambridge, MA).

Feynman, R. P. and Gell-Mann, M. (1958), Theory of the Fermi Interaction, *Physical Review* **109**, 193–8.

Field, J. V. (1988), *Kepler's Geometrical Cosmology* (University of Chicago Press, Chicago).

Fine, A. (1982a), Hidden Variables, Joint Probability, and the Bell Inequalities, *Physical Review Letters* **48**, 291–5.

Fine, A. (1982b), Antinomies of Entanglement: The Puzzling Case of the Tangled Statistics, *The Journal of Philosophy* **79**, 733–47.

Fine, A. (1986a), *The Shaky Game: Einstein, Realism and the Quantum Theory* (University of Chicago Press, Chicago).

Fine, A. (1986b), Unnatural Attitudes: Realist and Instrumentalist Attachments to Science, *Mind* **95**, 149–79.

Finocchiaro, M. A. (1989), *The Galileo Affair* (University of California Press, Berkeley, CA).

Fock, V. (1950), *The Theory of Space Time and Gravitation* (Pergamon Press, New York).

Ford, J. (1983), How Random is a Coin Toss?, *Physics Today* **36** (4), 40–7.

Frank, P. (1947), *Einstein: His Life and Times* (Alfred A. Knopf, New York).

Frank, P. (1949), Einstein's Philosophy of Science, *Reviews of Modern Physics* **21**, 349–55.

Frank, P. (1957), *Philosophy of Science* (Prentice-Hall, Englewood Cliffs, NJ).

Franklin, A. (1976), *The Principle of Inertia in the Middle Ages* (Colorado Associated University Press, Boulder, CO).

Friedman, M. (1992), *Kant and the Exact Sciences* (Harvard University Press, Cambridge, MA).

Frisch, D. H. and Smith, J. H. (1963), Measurement of the Relativistic Time Dilation Using μ-Mesons, *American Journal of Physics* **31**, 342–55.

Freud, S. (1965), *New Introductory Lectures on Psychoanalysis* (W. W. Norton & Co., New York).

Galilei, G. (1946), *Dialogues Concerning Two New Sciences* (Northwestern University Press, Evanston, IL).

Galilei, G. (1967), *Dialogue Concerning the Two Chief World Systems – Ptolemaic & Copernican* (University of California Press, Berkeley, CA).

Gardner, M. R. (1983) 'Realism and Instrumentalism in Pre-Newtonian Astronomy' in Earman, pp. 201–65.

Gell-Mann, M. (1981), 'Questions for the Future' in Mulvey, pp. 169–86.

Gershenson, D. E. and Greenberg, D. A. (eds.) (1964), *The Natural Philosopher*, Vol. 3

(Blaisdell Publishing Co., New York).
Gillispie, C. G. (1960), *The Edge of Objectivity* (Princeton University Press, Princeton, NJ).
Gillispie, C. G. (1970–6), *Dictionary of Scientific Biography*, 14 Vols. (Charles Scribner's Sons, New York).
Gingerich, O. (ed.) (1977), *Cosmology + 1* (W. H. Freeman and Company, San Francisco).
Gleick, J. (1987), *Chaos: Making a New Science* (Viking Press, New York).
Godzinski, Z. (1991), Investigations of Light Interference at Extremely Low Intensities, *Physics Letters A* **153**, 291–8.
Goldberg, A., Schey, H. M. and Schwartz, J. L. (1967), Computer-Generated Motion Pictures of One-Dimensional Quantum-Mechanical Transmission and Reflection Phenomena, *American Journal of Physics* **35**, 177–86.
Goldberg, S. (1970–1), The Abraham Theory of the Electron: The Symbiosis of Experiment and Theory, *Archive for History of Exact Sciences* **7**, 7–25.
Goldreich, P. (1972), Tides and the Earth–Moon System, *Scientific American* **226** (4), 43–52.
Goldstein, H. (1950), *Classical Mechanics* (Addison-Wesley Publishing Co., Reading, MA).
Goodman, N. (1946), A Query on Confirmation, *Journal of Philosophy* **43**, 383–5.
Goodman, N. (1965), *Fact, Fiction, and Forecast*, 2nd edn (Bobbs-Merrill Co., Indianapolis, IN).
Gott, J. R., Gunn, J. E., Schramm, D. N. and Tinsley, B. M. (1977), 'Will the Universe Expand Forever?' in Gingerich, pp. 82–93. (Appeared originally in *Scientific American* **234** (3) (1976), 62–79.)
Gould, S. J. (1989), *Wonderful Life: The Burgess Shale and the Nature of History* (W. W. Norton & Co., New York).
Green, A. W. (1966), *Sir Francis Bacon* (Twayne Publishers, New York).
Guckenheimer, J. and Holmes, P. (1983), *Nonlinear Oscillations, Dynamical Systems, and Bifurcations of Vector Fields* (Springer-Verlag, New York).
Guye, C.-E. and Lavanchy, C. (1915), Vérification expérimentale de la formule de Lorentz–Einstein par les rayons cathodiques de grand vitesse, *Comptes Rendus* **161**, 52–5.
Hacking, I. (1975), *The Emergence of Probability* (Cambridge University Press, London).

Hafele, J. C. (1972), Relativistic Time for Terrestrial Circumnavigations, *American Journal of Physics* **40**, 81–5.
Hafele, J. C. and Keating, R. E. (1972), Around the World Atomic Clocks, *Science* **177**, 166–70.
Haldane, E. S. and Ross, G. R. T. (eds.) (1977), *The Philosophical Works of Descartes*, Vol. I (Cambridge University Press, Cambridge).
Hanle, P. A. (1979), Indeterminacy Before Heisenberg: The Case of Franz Exner and Erwin Schrödinger, *Historical Studies in the Physical Sciences* **10**, 225–69.
Harding, S. G. (ed.) (1976), *Can Theories Be Refuted?* (D. Reidel Publishing Co., Dordrecht).
Harman, P. M. (1982), *Energy, Force and Matter: The Conceptual Development of Nineteenth-Century Physics* (Cambridge University Press, New York).
Harrison, E. R. (1981), *Cosmology, The Science of the Universe* (Cambridge University Press, Cambridge).
Harrison, E. R. (1985), *Masks of the Universe* (Macmillan Publishing Co., New York).
Harrison, E. R. (1986), Newton and the Infinite Universe, *Physics Today* **39** (2), 24–32.
Harvey, W. (1952), *A Second Disquisition to John Riolan* in Hutchins, Vol. 28.
Heath, T. L. (ed.) (1896), *Apollonius of Perga: Treatise on Conic Sections* (Cambridge University Press, Cambridge).
Heath, T. L. (ed.) (1897), *The Works of Archimedes* (Cambridge University Press, Cambridge).
Heath, T. L. (1981a), *Aristarchus of Samos* (Dover Publications, New York).
Heath, T. L. (1981b), *A History of Greek Mathematics*, Vols. I and II (Dover Publications, New York).
Heilbron, J. L. (1985), Review of Jagdish Mehra and Helmut Rechenberg's '*The Historical Development of Quantum Theory*', *Isis* **76**, 388–93.
Heilbron, J. L. (1988), 'The Earliest Missionaries of the Copenhagen Spirit' in Ullmann-Margalit, pp. 201–33.
Heilbron, J. L. and Kuhn, T. S. (1969), The Genesis of the Bohr Atom, *Historical Studies in the Physical Sciences* **1** (University of Pennsylvania Press, Philadelphia), pp. 211–90.

Bibliography

Heisenberg, W. (1927), Über den anschaulichen Inhalt der quantentheoretischen Kinematik und Mechanik, *Zeitschrift für Physik* **43**, 172–98. (Appears in English translation as 'The Physical Content of Quantum Kinematics and Mechanics' in Wheeler and Zurek (1983, 62–84).)

Heisenberg, W. (1958), *Physics and Philosophy* (Harper & Row, New York).

Heisenberg, W. (1971), *Physics and Beyond* (Harper & Row, New York).

Hempel, C. G. (1945), Studies in the Logic of Confirmation, *Mind* **54**, 1–26 and 97–121. (Also reprinted in Hempel (1965), pp. 3–46.)

Hempel, C. G. (1965), *Aspects of Scientific Explanation* (The Free Press, New York).

Hendry, J. (1984), *The Creation of Quantum Mechanics and the Bohr–Pauli Dialogue* (D. Reidel Publishing Co., Dordrecht).

Herivel, J. (1965), *The Background to Newton's Principia* (Oxford University Press, Oxford).

Herschel, J. F. (1830), *A Preliminary Discourse on the Study of Natural Philosophy* (Longman, Rees, Orme, Brown & Green, London).

Hertz, H. (1900), *Electric Waves* (Macmillan Publishing Co., London).

Holland, P. R. (1993), *The Quantum Theory of Motion* (Cambridge University Press, Cambridge).

Holton, G. (1968), Mach, Einstein and the Search for Reality, *Daedalus* **97** (2), 636–73.

Holton, G. (1973), *Thematic Origins of Scientific Thought* (Harvard University Press, Cambridge, MA).

Holton, G. (1978), *The Scientific Imagination: Case Studies* (Cambridge University Press, Cambridge).

Holton, G. (1996), *Einstein, History, and Other Passions* (Addison-Wesley Publishing Co., Reading, MA).

Hon, G. (1996), 'Gödel, Einstein, Mach: Completeness of Physical Theory' in Schimanovich *et al.* (to be published).

Howard, D. (1997), 'A Peek Behind the Veil of Maya: Einstein, Schopenhauer, and the Historical Background of the Conception of Space as a Ground for the Individuation of Physical Systems' in Earman and Norton (to be published).

Howson, C. (ed.) (1976), *Method and Appraisal in the Physical Sciences* (Cambridge University Press, Cambridge).

Hughes, R. J. (1990), The Bohr–Einstein 'Weighing-of-Energy' Debate and the Principle of Equivalence, *American Journal of Physics* **58**, 826–8.

Hume, D. (1902), *An Enquiry Concerning Human Understanding* (Oxford University Press, Oxford).

Hutchins, R. M. (ed.) (1952), *Great Books of the Western World* (Encyclopaedia Britannica, Chicago).

Huygens, C. (1920), *Oeuvres Complètes de Christiaan Huygens*, Vol. 14 (Martinus Nijhoff, The Hague).

Huygens, C. (1934), *Oeuvres Complètes de Christiaan Huygens*, Vol. 18 (Martinus Nijhoff, The Hague).

Huygens, C. (1952), *Treatise on Light* in Hutchins, Vol. 34.

Jackson, J. D. (1975), *Classical Electrodynamics*, 2nd edn (John Wiley & Sons, New York).

Jaki, S. L. (1966), *The Relevance of Physics* (University of Chicago Press, Chicago).

Jammer, M. (1957), *Concepts of Force* (Harvard University Press, Cambridge, MA).

Jammer, M. (1961), *Concepts of Mass* (Harvard University Press, Cambridge, MA).

Jammer, M. (1969), *Concepts of Space*, 2nd edn (Harvard University Press, Cambridge, MA).

Jammer, M. (1974), *The Philosophy of Quantum Mechanics* (John Wiley & Sons, New York).

Jammer, M. (1989), *The Conceptual Development of Quantum Mechanics*, 2nd edn (Tomash Publishing Co., New York).

Jefferson, T. (1952), *The Declaration of Independence* in Hutchins, Vol. 43.

Jensen, R. V. (1987), Classical Chaos, *American Scientist* **75** (2), 168–81.

The Jerusalem Bible (1966) (Doubleday & Co., Garden City, NY).

Jevons, W. S. (1958), *The Principles of Science: A Treatise on Logic and Scientific Method* (Dover Publications, New York).

Jowett, B. (ed.) (1892), *The Dialogues of Plato*, Vols. I–V (Macmillan Publishing Co., New York).

Kamefuchi, S., Ezawa, H., Murayama, Y., Namiki, M., Nomura, S., Ohnuki, Y. and Yojima, T. (eds.) (1984), *Proceedings of the International Symposium on the Foundations of Quantum Mechanics in the Light of New Technology* (Physical Society of Japan, Tokyo).

Kaufmann, W. (1902), Die elektromagnetische Masse des Elektrons, *Physikalische Zeitschrift* **4**, 54–7.

Kaufmann, W. (1905), Über die Konstitution des Elektrons, *Sitzungsberichte der Königlich Preussischen Akademie der Wissenschaften* **45**, 949–56.

Kellert, S. H. (1993), *In the Wake of Chaos* (University of Chicago Press, Chicago).

Kelvin, Lord (1901), Nineteenth Century Clouds Over the Dynamical Theory of Heat and Light, *Philosophical Magazine* **2**, 1–40. (Reprinted in Kelvin (1904), pp. 486–527).

Kelvin, Lord (1904), *Baltimore Lectures on Molecular Dynamics and the Wave Theory of Light* (C. J. Clay & Sons, London).

Kepler, J. (1937), *Astronomia Nova* in Caspar, Vol. 3.

Kepler, J. (1952a), *Epitome of Copernican Astronomy* in Hutchins, Vol. 16.

Kepler, J. (1952b), *Harmonies of the World* in Hutchins, Vol. 16.

Kepler, J. (1992), *New Astronomy* (Cambridge University Press, Cambridge).

Keynes, J. M. (1963), 'Newton, the Man' in *Essays in Biography* (W. W. Norton & Co., New York), pp. 310–23.

Klein, M. J. (1962), Max Planck and the Beginnings of the Quantum Theory, *Archive for History of Exact Sciences* **1**, 459–79.

Klein, M. J. (1964), 'Einstein and the Wave–Particle Duality' in Gershenson and Greenberg, pp. 1–49.

Kockelmans, J. J. (ed.) (1968), *Philosophy of Science* (The Free Press, New York).

Koestler, A. (1959), *The Sleepwalkers* (Macmillan Publishing Co., New York).

Kourany, J. A. (ed.) (1987), *Scientific Knowledge* (Wadsworth Publishing Co., Belmont, CA).

Koyré, A. (1957), *From the Closed World to the Infinite Universe* (Johns Hopkins University Press, Baltimore, MD).

Kuhn, T. S. (1957), *The Copernican Revolution* (Harvard University Press, Cambridge, MA).

Kuhn, T. S. (1970), *The Structure of Scientific Revolutions*, 2nd edn (University of Chicago Press, Chicago).

Kuhn, T. S. (1978), *Black-Body Theory and the Quantum Discontinuity: 1894–1912* (Clarendon Press, Oxford).

Kuntz, P. G. (ed.) (1968), *The Concepts of Order* (University of Washington Press, Seattle, WA).

Lakatos, I. (1970), 'Falsification and the Methodology of Scientific Research Programmes' in Lakatos and Musgrave, pp. 91–196.

Lakatos, I. (1976), 'History of Science and Its Rational Reconstructions' in Howson, pp. 1–39.

Lakatos, I. (1978), *The Methodology of Scientific Research Programmes* (Cambridge University Press, Cambridge).

Lakatos, I and Musgrave, A. (eds.) (1970), *Criticism and the Growth of Knowledge* (Cambridge University Press, London).

Lakatos, I. and Zahar, E. (1978), 'Why Did Copernicus' Research Program Supersede Ptolemy's?' in Lakatos (1978), pp. 168–92.

Landau, L. D. and Lifshitz, E. M. (1977), *Quantum Mechanics: Non-Relativistic Theory*, 3rd edn (Pergamon Press, Oxford).

Laplace, P. S. (1886), *Théorie Analytique des Probabilités* in *Oeuvres Complètes de Laplace*, Vol. VII (Gauthier-Villars, Paris).

Laplace, P. S. (1902), *A Philosophical Essay on Probabilities* (John Wiley & Sons, New York).

Laudan, L. (1981), *Science and Hypothesis* (D. Reidel Publishing Co., Dordrecht).

Laudan, L. (1984), *Science and Values* (University of California Press, Berkeley, CA).

Laudan, L. and Leplin, J. (1991), Empirical Equivalence and Underdetermination, *The Journal of Philosophy* **88** (9), 449–72.

Leighton, R. B. (1959), *Principles of Modern Physics* (McGraw-Hill Book Co., New York).

Leplin, J. (ed.) (1984), *Scientific Realism* (University of California Press, Berkeley, CA).

Livingston, D. M. (1973), *The Master of Light: A Biography of Albert A. Michelson* (Charles Scribner's Sons, New York).

Longair, M. (1984), *Theoretical Concepts in Physics* (Cambridge University Press, Cambridge).

Lorentz, H. A. (n.d.), 'Michelson's Interference Experiment' in Lorentz *et al.*, pp. 3–7.

Lorentz, H. A. (1952), *The Theory of Electrons* (Dover Publications, New York).

Lorentz, H. A., Einstein, A., Minkowski, H. and Weyl, H. (n.d.), *The Principle of Relativity* (Dover Publications, New York).

Lovejoy, A. O. (1936), *The Great Chain of Being* (Harvard University Press, Cambridge, MA).

Bibliography

Lucretius (1952), *On the Nature of Things* in Hutchins, Vol. 12.

Luria, S. E. (1973), *Life – The Unfinished Experiment* (Charles Scribner's Sons, New York).

Mach, E. (1960), *The Science of Mechanics*, 6th edn (Open Court Publishing Co., La Salle, IL).

MacKinnon, E. M. (1982), *Scientific Explanation and Atomic Physics* (University of Chicago Press, Chicago).

Magie, W. F. (1963), *A Source Book in Physics* (Harvard University Press, Cambridge, MA).

Mansouri, R. and Sexl, R. U. (1977a), A Test Theory of Special Relativity: I. Simultaneity and Clock Synchronization, *General Relativity and Gravitation* **8**, 497–513.

Mansouri, R. and Sexl, R. U. (1977b), A Test Theory of Special Relativity: II. First-Order Tests, *General Relativity and Gravitation* **8**, 515–24.

Mansouri, R. and Sexl, R. U. (1977c), A Test Theory of Special Relativity: III. Second-Order Tests, *General Relativity and Gravitation* **8**, 809–14.

Manuel, F. E. (1968), *A Portrait of Isaac Newton* (Harvard University Press, Cambridge, MA).

Maudlin, T. (1994), *Quantum Non-Locality and Relativity* (Basil Blackwell Publishers, Oxford).

Maudlin, T. (1996), 'Space–Time in the Quantum World' in Cushing *et al.*, pp. 285–307.

Maxwell, J. C. (1890), *The Scientific Papers of James Clerk Maxwell* (Cambridge University Press, Cambridge).

Maxwell, J. C. (1954), *A Treatise on Electricity and Magnetism* (Dover Publications, New York).

Maxwell, N. (1985), Are Probabilism and Special Relativity Incompatible?, *Philosophy of Science* **52**, 23–43.

McCauley, J. L. (1993), *Chaos, Dynamics, and Fractals* (Cambridge University Press, Cambridge).

McMullin, E. (ed.) (1967a), *Galileo, Man of Science* (Basic Books, New York).

McMullin, E. (1967b), 'Introduction: Galileo, Man of Science' in McMullin (1967a), pp. 3–51.

McMullin, E. (1968) 'Cosmic Order in Plato and Aristotle' in Kuntz, pp. 63–76.

McMullin, E. (1978a), *Newton on Matter and Activity* (University of Notre Dame Press, Notre Dame, IN).

McMullin, E. (1978b), 'Philosophy of Science and its Rational Reconstructions' in Radnitzky and Andersson, pp. 221–52.

McMullin, E. (1984), 'A Case for Scientific Realism' in Leplin, pp. 8–40.

Melchior, P. (1966), *The Earth Tides* (Pergamon Press, Oxford).

Meyerson, E. (1930), *Identity and Reality* (George Allen & Unwin, London).

Mill, J. S. (1855), *System of Logic* (Harper & Brothers, New York).

Miller, A.I. (1981), *Albert Einstein's Special Theory of Relativity: Emergence (1905) and Early Interpretation (1905–1911)* (Addison-Wesley Publishing Co., Reading, MA).

Monod, J. (1971), *Chance and Necessity* (Alfred A. Knopf, New York).

Moore, W. (1989), *Schrödinger: Life and Thought* (Cambridge University Press, Cambridge).

More, L. T. (1934), *Isaac Newton, A Biography* (Charles Scribner's Sons, New York).

Moulton, F. R. (1914), *An Introduction to Celestial Mechanics* (Macmillan Publishing Co., New York).

Mulvey, J. H. (ed.) (1981), *The Nature of Matter* (Oxford University Press, Oxford).

Nelson, E. (1966), Derivation of the Schrödinger Equation from Newtonian Mechanics, *Physical Review* **150**, 1079–85.

Nelson, E. (1967), *Dynamical Theories of Brownian Motion* (Princeton Unversity Press, Princeton, NJ).

Neugebauer, O. (1968), 'On the Planetary Theory of Copernicus' in Beer, pp. 89–103.

Neugebauer, O. (1969), *The Exact Sciences in Antiquity* (Dover Publications, New York).

Neumann, G. (1914), Die träge Masse schnell bewegter Elektronen, *Annalen der Physik* **45**, 529–79.

Newton, I. (1934), *Mathematical Principles of Natural Philosophy and His System of the World* (University of California Press, Berkeley, CA).

Newton, I. (1952), *Optics* in Hutchins, Vol. 34.

Pais, A. (1982), *Subtle is the Lord* (Clarendon Press, Oxford).

Park, D. (1988), *The How and the Why* (Princeton University Press, Princeton, NJ).

Pauli, W. (1981), *Theory of Relativity* (Dover

Publications, New York).
Peacock, G. (ed.) (1855), *Miscellaneous Works of the Late Thomas Young*, Vol. I. (John Murray Publishers, London).
Pederson, O. (1993), *Early Physics and Astronomy*, rev. edn (Cambridge University Press, Cambridge).
Peres, A. (1978), Unperformed Experiments Have No Results, *American Journal of Physics* **46**, 745–7.
Petruccioli, S. (1993), *Atoms, Metaphors and Paradoxes* (Cambridge University Press, Cambridge).
Pickering, A. (1984), *Constructing Quarks: A Sociological History of Particle Physics* (University of Chicago Press, Chicago).
Planck, M. (1906a), Das Prinzip der Relativität und die Grundgleichungen der Mechanik, *Verhandlungen der Deutschen Physikalischen Gesellschaft* **8**, 136–41.
Planck, M. (1906b), Die Kaufmannschen Messungen der Ablenkbarkeit der β-Strahlen in ihrer Bedeutung für die Dynamik der Elektronen, *Verhandlungen der Deutschen Physikalischen Gesellschaft* **8**, 418–32.
Planck, M. (1907), Nachtrag zu der Besprechung der Kaufmannschen Ablenkungsmessungen, *Verhandlungen der Deutschen Physikalischen Gesellschaft* **9**, 301–5.
Planck, M. (1949), *Scientific Autobiography and Other Papers* (Philosophical Library, New York).
Plato (1892), *Timaeus* in Jowett, Vol. III, pp. 437–515.
Poincaré, H. (1952), *Science and Hypothesis* (Dover Publications, New York).
Popper, K. R. (1965), *The Logic of Scientific Discovery* (Harper & Row, New York).
Popper, K. R. (1968), *Conjectures and Refutations* (Harper & Row, New York).
Pound, R. V. and Rebka, G. A. (1960), Apparent Weight of Photons, *Physical Review Letters* **4**, 337–41.
Prigogine, I. and Stengers, I. (1984), *Order Out of Chaos* (Bantam Books, New York).
Przibram, K. (ed.) (1967), *Letters on Wave Mechanics* (Philosophical Library, New York).
Ptolemy (1952), *The Almagest* in Hutchins, Vol. 16.
Quine, W. V. (1951), Two Dogmas of Empiricism, *The Philosophical Review* **60**, 20–43.

Quine, W. V. (1990), *Pursuit of Truth* (Harvard University Press, Cambridge, MA).
Quirino, C. (1963), *Philippine Cartography* (N. Israel, Amsterdam).
Radnitzky, G. and Andersson, G. (eds.) (1978), *Progress and Rationality in Science* (D. Reidel Publishing Co., Dordrecht).
Rasband, S. N. (1990), *Chaotic Dynamics of Nonlinear Systems* (John Wiley & Sons, New York).
Redhead, M. L. G. (1987), *Incompleteness, Nonlocality, and Realism* (Clarendon Press, Oxford).
Reitz, J. R. and Milford, F. J. (1967), *Foundations of Electromagnetic Theory*, 2nd edn (Addison-Wesley Publishing Co., New York).
Rohrlich, F. and Hardin, L. (1983), Established Theories, *Philosophy of Science* **50**, 603–17.
Ronan, C. A. (1974), *Galileo* (G. P. Putnam's Sons, New York).
Rose, V. (1886), *Aristotelis Fragmenta* (B. G. Teubner, Leipzig).
Rosen, E. (ed.) (1959), *Three Copernican Treatises* (Dover Publications, New York).
Rosenfeld, L (1965), Newton and the Law of Gravitation, *Archive for History of Exact Sciences* **2**, 365–86.
Ross, W. D. (ed.) (1942), *The Student's Oxford Aristotle*, Vols. I–VI (Oxford University Press, New York).
Russell, B. (1917), *Mysticism and Logic*, 2nd edn (George Allen & Unwin, London).
Sachs, M. (1971), A Resolution of the Clock Paradox, *Physics Today* **24** (9), 23–9. (Replies in The Clock 'Paradox' – Majority View, *Physics Today* **25** (1) (1972), 9–15, 47–51.)
Sambursky, S. (1975), *Physical Thought From the Presocratics to the Quantum Physicists* (Pica Press, New York).
de Santillana, G. (1955), *The Crime of Galileo* (University of Chicago Press, Chicago).
Schilpp, P. A. (ed.) (1949), *Albert Einstein: Philosopher–Scientist*; Library of Living Philosophers, Vol. 16 (Open Court Publishing Co., La Salle, IL).
Schimanovich, W., Buldt, B., Köhler, E. and Weibel, P. (eds.) (1996), *Wahrheit und Beweisbarkeit. Leben und Werk Kurt Gödels* (Hölder–Pichler–Tempsky, Wien) (to be published).
Schrödinger, E. (1935), Die Gegenwärtige Situation in der Quantenmechanik, *Die*

Bibliography

Naturwissenschaften **23**, 807–12, 823–8, 844–9. (Appears in English translation as 'The Present Situation in Quantum Mechanics' in Wheeler and Zurek (1983, 152–67).)

Schrödinger, E. (1944), *What Is Life?* (Cambridge University Press, Cambridge).

Schuster, H. G. (1988), *Deterministic Chaos*, 2nd rev. edn (VCH, New York).

Segre, M. (1980), The Role of Experiment in Galileo's Physics, *Archive for History of Exact Sciences* **23**, 227–52.

Segre, M. (1989), Galileo, Viviani and the Tower of Pisa, *Studies in History and Philosophy of Science* **20**, 435–51.

Selleri, F. (1990), *Quantum Paradoxes and Physical Reality* (Kluwer Academic Publishers, Dordrecht).

Selleri, F. (1994), 'Theories Equivalent to Special Relativity' in Barone and Selleri, pp. 181–92.

Shimony, A. (1984), 'Controllable and Uncontrollable Non-Locality' in S. Kamefuchi et al., pp. 225–30.

Shipman, H. L. (1976), *Black Holes, Quasars, and The Universe* (Houghton Mifflin Co., Boston, MA).

Smith, A. J. (ed.) (1974), *John Donne, The Complete English Poems* (Allen Lane, London).

de Sitter, W. (1917), On Einstein's Theory of Gravitation and its Astronomical Consequences, *Monthly Notices of the Royal Astronomical Society* **78**, 3–28.

Stapp, H. P. (1989), 'Quantum Nonlocality and the Description of Nature' in Cushing and McMullin, pp. 154–74.

Stapp, H. P. (1990), Comments on 'Nonlocal Influences and Possible Worlds', *British Journal for the Philosophy of Science* **41**, 59–72.

Stein, H. (1991), On Relativity Theory and Openness of the Future, *Philosophy of Science* **58**, 147–67.

Stewart, I. (1989), *Does God Play Dice?* (Basil Blackwell Publishers, London).

Tangherlini, F. R. (1961), An Introduction to the General Theory of Relativity, *Supplemento del Nuovo Cimento* **20**, 1–86.

Taylor, E. F. and Wheeler, J. A. (1966), *Spacetime Physics* (W. H. Freeman and Company, San Francisco).

ter Haar, D. (1967), *The Old Quantum Theory* (Pergamon Press, London).

Thayer, H. S. (ed.) (1953), *Newton's Philosophy of Nature* (Hafner Press, New York).

Thompson, J. M. T. and Stewart, H. B. (1986), *Nonlinear Dynamics and Chaos* (John Wiley & Sons, New York).

Tolstoy, I. (1982), *James Clerk Maxwell* (University of Chicago Press, Chicago).

Truesdell, C. (1960–2), A Program Toward Rediscovering the Rational Mechanics of the Age of Reason, *Archive for History of Exact Sciences* **1**, 3–36.

Ullmann-Margalit, E. (ed.) (1988), *Science in Reflection* (D. Reidel Publishing Co., Dordrecht).

Valentini, A. (1996), 'Pilot-Wave Theory of Fields, Gravitation and Cosmology' in Cushing et al., pp. 45–66.

Valentini, A. (1997), *On the Pilot-Wave Theory of Classical, Quantum and Subquantum Physics* (Springer-Verlag, Berlin) (to be published).

van der Waerden, B. L. (ed.) (1967), *Sources of Quantum Mechanics* (North-Holland Publishing Co., Amsterdam).

van Fraassen, B. C. (1980), *The Scientific Image* (Oxford University Press, Oxford).

vann Woodward, C. (1986), Gone with the Wind, *The New York Review of Books*, **33** (12), 3–6.

Vogt, E. (1996), Elementary Derivation of Kepler's Laws, *American Journal of Physics* **64**, 392–6.

von Neumann, J. (1955), *Mathematical Foundations of Quantum Mechanics* (Princeton University Press, Princeton, NJ).

Waldrop, M. M. (1991), Cosmologists Begin to Fill in the Blanks, *Science* **251**, 30–1.

Watkins, J. (1978), 'The Popperian Approach to Scientific Knowledge' in Radnitzky and Andersson, pp. 23–43.

Watson, J. D. (1970), *Molecular Biology of the Gene*, 2nd edn (W. A. Benjamin, New York).

Weimer, W. (1975), The Psychology of Inference and Expectation: Some Preliminary Remarks, *Minnesota Studies in the Philosophy of Science*, Vol. VI (University of Minnesota Press, Minneapolis, MN), pp. 430–86.

Weinberg, S. (1977), *The First Three Minutes* (Basic Books, New York).

Weinstock, R. (1982), Dismantling a Centuries-Old Myth: Newton's *Principia* and Inverse-Square Orbits, *American Journal of Physics* **50**, 610–17.

Weisskopf, V. F. (1960), The Visual Appearance

of Rapidly Moving Objects, *Physics Today* **13** (9), 24–7.

Westfall, R. S. (1962), The Foundations of Newton's Philosophy of Nature, *British Journal for the History of Science* **1**, 171–82.

Westfall, R. S. (1971), *Force in Newton's Physics* (American Elsevier, New York).

Westfall, R. S. (1980a), Newton's Marvelous Years of Discovery and Their Aftermath: Myth versus Manuscript, *Isis* **71**, 109–21.

Westfall, R. S. (1980b), *Never At Rest. A Biography of Isaac Newton* (Cambridge University Press, Cambridge).

Westfall, R. S. (1989), *Essays on the Trial of Galileo* (University of Notre Dame Press, Notre Dame, IN).

Wheeler, J. A. and Zurek, W. H. (eds.) (1983), *Quantum Theory and Measurement* (Princeton University Press, Princeton, NJ).

Whitaker, A. (1995), *Einstein, Bohr and the Quantum Dilemma* (Cambridge University Press, Cambridge).

White, H. E. (1934), *Introduction to Atomic Spectra* (McGraw-Hill Book Co., New York).

Whitehead, A. N. (1967), *Science and the Modern World* (The Free Press, New York).

Whiteside, D. T. (1964), Newton's Early Thoughts on Planetary Motion: A Fresh Look, *British Journal for the History of Science* **2**, 117–37.

Whiteside, D. T. (ed.) (1967–76), *The Mathematical Papers of Isaac Newton*, Vols I–VII (Cambridge University Press, Cambridge).

Whitrow, G. (ed.) (1973), *Einstein: The Man and His Achievement* (Dover Publications, New York).

Whittaker, E. (1973), *A History of the Theories of Aether and Electricity*, Vols. I and II (Humanities Press, New York).

Wylie, F. E. (1979), *Tides and the Pull of the Moon* (Stephen Greene Press, Brattleboro, VT).

Young, T. (1845), *A Course of Lectures on Natural Philosophy*, Vol. I (Taylor & Walton, London).

Zahar, E. (1976), 'Why Did Einstein's Programme Supersede Lorentz's?' in Howson, pp. 211–75.

Zahar, E. (1978), '"Crucial" Experiments: A Case Study' in Radnitzky and Andersson, pp. 71–97.

Zahar, E. (1989), *Einstein's Revolution: A Study in Heuristic* (Open Court Publishing Co., La Salle, IL).

Ziman, J. (1978), *Reliable Knowledge* (Cambridge University Press, Cambridge),

Author index

This index contains only citation references, listed by author. Proper names appearing in the text, or those actually discussed as such in the *Notes*, appear in the subject index that follows. *Note* numbers are given in parentheses. Multiple *Note* references for a page are given in the order in which they appear on that page.

Adler, C. G., 380 (16, 19)
Albert, D. Z , 394 (26)
Andersson, G., 377
Archimedes, 379 (3)
Aristotle, 28, 378 (2, 3, 3, 5, 10), 379 (13–17), 383 (10)
Armitage, A., 379 (17), 380 (7)
Aspect, A., 394 (24)

Bacon, F., 379 (12), 381 (8), 396 (42)
Baggott, J. E., 289, 304, 315, 330, 344, 394 (26)
Ball, R. S., 131
Becker, R. A., 384 (9)
Bell, J. S., 318, 330
Beller, M., 289, 390 (22), 391 (34, 35), 395 (25)
Ben-Menahem, Y, 395 (13, 15)
Bernstein, J., 240, 389 (14)
Bestelmeyer, A., 387 (32)
Biagioli, M., 147
Bitbol, M., 289
Blackwell, R. J., 380 (11)
Blake, R. M., 38
Bloch, F., 384 (27)
Bloor, D., 396 (29)
Bohm, D., 344, 393 (22), 394 (8, 10–12, 15–17), 396 (35)
Bohr, N., 271, 317, 390 (14, 17), 391 (33), 393 (2)
Boorse, H. A., 386 (1), 399
Bork, A. M., 385 (2)
Born, M., 317, 391 (4), 392 (3), 393 (3–6), 396 (6)
Bose, S. K., 270, 389 (28, 29, 36, 41, 42)
Brackenridge, J. B., 131
Briggs, K., 179, 384 (11)
Brody, B. A., 39
Bromberg, J., 385 (2)
Bucherer, A. H., 387 (33)

Buchwald, J. Z., 194, 207, 385 (2)
Burtt, E. A., 376, 397 (21)
Butterfield, H., 355, 396 (38, 39, 41)
Butterfield, J. N., 393 (19)

Cardwell, D. S. L., 385 (13)
Casper, B. M., 380 (16, 18)
Cassidy, D. C., 391 (35), 395 (25)
Chang, H., 251, 388 (13)
Clagett, M., 86
Clark, R. W., 240, 387 (1), 388 (5, 6, 10), 390 (10)
Clauser, J. F., 271, 393 (17, 18)
Clemence, G. M., 389 (17)
Clifton, R. K., 393 (19)
Cohen, I. B., 113, 380 (4), 381 (15–17), 382 (19, 22–4)
Cohen, M. R., 379 (16), 380 (2, 3), 398
Cooper, L., 86, 378 (8), 380 (2, 14)
Copernicus, N., 379 (14, 17), 380 (1–8)
Coulter, B. L., 380 (16, 19)
Crowe, M. J., 58, 270
Crutchfield, J. P., 179
Cushing, J. T., 113, 131, 222, 289, 330, 344, 355, 380 (5), 382 (26, 27, 4, 8, 11, 18), 386 (5, 6, 8, 10, 14, 15, 23, 24), 387 (27, 40), 390 (19), 391 (29, 36, 14), 392 (7, 14), 393 (15, 23), 394 (2, 10, 12, 13, 19, 22), 395 (1, 22, 26–9, 32), 396 (33, 39, 40)

Dalibard, J., 394 (24)
Darrigol, O., 289
Darwin, G., 131
Daston, L., 179
Dawkins, R., 377
Debs, T. A., 251, 388 (13)
Defant, A., 131
Densmore, D., 131

Descartes, R., 378 (5, 12), 380 (10), 383 (11, 12)
Diacu, F., 179
Dickson, M., 393 (19)
Dijksterhuis, E. J., 378 (11)
Dirac, P. A. M., 391 (7–10), 392 (17, 19, 16–19)
Dobbs, B. J. T., 102
Donne, J., 383 (3)
Drabkin, I. E., 379 (16), 380 (2, 3), 398
Drake, S., 86, 147, 383 (3, 4, 6, 8, 9, 11, 17, 5)
Dresden, M., 179, 385 (20)
Dugas, R., 378 (7)
Duhem, P., 58, 379 (11), 395 (3–6), 396 (26)
Dürr, D., 395 (29)
De Duve, C., 377, 397 (46)

Earman, J., 179
Eddington, A. S., 389 (13, 40)
Edwards, P., 398
Einstein, A., 1, 223, 271, 317, 357, 377, 387 (31, 3, 4, 7, 8, 10, 11, 15, 17, 18), 388 (6, 9, 11, 12, 2–4), 389 (11, 12, 27), 391 (28), 393 (1, 7, 9, 10, 12), 395 (16, 30), 396 (34, 2, 5, 7–9, 12)
Eisberg, R. M., 391 (6, 15)
Ellis, G. F. R., 270
The New Encyclopaedia Britannica, 398
d'Espagnat, B., 315

Fahie, J. J., 383 (2)
Faraday, M., 385 (5–10)
Faye, J., 390 (21)
Feinberg, G., 380 (15, 17)
Feynman, R., 304, 317, 387 (37), 392 (20), 397 (44)
Field, J. V., 73

Author index

Fine, A., 330, 344, 377, 394 (21, 25), 395 (28), 396 (41)
Finocchiaro, M. A., 147, 383 (25, 32, 1)
Fock, V., 387 (22), 388 (17), 389 (12)
Ford, J., 179, 384 (19)
Frank, P., 13, 240, 387 (2, 5, 6), 396 (1, 22)
Franklin, A., 86, 380 (12)
Freud, S., 396 (42)
Friedman, M., 396 (20)
Frisch, D. H., 387 (21)

Galilei, G., 41, 87, 380 (4–8, 13, 20–2), 383 (26, 28–31, 11, 12), 384 (26)
Gardner, M. R., 379 (13)
Gell-Mann, M., 317, 387 (37)
Gillispie, C. C., 58, 398
Gingerich, O., 270
Gleick, J., 179
Godzinski, Z., 392 (21)
Goldberg, A., 391 (3)
Goldberg, S., 222, 386 (7)
Goldreich, P., 131
Goldstein, H., 390 (13), 391 (29)
Goldstein, S., 344, 395 (29)
Goodman, N., 396 (28)
Gott, J. R., 389 (28, 29, 32)
Gould, S. J., 378 (10), 396 (17)
Green, A. W., 28
Guckenheimer, J., 179, 384 (10)
Gunn, J. E., 389 (28, 29, 32)
Guye, C.-E., 387 (36)

Hacking, I., 179
Hafele, J. C., 388 (18)
Hanle, P. A., 395 (20)
Hardin, L., 383 (1)
Harding, S. G., 395 (2)
Harman, P. M., 194
Harrison, E. R., 270, 388 (8), 389 (19, 28, 32, 33, 38, 41, 42, 45)
Harvey, W., 379 (1)
Heath, T. L., 379 (2, 15–17), 380 (15), 382 (5)
Heilbron, J. L., 289, 390 (12, 15), 391 (37–9), 395 (23, 24)
Heisenberg, W., 317, 392 (16, 2), 396 (17)
Hempel, C. G., 396 (27)
Hendry, J., 391 (32)
Herivel, J., 381 (11)
Herschel, W., 378 (6)
Hertz, H., 357
Hiley, B. J., 344, 394 (12), 396 (35)
Holland, P. R., 344, 394 (12)

Holmes, P., 179, 384 (10)
Holton, G., 14, 377, 378 (8), 389 (14), 397 (36)
Hon, G., 388 (7)
Howard, D., 384 (24)
Howson, C., 377, 396 (38)
Hughes, R. J., 392 (4)
Hume, D., 379 (5)
Hutchins, R. M., 398
Huygens, C., 381 (5), 382 (6), 383 (8)

Infeld, L., 388 (3)

Jackson, J. D., 207, 385 (1, 9), 386 (4), 388 (6, 7)
Jaki, S. L., 28, 162, 179
Jammer, M., 162, 289, 315, 381 (6–8), 383 (18), 384 (25), 390 (4, 7, 16, 19, 20), 391 (29), 392 (1), 393 (8)
Jefferson, T., 378 (1)
Jensen, R. V., 179, 384 (14, 16)
The Jerusalem Bible, 383 (7, 13)
Jevons, W. S., 1

Kaloyerou, P. N., 394 (12), 396 (35)
Kaufmann, W., 386 (16, 18)
Keating, R. E., 388 (18)
Kellert, S. H., 179
Kepler, J., 380 (10–14), 383 (27)
Keynes, J. M., 102
Klein, M. J., 390 (7–9), 391 (27, 30, 31)
Kockelmans, J. J., 377
Koestler, A., 58, 73, 380 (10)
Kourany, J. A., 377
Koyré, A., 163, 383 (14), 389 (20)
Kuhn, T. S., 58, 73, 377, 379 (1, 4, 5, 15), 380 (7), 390 (6, 7, 9, 12, 15), 396 (30–3)

Lakatos, I., 377, 379 (9), 380 (1), 396 (38)
Landau, L. D., 318
Laplace, P.-S., 384 (6, 7)
Laudan, L., 377, 395 (14), 396 (19, 39)
Lavanchy, C., 387 (36)
Leighton, R. B., 390 (2), 391 (15)
Leplin, J., 395 (14)
Lifshitz E. M., 318
Livingston, D. M., 207
Longair, M., 102
Lorentz, H. A., 385 (6, 8), 388 (4)
Lovejoy, A. O., 383 (4)
Lucretius, 378 (9), 383 (9)
Luria, S. E., 396 (16)

Mach, E., 381 (28), 384 (19)
MacKinnon, E. M., 315
Magie, W. F., 398
Mansouri, R., 387 (14)
Manuel, F. E., 102, 381 (7)
Maudlin, T., 393 (19, 20, 21), 394 (20, 27)
Maxwell, J. C., 385 (12, 15–17, 1, 3, 9)
Maxwell, N., 388 (20)
McCauley, J. L., 179
McMullin, E., 13, 86, 102, 330, 380 (9), 385 (2), 393 (23), 395 (19), 396 (18)
Melchior, P., 131, 383 (23)
Meyerson, É., 384 (3)
Milford, F. J., 207
Mill, J., 379 (6)
Miller, A. I., 222, 240, 386 (2, 3, 7, 8), 388 (16)
Minkowski, H., 388 (4)
Monod, J., 377, 396 (15), 397 (43, 45)
Moore, W., 317
More, L. T., 102, 381 (6, 9)
Motz, L., 386 (1), 399
Moulton, F. R., 383 (20)
Musgrave, A., 377

Nelson, E., 395 (29)
Neugebauer, O., 58, 380 (9)
Neumann, G., 387 (34, 35)
Newton, I., 87, 379(2), 381 (9–13, 19–22, 24–7, 2, 3, 10, 13–15), 382 (16, 17, 21, 25, 28–32, 34, 1–3, 9, 10, 13, 15–17), 383 (33, 34, 39, 40, 15–17), 384 (20–2, 2, 4, 5,7, 8), 389 (18, 21, 22, 24)

Pais, A., 240, 388 (2)
Pannekoek, A., 399
Park, D., 162
Pauli, W., 390 (23)
Peacock, G., 383 (41)
Pedersen, O., 383 (21, 22, 24)
Peres, A., 393 (25)
Petruccioli, S., 304
Pickering, A., 396 (29)
Planck, M., 378 (7), 379 (3), 386 (20–2), 387 (29), 390 (11), 391 (11)
Plato, 378 (2), 379 (10)
Podolsky, B., 393 (9, 10, 12)
Poincaré, H., 381 (29), 396 (24, 25)
Popper, K., 1, 38, 379 (7)
Pound, R. V., 389 (16)
Prigogine, I., 179
Przibram, K., 392 (13)
Ptolemy, C., 379 (6–8, 12), 383 (5)

413

Author index

Quine, W. V., 2, 395 (7–12)
Quirino, C., 383 (38)

Radnitzky, G., 377
Rasband, S. N., 179, 384 (10, 14, 16)
Rebka, G. A., 389 (16)
Redhead, M. L. G., 251, 330, 388 (13), 393 (19)
Reitz, J. R., 207
Roger, G., 395 (24)
Rohrlich, F., 383 (1)
Ronan, C. A., 86, 147
Rose, V., 378 (1)
Rosen, N., 393 (9, 10, 12)
Rosenfeld, L., 113
Russell, B., 133, 181, 395 (21)

Sachs, M., 388 (15)
Sambursky, S., 399
de Santillana, G., 147, 383 (10, 11)
Schey, H. M., 391 (3)
Schilpp, P. A., 240
Schramm, D. N., 389 (28, 29, 32)
Schrödinger, E., 317, 377, 392 (12), 396 (13)
Schuster, H. G., 179, 384 (10, 14)
Schwartz, J. L., 391 (3)
Segre, M., 86
Selleri, F., 355, 387 (12, 13), 395 (26)

Sexl, R. U., 387 (14)
Shimony, A., 271, 393 (17, 18), 394 (19)
Shipman, H. L., 270, 389 (28, 29, 39)
Smith, A. J., 383 (2)
Smith, J. H., 387 (21)
Stapp, H. P., 393 (19)
Stein, H., 388 (20)
Stengers, I., 179
Stewart, H. B., 179
Stewart, I., 179

Tangherlini, F. R., 387 (14)
Taylor, E. F., 240, 251
ter Haar, D., 390 (7)
Thayer, H. S., 382 (18), 383 (6, 7), 384(23), 389 (23, 25)
Thompson, J. M. T., 179
Thomson, W. (Lord Kelvin), 223
Tinsley, B. M., 389 (28, 29, 32)
Tolstoy, I., 207
Truesdell, C., 381 (23)

Valentini, A., 344, 391 (36), 395 (29)
van der Waerden, B. L., 391 (28)
van Fraassen, B. C., 395 (18)
von Neumann, J., 317
van Woodward, C., 396 (37)
Vogt, E., 382 (19)

Waldrop, M. M., 389 (36)
Watkins, J., 379 (4)
Watson, J. D., 396 (14)
Weimer, W., 396 (37)
Weinberg, S., 270
Weisskopf, V. F., 387 (20)
Westfall, R. S.,102, 113, 147, 380 (1–3), 381 (4, 11, 12), 382 (22–4, 7), 383 (13–15)
Weyl, H., 388 (4)
Wheeler, J. A., 240, 251, 392 (16, 12)
Whitaker, A., 315
White, H. E., 378 (9)
Whitehead, A. N., 383 (16), 389 (13)
Whiteside, D. T., 380 (4), 382 (20, 7)
Whitrow, G., 388 (10)
Whittaker, E., 194, 385 (11, 14, 7), 399
Williams, R. M., 270
Wylie, F. F., 131

Young, T., 383 (41)

Zahar, E., 379 (9), 385 (5), 386 (19), 387 (16)
Zanghi, N., 395 (29)
Ziman, J., 396 (3, 4, 34, 35)
Zurek. W. H., 392 (16, 12)

Subject index

Note numbers are given in parentheses.

aberration, 185
Abraham, M., 208, 209, 233; model of electron, 209, 218, 219, 221, 222, 387 (41)
Academy in Athens, 4
action at a distance: in electrostatics, 183; in gravity, 109, 184. *See also* Faraday; nonlocality; Thomson, W.
actuality. *See* Aristotle
Adams, J., 124, 152, 153
ad hoc: defined, 22; examples of such explanations, 53, 65, 154, 218, 220, 277, 280, 281, 311, 335, 348, 353
Adler, A., 33, 38
Adler, F., 227
aether: as an absolute reference frame, 160, 196; models of, 183–94. *See also* Aristotle; Descartes; Einstein; Lorentz; Maxwell; Michelson–Morley experiment; Newton; Poincaré; Thomson, W.
Aeolus, 17
Airy, G., 153
alchemy, 89, 90, 91.
Alexander the Great, 15
Alexandria, 47
Alfonso X of Castile, 54
al-Kindi, 124
Almagest. *See* Ptolemy
Alpha Centauri, 49, 261, 268
Alpha Lyrae. *See* Vega
alpha-particle scattering, 276
alternative interpretation of quantum mechanics. *See* Bohm interpretation
Ampère's law, 205, 206
Amyntas II of Macedon, 15
Anaxagoras, 165
Anaximander, 17, 165
Anaximenes, 17, 165

Andromeda Nebula, 10, 268
angular momentum, 96. *See also* Bohr
anima motrix. *See* Kepler
animist theory, 375
antecedent, 6, 7, 35
antirealism, 348, 349
annus mirabilis. *See* Newton
aphelion, 70, 87, 122
Apollonius of Perga, 71, 85, 382 (5)
a priori: defined, 378 (4); examples of such reasoning, 1, 19, 25, 64, 81, 112, 160, 165–7, 226, 366–8, 374
Arago, D., 185
Archimedes, 48
Ares, 17
Aries, 46
Aristarchus, 48, 50, 60, 63, 379 (16); and sizes of and distances to the sun and moon, 56–8
Aristotle, 3, 4, 15, 16, 35, 55; on the aether, 18; and astronomy, 55; and Bacon, 24, 25; on first principles, 4, 5, 365; on form and matter, 4; *On the Heavens*, 18; on knowledge, 5; and materialistic philosophy, 21; on motion, 19–22, 25–7, 50, 165; on nature, 18, 19; *Physics*, 5; on potentiality and actuality, 18; on sense experience, 4, 16; on the shape of the earth, 18; on space, 155, 156; on speed and weight, 20, 27; on the universe as organism, 17–19. *See also* Galilei
astronomy, 55. *See also* instrumentalism; Plato; Ptolemy; realism
Athena, 74
axiomatic, 34
Augustine, St., 90
Avempace, 75
Averroes, 75

Avicenna, 75
Ayscough, H., 89

Babylonians, 55
Bacon, F., 15, 22–4, 35; and Aristotle, 24, 25; on induction, 24; ladder of axiom, 24, 25; *The New Organon*, 23; on science, 24, 25, 30, 93, 166, 354. *See also* induction
Bacon–Descartes ideal, 35
Bacon, N., 23
Bacon, R., 29
Balashov, Y., xvi, 395 (2), 396 (38)
Balmer, J., 275; formula, 280, 282, 390 (15)
Barberini, F., 144
Barberini, M. (Urban VIII), 142–5.
Barrow, I., 90, 157, 366
Bede the Venerable, 124
Bell, J., 288, 324; on the Bohm interpretation, 318. *See also* Bell's theorem
Bell's inequality. *See* Bell's theorem
Bell's theorem, 324–30, 332, 340, 393 (17, 19); and Bohm's theory, 337; 340; and experiment, 271, 340, 395 (24). *See also* EPR paradox; hidden variables
Bellarmine, R., 135, 140, 145; his charge to Galileo, 142, 143; on the Copernican system and Galileo, 137, 138.
bending of light. *See* light
Bentley, R., 184, 260
Berkeley, G., 159
Bernoulli, J., 93
Bessel, F., 49, 152
Besso, M., 227, 229, 233, 254
Bestelmeyer, A., 218
Bible, 3, 60, 135–8, 142, 146, 147; purpose of, 140, 141. *See also* Scripture

Subject index

bifurcation, 176
big bang theory. *See* expanding universe
Big Dipper, 45
binary stars, 49
binomial theorem. *See* Newton
Biot–Savart law, 195, 205, 210
blackbody radiation, 273, 390 (1); and expanding universe, 268; and quantum theory, 273, 277, 280, 281, 285, 291. *See also* Einstein; Planck
black hole. *See* relativity, the general theory of
Bloch, F., 162
block universe. *See* simultaneity
blueshift, 242, 251, 265
Bohm, D., 318, 331, 334, 335, 351, 353; on the value of an alternative interpretation, 336–8; version of the EPR paradox, 325, 326. *See also* Bohm interpretation
Bohm interpretation, 331, 334–8, 342, 343, 345, 349, 394(1), 395 (26, 29, 30), 396 (27); and the classical limit, 335; empirical equivalence with the Copenhagen interpretation, 333–5; meaning of the wave function in, 335; and measurement, 335, 337; nonlocality of, 332, 337, 340; quantum potential, 337, 343; and quantum probability, 335; and realism, 349; and relativity, 337, 353, 395 (19), 396 (28); status of the uncertainty relation in, 337. *See also* de Broglie; chaos; determinism; historical contingency; measurement problem; stochastic mechanics; trajectory
Bohr, N., 276, 305, 350; on the context of observation, 271, 306, 320, 323, 371; and the Copenhagen interpretation, 271, 283, 286–9; and Høffding, 283; on measurement, 271, 306, 320; and positivism/operationalism, 287; quantization of angular momentum, 280, 282, 390 (17); quantum postulate, 306, 317; and reality, 271, 320, 360; semiclassical model, 278–82, 390 (13). *See also* Bohr–Einstein confrontations; causality; complementarity; wave–particle duality

Bohr–Einstein confrontations, 307–9
Boltzmann, L., 226, 277; constant, 274. *See also* equipartition theorem
Bolyai, J., 262
Bonaparte, N., 169
Born, M., 283, 284, 307, 318, 320, 321, 350; and the Copenhagen interpretation, 283, 287, 288, 296; and the meaning of the wave function, 289, 300; probability interpretation, 293–5, 299, 300, 317, 320. *See also* causality
Bose, S., xvi
Boussinesq, J., 186
Boyle, R., 90, 366; law, 378 (11)
brachistochrone, 380 (5)
Bradley, J., 49, 185
Brahe, T., 65–7, 139, 149; Tychonic model, 139
de Broglie, L. 285–9, 318, 351, 352, 391 (29), 396 (27); double solution, 288; duality between waves and matter, 291; pilot-wave theory, 288, 318, 321, 331, 350; relation, 291; and the Solvay Congress, 288, 321. *See also* Bohm interpretation; historical contingency
de Broglie–Bohm program, 353. *See also* Bohm interpretation; de Broglie
Brownian motion, 227, 352
Bucherer, A., 208–10, 219, 233; model of electron, 209, 210, 218, 219, 387 (41)
Buridan, J., 29, 75
Butterfield, H., 354
butterfly effect, 173

calculus, 29, 78, 90–3, 96, 103, 110, 111, 117, 381 (14, 15, 17). *See also* Newton
calendar, 44; reform, 60
Caccini, T., 135, 137
Caesar, J., 44
Calvin, J., 60
Carnap, R., 160, 361, 367
Castelli, C., 135–7
Castellino, F., xvii
cat paradox. *See* Schrödinger
Cauchy, A.-L., 186
causality: Born on, 317; Bohr on, 271, 306; Dirac on, 298; and Einstein, 285, 352, 391 (26); and explanation, 339; Heisenberg on, 299, 341; Hume on, 31; and quantum mechanics, 283, 290, 298, 305, 325, 331, 333, 338, 340; and relativity, 273; statistical, 334, 351. *See also* Bell's theorem; Bohm interpretation; cause; determinism; indeterminism
cause: of acceleration, 41, 77, 79; common, 340; and effect, 31, 32, 36, 37, 75, 94, 273, 298; final, 18, 19, 374, 375; of gravity, 87, 109; of the tides, 124–6. *See also* Aristotle; causality; motion; teleonomy
Cavendish, H., 108, 153
celestial sphere, 45, 46
center of curvature. *See* radius of curvature
central force. *See* force
centrifugal force. *See* force
centripetal force. *See* force
Challis, J., 152, 153
chaos, 124, 164, 171, 173, 305, 333, 385 (20); and Bohm's theory, 334, 337; examples of, 173; and nonlinearity, 174; and prediction, 171, 173, 178; and sensitivity to initial conditions, 173, 176, 178, 384 (12); and randomness, 176, 178, 384 (18). *See also* maps
Christina of Lorraine, 136
circular motion: and the ancients, 18, 50, 54, 138, 148, 160; and Copernicus, 59, 166; and Cusa, 139; and Galileo, 41, 103; and Kepler, 67; and Ptolemy, 51–3, 65
Clarke, S., 159
classical mechanics, 98, 99; logical structure of, 100–2; traditional view on, 171, 172
Clavius, C., 139, 142, 144
clockwork universe, 170–3
coexistence. *See* simultaneity
coherence (of a theory), 150–4, 282, 348; horizontal, 148, 151; vertical, 148
Coke, E., 23
collapse of the wave function. *See* measurement problem; wave function
Columbus, C., 60
commutative. *See* operator
compass directions. *See* sun
complementarity, 271, 283, 306, 333. *See also* wave–particle duality
completeness, 305–7, 311–13, 317, 319–24, 393 (14)

Compton, A., 285
confirmation, 30, 35, 93; paradox of, 369, 370
conic section, 71–3, 96, 115, 118, 120, 121; ellipse, 68, 71; hyperbola, 73; parabola, 73, 85
consequent, 6, 35
constant conjunction, 31, 36, 55. *See also* Hume
constellation, 46; of the zodiac, 47, 49
contextuality, 306, 320
continental drift, 362
constructive theory, 233, 339
convention (in science). *See* Poincaré
Copenhagen interpretation, 290–304, 331–4, 345, 349, 355, 394 (1), 395 (26); and the classical limit, 334; the road to, 282–4, 286–9. *See also* Bohm; Bohr; Born; Dirac; Heisenberg; historical contingency; indeterminism; measurement; measurement problem; Pauli; trajectory
Copernican model. *See* Copernicus; heliocentric model
Copernicus, N., 60, 61, 66; Aristotelian commitments of, 59; *De Revolutionibus*, 55; and the equant, 59, 65, 166; model of the universe, 34, 61–6, 137, 149, 166; and realism, 55, 56, 139; and the size of the universe, 48; on the sizes of the planetary orbits, 63, 64. *See also* heliocentric model
correlation, 326–30; calculation of for the EPRB experiment, 329, 330
Cosimo II, 76, 136
cosmic background radiation. *See* blackbody radiation
cosmological principle, 264
Cotes, R., 151
Coulomb's law, 183, 195, 205
Counter-Reformation, 61
covering-law model. *See* deductive-nomological model
Crick, F., 364
Curie, M., 211
Curie, P., 211
curvature. *See* radius of curvature; space
Cusa, Nicholas of, 139
Cushing, N., xvii
Cygnus X-1, 269

61 Cygni, 49

dark matter, 267
Darwin, C., 32, 276
Darwin, C. G., 276
deduction, 5, 7, 24, 29, 35, 101
deductive-nomological model, 338
deferent, 51, 52, 62, 65
Delbrück, M., 363, 364
Demeter, 17
Democritus, 17, 155, 165, 384 (1)
De Revolutionibus. *See* Copernicus
Descartes, R., 7, 25, 29, 35, 90; on the aether, 103; on certain knowledge, 7–9; on first principles, 7, 8; and inertia, 80, 81, 98, 103; on intuition, 7, 12, 13; on light, 183; on matter and space, 156; and the plenum, 183; and the warrant for an hypothesis, 24
determinate, 324, 394 (16)
determinism, 164, 167–9, 176–8, 250, 396 (21); and Bell's theorem, 325; in Bohm's theory, 318, 331, 337, 342, 349; in de Broglie's theory, 321; in the Copenhagen interpretation, 298, 305, 317; warrant for a belief in, 170–2, 178. *See also* causality
deterministic chaos. *See* chaos
Devereux, R. (Earl of Essex), 23
Dialogue Concerning the Two Chief World Systems. *See* Galilei
Dialogues Concerning Two New Sciences. *See* Galilei
Dirac, P., 286, 287, 307; and the Copenhagen interpretation, 296; on large vs. small, 296–8; on measurement, 313–15; on superpositions, 301, 302, 314; transformation theory, 352, 396 (31). *See also* causality
directrix, 71, 118
discontinuity, 279, 282, 283
discrete maps. *See* maps
displacement current. *See* Maxwell
distances to the sun and moon. *See* Aristarchus
Donne, J., 149
Doppler, C., 241
Doppler shift, 241, 242, 265, 388 (4); blueshift, 242; redshift, 241; relativistic, 241, 242. *See also* Hubble, law
double-slit experiment, 301–4, 392 (21)

double solution. *See* de Broglie
duality. *See* wave–particle duality
Duhem, P., 76, 345, 346, 368. *See also* Duhem–Quine thesis
Duhem–Quine thesis, 345, 347, 395 (2). *See also* underdetermination
de Duillier, N. 92
dynamics, 77, 97

earth-centered model. *See* geocentric model
eccentric circle, 51, 65
eccentricity, 71; values of for the planets, 70
ecliptic, 45, 49, 68
economy. *See* simplicity
Eddington, A., 38, 266
effect. *See* cause
Ehrenfest's theorem, 394 (14)
Ehrenhaft, F., 10
eigenstate, 309, 322, 392(6, 9). *See also* operator; wave function
eigenvalue, 290, 309, 322, 392 (6, 9), 394 (5). *See also* operator; wave function
Einstein, A., 3, 33, 38, 101, 225–8, 255–7, 278, 289; on the aether, 204, 230, 232; and de Broglie, 285, 286, 352; Brownian motion, 227, 352; cosmological constant, 262, 266; cosmological principle, 264; Doppler shift, 241; early thought experiments, 229–30, 239, 240; hidden variables theory, 396 (32); and Hubble's law, 266; and the Kaufmann experiments, 215, 220, 233; Lorentz transformations, 242; mass–energy equivalence, 245, 246; on measurement, 321; on the nature of space, 160–2, 384 (25); and nonlocality, 320, 353; philosophical commitments of, 285, 287, 352; on physics, 1, 360, 361; and Planck's law, 244, 285; and positivism, 228, 287, 357, 367; postulates for special relativity, 231, 232; quanta, 285; and quantum mechanics, 271, 285, 288, 305, 307, 308, 317; on reductionism, 361; and Schrödinger, 286; on science and its goals, 1, 352, 359–61; on scientific knowledge, 1, 8; and separability, 353; on simultaneity, 231; skepticism about classical physics, 228–31; on underdetermination, 348;

417

Subject index

Einstein, A. (*cont.*)
 unified field theory, 228, 255, 285. *See also* Bohr–Einstein confrontations; causality; Einstein–Lorentz theory; EPR paradox; expanding universe; historical contingency; Kaufmann, experiments; Michelson–Morley experiment; relativity, the general theory of; relativity, the special theory of
Einstein–Lorentz theory, 209, 210, 216, 217
Einstein–Podolsky–Rosen paradox. *See* EPR paradox
electrodynamic vs. mechanical world views, 210
electromagnetic theory, 181, 198; and quanta, 278, 279, 291. *See also* Maxwell
electromagnetic wave, 181, 183; energy of, 210, 243; relation to E and B fields, 191, 195, 206. *See also* Maxwell
electron; discovery of, 208; electromagnetic mass of, 210; charge of, 10
element of reality, 319, 320, 322. *See also* EPR paradox
elements (the four basic), 18, 20, 26, 81, 155, 156, 165
Elizabeth I of England, 22
Ellingson, A., xvi
ellipse. *See* conic section
e/m, determination of, 218, 219. *See also* Kaufmann, experiments
empirical adequacy. *See* explanation vs. understanding
empiricist, 9, 24
energy quantization. *See* quantum mechanics
entanglement. *See* wave function
entropy, 363–5
epicycle, 51, 65
epistemological, 172, 288, 339. *See also* ontology
EPR paradox, 311, 319–21; analysis of the EPR argument, 322–4, 393 (14); Bohm version of (EPRB), 325, 326; Bohr's response to, 320. *See also* Bell's theorem; cat paradox; completeness; explanation vs. understanding
EPRB. *See* EPR paradox
equant. *See* Copernicus; Ptolemy
equinox, 44
equipartition theorem, 223, 273–5

equivalence principle. *See* relativity, the general theory of
Eratosthenes, 47, 56, 379 (2)
escape velocity, 264, 389 (31)
Euclid, 4, 8
Euclidean. *See* space
Eudoxus, 50, 55, 148
Euler, L., 99, 166, 167
expanding universe: cosmic background radiation, 268; critical density, 263, 267; dark matter in, 267; horizon in, 266, 270, 390 (44); initial big bang, 263, 266; as a model, 267–9. *See also* Hubble, law; Olbers' paradox; redshift
expansion of space. *See* expanding universe; redshift
experiment (the role of in science), 9, 10, 29, 30–3, 140, 371, 373; Bacon on, 24, 166; Einstein on, 357; Galileo on, 78, 82, 380 (8); Kepler on, 67, 166; Newton on, 94; vs. observation, 16, 17. *See also* scientific method
explanation vs. understanding, 338–40; and EPRB, 339, 340; and quantum mechanics, 317, 341, 342
extremum. *See* maximum/minimum principle

Fallon, S. M., xvi
falsification, 32, 33, 38, 366. *See also* Popper; refutation
Faraday, M., 93, 181; on the aether vs. action at a distance, 187; law of induction, 205; lines of force, 187, 192. *See also* Maxwell
Fates, 17
de Fermat, P., 183
fertility, 282, 349, 350
Feynman, R., 387 (37), 392 (20), 397 (44); on quantum mechanics, 317
field theory, 187, 284; unified, 228, 255, 285. *See also* Maxwell
first principles, 4, 5, 7–9, 34, 365. *See also* hypothesis
fixed point, 175
fixed stars. *See* celestial sphere
FitzGerald, G., 202, 203
Fizeau, A.-H., 185
fluent, 78, 90. *See also* calculus
fluxions, 78, 90, 91, 96. *See also* calculus
focus, 67, 68, 118
formalism vs. interpretation, 287,

333, 334, 336, 348, 394 (1). *See also* theory
force, 95, 96, 100, 101; central, 96, 111, 115, 116, 120; centrifugal, 105, 111; centripetal, 105, 111; impressed (impetus), 74, 75. *See also* Newton
Foscarini, P., 137
Foucault, J.-B., 185
Frank, P., 160, 359, 367
Frederick II of Denmark, 66
Fresnel, A.-J., 185, 223
Freud, S., 33, 38, 374, 375
Friedmann, A., 262. *See also* relativity, the general theory of

Galilei, G., 18, 22, 23, 35, 74, 76, 77, 102, 103, 160; on Aristotle and motion, 81–4; and Bellarmine, 142, 143; *Dialogue Concerning the Two Chief World Systems*, 76; *Dialogues Concerning Two New Sciences*, 77; and experiment, 78, 82, 380 (8); on gravity, 87; on inertia, 41, 80, 81, 103, 150; on the interpretation of Scripture, 140, 146, 147; and the Jesuits, 136, 142, 144; and the Leaning Tower of Pisa, 76, 83; *Letter to the Grand Duchess*, 135, 137, 140–2, 146, 147; on mathematics, 150, 367; on naturally accelerated motion, 41, 76–8; on projectile motion, 77, 79, 85, 86; and realism, 139; and scientific method, 29, 30, 166; and simplicity, 41, 78, 380 (6); on the tides, 125, 126, 143, 383 (25); and the telescope, 76, 150; and Urban VIII, 142–5. *See also* Bellarmine
Galle, J., 124, 153
Gandhi, M., 376
Gauss, C., 160
Geiger, H., 276
general relativity. *See* relativity, the general theory of
genetic code, 364
geocentric model, 34, 50–4, 76, 136, 141, 148, 267. *See also* Copernicus; Galilei; Ptolemy
geodesic, 255, 389 (41), 390 (43)
geometry, 4. *See also* space
Gilbert, W., 22, 23, 366
goals of science. *See* science
Gödel, K., 160, 254, 367
Gould, S., 365
Grassi, H., 143, 144
gravitation, 6, 87, 90, 94, 97, 103,

Subject index

108–12, 115, 154; *See also* relativity, the general theory of
Green, G., 188
Gregory the Great, 44
Grossmann, M., 226, 227
guidance condition, 288, 331, 334, 343, 395 (30). *See also* Bohm interpretation

half-life, 235, 393 (15)
Halley, E., 92, 129
Hamilton–Jacobi theory, 286, 352, 391 (29), 396 (27)
Harvey, W., 22, 30, 366
heavenly vs. terrestrial laws, 18, 87, 103, 139.
hegemony, 286, 287, 350
Heisenberg, W., 162, 284, 288, 318, 320, 350; Copenhagen interpretation, 283, 307, 317; on formalism and interpretation, 287, 348; matrix mechanics, 284, 286, 287, 352; and the meaning of the wave function, 289, 300; and measurement, 300, 317; and positivism/operationalism, 287; *potentia*, 342; on probability in quantum mechanics, 317. *See also* causality; uncertainty relation
heliocentric model, 34, 60–5, 76, 136, 141, 149; and sizes of the planetary orbits, 63; and Mercury–Venus problem, 63, 64. *See also* Aristarchus; Copernicus; Galilei
Helmholtz, H. von, 226
Hempel, C., 367, 369, 370, 397 (23)
Henderson, T., 49
Heraclitus, 165
Hermetic tradition, 90, 91
Herschel, J., 9, 10, 152
Herschel, W., 49; discovery of Neptune, 124, 152; nebulae, 10
Hertz, H., 181, 226, 357, 367, 368
hidden variables, 305, 321, 325, 394 (20); Bell on, 318, 324; Bohm on, 336; Einstein on, 396 (32); von Neumann on, 288, 317. *See also* locality; nonlocality
Hipparchus, 52, 74, 75, 379 (16)
historical contingency: and Bohm vs. Copenhagen interpretations, 345, 351–3, 396(22); and the Copenhagen interpretation, 283, 350; in evolution, 378 (10); and the goals of science, 373; and retrospective reconstructions 354;

in technology, 11, 12; and theory acceptance, 351
Hobbes, T., 145
Høffding, H., 283
Homer, 17, 346
homocentric-sphere model, 50, 55, 148
homogeneous, 155, 237, 264
Hooke, R., 91; on light, 183; on the planetary orbits, 92, 110.
horror vacui, 77, 148
Howard, T. (Earl of Suffolk), 23
Hubble, E., 265; law, 265, 267–70; radius, 266, 270
Hume, D., 9, 25, 31, 34–7, 55, 229, 361, 366. *See also* induction
Huygens, C.: and absolute space, 159; on centripetal force, 105; and curvature, 118, 382 (6); on inertia, 80, 81; on light, 184
Hydra II, 268
hyperbola. *See* conic section
hypothesis (in science), 11, 24, 30, 32–4, 93, 369. *See also* first principles; Newton; warrant
hypothetical proposition, 5, 6, 378 (4)
hypothetico-deductive method, 33, 34, 94, 151, 366.

impetus theory. *See* force
impressed force. *See* force
incompleteness. *See* completeness
indeterminism, 282, 283, 396 (21); in quantum mechanics, 287, 288, 298, 305, 318, 305, 331, 332, 334, 337, 341, 342. *See also* causality; determinism
Index (forbidden books), 61, 136, 137
induction, 1, 9, 24, 25, 29, 31–5, 37
inertia, 41, 79–81, 87, 98, 100–3, 150
inertial frame, 99, 160, 196, 204, 254, 337; and special relativity, 229, 231, 232, 237, 238; and the twin paradox, 246, 247
Inquisition, 76, 137, 142
institutional authority. *See* intellectual freedom vs. institutional authority
instrumentalism, 55, 139, 140, 144. *See also* logical positivism; operationalism; pragmatism
intellectual freedom vs. institutional authority, 135, 137, 140, 144. *See also* Bellarmine
interference: of light, 185, 201, 202;

of the tides, 129–32; of wave functions, 291, 301–4, 313, 314
internal vs. external factors in science. *See* science
interpretation (of a scientific theory). *See* formalism vs. interpretation
interpretations of quantum mechanics. *See* Bohm interpretation; Copenhagen interpretation
instrumentalism, 55. *See also* logical positivism; operationalism; pragmatism; to save the phenomena; truth
intuition, 7, 11–13, 35, 366
inverse-square law. *See* Newton
isotropic, 155, 264.

James I of England, 22, 23
Jeans, J., 274
Jefferson, T., 3
Jesuits, 136, 142, 144
Jones, G., xvi, xvii
Jordan, P., 283, 287
Joshua, 141
Jupiter, 49, 97, 124, 152; moons of, 104, 135, 136, 151

Kamminga, H., 396 (21)
Kant, I., 366
Kaufmann, W., 208, 209; on electromagnetic mass, 214; experiments, 209–15, 233, 346, 366; on the Einstein–Lorentz theory, 215; methodological lessons from, 220, 386 (19). *See also* Planck
Kennedy, R., xvi
Kepler, J., 22, 55, 66, 67, 166; *anima motrix*, 67, 141; and causes, 56, 367; laws of planetary motion, 66–71; and mathematical harmonies, 367; orbit of Mars, 67; and realism, 139; and the regular solids, 66, 155; on the tides, 125
kicked rotator, 176, 177
Kierkegaard, S., 283
kinematics, 77
Kirchhoff, G., 226, 367, 368
Kneller, G., 89
Kuhn, T., 370, 371
Kurlbaum, F., 273

Lagrange. J.-L., 123, 160, 166
Lakatos, I., 397 (38)
Lamb, H., 189
Landau, L., 318

419

Subject index

Laplace, P.-S., 123, 152, 160, 269; on determinism, 169; and the discovery of Neptune, 152
Larmor, J., 189
Lateran Council, 60
law. *See* theory
laws of motion. *See* Kepler; Newton
law vs. theory, 378 (11)
Laudan, L., 397 (39)
Laue, M. von, 247
Leaning Tower of Pisa. *See* Galilei
Leibniz, G., 92, 93, 166, 381 (14); and the nature of space, 159, 161; principle of sufficient reason, 166; and priority dispute with Newton, 92, 381 (17); and an infinite universe, 261
Lemaître, G., 263, 266
Lenard, P., 228
length contraction, 233–5, 248, 387 (20). *See also* Lorentz–FitzGerald contraction hypothesis
Lesage, G.-L., 184
Letter to the Grand Duchess. See Galilei
Leucippus, 165, 384 (1)
Leuconia, 129, 383 (38)
Leverrier, U., 124, 153, 154
life (as biochemistry), 363–5
light: bending of, 256, 257; corpuscular and wave theories theory of, 183–5; speed of (c), 151, 152, 185, 195, 196, 205. *See also* photon; relativity, the general theory of
Lifshitz, E., 318
Lincean Academy, 142, 143
lines of force. *See* Faraday
Lobachevski, N., 262
Local Group, 268
locality, 320, 325; and Bell's theorem, 327, 328; Einstein on, 321, 352; and the EPR paradox, 323, 324. *See also* Bell's theorem; nonlocality
logic, 5–7; inductive, 35
logical empiricism. *See* logical positivism
logical positivism, 160, 228, 283, 357, 367. *See also* instrumentalism; operationalism
logistic map, 174–8
Lord Kelvin. *See* Thomson, W.
Lord Rayleigh. *See* Strutt
Lorentz, H., 203, 208, 229; and the aether, 203, 204; on Einstein's theory of relativity, 204; force law, 205, 244, 385 (10); and the Kaufmann experiments, 215; model of the electron, 209, 222; molecular force hypothesis, 203; theory of electrons, 203, 204, 218
Lorentz–FitzGerald contraction hypothesis, 203, 209, 235
Lorentz transformation, 203, 231, 232, 237–9; velocity addition law, 239
Love, A., 189
Lowell, P., 153
Lucretius, 20, 155, 162
Lummer, O., 273
Luria, S., 365
Luther, M., 60

MacCullagh, J., 186
Mach, E., 160; on absolute space, 158–60, 254, 367; on classical mechanics, 101, 368; and Einstein, 160, 229; and positivism, 360, 367; principle, 254; on science, 167, 367
Mach's principle. *See* Mach
Madelung, E., 353
Maestlin, M., 66
maps, 174, 384 (13); examples of, 174–8. *See also* chaos
Maric, M., 227
Mars, 49, 61, 67; god, 17
Marsden, E., 276
Marx, K., 33, 38
mass, 95, 98, 100, 112, 113; inertial vs. gravitational, 112, 113, 252, 388 (1)
mass–energy equivalence. *See* relativity, the special theory of
materialistic philosophy, 21, 164. *See also* mechanistic philosophy
matter and space, 90, 149. *See also* Descartes; relativity, the general theory of
matrix mechanics. *See* quantum mechanics
maximum/minimum principle, 166, 167, 183. *See also* teleonomy
Maupertuis, P.-L. de, 166, 167
Maxwell, J., 93, 181, 183, 229, 282; on action at a distance, 193, 194; on the aether, 192–4; displacement current, 197, 198, 205, 206, 385 (2); equations, 195, 198, 205, 206, 229, 243, 357, 385 (10); on lines of force, 190, 192–4; models of the aether, 190, 191; speed of light, 195, 196, 205
Maxwell–Boltzmann law, 223, 335. *See also* equipartition theorem
McGlinn, W., xvi
McMullin, E., xvi
measurement: Bohm on, 337; Bohr on, 306, 320; Dirac on, 302, 313–15; Heisenberg on, 300, 317; Pauli on, 284
measurement problem, 309–11, 323, 333–5, 341, 342, 348; in Bohm's theory, 335, 337, 394 (13, 18). *See also* measurement
mechanics. *See* Aristotle; Galilei; motion; Newton
mechanistic philosophy, 17, 21, 90, 366. *See also* materialistic philosophy
medium. *See* Aristotle; motion
Menaechmus, 71
Mercury, 49, 63, 64; perihelion shift of, 153, 154, 258, 259
Mercury–Venus problem, 50, 63, 64
Merton theorem, 76
Meyerson, É., 167, 352
Michelson, A., 199, 204, 234
Michelson–Morley experiment, 199–202, 385 (4); and Einstein, 227, 232
Middle Ages, 19, 20, 75, 76, 139, 366
Miletus, 164
Milky Way, 268
Mill, J., 25, 31, 34, 32, 35, 37. *See also* induction
Millikan, R., 10
Milne, E., 264
Minkowski, H., 226
models of the universe. *See* Aristarchus; Brahe; clockwork universe; Copernicus; expanding universe; geocentric model; Galilei; heliocentric model; homocentric-sphere model; Ptolemy; relativity, the general theory of; stability of the classical universe; steady-state theory; two-sphere model
momentum, 80, 98, 221, 244, 291, 343
Monod, J., 364, 397 (43); on the basis of life, 364, 365; on science vs. traditional values, 375, 376, 397 (46), 397 (46)
moon, 18, 44, 49, 65; and Galileo's observations, 135; and Newton's law of gravity, 90, 91, 107, 108,

110; size of and distance to, 56–8; and the tides, 124–31. *See also* Jupiter; Saturn
More, H., 90, 157, 366
Morley, E., 199, 234
motion: cause of, 21, 98; forced (unnatural), 18, 19, 74; Lucretius on, 20; natural, 18, 19, 21, 25–7, 81; naturally (uniformly) accelerated, 41, 76–8; projectile, 79. *See also* Aristotle; circular motion; Galilei; Newton
muon, 235–7
μ meson. *See* muon

Nash, N., xvi
natural motion. *See* motion
natural place, 18, 19, 27
naturally accelerated motion. *See* motion
nature. *See* Aristotle
Navier, C.-L., 186
Neoplatonism, 90
Neptune; discovery of, 124, 152, 153
neutrino, 235
Newcomb, S., 124
The New Organon. See Bacon
Newton, I. (*père*), 89
Newton, I., 22, 30, 35, 89–93, 145, 149, 223; on absolute space, 96, 157–9, 161, 383 (18); on absolute time, 96; on action at a distance, 109; on the aether, 90, 95, 184; and alchemy, 89–91; *annus mirabilis*, 90; binomial theorem, 90; calculus, 78, 90–2, 96, 110, 381 (14, 15, 17); centripetal force, 105–7; and curvature, 118, 382(5); gravitation, 87, 90, 97, 103, 108–12, 115; and Halley, 92, 129; on hypotheses, 87, 94, 95, 109, 157; on inertia, 80, 81; on interference, 129, 130; inverse-square law, 87, 104–7, 115, 122; and Kepler's laws, 94, 96, 97, 103–7, 114, 115, 117–23, 382 (14); laws of motion, 87, 96, 97–100, 388 (10); on light, 184; optics, 90, 91; and perturbations, 123; and philosophy of science, 93–5; *Principia*, 81, 95–7; and priority dispute with Leibniz, 92, 381 (17); Queries (*Optics*), 95, 145, 159, 168, 260; *Rules of Reasoning*, 33, 94, 97, 366; on space and God, 157; and the stability of the universe, 168, 169, 260; on the tides, 126, 129, 130. *See also* force; Hooke

Nicomachus, 15
nominalism, 166
noncommutative. *See* operator
non-Euclidean. *See* space
nonlocality: and Bell's theorem, 325, 334, 393 (19); and Bohm's theory, 332, 337, 342, 349; Bohr on, 320; Einstein on, 288, 321, 352, 353; and EPR, 393 (14). *See also* locality; nonseparability
nonseparability, 334, 341, 353. *See also* wave function, entangled
normal science. *See* paradigm model of science
North star. *See* Polaris
no-signaling theorem, 337, 353, 395 (19)
nuclear atom. *See* Bohr; Rutherford; Thompson, J.

objective reality. *See* realism
objectivity of science: belief in, 30, 164, 374, 375; questioned, 34, 350, 372
observable. *See* operator
observation. *See* measurement; measurement problem
observation vs. experimentation. *See* experiment
ocean tides. *See* tides
Ockham, William of, 75; razor, 29, 166
Olbers, H., 261; paradox, 261, 267, 389 (37)
Oldenburg, H., 105, 151
old quantum theory, 281
On the Heavens. See Aristotle
ontology, 172, 334, 337, 341, 347, 348, 352. *See also* epistemological
operationalism, 284, 287, 391 (24). *See also* instrumentalism; logical positivism; pragmatism
operator: noncommutative, 284, 299, 323, 391 (25,12); and physical observables, 284, 309, 319, 322, 333, 392 (5), 394(5). *See also* wave function
optics. *See* Hooke; Huygens; Newton; Young
Oresme, N., 29, 75
organism, 17–19, 366
Orion, 10
Osiander, A., 56

Pappus, 71
parabola. *See* conic section; Galilei
paradigm model of science, 370–2;

normal science, 370; revolutionary science, 370
paradox of confirmation, 369, 370
paradox of the ravens. *See* paradox of confirmation
parallax. *See* stellar parallax
patronage, 144, 145
Pauli, W., 318, 284, 350; and Bohm's theory, 351; and the Copenhagen interpretation, 283, 287; on measurement, 284; and positivism/operationalism, 284, 287; and the Solvay Congress, 288
Peirce, C., 367
Penzias, A., 268
Peres, A., 328
perihelion, 70, 262; shift for Mercury, 153, 154, 258, 259
periodic, 123
perturbations: and the discovery of new planets, 124, 152–4; in classical gravitational theory, 97, 123, 124, 126, 169; and Mercury, 153, 154, 258; stability against, 123, 171. *See also* chaos
phase space, 176, 384 (12)
Philip II of Macedon, 15
Philippines, 129
Philoponus, J, 74, 75
philosophy (etymology of), 3, 4
philosophy of science. *See* scientific method
photon: and Bohr, 279, 288; and de Broglie, 291; and complementarity, 333; Dirac on, 296, 301, 302, 313, 314; and Einstein, 285, 308, 309; and relativity, 244, 258, 267; and Schrödinger, 286
physics (etymology of), 4
Physics. See Aristotle
pilot-wave theory. *See* Bohm; de Broglie
place (concept of), 156. *See also* natural place
Planck, M., 9, 34, 208; analysis of Kaufmann's experiments, 215–18; blackbody radiation formula, 277, 280, 281, 285, 390 (7, 9); on causality, 298; constant, 258, 278, 301; law, 244, 258, 276–8, 281, 282; 285, 291; on the nature of science, 9, 10, 35, 360, 361; relativistic version of Newton's second law, 216, 244, 386 (20)

421

Subject index

planets, 49–51; astronomical data on, 70; distances to, 50, 63. *See also names of individual planets*
plate tectonic theory, 362
Plato, 3, 15; and astronomy, 50, 54, 55; *Forms*, 4, 35; on knowledge, 4; and materialistic philosophy, 21; on opinion vs. science, 4; to save the phenomena, 54–6, 138; on space, 155, 156; view of the universe, 17. *See also* Socrates
plenum, 27, 183
Pliny the Elder, 124, 379 (2)
Pluto, 153
Podolsky, B., 319
Poincaré, H.: on the aether, 204; conceptual influence of, 283; on convention in science, 101, 368; influence on Einstein, 229, 360; on the logical structure of classical mechanics, 101, 368; principle of relativity, 204, 252; on the stability of the solar system, 124, 173
Poisson, S.-D., 123, 160, 186
Polaris, 45
polarization, 183, 185, 189, 385 (1)
Pole Star. *See* Polaris
Popper, K.: on falsification, 1, 32–4, 37–9, 397 (38); and the rationality of science, 272; on the social milieu, 284
Poseidon, 17
Poseidonios of Rhodes, 124
positivism. *See* logical positivism
postulate of objectivity, 375, 397 (43). *See also* objectivity of science
potentiality, 18
potentia. *See* Heisenberg
pragmatism, 367. *See also* instrumentalism; logical positivism; operationalism
prediction, 30, 35, 220, 366, 369; Bacon on, 24, 93; Einstein on, 360; Newton on, 32, 93; Popper on, 32, 33, 38. *See also* chaos; falsification
Principia. *See* Newton
principle of relativity, 204. *See also* Einstein
principle of sufficient reason, 166
principle theory. *See* constructive theory
Pringsheim, E., 273
probability (quantum). *See* indeterminism; wave function
projectile motion. *See* motion
proper length, 234
proper time, 233, 246, 247
Ptolemaic model. *See* geocentric model; Ptolemy
Ptolemy, C., 51, 55, 66; *Almagest*, 52; on astronomy, 138, 139; equant, 52, 53; model of the universe, 34, 51–4, 62, 65. *See also* geocentric model
Pythagoras, 165; theorem, 162
Pythagorean brotherhood, 165

quanta, 227, 277, 278, 285. *See also* photon
quantum mechanics: matrix mechanics vs. wave mechanics, 282, 284–7; routes to, 282–6, 391 (19); a simple example of energy quantization, 294–6. *See also* Bohm interpretation; Bohr; Born; de Broglie; completeness; Copenhagen interpretation; Dirac; double-slit experiment; Einstein; EPR paradox; Heisenberg; indeterminism; Pauli; Schrödinger; spin; trajectory; uncertainty relation; understanding; wave function
quantum postulate. *See* Bohr
quantum potential. *See* Bohm interpretation
quantum theory. *See* Bohr; complementarity; Planck; quantum mechanics
quasar, 268
Queries. *See* Newton (*Optics*)
Quine, W., 33, 35, 346, 373. *See also* Duhem–Quine thesis; underdetermination

radius of curvature, 118, 120, 382 (5–7)
Raleigh, W., 23
randomness. *See* chaos
random walk. *See* chaos
rationalist, 9, 24
Rayleigh–Jeans law, 274, 277, 390 (7)
realism: as the existence of an objective reality, 4, 271, 312, 317, 321, 325, 352, 353, 357, 360; scientific, 55, 138–40, 144, 341, 345, 347–51. *See also* instrumentalism; underdetermination
redshift: Doppler, 241, 251, 265, 270; and the expansion of space, 265, 269, 270; gravitational, 258
reduction of state. *See* collapse of the wave function; measurement problem
reductionism: defined, 347; and biology, 363–5; and chemistry, 362; and geology, 362; and physics, 361, 362, 365
Reichenbach, H., 361, 367, 397 (23)
Reformation, 60
refutation, 32, 33, 35, 345. *See also* falsification; Popper
relativity, the general theory of, 113, 252–70, 339; black hole, 269; Einstein solution, 262; equivalence principle, 252–4, 388 (2); experimental tests of, 256–9, 389 (13); Friedmann solution, 262, 263; and geometrization of gravity, 255; Gödel solution, 254; and Mach's principle, 254; and perihelion precession, 154; and Popper, 33, 38; de Sitter solution, 254, 389 (8); and solar eclipse expeditions, 257; structure of space in, 255, 256. *See also* expanding universe; space; space–time
relativity, the special theory of, 160, 198, 204, 220, 221, 225–51; postulates for, 231, 232; precursors to, 202–4; and equivalence to Lorentz's theory, 209, 210; mass–energy equivalence, 242–6; and tension with quantum mechanics, 324, 325, 337, 353. *See also* Bohm interpretation; Doppler shift; Einstein; Kaufmann, experiments; length contraction; Lorentz; Lorentz transformation; time dilation; twin paradox
relativity theory. *See also* relativity principle; the special and general theories of
religion and science. *See* science
Renaissance, 19, 20, 29, 46, 55, 60
retroduction. *See* hypothetico-deductive method
retrograde motion, 51, 61, 62
revolutionary science. *See* paradigm model of science
Rheticus, G., 60
Riemann, B., 161, 186, 262
Roemer, O., 151, 383 (8)
Roosevelt, F., 228

Subject index

Rosen, N., 319, 391 (38)
Rosenfeld, L. 318
Rossi, B., 236
Rubens, H., 273
Rudolph II of Prussia, 66
Rutherford, E., 276, 278, 280; model of atom, 276
Rydberg, J., 275; constant, 275, 280; formula, 275

Sagredo, G., 84
Salviati, F., 80, 82, 84, 103, 143
Saturn, 49, 50, 66, 97, 124, 152; moons of, 104
Schlick, M., 160, 367
Schrödinger, E., 284–9; cat paradox, 311–13, 393 (12); on determinism vs. indeterminism, 349; equation, 291, 292, 295, 343, 392 (10), 394 (4); and the explanation of life, 363, 364; interpretation of the wave function, 293; wave mechanics, 285, 321, 391 (29). See also historical contingency; quantum mechanics
science: element of belief in, 30; geometry as a model of, 8; goals of, 181, 359–76; internal vs. external factors in, 353–5; natural, 5; and opinion, 4; and religion, 144, 145, 164; theological underpinnings of, 145, 170; as truth, 4. See also Galilei; Monod; Newton; scientific method
scientific knowledge, 11, 34–6, 359–76. See also objectivity of science; scientific method
scientific method, 1, 2, 9, 25, 29, 30, 93, 357; styles of inference in, 365–8. See also Aristotle; Bacon; Descartes; Duhem; Einstein; experiment; Galilei; Mach; Monod; Newton; Planck; Popper
scientific realism. See realism
scientific research program, 397 (38)
scintillation counter, 236
Scripture, 60, 137–44, 146, 147, 149. See also Bible
seasons, 44
secular (motion), 123
semiclassical model. See Bohr
Seleucus of Babylon, 124
sense experience, 30, 361; Aristotle on, 4, 5, 16; Mach on, 367; Planck on, 35; Plato and, 4. See also logical positivism
separable (equations), 171

separability. See nonseparability
Shakespeare, W., 22, 65
Shapley, H., 265
signaling. See no-signaling theorem
Simplicio, 80, 84, 143, 144
simplicity (of explanation), 50, 78, 161, 164–7
Simplicius, 54, 139, 143
simultaneity, 231, 232, 238; absolute, 231, 387 (13); and the block universe, 250, 388 (20); and coexistence, 248–50; conventionality of, 231, 337, 388 (14). See also Einstein
de Sitter, W., 254. See also relativity, the general theory of
size of the earth. See Eratosthenes
sizes of the planetary orbits. See Copernicus
sizes of the sun and moon. See Aristarchus
Slipher, V., 265
Smith, B., 89
Socrates, 3, 4, 16; on materialistic philosophy, 17
solstice, 44
Solvay Congress, 288, 306, 307, 321, 351
Sommerfeld, A., 254, 281, 283
space: absolute, 96, 157–9; 160, 196; curvature of, 255, 256, 262, 263; Euclidean, 8, 260, 262, 368; and God, 156, 157; non-Euclidean, 8, 162, 262, 368; physical vs. mathematical, 160–2; views on, 155–7. See also expanding universe; matter and space; Newton; plenum; relativity, general theory of; space–time
space–time: as a background arena in quantum mechanics, 271, 290, 299, 306, 308, 317, 321, 331–3; in classical physics, 232; in general relativity, 255, 262, 340; in special relativity, 237, 238. See also space; time
special relativity. See relativity, the special theory of
spectra, 11, 265, 275, 278
speed of light (c). See light, speed of
spin: and Bohm's theory, 353; and the EPRB experiment, 326, 329, 330, 340; in quantum mechanics, 281, 288, 309, 310, 333, 348, 392 (9)
stability of the classical universe, 260, 262

standard map, 176
stadia, 379 (2)
standing waves, 273, 292
Stapp, H. P., 393 (19)
state, 290, 298, 309–11, 313, 323. See also completeness; wave function
state vector. See wave function
statistical mechanics, 378 (11)
steady-state theory, 268, 269
stellar aberration, 49, 185
stellar parallax, 48; measurement of, 49, 265
Stern–Gerlach apparatus, 309
stochastic mechanics, 352, 396 (29)
Strutt, W. (Lord Rayleigh), 189, 204, 274
Struve, F., 49
sun: and the bending of light, 253, 254, 256, 257; and the celestial sphere, 45, 46; and the compass directions, 43; and Galileo, 141, 142; in the geocentric model, 45, 46, 49; in the heliocentric model, 59; and Hume on induction, 31; in modern models of the universe, 267, 268; size of and distance to, 56–8; and the seasons, 44; and the tides, 383 (37)
sun-centered model. See heliocentric model
superposition, 291, 301, 303, 310–15. See also measurement problem
Syene (Aswan), 47
syllogism, 5

teleonomy, 19, 166, 167, 364, 365, 375. See also cause; maximum/minimum principle
telescope, 10, 57, 76, 150, 185, 256
Thales, 17, 164
theory (law), 12, 30–2, 35, 148, 378 (11), 394 (1). See also formalism vs. interpretation
theory-ladenness, 387 (38)
Thomson, J., 189, 208; model of atom, 276
Thomson, W. (Lord Kelvin), 188–90, 274, 276; on the aether, 223
tides, 114, 124–31, 143, 151
time: absolute, 96; cosmic, 389 (41); as an independent variable, 78, 90. See also time dilation
time dilation, 233, 234, 388 (5); experimental verification of, 235–7
Tombaugh, C., 153
Tonkin, Gulf of, 129

423

Subject index

to save the phenomena, 54–6, 138, 139. *See also* instrumentalism; truth
Torricelli, E., 77
trajectory, 79, 117, 118, 167, 173; in quantum mechanics, 285, 290, 318, 331, 343
triadic model of science, 372, 373, 397 (39)
trigonometry, 379 (16)
truth (as a goal of science), 34–6, 281, 366, 373; Aristotle on, 4; Bacon on. 24; and Copernicus' theory, 56; Descartes on, 7, 12; Einstein on, 357, 361; Galileo on, 78; Kuhn on, 371; Newton on, 166; Planck on, 10; Plato on, 4; Popper on, 1; and simplicity, 50, 161. *See also* instrumentalism; to save the phenomena; simplicity
turbulence, 173
twin paradox, 246–8; 251, 388 (15); and general relativity, 248, 388 (17)
two-sphere model, 48, 49, 52, 148
Tyache, S., 383 (38)

ultraviolet catastrophe, 274
uncertainty; classical, 168; and Kierkegaard, 283. *See* also uncertainty relation
uncertainty principle. *See* uncertainty relation
uncertainty relation, 286, 287, 298–301, 333, 392 (13, 16); in Bohm's theory, 331, 333, 335, 337; and Einstein, 308, 309
underdetermination, 220, 345–8, 395 (2, 14). *See also* scientific realism
understanding. *See* explanation vs. understanding
uniformly accelerated motion. *See* motion
universe. *See* models of the universe
unnatural motion. *See* motion
Uranus, 124, 152
Urban VIII. *See* M. Barberini

vacuum: in Aristotelian physics, 18, 26, 27, 77, 148, 183; in early electromagnetic theory, 196, 197; in the Kaufmann experiments, 211, 217; vs. a plenum, 183; solution in general relativity, 254 *See also* void
variational principle. *See* maximum/minimum principle
Vega (Alpha Lyrae), 49
Venus, 49, 63, 64, 154
verification: Bacon on, 24; Popper on, 33, 38
Vienna Circle, 160, 367, 397 (23)
da Vinci, L., 29
vitalist theory, 375
void, 21, 27, 50, 75, 155, 183, 184. *See also* vacuum
Virgo, 268
Vitruvius, 379 (4)
Volta, A., 306
Voltaire, F.-M., 166
von Neumann, J., 288, 318
Vulcan, 154

warrant (for an hypothesis): Bacon on, 24; for a belief in determinism, 170–3; for the Copenhagen interpretation, 288; Einstein on, 1; Galileo on, 78; Newton on, 33, 94; various types, 34, 365–9
Watson, J., 364
wave: longitudinal, 184, 385 (3); standing, 273, 390 (3); transverse, 185, 186, 385 (4). *See also* aether; electromagnetic wave; interference; wave function; wave–particle duality
wave equation. *See* Maxwell; Schrödinger
wave function: Bohm on, 336; in Bohm's theory, 334, 335; Born on, 300, 317; collapse of, 309–11, 313; defined, 290; entangled, 311, 323, 334; example of, 291–3; Heisenberg on, 300; Planck on, 298; and probability, 293; Schrödinger on, 286, 293; 313. *See also* completeness; quantum mechanics; superposition; measurement problem
wave mechanics. *See* quantum mechanics
wave-particle duality, 285, 286, 291. *See also* complementarity
weight: Aristotle on, 20, 25–7, 379 (18); Descartes on, 156; of light, 256; Newton on 112; speed as a measure of, 20, 26; the tides and, 128. *See also* motion
Wegener, A., 262
Weltanschauung (world view), 374; and science, 375
Whewell, W., 339
whig interpretation of history, 354, 355
Whitehead, A., 145, 170
Whiteside, D., 381 (14)
Wien, W., 277
Wilson, R., 268
world view. *See Weltanschauung*
Wren, C., 92

year, 44
Young, T., 130, 185, 223

Zeus, 74
zodiac. *See* constellation